ELECTROMAGNETIC INTERACTIONS AND HADRONIC STRUCTURE

The nuclei of atoms are constructed from protons and neutrons that, in turn, are constructed from more fundamental particles: quarks. The distribution of quarks within protons and neutrons, their interactions, and how they define the properties of protons and neutrons, and hence nuclei, are subjects of major research worldwide. This study requires the use of beams of high-energy electrons. Understanding of proton structure at high energies has been greatly expanded by the study of generalized parton distributions and of nucleon spin structure. Photons can separate the roles of quark and gluonic degrees of freedom within hadrons and hence filter glueballs and hybrid mesons. At high energies, both photon and nucleon structure can be probed. The former manifests itself in diffractive photoprocesses, where there is a uniquely rich interplay between perturbative and non-perturbative effects. This book provides an authoritative overview on the subject, and sections on chiral perturbation theory, crucial in understanding soft pions and soft photons near threshold, and duality ideas, equally crucial at intermediate energies, are included.

The emphasis throughout the book is on phenomenology, and the book concentrates on describing the main features of the experimental data and the theoretical ideas used in their interpretation. Written by leading experts in the field, both theoretical and experimental, this is an essential reference for graduate students and researchers in the field of particle physics and electromagnetic interactions.

FRANK CLOSE is Professor of Physics at the Rudolph Peierls Centre for Theoretical Physics, Oxford University. He gained a D.Phil at Oxford University, and has worked at Stanford Linear Accelerator Center (SLAC), CERN, and the Rutherford Appleton Laboratory. Professor Close has been awarded the Kelvin Medal from the Institute of Physics and is the author of many books aimed at both a professional and a lay audience, most recently *The New Cosmic Onion* (Taylor & Francis, 2006).

SANDY DONNACHIE is Honorary Research Professor in the School of Physics and Astronomy at the University of Manchester. Since gaining his Ph.D. from the University of Glasgow, Professor Donnachie has held positions at the University College London, CERN, the University of Glasgow and Daresbury National Laboratory. He was awarded the Glazebrook Medal from the Institute of Physics and was co-author of *Pomeron Physics and QCD* (Cambridge University Press 2002).

GRAHAM SHAW is a senior lecturer in the School of Physics and Astronomy at the University of Manchester. He gained his Ph.D. from the University of London. He has held positions at the Rutherford Laboratory, Daresbury National Laboratory and Columbia University. His many books include the well-known texts on *Quantum Field Theory* and *Particle Physics* (Wiley, latest editions 1993 and 1997).

CAMBRIDGE MONOGRAPHS ON PARTICLE PHYSICS NUCLEAR PHYSICS AND COSMOLOGY 25

General Editors: T. Ericson, P. V. Landshoff

ELECTROMAGNETIC INTERACTIONS AND HADRONIC STRUCTURE

Edited by

FRANK CLOSE
University of Oxford, UK

SANDY DONNACHIE
University of Manchester, UK

GRAHAM SHAW
University of Manchester, UK

CAMBRIDGE
UNIVERSITY PRESS

CAMBRIDGE UNIVERSITY PRESS
Cambridge, New York, Melbourne, Madrid, Cape Town, Singapore, São Paulo, Delhi

Cambridge University Press
The Edinburgh Building, Cambridge CB2 8RU, UK

Published in the United States of America by Cambridge University Press, New York

www.cambridge.org
Information on this title: www.cambridge.org/9780521115940

© Cambridge University Press 2007

First published 2007
This digitally printed version 2009

A catalogue record for this publication is available from the British Library

ISBN 978-0-521-84420-8 hardback
ISBN 978-0-521-11594-0 paperback

Contents

List of contributors

M C Birse, School of Physics and Astronomy, University of Manchester, Manchester M13 9PL, UK

V D Burkert, Jefferson Laboratory, 12000 Jefferson Avenue, Newport News, VA 23606, USA

F E Close, Department of Theoretical Physics, University of Oxford, 1 Keble Road, Oxford OX1 3NP, UK

M Diehl, DESY, 22063 Hamburg, Germany

A Donnachie, School of Physics and Astronomy, University of Manchester, Manchester M13 9PL, UK

R Ent, Jefferson Laboratory, 12000 Jefferson Avenue, Newport News, VA 23606, USA

J Forshaw, School of Physics and Astronomy, University of Manchester, Manchester M13 9PL, UK

Yu S Kalashnikova, ITEP, 117259 Moscow, Russia

T-S H Lee, Physics Division, Argonne National Laboratory, Argonne, Illinois 60439, USA

J McGovern, School of Physics and Astronomy, University of Manchester, Manchester M13 9PL, UK

G A Miller, Department of Physics, University of Washington, Seattle, WA 98195-1560, USA

P J Mulders, Department of Physics and Astronomy, Vrije Universiteit, De Boelelaan 1081, NL-1081 HV Amsterdam, The Netherlands

G Shaw, School of Physics and Astronomy, University of Manchester, Manchester M13 9PL, UK

Preface

Within the Standard Model of particle physics, it is the strong phase of QCD that is least understood, and the electromagnetic interaction that is best understood. It is therefore natural to use the electromagnetic interaction as a relatively gentle probe of the internal structure of hadrons and of other aspects of non-perturbative strong interactions.

This approach is hardly new: electron scattering was first used to measure the charge distribution within the proton some 50 years ago. However, its importance has been enormously enhanced by the recent development of many experimental facilities dedicated to electromagnetic interactions, and the realization that other laboratories can access electromagnetic interactions in novel ways. These facilities are primarily at low and medium energies, which probe the small-Q^2 kinematic regions, and upgrades are planned. In contrast, existing high energy electron accelerators, such as HERA, are soon to close, and even B-factories in e^+e^- annihilation are coming to the end of their lives. The focus of electromagnetic and hadron physics will be on QCD in the strong interaction regime ('strong QCD') as distinct from perturbative QCD physics.

The physics of strong QCD falls mainly into two areas: hadron spectroscopy at low energies and non-perturbative aspects of high-energy processes. In spectroscopy, the ability to tune the virtuality of the photon in electroproduction enables the spatial structure of baryons to be explored; and since photons only couple directly to charged particles, they are a vital tool in separating the roles of quark and gluonic degrees of freedom within hadrons, and hence in filtering glueballs and hybrid mesons. At high energies, there has been considerable progress in understanding both photon and nucleon structure. The former manifests itself in diffractive photoprocesses, where there is a uniquely rich interplay between perturbative and non-perturbative effects, and intermediate colour dipoles play the dominant role; while the study of proton structure at high energies has been greatly expanded by the study of skewed or generalized parton distributions and of nucleon spin structure. In addition we have included reviews of chiral perturbation theory, which plays a crucial role in understanding both soft pions and soft photons in the threshold region; and of duality ideas, which play an equally crucial role at intermediate energies.

In view of the enormous and continuing interest stimulated by these developments, a review of the present state of knowledge is both timely and useful. Because of the wide range of the material, a cooperative format has been adopted, similar to that used in a review of early work on the same topic published many years ago.[*] Again, the emphasis throughout is primarily on phenomenology, concentrating on describing the main features of the experimental data and the theoretical ideas used in their interpretation. As such we hope it will be of interest and use to all physicists interested in hadron physics, including graduate students.

For ease, laboratories and experiments will be referred to by their common acronyms or abbreviated forms. For reference the most common of these are the Thomas Jefferson National Accelerator Facility (JLab), LEGS at the Brookhaven National Laboratory (BNL), BaBar at the Stanford Linear Accelerator Center (SLAC), CLEO at Cornell University and MIT-Bates at MIT in the USA; MAMI at Mainz, ELSA at Bonn and HERMES at DESY in Germany; GRAAL at Grenoble in France; FENICE at Frascati in Italy; the COMPASS experiment at CERN; VEPP at Novosibirsk in Russia; and LEPS at Spring-8 and BELLE in Japan.

Finally, we are grateful to all the participating authors, both for their contributions and for their cooperation with the editors in obtaining a well-balanced presentation.

Frank Close
Sandy Donnachie
Graham Shaw

---

[*] *Electromagnetic Interactions of Hadrons*, A Donnachie and G Shaw eds, Plenum Press, New York, Vol I (1978) and Vol II (1978).

1

Quark models of hadrons and issues in quark dynamics

F E Close

1.1 Chromostatics

The discovery of quarks in inelastic electron scattering experiments, following their hypothesized existence to explain the spectroscopy of hadrons, led rapidly to the quantum chromodynamic (QCD) theory and the Standard Model, which has underpinned particle physics for three decades. Today, all known hadrons contain quarks and/or antiquarks.

The QCD Lagrangian implies that gluons also exist, and the data for inelastic scattering at high energy and large momentum transfer confirm this. What is not yet established is the role that gluons play at low energies in the strong interaction regime characteristic of hadron spectroscopy. QCD implies that there exist 'glueballs', containing no quarks or antiquarks, and also quark–gluon hybrids. The electromagnetic production of hybrids is one of the aims of JLab. Glueballs, on the other hand, are not expected to have direct affinity for electromagnetic interactions; hence hadroproduction of a meson that has suppressed electromagnetic coupling is one of the ways that such states might be identified.

Quarks are fermions with spin $\frac{1}{2}$ and baryon number $\frac{1}{3}$. A baryon, with half-integer spin, thus consists of an odd number of quarks (q) and/or antiquarks ($\bar{q}$), with a net excess of three quarks. Mesons are bosons with baryon number zero, and so must contain the same number of quarks and antiquarks.

The simplest configuration to make a baryon is thus three quarks, qqq; a meson most simply is $q\bar{q}$. Within this hypothesis well over two hundred hadrons listed by the Particle Data Group (PDG) [1] can be described. The question of whether there are hadrons whose basic constitution is more complicated than these, such as mesons made of $qq\bar{q}\bar{q}$ or baryons made of $qqqq\bar{q}$, is an active area of research,

Electromagnetic Interactions and Hadronic Structure, eds Frank Close, Sandy Donnachie and Graham Shaw. Published by Cambridge University Press. © Cambridge University Press 2007

which we shall summarize later. First we examine what property of the attractive forces causes such combinations to occur and then discuss how the multitude of hadrons are described.

The fundamental theory of the strong forces between quarks is QCD. The details of this theory and rules of calculation may be found in dedicated texts such as [2]. A quark carries any of three colours – which we label *RBG*. They are the charges that are the source of the force between quarks. The rules of attraction and repulsion are akin to those of electrostatics where like repel and unlike attract. Associate positive charges with quarks and negative with antiquarks. The attraction of plus and minus then naturally leads to the $q\bar{q}$ configurations, the mesons, for which the colour charges have counterbalanced.

In quantum electrodynamics, QED, the electromagnetic force is transmitted by photons; analogously, in QCD the forces between quarks are transmitted by gluons. This far is analogous to the formation of electrically neutral atoms. The novel feature arises from the three different colour charges. Two identical colours repel one another but two different, namely *RG*, *RB* or *BG*, can mutually attract. A third quark can be mutually attracted to the initial pair so long as its colour differs from that pair. This leads to attraction between three different colours: *RBG*. A fourth quark must carry the same colour charge as one that is already there and will be repelled by that, meanwhile being attracted to the dissimilar pair.

The above pedagogic illustration needs better specification. The rules of attraction and repulsion depend on the symmetry of the pair under interchange. Thus symmetric combinations repel, antisymmetric attract. Two identical colours, being indistinguishable, are trivially symmetric. Two differing colours can be either symmetric or antisymmetric:

$$[RB]_S \equiv \sqrt{\tfrac{1}{2}}(RB + BR), \tag{1.1}$$

$$[RB]_A \equiv \sqrt{\tfrac{1}{2}}(RB - BR). \tag{1.2}$$

Thus any pair of quarks in a baryon is in an antisymmetric symmetry state for the saturation of the attractive forces. The full wave function for the colour of a three-quark baryon is thus

$$\sqrt{\tfrac{1}{6}}((RB - BR)G + (BG - GB)R + (GR - RG)B). \tag{1.3}$$

The three colours form the basic **3** representation of SU(3); the three 'negative' colours of antiquarks are then a $\bar{\mathbf{3}}$. The rules of combining representations give

$$\mathbf{3} \otimes \bar{\mathbf{3}} = \mathbf{1} \oplus \mathbf{8}; \quad \mathbf{3} \otimes \mathbf{3} \otimes \mathbf{3} = \mathbf{1} \oplus \mathbf{8} \oplus \mathbf{8} \oplus \mathbf{10}. \tag{1.4}$$

It is the colour-singlet representations that are the formally correct SU(3) expressions of the above heuristic combinations. Hadrons are thus colour singlets of colour SU(3). Building a relativistic quantum gauge field theory of colour SU(3) leads to QCD. The baryons are thus in the antisymmetric representation of colour SU(3). The above argument shows how this is a natural consequence of the attractive colour forces in chromostatics. The antisymmetry under interchange of colour labels, combined with the Pauli principle that requires fermions, such as quarks, to be antisymmetric under the exchange of all their quantum numbers, leads to essential constraints on the pattern of hadrons and their properties.

1.2 Mesons as bound states from $b\bar{b}$ to light flavours

For heavy-flavour mesons, such as $c\bar{c}$ and $b\bar{b}$, a non-relativistic potential model description of meson spectroscopy is realized phenomenologically and may be justified theoretically. The ground state $1S$, the first excited state with orbital excitation $L = 1$, denoted by $1P$, and the radial excitation of the S-state, $2S$ with the $1D$ level being slightly higher than this, are qualitatively in accord with the pattern of a linear potential $V(r) = Kr$. Here r is the radial separation of the q and $\bar{q}$, and K (known as the string tension) has dimensions of $(\text{Energy})^2$. Empirically $K \sim 1$ GeV/fm ~ 0.18 GeV^2.

The qualitative features of the spectrum of states survive for all flavours (with some exceptions, such as 0^{++}, which we discuss later). This has enabled a successful phenomenology to be built in applying the non-relativistic constituent-quark model to light flavours even though the a priori theoretical justification for this remains unproven. A widely-used approach has been to approximate the dynamics to that of the harmonic oscillator, with Gaussian wave functions of form $\exp(-r^2 \beta_M^2/2)$ multiplied by the appropriate polynomials and β treated as a variational parameter in the Hamiltonian H for each of the $1S, 1P, 2S, 1D, \ldots$ states.

There are also spin-dependent energy shifts among states with the same overall L, the forms of which phenomenologically share features with those generated by the Fermi–Breit Hamiltonian in QED. This is widely interpreted as evidence for analogous chromomagnetic effects in QCD.

The $q\bar{q}$ picture is only literally true for states that are stable. If the number of colours $N_c \to \infty$, the amplitudes for $q\bar{q} \to q\bar{q} + q\bar{q} \sim 1/N_c \to 0$. In the real world $N_c = 3$ and the coupling to meson channels must distort the simple $q\bar{q}$ picture. A particular example occurs for $c\bar{c}$ where the $D\bar{D}$ thresholds cause non-trivial admixtures of $c\bar{c}u\bar{u}$ and $c\bar{c}d\bar{d}$ in the 'primitive' $c\bar{c}$ wave functions of the $J^{PC} = (0, 1, 2)^{++}$ χ states [3–5] and the 1^{--} $\psi(3685)$ which are all below the $D\bar{D}$ threshold. In the simple $c\bar{c}$ non-relativistic-potential picture, electromagnetic transitions among these states, such as

$$\psi(3685) \to \gamma\chi_J; \chi_J \to \gamma\psi, \qquad (1.5)$$

are electric dipole, E1, transitions. In this case if $\psi(3685) \equiv {}^3S_1$, then apart from phase-space effects the relative widths $\psi(3685) \to \gamma\chi_J$ would be in the ratio $2^+ : 1^+ : 0^+ = 5 : 3 : 1$, whereas for a ${}^3D_1$ initial state these ratios become $2^+ : 1^+ : 0^+ = 20 : 15 : 1$ [6,7]. While data are qualitatively in accord with the predictions of such a model for the ${}^3S_1$ case, future precision data can reveal the presence of relativistic effects, ${}^3D_1$ components and of $D\bar{D}$ mixing in the $c\bar{c}$ states. This is a particular example where electromagnetic transitions can give precision information on hadron wave functions and dynamics. More discussion of this can be found in chapter 4.

An example of this is the conundrum of the state $X(3872)$ [8]. This charmonium state is degenerate with the neutral $D\bar{D}^*$ threshold and as such is suspected of having $u\bar{u}$ admixture in its $c\bar{c}$ wave function. The $u\bar{u}$ and absence of $d\bar{d}$ will lead to significant isospin violation in its decays. If this state is 1^{++}, then one may anticipate such a light flavour asymmetry at a small level in the wave function of the $\chi_1(3500)$. High-statistics data on the hadronic decays of the $\chi_1(3500)$ could reveal if this is the case.

Such subtle effects could occur more widely in the charmonium states. The basic idea is as follows. The mass difference between $d\bar{d}$ and $u\bar{u}$, although small and widely neglected in analysis, can have measurable effects when the dynamics involves the differences among various energies. For example, the mixing of $d\bar{d}$ and $u\bar{u}$ in the χ states may be driven by $M(D\bar{D}) - M(\chi)$. For $u\bar{u}$ mixing into the χ_0 for example, it will be the neutral threshold $D^0\bar{D}^0$ that is relevant, while for the $d\bar{d}$ mixing it is D^+D^-. The difference in the energy gaps in the two cases is $\sim 5\%$; for states that are nearer the threshold such flavour-dependent effects can become highly significant. In the case of the $c\bar{c}$ state $X(3872)$ [8] one has almost perfect degeneracy with the $D^0\bar{D}^{*0}$ threshold such that admixture of $u\bar{u}$ is expected to dominate $d\bar{d}$ utterly [9,10].

High-statistics data on χ hadronic decays should be studied to see if there is an asymmetry between the neutral and charged particles in the final states, which would show a failure of simple isoscalar decay. These data can be taken in the e^+e^- facilities CLEO-c at Cornell University and BES in Beijing.

These mixing effects may be studied in precision data for heavy flavours and the resulting insights applied to light flavours. In the latter we already have qualitative understanding of where the limits of the $q\bar{q}$ model occur. The strategy is to quantify these en route to a more mature dynamical picture of the light flavoured hadrons. This has interest in its own right but also is needed when building Monte Carlo models for the decays of B and D heavy flavours into channels involving light hadrons.

1.2.1 Linear plus Coulomb potential

The phenomenological picture of the linear potential deduced from the pattern of the energy levels gives insight into the nature of the force fields acting on the constituents. As the force field $\sim dV(r)/dr$, then $V(r) \sim Kr$ implies the field is constant as a function of the distance r between the colour sources. This immediately contrasts with the behaviour in QED where $V(r) \sim 1/r$ implies that $E(r) \sim 1/r^2$, whereby the fields spread freely into all directions of three-dimensional space. The message for QCD is that the field lines concentrate along the line connecting the colour source q and sink $\bar{q}$. Thus the 'linear' potential is aptly named!

This is in accord with the picture that emerges from lattice QCD [11,12]. The potential is predicted to be linear and by implication the field lines collectively form a tube-like configuration. This has led to 'flux-tube models' [13,14] of $q\bar{q}$. These models underpin the potential, which is all that is needed for many calculations. However, they and the lattice computations also imply that the flux-tube provides an independent degree of freedom, which can be excited. The resulting states that form when the flux-tube is excited in the presence of $q\bar{q}$ are known as 'hybrids' [13–18]. We consider their dynamics later.

For the conventional $q\bar{q}$ states one views the flux-tube as the source of the linear potential, at least at distances comparable to the confinement scale ($r \sim 1$ fm). At short range, say $r \sim 0.1$ fm, QCD theory implies that the colour force is transmitted by gluons, which act independently of one another analogously to the way that photons behave in QED. This gives a Coulombic contribution to the potential as $r \to 0$. The exchange of a single gluon gives perturbative corrections to the simple potential, generating analogues of the spin-dependent hyperfine shifts that are familiar in QED.

The effective potential arising from QCD is thus taken as [3,7]

$$V(r) = Kr - \frac{4}{3}\frac{\alpha_s}{r} + C \qquad (1.6)$$

(the factor $\frac{4}{3}$ is a normalization factor arising from the SU(3) colour matrices at the quark vertices). In calculations it is often useful to make a Gaussian approximation to the wave functions, which may be found variationally from the Hamiltonian

$$H = \frac{p^2}{\mu} + Kr - \frac{4}{3}\frac{\alpha_s}{r} + C, \qquad (1.7)$$

where $\mu = m_q m_{\bar{q}}/(m_q + m_{\bar{q}})$ is the reduced mass, with standard quark-model parameters $m_q = 0.33$ GeV for u and d quarks and 0.45 GeV for s quarks, $K = 0.18$ GeV^2 and $\alpha_s \sim 0.5$.

Not only are the patterns of the energy levels preserved as one goes from heavy to lighter flavours, but many of the energy gaps are quantitatively approximately independent of flavour mass. Thus the gap between the 1^{--} $q\bar{q}$ in the $1S$ to $2S$ levels is 563 MeV for $b\bar{b}$ ($\Upsilon(10023)-\Upsilon(9460)$) and 589 MeV for $c\bar{c}$ ($\psi(3686)-\psi(3097)$). For the light flavours the analogous gap between $\rho(1460)$ and $\rho(770)$ is only some 10% larger while the absolute mass scales have changed by over an order of magnitude.

For a constituent of mass m in a potential that behaves as $V(r) \sim r^N$, this gap would vary as $m^{-N/(N+2)}$, hence $\sim m$ for the Coulomb potential and $\sim m^{-1/3}$ for linear. Mass independence would ensue for a potential $V(r) \sim \ln r$, which is approximately how the linear + Coulomb appears in the region of r most sensitive to the bound states.

However, there are also clear mass-dependent effects, notably in the splittings between the $^3S_1-{}^1S_0$ levels ($1^{--}-0^{-+}$). These vary from over 600 MeV for $\rho(770) - \pi(140)$, through 400 MeV for $K^*(890)-K(490)$ to significantly less for $\psi(3095)-\eta_c(2980)$ (we adopt the naming conventions for particles of the PDG [1]).

1.2.2 Hyperfine shifts

Although the mass gaps between successive orbital excitation levels of the effective potential are empirically approximately flavour-independent, there is a marked flavour dependence of the splitting between the S-wave states of differing total spin. Specifically this concerns the 1^- and 0^- states of $q\bar{q}$ and the $J^P = \frac{3}{2}^+$ and $\frac{1}{2}^+$ baryons.

Early evidence that mesons and baryons are made of the same quarks was provided by the remarkable successes of the Sakharov–Zeldovich constituent quark model [19], in which static properties and low-lying excitations of both mesons and baryons are described as simple composites of asymptotically free quasiparticles with a flavour-dependent linear mass term and hyperfine interaction,

$$M = \sum_i m_i + \sum_{i>j} \frac{\sigma_i \cdot \sigma_j}{m_i \cdot m_j} \cdot v^{hyp}, \qquad (1.8)$$

where m_i is the effective mass of quark i, σ_i is a quark spin operator and v_{ij}^{hyp} is a hyperfine interaction. This form has analogy with the source of hyperfine splitting in atomic hydrogen and suggests for hadrons that there is a QCD source in single-gluon exchange. As in the QED case, this (chromo)magnetic interaction is inversely proportional to the constituent masses.

In QCD the colour couplings are proportional to $\lambda_i \cdot \lambda_j$, with λ the SU(3) matrices [20]. The spin-dependent $\sigma_i \cdot \sigma_j$ term then causes the lowest-spin combinations to be further attracted, their high-spin analogues suffering a relative repulsion. This leads to a strong chromomagnetic attraction between a u and a d flavour when the ud diquark is in the $\bar{3}$ of the colour SU(3) and in the $\bar{3}$ of the flavour SU(3) and has $I = 0, S = 0$.

The relative magnitudes of the spin-dependent shifts for a common set of flavours depend on the net colour of the interacting constituent pair. Since the colour expectation values are

$$\langle \lambda_1 \cdot \lambda_2 \rangle_{qq(3)} = \tfrac{1}{2} \langle \lambda_1 \cdot \lambda_2 \rangle_{q\bar{q}(1)}, \tag{1.9}$$

the relative shifts for colour-singlet mesons and baryons are

$$J = 0 \to -3; J = 1 \to +1; J = \tfrac{3}{2} \to +\tfrac{3}{2}; J = \tfrac{1}{2} \to -\tfrac{3}{2}, \tag{1.10}$$

whence $m(\Delta) > m(N)$ and $m(\rho) > m(\pi)$. These need to be rescaled by the appropriate masses following (1.8) when comparing the flavour dependence of the energy shifts [20,21], such as for $m(\Sigma^*) > m(\Sigma)$ and $m(K^*) > m(K)$.

These spin- and flavour-dependent energy shifts are manifested not only in the different masses of hadrons with different spins, but also cause the splitting of $\Sigma - \Lambda$ baryons. This is because the ud in $\Sigma_Q(Qud)$ have, by the Pauli correlation, $S = 1$ and are hence pushed up in energy relative to their counterparts in the $\Lambda_Q(Qud)$, which are in $S = 0$. Details are in [21].

These attractive forces can generate correlations among pairs of quarks and/or antiquarks, which are manifested as spin-dependent effects in inelastic scattering and in the appearance of colour-singlet hadrons with content $qq\bar{q}\bar{q}$ or $qqqq\bar{q}$. These attractive correlations arise when a qq or $\bar{q}\bar{q}$ are antisymmetric in each quantum number, thus qq in colour $\bar{3}$ ($\bar{q}\bar{q}$ in 3), spin-zero and flavour singlet. This has significant implications for the structure of mesons with $J^{PC} = 0^{++}$ below 1 GeV [22,23]. It also can lead to the possibility of 'pentaquark' states $(ud)(ud)\bar{Q}$, where the (ud) denotes a correlated pair and the antiquark has a distinct flavour $Q = s, c, b$. A particular example of the latter would be $udud\bar{s}$, which would form a baryon with positive strangeness, which is thus manifestly exotic in that it cannot be formed from any combination of qqq.

1.3 Flavour mixings

Any flavour of quark Q and its antiquark $\bar{Q}$ when attracted together form a state with no net flavour, in particular having zero electric charge and strangeness. How then are we to tell what combination of flavours occurs in any given physical state? We begin with some theoretical expectations.

Consider two heavy flavours, say $b\bar{b}$ and $c\bar{c}$. The mass matrix will have on its diagonal $2m_b$ and $2m_c$, and if this was the whole story these would be the physical eigenstates. But there is also the possibility that these neutral states can annihilate through some common channel, for example gluons. Let the strength of this annihilation be A. It may connect $b\bar{b}$ to $c\bar{c}$ and it can also connect either of these to itself. The matrix thus becomes

$$\begin{pmatrix} 2m_b + A & A \\ A & 2m_c + A \end{pmatrix}. \qquad (1.11)$$

The eigenstates depend on the relative size of $(m_b - m_c)/A$. If this is large the eigenstates are $b\bar{b}$ and $c\bar{c}$; this is indeed the case if we identify the c, b with the physical charm and bottom quarks where the vector mesons, for example, are the $\psi(c\bar{c})$ and $\Upsilon(b\bar{b})$. If it is small, which would be the case if we chose u,d instead of c,b, they tend to the equally mixed states $(u\bar{u} \pm d\bar{d})/\sqrt{2}$, which are the familiar isospin eigenstates. The $I = 1$ state $(u\bar{u} - d\bar{d})/\sqrt{2}$ decouples from the annihilation A channel, while the $I = 0$ $(u\bar{u} + d\bar{d})/\sqrt{2}$ couples with an enhanced amplitude $\sqrt{2}$ times that of an individual flavour and the mass gap is proportional to A.

Thus we have a qualitative understanding of why the $c\bar{c}$ and $b\bar{b}$ spectroscopies are distinct (or 'ideal') while their u,d counterparts are mixed into the 'isospin' basis. Now consider the latter systems but in the presence of the strange quark, which can form $s\bar{s}$ states.

Consider the limit where $m_u \sim m_d$ but $m_s - m_d >> A$. The eigenstates will then be the same $u\bar{u} \pm d\bar{d}$ as above with the third state being $s\bar{s}$. This is realised in the vector mesons where the isoscalar mesons are $\omega = (u\bar{u} + d\bar{d})/\sqrt{2}$ and $\phi = s\bar{s}$. The evidence for this will be described shortly; the implication of it is that A is small for the vector meson channel [20].

Now consider the limit where $m_s > m_d$ and $A > m_s - m_d$. The eigenstates are now orthogonal mixtures of $u\bar{u} + d\bar{d}$ and $s\bar{s}$. This is what is observed for the 0^{-+} mesons where $\eta(550)$ and $\eta'(960)$ are mixtures of these flavours. Specifically, it was found in [20] that $A(0^{-+})$ is in the range $80-600$ MeV while $A(1^{--})$ is $5-7$ MeV, in both cases there being a slight hint that the magnitude falls with increasing energy. The annihilation contribution thus seems to be much smaller for the vector mesons than for the pseudoscalar and a question for dynamics is why?

Determining which J^{PC} states are 'ideal' (that is like the 1^{--}) and which are strongly mixed is one of the unresolved issues in the spectroscopy of light flavours. Answering it may help to identify the dynamics that controls this mixing. Electromagnetic interactions can play a significant role in addressing these issues as we now illustrate by showing how they have already been seminal in the case of the 1^{--} and 2^{++} multiplets, at least.

1.3.1 The 1^{--} and 2^{++} nonets

That the vector and tensor multiplets are near ideal can be seen from the pattern of their masses. One $I = 0$ state has mass similar to that of the isovector, while the other $I = 0(s\bar{s})$ is heavier with the strange $K(u\bar{s})$ midway between them. As the $I = 1$ state contains only u and d flavours, this pattern suggests that the lighter isoscalar is $n\bar{n} \equiv (u\bar{u} + d\bar{d})/\sqrt{2}$, while the heavier is $s\bar{s}$. Examples of such nonets are

$$1^{--} : \rho(770) \sim \omega(780); K^*(890); \phi(1020), \qquad (1.12)$$
$$2^{++} : a_2(1320) \sim f_2(1270); K^*(1430); f_2(1525). \qquad (1.13)$$

This pattern of flavours is also confirmed by the strong decays, in the approximation that the dominant hadronic decay is driven by the creation of $q_k \bar{q}_k$ in the field lines between the original $q_i \bar{q}_j$ (where i,j,k denote the flavour labels) such that

$$q_i \bar{q}_j \rightarrow q_i \bar{q}_k q_k \bar{q}_j \rightarrow (q_i \bar{q}_k) + (q_k \bar{q}_j). \qquad (1.14)$$

Thus $s\bar{s}$ can decay to $s\bar{u} + u\bar{s}$, which is $\equiv K\bar{K}$, but it does not decay to $\pi\pi$. This rule underpins the suppressed decays of the ϕ and $f_2(1525)$ to $\pi\pi$. The relative strengths of the electromagnetic couplings of these states also fit with this ideal picture.

For 1^{--} one has the direct coupling $q\bar{q}(1^{--}) \rightarrow \gamma \rightarrow e^+e^-$. Thus the leptonic widths, after phase-space corrections, give a measure of the flavour contents. This is discussed further in chapter 5. The amplitude is proportional to the sums of electric charges of the $q\bar{q}$ contents weighed by their amplitude and phases. Thus for the relative squared amplitudes

$$\rho(d\bar{d} - u\bar{u})/\sqrt{2} : \omega(u\bar{u} + d\bar{d})/\sqrt{2} : \phi(s\bar{s}) = 9 : 1 : 2, \qquad (1.15)$$

which may be compared with their e^+e^- widths in keV

$$\Gamma^{e^+e^-} [\rho : \omega : \phi] = 6.8 \pm 0.11 : 0.60 \pm 0.02 : 1.28 \pm 0.02. \qquad (1.16)$$

The differences in phase space are only small and so do not affect the arguments here. However, it is noticeable that the ratios seem to apply to the widths in that they hold also for the $\Gamma^{e^+e^-} (\psi(c\bar{c}); \Upsilon(b\bar{b}))$

$$\Gamma^{e^+e^-} [\psi(c\bar{c}) : \Upsilon(b\bar{b})] = 5.26 \pm 0.37 : 1.32 \pm 0.05, \qquad (1.17)$$

which are experimentally in accord with $4\Gamma^{e^+e^-} (\phi(s\bar{s}))$ and $\Gamma^{e^+e^-} (\phi(s\bar{s}))$ respectively.

For future reference, it is useful to show these states in the **1–8** basis of $SU(3)_F$:

$$\omega_1 \equiv (u\bar{u} + d\bar{d} + s\bar{s})/\sqrt{3}; \quad \omega_8 \equiv (u\bar{u} + d\bar{d} - 2s\bar{s})/\sqrt{6} \quad (1.18)$$

whereby (denoting the light quark component by $n\bar{n} \equiv (u\bar{u} + d\bar{d})/\sqrt{2}$)

$$\phi(s\bar{s}) = \sqrt{\tfrac{1}{3}}\omega_1 - \sqrt{\tfrac{2}{3}}\omega_8,$$

$$\omega(n\bar{n}) = \sqrt{\tfrac{2}{3}}\omega_1 + \sqrt{\tfrac{1}{3}}\omega_8. \quad (1.19)$$

The electromagnetic couplings of the 2^{++} states also confirm their tendency towards ideal flavour states. Here the decays to $\gamma\gamma$ have amplitudes proportional to the sum of the squares of the electric charges of the quarks weighted by their relative phases. Thus for the relative squared amplitudes

$$a_2(d\bar{d} - u\bar{u})/\sqrt{2} : f_2(u\bar{u} + d\bar{d})/\sqrt{2} : f_2(s\bar{s}) = 9 : 25 : 2, \quad (1.20)$$

which may be compared with their $\gamma\gamma$ widths in keV

$$\Gamma(a_2(1320) \to \gamma\gamma) : \Gamma(f_2(1270) \to \gamma\gamma) : \Gamma(f_2(1525) \to \gamma\gamma)$$
$$= 100 \pm 8 : 261 \pm 30 : 9.3 \pm 1.5. \quad (1.21)$$

The $a_2(1320) : f_2(1270)$ ratio is in excellent agreement with this; the $f_2(1525)$ is about a factor 2 smaller. The agreement between the mass-degenerate $a_2 : f_2$ states is in accord with ideal flavour states and then, if the heavier f_2 is $s\bar{s}$, its strange quark masses will suppress its magnetic contribution to the $\gamma\gamma$ amplitude and thus be consistent with the reduced strength.

To the extent that the vector mesons are ideal states, the radiative transitions of $C = +$ states $(C = +) \to \gamma V (= \rho : \omega : \phi)$ may be used to determine the flavour contents of the initial $C = +$ states (see chapter 4). As flavour is conserved in electromagnetic transitions to leading order, decays to $\gamma\rho$ weigh the $n\bar{n}$ component of the initial $C = +$ state, and those to $\gamma\phi$ weigh its $s\bar{s}$ component. This can be used as a further measure of the flavours of tensor mesons, where the reduced widths (phase space removed) would be expected to satisfy

$$\frac{\Gamma_R(f_2 \to \rho\gamma)}{\Gamma_R(f_2' \to \phi\gamma)} = \frac{9}{4}. \quad (1.22)$$

Empirically there is only an upper limit on these transitions. Obtaining their magnitudes is thus important, both as a check of this flavour filter and also for comparison with the analogous transition magnitudes for $f_{0,1} \to \gamma V$ as these can test the single-quark transition hypothesis for radiative transitions [24].

Complementary to this is the question of to what extent these ideal flavour states are realized for excited vector mesons such as $\rho(1460)$, $\rho(1700)$, $\omega(1420)$, $\omega(1650)$

and $\phi(1680)$. The decays of these $\rho^* : \omega^* : \phi^* \to 2^{++}\gamma$ have been discussed in [7] and chapter 4 as a potential way of addressing these issues and of discriminating the $^3D_1$ and 2^3S_1 (radial excitation) content of these states. These issues are also important in helping disentangle any hybrid vector meson presence in the wave functions of these states [7,25].

The radiative decays into the ground state ρ, ω, ϕ are potentially especially interesting for the enigmatic scalar mesons, the isoscalar members of which are hypothesized to be mixtures of $n\bar{n}$, $s\bar{s}$ and also pure glue ('glueball') configurations. By studying the relative rates for $0^{++} \to \gamma V(\rho : \omega : \phi)$ the flavour content of a set of scalar mesons may be weighed and then unitarity be checked to determine how much 'inert' (glue) presence is in the wave functions.

1.3.2 Allowed J^{PC}

The spin-$\frac{1}{2}$ q and $\bar{q}$ couple their spins to a total of 0 (singlet) or 1 (triplet). In the non-relativistic picture these couple with the orbital angular momentum L to give the total spin of the system $\vec{J} = \vec{L} + \vec{S}$. This is analogous to the atomic physics conventions for positronium and is how we combine the spins for quarks of the same or similar mass. In the more extreme asymmetric case of heavy Q and light $\bar{q}$ we follow the scheme more familiar in the hydrogen atom. The spin of the light (anti)quark s_q is first coupled to the orbital angular momentum l to give the total effective state j_q. This is then combined with the spin S_Q of the heavy quark. One can of course use either coupling scheme as the one is linearly related to the other. However, their utility varies.

It is when one has more than one state with the same overall spin (for example the $J^P = 1^+$ mesons formed from the $q\bar{q}$ with $L = 1$ and $S = 0, 1$) that one or the other scheme becomes more appropriate. For q and $\bar{q}$ of the same flavour the meson is an eigenstate of C. Since $C = (-1)^{(L+S)}$, the charge conjugation can distinguish the triplet (1^{++}) and singlet (1^{+-}) states. Thus the $\vec{J} = \vec{L} + \vec{S}$ basis is appropriate. For flavoured states, such as the K,D,B there is no C eigenvalue. The physical states are in general mixtures of the singlet and triplet basis states:

$$K_1(1273) = \cos\theta|^1P_1\rangle + \sin\theta|^3P_1\rangle,$$
$$K_1(1402) = -\sin\theta|^1P_1\rangle + \cos\theta|^3P_1\rangle. \tag{1.23}$$

It is important to be careful about the convention. Barnes *et al* [26] uses the mixing angle formula for kaons ($n\bar{s} = K_1$), whereas Blundell *et al* [27] apply it to antikaons ($s\bar{n}$). The opposite phases in the two conventions arise because the charge conjugation operator gives opposite phases when applied to $|^1P_1\rangle$ and $|^3P_1\rangle$ states.

The precise mechanism of this mixing of different $S_{q\bar{q}}$ states in the K-system is an open question. In the heavy-quark limit the angle in the above convention is $\theta = \tan^{-1}(1/\sqrt{2}) \sim +35.3°$. In the single-gluon exchange of QCD, the fact that $m_s \neq m_{u,d}$ causes a non-conservation of $q\bar{q}$ spin in the spin–orbit interaction. However, with a realistic strange quark mass the mixing angle is only $\theta = +5°$, which does not appear large enough to explain the observed mixing angle (see footnote number 90 in [26] for further discussion of this point). Contrast this with the hypothesis of Lipkin [28] who notes that the mixing might be determined by the coupling of the two $|K_1\rangle$ states through their decay channels. With sufficiently strong decay couplings the physical resonances are driven into near 'mode eigenstates' [26], which explains the separation into 'ρK' and 'πK^*' resonances. This model suggests a singlet-triplet mixing angle of $\sim 45°$, which is in accord with an analysis of the hadronic decays in [26]. This mixing pattern needs to be more precisely determined. Radiative transitions $K_1 \to K\gamma$ can help here. The $E1$ transition in leading order conserves the $q\bar{q}$ spin and so one would anticipate the $^1P_1$ component to dominate. Detailed estimates are needed in order to exploit the opportunities presented by radiative transitions in the kaonic system. This may become possible experimentally in $e^+e^- \to \psi \to \bar{K}K_1 \to \bar{K}K\gamma$ at a high-luminosity ψ factory such as CLEO-c or BES.

For the neutral mesons $q\bar{q}$, C is an eigenvalue. However, not all correlations of J^{PC} are accessible. As an antiquark has the opposite intrinsic parity to a quark, the parity of $q\bar{q}$ with relative angular momentum L is $P = (-1)^{L+1}$. Their charge conjugation eigenvalue in a state of total spin S and angular momentum L is $C = (-1)^{L+S}$. Thus the pattern of J^{PC} for the lowest levels is as follows:

$$S = 0 : 0^{-+}, 1^{+-}, 2^{-+},$$
$$S = 1 : 1^{--}, (0, 1, 2)^{++}, (1, 2, 3)^{--}. \tag{1.24}$$

Thus we see that $J^{\pm,\pm}$ correlations occur for any J with the exception of 0^{--}. In the series $J^{\pm\mp}$ the $J =$ odd correlates with positive parity and $J =$ even correlates with negative parity. We do not have the sequence $0^{+-}, 1^{-+}, 2^{+-} \ldots$. Thus we have the concept of 'exotic' states

$$0^{--}; \quad \text{and} \quad 0^{+-}, 1^{-+}, 2^{+-}, \ldots$$

A meson observed with any of the above exotic combinations cannot be described as a non-relativistic $q\bar{q}$ state. Thus looking for charge-neutral mesons with exotic J^{PC} is a strategy for isolating glueballs or hybrids ($q\bar{q}$ states where the gluonic degrees of freedom are excited) as these can form the 'exotic' J^{PC} combinations.

1.4 Baryons

A baryon consists most simply of three quarks. For spectroscopy and the low-energy phenomenology of baryon resonances and transitions, such a picture gives a remarkably good description though precision data are beginning to show this may not be the whole story. The elastic form factors of the nucleons and the transition to Δ begin to show the effects arising from π components in the wave function. These are discussed in chapters 2, 3 and 5. Furthermore, when viewed at high resolution the proton and neutron appear as swarms of quarks, their antimatter counterparts, antiquarks, and gluons – the quantum bundles that bind these constituents. The $q\bar{q}$ meson and qqq baryon configurations are the simplest combinations for which the attractive forces of QCD saturate, leading to stable hadrons. First we consider the qqq picture of baryons in 'strong QCD' and then the richer details of nucleons as revealed in deep inelastic scattering in the pQCD regime. In the latter there is evidence that there is a 'sea' of $q\bar{q}$ pairs resolved at small distances. There is a wider problem here: QCD does not forbid configurations such as $qqq\bar{q}$ or $qqqq\bar{q}$ to occur in the strong QCD regime. The question about such multiquark states is not one of existence but of observability. For example many are expected to have widths that are too broad for the states to be seen.

1.4.1 Role of the Pauli principle

Quarks, being fermions, obey the Pauli exclusion principle, which implies that when two quarks are interchanged, their total wave function must be antisymmetric. This has no impact on $q\bar{q}$ mesons but is essential in baryons. As the qqq is already antisymmetric in the colour degree of freedom, it must be globally symmetric in the space–spin–flavour product.

In the S-state the spatial wave function is symmetric and so we seek symmetry in flavour and spin. This immediately implies that two identically flavoured quarks, such as uu or dd, must be symmetric in spin, hence in spin 1. If $\uparrow\downarrow$ refer to S_z projections, then for a pair we have three symmetric combinations for $S = 1$

$$S = 1: \quad \uparrow\uparrow; \ (\uparrow\downarrow + \downarrow\uparrow)/\sqrt{2}; \ \downarrow\downarrow \tag{1.25}$$

and one antisymmetric for $S = 0$

$$S = 0: \quad (\uparrow\downarrow - \downarrow\uparrow)/\sqrt{2}. \tag{1.26}$$

If we replace $\uparrow, \downarrow$ respectively by u,d we can write the analogous states for isospin $I = 1$ (symmetric) and $I = 0$ (antisymmetric). In the S-wave baryons the overall symmetry leads to the correlation that pairs with $I = 0$ must have $S = 0$, and those

ϕ_λ	ϕ_ρ
P: $[(ud+du)u-2uud]/\sqrt{6}$	$(ud-du)u/\sqrt{2}$
N: $-[(ud+du)d - 2ddu]/\sqrt{6}$	$(ud-du)d/\sqrt{2}$

Table 1.1. Mixed symmetry combinations of u and d flavours for proton and neutron.

with $I = 1$ have $S = 1$. Thus in the $\Sigma(uds)$ and $\Lambda(uds)$, the (ud) are in $I = 1, 0$ respectively and hence are coupled to $S = 1, 0$. This implies that the spin of a polarized Λ is carried by that of the strange quark. Also this immediately generalizes to $\Lambda_{c,b}$ where the polarization will be carried by the c,b flavour. These correlations underpin the mass-difference of $\Sigma > \Lambda$. They also drive the phenomenology of the spin polarization asymmetry in inelastic electron scattering.

The above are easy to see for a pair of quarks. The extension to three is more involved. As regards flavour, the uds combinations lead to ten that are flavour symmetric. Write them (abc) to denote the totally-symmetric state $(ab + ba)c + (ac + ca)b + (bc + cb)a$, suitably normalized. The baryons with the corresponding quantum numbers are:

$$(uuu); (uud); (ddu); (ddd) \equiv \Delta^{++,+,0,-},$$
$$(uus); (uds); (dds) \equiv \Sigma^{(*)+,0,-},$$
$$(uss); (dss) \equiv \Xi^{(*)0,-},$$
$$(sss) \equiv \Omega^-. \tag{1.27}$$

As these are totally symmetric in flavour they must also be symmetric in spin. This implies they have spin $\frac{3}{2}$; the totally symmetric spin states with $S_z = +\frac{3}{2}$; $+\frac{1}{2}; -\frac{1}{2}; -\frac{3}{2}$ being

$$\uparrow\uparrow\uparrow; \quad (\uparrow\uparrow\downarrow + \uparrow\downarrow\uparrow + \downarrow\uparrow\uparrow)/\sqrt{3}; \quad (\downarrow\downarrow\uparrow + \downarrow\uparrow\downarrow + \uparrow\downarrow\downarrow)/\sqrt{3}; \quad \downarrow\downarrow\downarrow.$$

Thus it is the overall symmetry that causes the **10** to have spin $\frac{3}{2}$.

States where all flavours are identical can only be written in a symmetric form. States where one or more is distinct can be written with antisymmetric parts. Only when all three are distinct can a totally antisymmetric combination be written. Combinations like uud can be written in 'mixed' symmetry forms denoted $\phi_{\lambda,\rho}$, which are listed in table 1.1. These are respectively antisymmetric (ρ) and symmetric (λ) under the exchange $1 \leftrightarrow 2$ while they go into linear combinations of one another under $1, 2 \leftrightarrow 3$. Thus we say they transform with mixed symmetry $M_S(\lambda)$ and $M_A(\rho)$ respectively. There is no agreed convention and we will use $M_{S,A}$ or the λ and ρ notation throughout this book. The corresponding states for other members of the octet follow by the cyclic replacements $u \to d \to s$ starting from the P forms of the table. We have illustrated the result for the N.

In this simple picture of qqq, baryons can only have strangeness $0, -1, -2$ or -3. In particular no positive strangeness baryons can be formed. The possibility of such an exotic baryon is discussed in section 1.8.

1.5 Spin–flavour correlations

While the flavour content and relative phases of wave functions are probed by electric couplings, the correlations of flavour and spin are most directly probed by the magnetic interaction. Immediate examples are the relative sizes of the magnetic moments of baryons, in particular the neutron and proton.

The Pauli correlation of three nucleons in the nuclei ^{3}He and ^{3}H implies that the two identical fermions (for example pp in ^{3}He) have net spin $S = 0$ and hence do not contribute to the magnetic moment $\vec{\mu}$ (in the approximation that all are in the S-state). Thus $\mu({}^3\text{He}) : \mu({}^3\text{H}) = \mu(n) : \mu(p)$, which is true to $\sim 20\%$, any discrepancies being due to higher angular momentum components and/or transient pion exchanges. For the p and n made of three constituent quarks, the colour degree of freedom causes the Pauli correlation of like flavours to be reversed. Thus uu in the proton (uud) has net $S = 1$. Hence the S-state wave function involves the coupling of $S = \frac{1}{2} \otimes S = 1 \to S = \frac{1}{2}$. The resulting Clebsch–Gordan coefficients then imply that for a proton with $S_z = \uparrow$

$$|p \uparrow\rangle = \sqrt{\tfrac{2}{3}} |u \uparrow u \uparrow d \downarrow\rangle + \sqrt{\tfrac{1}{3}} |u \uparrow u \downarrow d \uparrow\rangle \qquad (1.28)$$

(the constitution of the neutron follows immediately upon replacing $u \leftrightarrow d$). The spin-weighted probabilities are then $P(u \uparrow) = \frac{5}{3}$, $P(u \downarrow) = \frac{1}{3}$, $P(d \uparrow) = \frac{1}{3}$ and $P(d \downarrow) = \frac{2}{3}$. The expectation value of $I_z \sigma_z$, which gives the magnitude of g_A/g_V, is $5/3$ in this simple state. Also the ratio of magnetic moments is given by the expectation value of the charge-weighted sum of the quark spin-projection σ_z. The above wave function then implies that the ratio of proton and neutron magnetic moments is $-\frac{3}{2}$. The assumption that the magnetic moments of quarks are uniformly proportional to their charges is profound and merits deeper understanding. The agreement with data is good to within $\sim 3\%$, supporting the S-state constituent qqq model description.

This is better even than the nuclear case and qualitatively suggests that pion-exchange currents and $qqq q\bar{q}$ components cannot play a major role in this constituent wave function. However, precision data on charge radii and inelastic scattering show that things cannot be so simple (see chapters 2 and 3). Extending the analysis to the baryon octet, assuming that the intrinsic magnetic moment of a strange quark and a down quark are in the ratio $\mu(s)/\mu(d) = m(d)/m(s)$, gives overall reasonable agreement though the data, especially on the magnetic moments of Ξ, merit better precision in order to constrain the wave functions. Thus precision data on the magnetic moments of the octet and on the magnetic dipole $M1$ transition $\gamma N \to \Delta(1232)$ are needed, as are possible electric quadrupole $E2$ contributions, which are absent in the most naive qqq **56** wave functions but present when pion clouds or admixtures of other components are in the wave function, as suggested by pQCD, are needed. The role of a pion cloud can also be better constrained through measurements of charge radii and of the ratio $G_E(q^2)/G_M(q^2)$ of the electric and magnetic form factors.

The constituent wave function (1.28) is also manifested in spin-dependent asymmetries in deep inelastic polarized scattering. The essential assumption here is of incoherent scattering on the valence quarks at $x \geq 0.1$. In this case, consider for example a polarized photon with $J_z = +1$ incoherently scattering from the polarized quarks forming a final state with $J_z = +\frac{3}{2}$ or $+\frac{1}{2}$. If the total photoabsorption cross section in each of these configurations is denoted $\sigma(\frac{3}{2})$ and $\sigma(\frac{1}{2})$ respectively, the polarization asymmetry $A = (\sigma(\frac{3}{2}) - \sigma(\frac{1}{2}))/(\sigma(\frac{3}{2}) + \sigma(\frac{1}{2}))$ and is driven by $\sim \sum_{u,d} e_{u,d}^2 \sigma_z$. Thus

$$A(\gamma n) = 0; \quad A(\gamma p) = \tfrac{5}{9}, \tag{1.29}$$

where in this simple model $\frac{5}{9} \equiv \frac{1}{3} g_A/g_V$.

In inelastic electroproduction at modest values of Q^2 the valence quarks play a substantial role in determining these cross-sections over a large range of the kinematic variable x (which is roughly interpreted as the fraction of the momentum of the initial nucleon along the direction of the incident virtual photon that was carried by the struck quark). In this picture it is natural that the valence quark contribution peaks when $x \sim 0.2-0.3$ and in this region the above predictions appear to be realized. However, at very large Q^2 the 'pollution' from gluons and $q\bar{q}$ pairs dominates the valence quark probability and it is only for $x \geq 0.5$ that the valence quarks dominate the wave function. The region $x \to 1$ probes the rare extreme where a single quark carries all of the momentum. The QCD forces that elevate the energy of a qq pair with $S = 1$ (thus for example uu will carry more than the average) imply that as $x \to 1$ the probability to find u is greater than that to find d. The spin wave function in (1.28) implies that in a polarized nucleon, if a quark carries all of its momentum it also carries all of its spin polarization [29]. So for example for the neutron the asymmetry at large values of x, $A(x \to 1) \to 1$, which is radically different from the vanishingly small magnitude at smaller x. A similar result is expected for the proton but as the value is already rather large, any increase in magnitude as $x \to 1$ will be harder to establish.

While the prediction of the $A(x \to 1) \to 1$ is universal in pQCD-based models, the approach to this value with increasing x is model-dependent. A dedicated study at JLab with a 12 GeV upgrade may shed light on this. The formalism relevant for polarized leptoproduction is described in chapter 7.

1.6 SU(6) multiplets and excited multiplets

Taking the $SU(3)_F$ fundamental representation (u,d,s) and combining it with the SU(2) spin $(\uparrow, \downarrow)$ one can form a fundamental six-dimensional representation of SU(6): $u \uparrow, d \uparrow, s \uparrow, u \downarrow, d \downarrow, s \downarrow$. The formalism is especially useful when discussing baryons, as here the Pauli principle limits the possibilities under interchange of any pair of quarks such that only the overall antisymmetric wave function is allowed. As the colour-singlet representation qqq is already antisymmetric, then as noted earlier this implies that the product of spatial, spin and flavour degrees of freedom must be overall symmetric under interchange.

The SU(6) representations for qqq are $6 \otimes 6 \otimes 6 = 56_S \oplus 70_M \oplus 70_M \oplus 20_A$ (where the subscripts denote the symmetric, mixed symmetric and antisymmetric behaviour of the representation). The Pauli principle will then constrain the correlations between these states and their spatial state as follows.

The spatial ground state, $L = 0$, is symmetric. Thus here the flavour-spin must be symmetric, namely **56**. Decomposed into $SU(3) \otimes SU(2)$ this becomes **8,2** $\oplus$ **10,4**. The **2** corresponds to the $2S + 1$ states of spin, hence $S = 1/2$; analogously the **4** corresponds to $S = 3/2$. Thus we see that the correlation of an octet with spin $\frac{1}{2}$ and a decuplet of spin $\frac{3}{2}$ is forced on us, in accord with data.

This simple assignment of N and $\Delta(1232)$ has immediate constraints on their electromagnetic properties. The magnetic moments of the neutron and proton are predicted to be in the ratio of $\mu_p/\mu_n = -\frac{3}{2}$, which is within 3% of the actual value. The amplitude for $\gamma N \to \Delta$ could in general be either magnetic dipole, $M1$, or electric quadrupole, $E2$ but in this model $E2 \equiv 0$; the magnitude of the $M1$ amplitude for the N to $\Delta(1232)$ transition relative to the magnetic moment of the proton is $\mu(N\Delta) : \mu = 2\sqrt{2}/3$. The ratio of electric and magnetic form factors of the proton, $G_E(Q^2)/G_M(Q^2)$, is predicted to be invariant with Q^2; for the neutron $G_E(Q^2) = 0$.

Although these results are qualitatively in agreement with experiment, it has long been known that for real photons there is a small non-zero $E2$ amplitude for $\gamma N \to \Delta(1232)$ [30] and that $\langle r^2 \rangle \sim dG_E^n/dQ^2 < 0$ [31]: see chapters 2 and 3. The latter is most simply described by giving the neutron a pion cloud: $n \to p\pi^-$ giving a negative outer charge due to the light π^-, whereas the compensating positive requires fluctuation to $\Delta^-\pi^+$, which is suppressed due to the higher mass of the Δ.

Modern precision data have confirmed and extended these results to higher Q^2, where the $E2$ amplitude remains very small compared to expectations from pQCD [32], and shown that G_E/G_M has a non-trivial dependence on Q^2 [33]. These data are clear indications that while the constituent-quark model is a good first approximation in certain circumstances, it is not the complete picture of hadrons. Further precision data on the complete set of electromagnetic transitions involving nucleons and $\Delta(1232)$ as a function of q^2 can show how to build the more complete picture of these states.

Now consider the excitation of a quark to the first excited level, a P-state, $L^P = 1^-$. For the three-quark system there are two independent degrees of internal spatial freedom. These are conventionally written

$$\sqrt{2}\vec{\rho} = \vec{r}_1 - \vec{r}_2; \quad \sqrt{6}\vec{\lambda} = 2\vec{r}_3 - \vec{r}_1 - \vec{r}_2 \equiv 3\vec{r}_3. \qquad (1.30)$$

These have mixed symmetry under interchange: they are respectively antisymmetric (ρ) and symmetric (λ) under the exchange $1 \leftrightarrow 2$ while they go into one another under $1, 2 \leftrightarrow 3$. Thus we say they transform with mixed symmetry $M_S(\lambda)$ and $M_A(\rho)$ respectively.

To form an overall symmetric space–SU(6) state with $L = 1$ thus requires that the SU(6) be in a mixed symmetry state. This is the **70** which under SU(3) $\otimes$ SU(2) becomes **8,2** $\oplus$ **8,4** $\oplus$ **10,2** $\oplus$ **1,2**. Finally combining these spins of $\frac{1}{2}$ or $\frac{3}{2}$ with the $L = 1$ gives negative-parity baryons in the following families, with candidate members identified:

$$^2\mathbf{8} : S_{11}(1550),\ D_{13}(1550),$$
$$^4\mathbf{8} : S_{11}(1700),\ D_{13}(1700),\ D_{15}(1700),$$
$$^2\mathbf{10} : S_{31}(1700),\ D_{33}(1700),$$
$$^2\mathbf{1}; S_{01}(1405),\ D_{03}(1520). \qquad (1.31)$$

These simple identifications are for illustration only. There are unresolved questions as to the mixing angles for the physical states S_{11} or D_{13} between the $^2\mathbf{8}$ and $^4\mathbf{8}$ bases and among the Λ states in $^2\mathbf{1}$, $^2\mathbf{8}$ and $^4\mathbf{8}$. Precision data on the photo- and electroproduction of these states can constrain these: see chapter 3.

While hyperfine splittings are clearly seen for the ground state $L = 0$ hadrons, when $L \neq 0$, spin–orbit splittings are expected. Empirically these are small, as is clear in (1.31), and explaining this continues to be a puzzle. In mesons the contributions to mass shifts arising from the short-range vector potential (single-gluon exchange) and from the long-range scalar potential are individually several hundreds of MeV but cancel.

The small splittings for mesons are therefore explicable. However, for baryons it is not so simple [34]. As this is a three-body system, the rest frame of the baryon in general differs from that of the interacting pair of quarks; transforming between these frames introduces a spin rotation that acts like a spin–orbit contribution to the Hamiltonian and which is not cancelled in general.

The spin dependence in meson and baryon spectroscopy in QCD is assumed to be due to gluon exchange. An alternative school has posited that π and η exchanges play a role in baryons [35]. This has been robustly criticized [36]. There has been a misconception that π exchange within baryons, being spinless, has no spin–orbit splittings and so may explain the empirical phenomenon. However, this is not so [36]. In all pictures Thomas precession occurs and, left to itself, would give *inverted* spin–orbit split multiplets (that is for states with the same value of L and different values of J, one finds that the state with the highest value of J lies lowest in mass, the splittings being several hundreds of MeV). The problem phenomenologically is to find a *cancellation* of the inevitable Thomas effects. In mesons this occurs but in baryons it fails, for the reasons above. The cause of the small spin–orbit splittings in baryons remains unresolved.

While it is possible to choose parameters such that N^* spin–orbit splittings are small, this fails in the case of the $\Lambda(1405)\frac{1}{2}^-$ and $\Lambda(1520)\frac{3}{2}^-$ where the parameters would suggest that the $\frac{1}{2}^-$ should actually be heavier than the $\frac{3}{2}^-$ [34]. The light mass of the $\Lambda(1405)$ may be an indicator that this state is affected by the KN threshold in S-wave and/or has $qqqq\bar{q}$ 'pentaquark' composition. Radiative couplings between the $\Lambda(1405)$ or $\Lambda(1520)$ and the ground state Λ/Σ can differentiate these pictures.

The single-gluon exchange perturbation also causes mixing among states. In this picture the $\Lambda(1405; 1800; 1670)$, which are all $\frac{1}{2}^-$, are mixtures of the **70**-plet $^2\mathbf{1}$, $^2\mathbf{8}$ and $^4\mathbf{8}$. The $\Lambda(1405)$ in particular is $0.9|^2\mathbf{1}\rangle + 0.4|^2\mathbf{8}\rangle$. Improved data on the electromagnetic couplings of these states and on their hadronic branching ratios are required to test these wave functions.

Electromagnetic transitions from the nucleon to N^* are direct and can test wave function mixings also. In the **70** one has $\frac{1}{2}^-$; $\frac{3}{2}^-$ states (known as S_{11}; D_{13} respectively), which can be either $^2\mathbf{8}$ or $^4\mathbf{8}$. The latter configuration cannot be photoproduced from protons in leading order. The physical states are predicted to be mixed ($N_{H,L}$ respectively referring to the heavier and lighter mass eigenstates),

$$|N_H\rangle = \cos(\theta)|^4\mathbf{8}\rangle + \sin(\theta)|^2\mathbf{8}\rangle,$$
$$|N_L\rangle = -\sin(\theta)|^4\mathbf{8}\rangle + \cos(\theta)|^2\mathbf{8}\rangle. \tag{1.32}$$

The masses are not well enough known to pin down $|\theta|$ with precision. Better data on $\gamma N \to S_{11} \to N\eta$ in particular are needed (see chapter 3).

This angle is a sensitive discriminator among models. The QCD inspired model of [37] predicts its magnitude for the S_{11} *parameter free*:

$$\theta = -\arctan[\tfrac{1}{2}(\sqrt{5} - 1)] \sim -32°.$$

Other models have more freedom. Hence precision data on the photo- and electro-production of these states together with their hadronic decay branching ratios (in particular into $N\eta$: $N\pi$) may significantly constrain the wave functions and expose the dynamics.

It is also possible to build negative-parity states by adding a $q\bar{q}$ to make $qqqq\bar{q}$ 'pentaquark' configurations. The QCD forces, however, appear to be repulsive for such configurations, unless they form the colour-singlet meson–nucleon configuration. If excitation to P-wave is included, however, it is possible for attractive forces to take over, leading to non-trivial pentaquark combinations such as $uudd\bar{s}$, which with positive strangeness is manifestly outside the states that are allowed within the qqq model.

One or more quarks may be excited to higher levels. The symmetry properties of the resulting spatial wave functions then determine the SU(6) representations allowed by Pauli. For the details of forming these configurations see a dedicated text such as [21]. If the ρ degree of freedom were to be 'frozen' in the ground state and only the λ degree of freedom were excited (thus qqq being effectively a quark–diquark model), then there would be a correlation between $L^p = (\text{even})^+ \equiv \mathbf{56}$ and $L^p = (\text{odd})^- \equiv \mathbf{70}$. If both ρ and λ can be excited, as in a 'genuine' qqq system, there is no such correlation and one can have $\mathbf{70}^+$, $\mathbf{56}^-$ and $\mathbf{20}$ dimensions for example. One of the challenges for baryon spectroscopy is to determine clearly which of these possibilities is realized in nature.

1.7 Role of Q^2 as a test of dynamics

Hadron form factors are the most basic observables that reflect the composite nature of hadrons. Indeed, the first indication that the proton is a composite object came from the measurements of the proton form factors in elastic electron–proton scattering [38]. In our modern QCD-based picture of hadrons, the high-Q^2 behaviour of elastic and transition form factors probes the high-momentum components of their valence quark wave functions. Of particular interest in this regard is understanding when the dynamics of valence quarks makes a transition from being dominated by the strong QCD of confinement to pQCD.

This transition should first occur in the simplest systems, in particular the elastic form factor of the pion seems the best hope for seeing this experimentally. Here the asymptotic behaviour is rigorously calculable in pQCD [39] and is $Q^2 F_\pi(Q^2) \to 8\pi\alpha_s f_\pi^2$, where $f_\pi = 133$ MeV is the π^+ axial-vector decay constant (see chapter 2). The approach to this asymptotic value is model-dependent. Experimental knowledge of $F_\pi(Q^2)$ is poor and a dedicated programme at JLab could give important insights.

The development of precise new measurement techniques has begun to give much more detailed images of the structure of the nucleon through its elastic and transition form factors. At small $Q^2 (< 1 \text{ GeV}^2)$ precise measurements of the electric form factor of the neutron, $G_E^n(Q^2)$, show values statistically different from zero. The neutron magnetic form factor remains less well determined. For both of these form factors few measurements exist for $Q^2 > 1 \text{ GeV}^2$. A non-zero value for the neutron electric form factor implies flavour-dependent correlations within the wave function. Two immediate possibilities for these are the hyperfine interaction between the quarks [40] and the pion cloud [41].

For the proton, polarization measurements [33] have shown that the electric form factor falls much faster than the magnetic one (see chapter 3). Furthermore, the $E2/M1$ ratio for the $N \rightarrow \Delta$ transition has remained near zero over the entire range of momentum transfer explored up to $Q^2 = 4 \text{ GeV}^2$ [32], in line with the most naive constituent-quark model, whereas the pQCD prediction for this number asymptotically is unity.

The $S_{11}(1535)$ may be considered the negative-parity partner of the nucleon. In the limit of exact chiral symmetry they would be degenerate. The properties of the S_{11} form factor reveal fundamental aspects of dynamical chiral symmetry breaking in QCD. Chiral symmetry is discussed in chapter 5 and applied to baryons in chapter 3.

At $Q^2 = 0$ the excitation of any given resonance with $J \geq \frac{3}{2}$ involves both electric and magnetic multipoles. The relative weight of these is determined by the SU(6) and orbital structure of the resonance wave function. These multipole magnitudes can be translated into predictions for the relative helicity amplitudes, where the resonance is photoproduced with $J_z = \frac{1}{2}$ or $\frac{3}{2}$. Thus for example the $D_{13}(1520)$ is found to be photoproduced from a proton target with $\sigma(\frac{1}{2}) \sim 0$ due to a destructive interference between the $E1$ and $M2$ multipoles. This tendency arises for the tower of resonances in the series $D_{13}, F_{15}, \ldots$ [42,43] and plays a role in generating the negative sign of the total in the Gerasimov–Drell–Hearn (GDH) integral for $Q^2 = 0$. See chapter 6 for a full discussion.

For $Q^2 \neq 0$ the magnetic multipoles are predicted in constituent models increasingly to dominate over the corresponding electric multipoles [42,43], leading to a reversal of the spin polarization. This causes $\sigma(\frac{1}{2}) > \sigma(\frac{3}{2})$ at large Q^2, in accord with the implications of the Bjorken sum rule and predictions of pQCD. An unresolved question is how rapidly this occurs; in the non-relativistic quark model (NRQM) the changeover from $\sigma(\frac{3}{2})$ to $\sigma(\frac{1}{2})$ is predicted to occur at $Q^2 \sim 0.1 - 0.2 \text{ GeV}^2$. This transition has yet to be quantified in data; its dependence on the resonance J^P is also unknown. A precise measure of this systematics is likely to give more detailed insight into dynamics than a global study of the Q^2 dependence of departures from the GDH integral.

The Q^2 dependence for small Q^2 can expose the threshold behaviour of transition form factors and thereby show how many 'orbital excitation gaps' occur between the wave functions of the nucleon and resonance state. Thus for example the excitation of $P_{11}(1440)$ needs precision data. Its internal quark structure has been variously suggested to be: (i) a radially excited qqq, in which case there would be a large threshold factor in the transition form factor to this radially excited state; (ii) a hybrid baryon, for which the qqq would be in the ground state and so there would be no threshold factors in the excitation form factor, but excitation of the gluonic degrees of freedom is not well understood theoretically; or (iii) a pentaquark ($qqqq\bar{q}$) for which a large suppression may be expected with increasing Q^2.

In general the transitions touch on broader questions of the duality between exclusive coherent excitation of N^* and the inclusive sum rules for incoherent inelastic scattering. This brings us to the question of quark–hadron duality.

1.7.1 Q^2 duality of form factors and deep-inelastic scattering (DIS)

Baryons, in particular the nucleon, appear as qqq states when viewed at low energy. By contrast, in highly inelastic scattering, such as in electron scattering with large momentum transfer Q^2, a richer structure of q, $\bar{q}$ and gluons is resolved. The transition between the low-energy 'constituent' picture and the short-distance 'parton' picture of pQCD is an area of active study both in experiment and theory. This is especially so at machines, such as at JLab, which can both probe the excitation of nucleon resonances as a function of Q^2 and also study the beginnings of the highly inelastic scattering where scale invariance appears in the data. One question among many is whether the coherent excitation of resonances disappears at large Q^2, thereby exposing the incoherent short-distance world of pQCD, or instead whether the former is in some sense dual to the latter [44,45].

Interest has been partly prompted by high-precision data from JLab [46] on the unpolarized F_2 structure function of the proton in the resonance region, which showed a striking similarity, when averaged over resonances, to the structure function measured at much higher energies in the deep-inelastic continuum. This is described in chapter 10. This phenomenon was first observed some time ago by Bloom and Gilman [44], who found that when integrated over the mass of the inclusive hadronic final state, W, the scaling structure function at high Q^2 smoothly averages that measured in the region dominated by low-lying resonances,

$$\int dW\ F_2^{expt}(W^2, Q^2) = \int dW\ F_2^{scaling}(W^2/Q^2) . \qquad (1.33)$$

The integrand on the left-hand side of (1.33) represents the structure function in the resonance region at low Q^2, while that on the right-hand side corresponds to the scaling function, measured in the deep-inelastic region at high Q^2. The latter is described by leading-twist pQCD as an incoherent sum over quark flavours, whereby the structure functions are proportional to $\sum e_i^2$; by contrast the former involves coherent excitation of resonances.

Global duality is said to hold after integration over all W in (1.33). This equality can be related to the suppression of higher-twist contributions to moments of the structure function [47], in which the total moment becomes dominated by the leading-twist (approximately Q^2-independent) component at some lower value of Q^2. Information on all coherent interaction dynamics is subsequently lost. A more local form of duality is also observed [46], in which the equality in (1.33) holds for restricted regions of W integration – specifically, for the three prominent resonance regions at $W \leq 1.8$ GeV. The duality between the resonance and scaling structure functions is also being investigated in other structure functions, such as the longitudinal structure function [48], and spin-dependent structure functions of the proton and neutron [49]. For spin-dependent structure functions, in particular, the workings of duality are more intricate, as the difference of cross sections no longer needs to be positive. An example is the contribution of the $\Delta(1232)$ resonance to the polarization-dependent structure function of the proton, g_1, which is large and negative at low Q^2, but may become positive at higher Q^2.

Early work within the SU(6) symmetric quark model [43,42,50] showed how the ratios of various deep inelastic structure functions at $x = Q^2/2M\nu \sim \frac{1}{3}$, for both spin-dependent and spin-independent scattering, could be dual to a sum over N^* resonances in **56** and **70** representations of SU(6). With the emergence of precision data, showing detailed and interesting x dependence as $x \to 1$, various questions arise:

- How do changes in ratios as $x \to 1$ relate to the pattern of N^* resonances identified in [43,50,51]?
- Are certain families (spin–flavour correlations) of resonances required to die out at large Q^2 in order to maintain duality? If so, can electroproduction of specific examples of such resonances test this?
- Can such a programme reveal the spin–flavour dependence of short distance forces in the QCD bound state?

A detailed discussion of these questions is in [45].

The duality between the simplest SU(6) quark parton model results [52] for ratios of structure functions with sums over the **56** and **70** coherent N^* excitations was described in [42,43,51]. An essential feature of these analyses was that SU(6) was exact and that exotics in the t-channel were suppressed.

Although the *s*-channel sum was shown to be dual for ratios of incoherent quantities [43,50,51], this alone did not explain why (or if) any individual sum over states scaled. The transition from resonances to scaling has been explored in microscopic models at the quark level. The phenomenological quark model duality of [43,50, 51] was shown [53] to arise in a simple model of spinless constituents. A simple model in which the hadron consisted of a point-like scalar 'quark' bound to an infinitely massive core by an harmonic oscillator potential was used [54] to explicitly demonstrate how a sum over infinitely narrow resonances can lead to a structure function which scales in the $Q^2 \to \infty$ limit.

Since the original quark-model predictions were made in the 1970s, the quantity and quality of structure function data have improved dramatically. We now know, for instance, that in some regions of x SU(6) symmetry is badly broken, the strongest deviations from the naive SU(6) expectations being prevalent at large values of x. The new data will set challenges for theories of quark–hadron duality. There are questions such as:

- Can duality survive locally in x, in principle?
- What do the observed variations in x require of N^* excitations if duality is to survive?
- In particular, what families (spin–flavour correlations, or SU(6) multiplets) are suppressed as $x \to 1$, or equivalently $Q^2 \to \infty$, for duality to hold?
- Does the excitation of low-lying prominent N^* resonances, belonging to such families, exhibit such behaviour?

If the $x \to 1$ systematics for N^* families are not matched by specific $N \to N^*$ transition form factors as a function of Q^2, then duality fails. If, however, they do match, then this can expose the patterns in the flavour–spin dependence of short-distance forces in the strong QCD limit.

1.8 Pentaquarks and exotic baryons

The original conception of the constituent-quark model, and of our modern picture, was based on the observation that hadrons exist with (apparently) unlimited amounts of spin, but with only very restricted amounts of electric charge and strangeness. In particular, all baryons (strongly interacting fermions such as the proton) seen hitherto in 60 years of research with cosmic rays or accelerator-based experiments, carry either no strangeness (like the proton and neutron) or **negative** amounts (like the Λ, Σ and Ω^-).

During 2003 a range of experiments claimed that a metastable particle known as the theta baryon may exist [55–59]. Described most simply: it is like a heavier version of the proton but it possesses **positive** strangeness in addition to its positive electrical charge and it is denoted as Θ^+. This made it utterly novel. As the absence of 'positive strangeness baryons' in part is what helped establish the quark model in the first place, the claims were indeed radical.

QCD allows more complicated clusters of quarks or antiquarks and there is good evidence for this. For example, when the proton is viewed at high resolution, as in inelastic electron scattering, its wave function is seen to contain configurations where its three 'valence' quarks are accompanied by further quarks and antiquarks in its 'sea'. The three-quark configuration is thus the simplest required to produce its overall positive charge and zero strangeness. The question thus arises whether there are baryons for which the minimal configuration cannot be satisfied by three quarks.

The most familiar examples of course are atomic nuclei, the deuteron for example requiring a minimum of six quarks; it is the attractive forces of QCD that conspire to cluster these into two triplets such that the neutron and proton remain identifiable within the nucleus (there are interesting questions as to whether the six quarks leave an imprint in the properties of the deuteron that go beyond this simple neutron and proton cluster). A baryon with a positive amount of strangeness would be another such example; in this case the positive strangeness could only be produced by the presence of a strange-antiquark $\bar{s}$, the overall baryon number requiring four further quarks to accompany it. Thus we would have three quarks accompanied by an additional quark and antiquark, making what is known as a 'pentaquark'.

Hitherto unambiguous evidence for such states in the data has been lacking, their absence having been explained by the ease with which they would fall apart into a conventional baryon and a meson. It has been estimated that they would survive for less than 10^{-24} seconds, with widths of several hundred MeV which is at the limits of what is detectable.

The surprise was that the claimed pentaquark had a width of less than 10 MeV, perhaps even ~1 MeV. While this created a challenge for theory, it was also an enigma for experiment. If the width was small, then the production cross section also should be small, yet experiments were claiming to see it produced at a rate comparable to that of conventional hadrons. There were other inconsistencies in the data (for a summary see [60–62]) until a dedicated high-statistics photoproduction experiment at JLab put an upper limit on the production that was less than some previously claimed signals. The pentaquark baryon seems to have been an artefact. However, the large number of theoretical papers addressed to this issue prepared to countenance the possibility of strong correlations in QCD (for example [63,64]) were testimony to how little is understood about the properties of strong QCD. For a detailed summary see [65].

1.8.1 States beyond $q\bar{q}$ and qqq in QCD

If the Θ^+ were to be confirmed it would become the first clear example of a simple hadronic state with exotic quantum numbers that forced one to go beyond qqq or $q\bar{q}$ in QCD. There is, however, evidence of states in the meson sector, albeit with conventional flavour quantum numbers, that appear to be in accord with predictions that the forces in QCD have a strong attraction in certain correlated states and thereby can give rise to multiquark configurations, such as $qq\bar{q}\bar{q}$.

To introduce these, let us begin with the well-understood heavy flavour sector where there are clearly established scalar mesons $c\bar{c}$ and $b\bar{b}$. They behave as canonical $^3P_0$-states which partner $^3P_{1,2}$ siblings. Their production (for example in radiative transitions from 2^3S_1-states) and decays (into 1^3S_1 or light hadrons) are all in accord with this. There is nothing to suggest that there is anything 'exotic' about such scalar mesons.

For light flavours too there are clearly identified $^3P_{1,2}$ nonets which call for analogous $^3P_0$ siblings. However, while all other J^{PC} combinations appear to be realized as expected (apart from well-known and understood anomalies in the 0^{-+} pseudoscalars), the light scalars empirically stand out as singular.

The interpretation of the nature of the lightest scalar mesons has been controversial for over 30 years. There is still no general agreement on where the $q\bar{q}$ states are, whether there is necessarily a glueball among the light scalars, and whether some of the too numerous scalars are multiquark, $K\bar{K}$ or other meson–meson bound states. These are fundamental questions of great importance in particle physics. The mesons with vacuum quantum numbers are known to be crucial for a full understanding of the symmetry breaking mechanisms in QCD, and presumably also for confinement.

Theory and data are now converging that QCD forces are at work but with different dynamics dominating below and above 1 GeV/c^2 mass. The experimental proliferation of light scalar mesons is consistent with two nonets, one in the 1 GeV region (a meson–meson nonet) and another one near 1.5 GeV (a $q\bar{q}$ nonet), with evidence for glueball degrees of freedom. At the constituent level these arise naturally from the attractive interquark forces of QCD. Below 1 GeV these give a strong attraction between pairs of quarks and antiquarks in S-wave leading to a nonet which is 'inverted' relative to the ideal nonets of the simple $q\bar{q}$ model. Conversely, above 1 GeV the states seeded by $^3P_0$ $q\bar{q}$ are present. The scalar glueball, predicted by lattice QCD in the quenched approximation, causes mixing among these states.

As pointed out by Jaffe [22] long ago, there is a strong QCD attraction among qq and $\bar{q}\bar{q}$ in S-wave, 0^{++}, whereby a low-lying nonet of scalars may be expected. As far as the quantum numbers are concerned these states will be like two 0^{-+} $q\bar{q}$

mesons in S-wave. In the latter spirit, Weinstein and Isgur [66] noticed that they could motivate an attraction among such mesons, to the extent that the $f_0(980)$ and $a_0(980)$ could be interpreted as $K\bar{K}$ molecules.

The relationship between these is being debated [67–71], but while the details remain to be settled, the rather compelling message of the data is as follows. Below 1 GeV the phenomena point clearly towards an S-wave attraction among two quarks and two antiquarks (either as $(qq)^{\bar{3}}(\bar{q}\bar{q})^3$, or $(q\bar{q})^1(q\bar{q})^1$, where superscripts denote their colour state), while above 1 GeV it is the P-wave $q\bar{q}$ that is manifested. There is a critical distinction between them: the 'ideal' flavour pattern of a $q\bar{q}$ nonet on the one hand, and of a $qq\bar{q}\bar{q}$ or meson–meson nonet on the other, are radically different; in effect they are flavoured inversions of one another. Thus whereas the former has a single $s\bar{s}$ heaviest, with strange in the middle and $I = 0$; $I = 1$ set lightest (ϕ; K; ω, ρ-like), the latter has the $I = 0$; $I = 1$ set heaviest ($K\bar{K}$; $\pi\eta$ or $s\bar{s}(u\bar{u} \pm d\bar{d})$) with strange in the middle and an isolated $I = 0$ lightest ($\pi\pi$ or $u\bar{u}d\bar{d}$) [22,66].

The phenomenology of the 0^{++} sector appears to exhibit both of these patterns with ~ 1 GeV being the critical threshold. Below 1 GeV the inverted structure of the four-quark dynamics in S-wave is revealed with $f_0(980)$; $a_0(980)$; κ and σ as the labels. One can debate whether these are truly resonant or instead are the effects of attractive long-range t-channel dynamics between the colour-singlet 0^{-+} members in $K\bar{K}$; $K\pi$; $\pi\pi$, but the systematics of the underlying dynamics seems clear.

For the region below 1 GeV, the debate centres on whether the phenomena are truly resonant or driven by attractive t-channel exchanges, and if the former, whether they are molecules or $qq\bar{q}\bar{q}$. The phenomena are consistent with a strong attraction of QCD in the scalar S-wave nonet channels. The difference between molecules and compact $qq\bar{q}\bar{q}$ will be revealed in the tendency for the former to decay into a single dominant channel – the molecular constituents – while the latter will feed a range of channels driven by the flavour spin Clebsch–Gordan coefficients. For the light scalars it has its analogue in the production characteristics.

The picture that is now emerging from both phenomenology [72–74] and theory [75] is that both components are present. As concerns the theory [75], think for example of the two-component picture as two channels. One, the quarkish channel (QQ) is somehow associated with the $(qq)_3(\bar{q}\bar{q})_3$ coupling of a two-quark–two-antiquark system, and is where the attraction comes from. The other, the meson–meson channel (MM) could be completely passive (for example no potential at all). There is some off-diagonal potential which flips that system from the QQ channel to MM. The way the object appears to experiment depends on the strength of the attraction in the QQ channel and the strength of the off-diagonal potential. The nearness of the f_0 and a_0 to the $K\bar{K}$ threshold suggests that the QQ component cannot be too dominant, but the fact that there is an attraction at all means that the QQ component cannot be negligible. So in this line of argument, a_0 and f_0 must be superpositions of four-quark states and $K\bar{K}$ molecules.

1.9 Gluonic excitations

Lattice QCD predictions for the mass of the lightest (scalar) glueball are now mature. In the quenched approximation the mass is ~1.6 GeV [77,76,78–80]. Flux-tube models imply that if there is a $q\bar{q}$ nonet nearby, with the same J^{PC} as the glueball, then $G - q\bar{q}$ mixing will dominate the decay [81]. This is found more generally [82] and recent studies on coarse-grained lattices appear to confirm that there is indeed significant mixing between G and $q\bar{q}$ together with associated mass shifts, at least for the scalar sector [83].

Furthermore, the maturity of the $q\bar{q}$ spectrum tells us that we anticipate the $0^{++} q\bar{q}$ nonet to occur in the 1.2–1.6 GeV region. There are the following candidates $a_0(\sim 1400);\ f_0(1370);\ K(1430);\ f_0(1500)$ and $f_0(1710)$.

One immediately notes that if all these states are real there is an excess, precisely as would be expected if the glueball predicted by the lattice is mixing in this region. Any such states will have widths and so will mix with a scalar glueball in the same mass range. It turns out that such mixing will lead to three physical isoscalar states with rather characteristic flavour content [78,72]. Specifically; two will have the $n\bar{n}$ and $s\bar{s}$ in phase ('singlet tendency'), their mixings with the glueball having opposite relative phases; the third state will have the $n\bar{n}$ and $s\bar{s}$ out of phase ('octet tendency') with the glueball tending to decouple in the limit of infinite mixing. There are now clear sightings of prominent scalar resonances $f_0(1500)$ and $f_0(1710)$ and probably also $f_0(1370)$. Confirming the resonant status of the last is one of the critical pieces needed to clinch the proof. The production and decays of these states are in remarkable agreement with this flavour scenario [72].

Precision data on scalar meson production and decay are consistent with this and the challenge now centres on clarifying the details and extent of such mixing.

A major question is whether the effects of the glueball are localized in this region above 1 GeV, as discussed by [72,23] or spread over a wide range, perhaps down to the $\pi\pi$ threshold [84]. This is the phenomenology frontier. There are also two particular experimental issues that need to be settled: to confirm the existence of $a_0(1400)$ and determine its mass and to determine whether the $f_0(1370)$ is truly resonant or is a t-channel exchange phenomenon associated with $\rho\rho$.

1.9.1 Hybrids and glueballs

While QCD implies that gluons can mutually couple, and thereby offers the prospect of bound states of pure glue, 'glueballs', we have no general insight into the spectroscopy and widths of these states. According to lattice QCD, the lightest glueball has $J^{PC} = 0^{++}$ and in the absence of mixing with $q\bar{q}$ or hadrons, has a mass ~ 1.6 GeV [78]. In this idealized scenario the next states are expected to be 2^{++} and 0^{-+} with masses ~ 2 GeV. A glueball with the same quantum numbers as the photon is not expected until $m \sim 4$ GeV. The lightest glueball with exotic J^{PC} is predicted to be 0^{+-} with $m \sim 3.0 \pm 0.8$ GeV.

Neither lattice QCD nor models give any clear guide as to the widths of glueballs. Ideas based on perturbative QCD and the Zweig rule have led to suggestions that they could be narrow. A lattice study of the 0^{++} gave the decay width into two pseudoscalars ~ 100 MeV, akin to that of conventional hadrons. There are even phenomenological analyses suggesting that the scalar glueball at least mixes strongly with mesons, affecting the spectrum of $I = 0$ states in the region 1.3 to 1.7 GeV [81,72] or even across a broader mass range [84].

As gluons are electrically neutral, how might photons aid in identifying their dynamics? Four areas come to mind: (i) the production of a $C = +$ glueball recoiling against a photon in $Q\bar{Q} \rightarrow \gamma G$; (ii) the measurement of the q^2 evolution of a glueball's structure functions, which will increase in strength as the gluons are resolved into $q\bar{q}$ pairs in contrary fashion to the evolution of 'normal' hadrons; (iii) the comparison of the production of a glueball in $gg \rightarrow G$ or in central production $pp \rightarrow pGp$, which are strong as in (i) above, with $\gamma\gamma \rightarrow G$, which is suppressed due to the glueball's electrical neutrality; (iv) looking for mixing of a glueball with the $I = 0$ members of a $q\bar{q}$ nonet that may be revealed by radiative transitions involving ideally mixed flavour states, for example $M_n(C = +) \rightarrow \gamma\rho$ and $M_s(C = +) \rightarrow \gamma\phi$ weigh the $n\bar{n}$ and $s\bar{s}$ flavour contents respectively of positive charge conjugation states [7,61].

In $\psi \rightarrow \gamma R$, pQCD implies that the rate is sensitive to $J^{PC}(R)$. Thus one has to allow for the J^{PC} before naively assuming that a large branching ratio implies $R \equiv G$. Close *et al* [85] have quantified this and argue that the production rate for a pure glueball with $m_G \sim 1.5$ GeV depends on its J^{PC} and total width thus

$$10^3 b(\psi \rightarrow \gamma G[0^{++}; 2^{++}; 0^{-+}]) \sim \Gamma(\text{MeV})/[96; 26; 50], \qquad (1.34)$$

whereas for a $q\bar{q}$ the scaled factors are ~ 5–10 times larger, for example for 0^{++} being $O(500$–$1000)$. Within this uncertainty therefore

$$10^3 b(\psi \rightarrow \gamma q\bar{q}[0^{++}; 2^{++}; 0^{-+}]) \sim \Gamma(\text{MeV})/[500$–$1000; 150$–$300; 250$–$500].$$
$$(1.35)$$

This result is only qualitative as it is necessary to take account of the SU(3) flavour wave functions of the $q\bar{q}$ (see [85]) but shows why $q\bar{q}$ with 2^{++} are prominent whereas those with 0^{++} are harder to isolate. With the advent of high-statistics data from CLEO-c and BES the rates as a function of J^{PC} need to be both better quantified experimentally and put on a more solid theoretical foundation.

Using the existing analysis of [85] as a benchmark against which improvements should be made we note the following.

(a) There does appear to be an enhanced rate in the 0^{-+} channel around 1.4 GeV consistent with the presence of a glueball, reinforcing early suggestions to this effect. Such a mass is lower than lattice QCD predicts and it is principally this feature that has led to reluctance to accept this as a prima facie glueball. Close *et al* [85] concluded that the $\eta(1440)$ data in $\psi \to \gamma\eta(1440)$ appeared to separate into two mesons. The lower-mass state $\eta_L(1410)$ has a strong affinity for glue, or mixing with the η' via the $U_A(1)$ anomaly, whereas the higher-mass state $\eta_H(1480)$ is consistent with being the $s\bar{s}$ member of a nonet, perhaps mixed with glue.

(b) In the 0^{++} channel there appears to be significant strength in the 1.5–1.7 GeV region where the $f_0(1500)$ and $f_0(1700)$ are seen. The $f_0(1710)$ at least appears at a strength consistent with at least half of its strength being gluonic. Resolving the presence of any further resonant scalars such as the role of $f_0(1370)$ will also be important in this programme.

A challenge for BES and CLEO-c will be to investigate these states at higher precision in ψ decays. In particular the decays of these mesons into $\gamma V(\equiv \rho, \omega, \phi)$ can probe their flavour and hence be done in concert with programme (iv) [7].

1.9.2 Hybrid production by photons

An essential step in the phenomenological study of the origin and nature of confinement is to identify the spectrum of hybrid mesons. Lattice QCD predicts such a spectroscopy [86], but there are no unambiguous signals against which these predictions can be tested.

A major stumbling block had been that while predictions for their masses [86,14], hadronic widths [89,88] and decay channels [89,88,87] were rather well agreed upon, the literature contained little discussion of their production rates in electromagnetic interactions. Meanwhile, a significant plank in the proposed upgrade of JLab has been its assumed ability to expose the predicted hybrid mesons in photo- and electroproduction.

Lattice QCD has demonstrated that a string-like chromoelectric flux-tube forms between distant static quarks [12]. In the simplest situation of a long tube with fixed $Q, \bar{Q}$ sources on its ends, a flux-tube has a simple vibrational spectrum corresponding to the excitation of transverse phonons in its string-like structure. The essential features of this gluonic spectrum are retained in the spectrum of real mesons with their flux-tube excited – the hybrid mesons. Implications for spectroscopy and hadronic decays in such a flux-tube model [14] have been extensively explored. However, until the late 1990s the only estimates of electromagnetic couplings in this model [90,91] were at best upper limits, in that they were based upon vector meson dominance of hadronic decays into modes including ρ and assumed that certain selection rules [88] against this mode were suspended in π exchange.

A calculation [92] has quantified the conjecture of Isgur [93] that electromagnetic excitation amplitudes of hybrids may be significant. In particular this was found to be expected for E1 transitions between conventional and hybrid mesons for electrically charged states. This could have implications for $\gamma p \to n \mathcal{H}^+$.

The lightest hybrids with exotic J^{PC} (0^{+-}, 1^{-+}, 2^{+-}) have their $q\bar{q}$ coupled to spin 1. As the photon is already a vector particle there has emerged a folklore that photons may therefore be a good source of hybrid production since in photoproduction it is only necessary then to 'tickle the flux tube' to have the possibility of an exotic J^{PC}. The more detailed calculations in [92,94,95] find no support for this belief. The excitation of the tube occurs as a combination of the photoexcitation of a quark which is displaced from the centre of mass of the $q\bar{q}$-flux-tube system. This leads to non-zero overlap with excited hybrid states but the net spin of the $q\bar{q}$ need not be trivially conserved.

Nonetheless, photoproduction of the exotic states is expected to be significant [92, 94,95]. It is found that the electric dipole transitions of the hybrid axial meson to $\pi^{\pm}\gamma$ and of the exotic $(0,2)^{+-}$ to $\rho\gamma$ give radiative widths that can exceed 1 MeV. This implies significant photoproduction rates in charge-exchange reactions $\gamma p \to H^+ n$. The exotic 2^{+-} may also be produced diffractively in $\gamma p \to H p$. The exotic 1^{-+} may be produced by π exchange in $\gamma N \to 1^{-+} N$. Depending on its total width, the resulting signal may compare with that for a_2 production [96]. The exotic 2^{+-} should be more readily produced [95]. It is predicted to have photoexcitation from π, with a strength that is comparable to that of the 1^{-+}, and in addition can be diffractively excited from a gluonic pomeron, $\gamma N \to 2^{+-} N$. Hence a test for hybrid production is to seek 2^{+-} and compare its production amplitudes with those of 1^{-+}.

There is also a spectroscopy of non-exotic J^{PC} and some of these are predicted to have significant photoproduction strength. In particular, the relative (reduced) widths of hybrid $a_{1H}(1^{++})$ and the conventional $b_{1Q}(1^{+-})$ are predicted in the flux-tube model to be

$$\frac{\Gamma_{E1}(a_{1H}^+ \to \pi^+\gamma)}{\Gamma_{E1}(b_{1Q}^+ \to \pi^+\gamma)} = \frac{72}{\pi^3} \frac{k}{m_n^2} \left| \frac{H\langle r\rangle_\pi}{b\langle r\rangle_\pi} \right|^2, \tag{1.36}$$

where k is the string tension $k \sim 1$ GeV/fm, m_n is the constituent-quark mass and the $\langle r \rangle$ factors are measures of the charge radii. This ratio should be studied in lattice QCD as it is the most direct measure of the price for exciting the gluonic degrees of freedom in strong QCD, at least if the flux-tube model is a guide to the dynamics.

Hybrid mesons with heavy flavours, in particular hybrid charmonium states, are anticipated. If a hybrid vector charmonium is above the threshold for decay to $D\bar{D}_1$, then it is predicted [88] that the dominant decay will be to this $D\bar{D}_1$ state and that decays to pairs of ground-state charmed mesons, such as $D\bar{D}$, $D_s\bar{D}_s$, $D^*\bar{D}^*$, are suppressed. It is thus interesting that a vector state $\Upsilon(4260)$ has been seen in $e^+e^- \to \psi\pi\pi$ [97] with hadronic width ~ 90 MeV and a leptonic width that has been inferred to be ~ 100 eV [98]. Its leptonic width, being much smaller than those of other vector mesons [1], and its significant decay width to $\psi\pi\pi$ testify to its unusual nature. It has been suggested that it may be a vector hybrid charmonium [99], which had been predicted [88,100], or a tetraquark state, $cs\bar{c}\bar{s}$ [99]. A test of these competing hypotheses is in their predictions of decay modes. With the tetraquark hypothesis the $\Upsilon(4260)$ will decay to $D_s\bar{D}_s$, while this mode is suppressed for a hybrid. As hybrid charmonium is predicted to occur around 4 GeV in mass and a vector state to couple strongly to $D\bar{D}_1$ [88], the fact that the $\Upsilon(4260)$ is near to the DD_1 threshold is already of interest.

In principle, the ψ in the $\psi\pi\pi$ final state can be polarized. As a hybrid vector meson has its $c\bar{c}$ coupled to spin 0, in contrast to conventional vector mesons where the $c\bar{c}$ are in spin 1, one may expect that the polarization will differ in these scenarios, though detailed model predictions are not yet agreed upon. Empirically one may study the analogous process at lower energies in the strange sector. The process $e^+e^- \to K\bar{K}_1 \to \phi\pi\pi$ has been seen; the polarization of the ϕ and the mass distribution of $\pi\pi$ should be measured here and compared with their counterparts in the charm sector $e^+e^- \to D\bar{D}_1 \to \psi\pi\pi$.

This is potentially an exciting example of how comparison of e^+e^- annihilation at energies appropriate to different flavours can enable the underlying dynamics to be explored.

From a dedicated programme in such photo- and electroproduction evidence for states with exotic quantum numbers should emerge. If they are not seen then there is something seriously amiss in our intuition. However, the observation of exotic quantum numbers does not of itself imply the production of hybrid states. One can form such combinations from $qq\bar{q}\bar{q}$. A detailed spectroscopy is required, in particular of an entire multiplet, to ascertain whether it forms a regular nonet (as in the simplest $q\bar{q}$ cases) or an inverted one (as for the 0^{++} states below 1 GeV). Only when this evidence is available can one decide with any confidence whether exotic mesons are $qq\bar{q}\bar{q}$ states (the latter case) or hybrids (the former). This brings us to the more general question of how reliable the simple constituent picture should be expected to be and what further tests might be envisaged.

1.9.3 When does the quark model work?

There is general agreement that the NRQM is a good phenomenology for $b\bar{b}$ and $c\bar{c}$ states below their respective flavour thresholds. Taking $c\bar{c}$ as an example we have S-states (η_c, ψ, $\psi(2S)$), P-states ($\chi_{0,1,2}$) and a D-state ($\psi(3772)$), the last just above the $D\bar{D}$ threshold. Their masses and the strengths of the E1 radiative transitions between $\psi(2S)$ and χ_J are in reasonable accord with their potential model status. In particular there is nothing untoward about the scalar states.

Do the same for the light flavours and one finds clear multiplets for the 2^{++} and 1^{++} states (though the a_1 is rather messy); it is when one comes to the scalars that suddenly there is an excess of states. An optimist might suggest that this is the first evidence that there is an extra degree of (gluonic) freedom at work in the light scalar sector. But there is more: there is a clear evidence of states that match onto either $qq\bar{q}\bar{q}$ or correspondingly meson–meson in S-wave.

Such a situation is predicted by the attractive colour–flavour correlations in QCD [22,23]. Establishing this has interest in its own right but it is also necessary to ensure that one can classify the scalar states and then identify the role of any glueball by any residual distortion in the spectrum. It is in this context that discovery of narrow states in the heavy flavour sector provide tantalizing hints of this underlying dynamics elsewhere in spectroscopy. If this is established it could lead to a more unified and mature picture of hadron spectroscopy.

The sharpest discoveries have involved narrow resonances: $c\bar{s}$ states, probably 0^+, 1^+, lying just below $D\bar{K}$, $D^*\bar{K}$ thresholds; and $c\bar{c}$ degenerate with the $D^o\bar{D}^{*o}$ threshold. These are superficially heavy-flavour states, but their attraction to these thresholds involves light quarks and links to a more general theme.

First note [5] that the $c\bar{c}$ potential picture gets significant distortions from the DD threshold region, such that even the $c\bar{c}$ χ states can have 10% or more admixtures of meson pairs, or four-quark states, in their wave functions. Also simple potential models of the D_s-states are inadequate to explain the 2.32 GeV and 2.46 GeV masses of these novel states as simply $c\bar{s}$ in some potential. Furthermore, the lattice seems to prefer the masses to be higher than actually observed, though the errors here are still large. In summary, there is an emerging picture that these data on the $\bar{D}_s$ sector (potentially 0^+ and 1^+ and the S-wave $D\bar{K}$ and $D^*\bar{K}$ thresholds) and the $c\bar{c}$ sector (with the S-wave $D^o\bar{D}^{*o}$ thresholds) confirm the suspicion that the simple potential models fail in the presence of S-wave continuum threshold(s).

Now let us examine this in the light flavoured sector. The multiplets where the quark model works best are those where the partial wave of the $q\bar{q}$ or qqq is lower than that of the hadronic channels into which they can decay. For example, the ρ is S-wave $q\bar{q}$ but P-wave in $\pi\pi$; as S-wave is lower than P-wave, the quark model wins; by contrast the σ is P-wave in $q\bar{q}$ but S-wave in $\pi\pi$ and in this case it is the meson sector that wins and the quark model is obscured.

A similar message comes from the baryons. The quark model does well for the Δ (S-wave in qqq but P-wave in πN); at the P-wave qqq level it does well for the D_{13}, which as its name implies is D-wave in hadrons, but poorly for the S_{11} which is S-wave in $N\eta$. The story repeats in the strange sector where the strange baryons with negative parity would be qqq in P-wave: the $D_{03}(1520)$ is fine but the $S_{01}(1405)$ is the one that seems to be contaminated with possible KN bound-state effects.

As an exercise I invite you to check this out. It suggests a novel way of classifying the Fock states of hadrons. Instead of classifying by the number of constituent quarks, list by the partial waves with the lowest partial waves leading. Thus for example

$$0^{++} = |0^-0^-(qq\bar{q}\bar{q})\rangle_S + |q\bar{q}\rangle_P + \cdots ,$$

while

$$1^{--} = |q\bar{q}\rangle_S + |0^-0^-(qq\bar{q}\bar{q})\rangle_P + \cdots$$

or

$$\Delta(1230) = |qqq\rangle_S + |\pi N(qqqq\bar{q})\rangle_P + \cdots .$$

This holds true for

$$2^{++} = |q\bar{q}\rangle_P + |0^-0^-(qq\bar{q}\bar{q})\rangle_D + \cdots$$

the relevant S-wave vector-meson pairs being below threshold. For the remaining P-wave $q\bar{q}$ nonet with $C = +$ we have a delicate balance

$$1^{++} = |q\bar{q}\rangle_P + |0^-1^-\rangle_S + \cdots ,$$

where the $\pi\rho$ S-wave distorts the $q\bar{q}$ a_1, as is well known; the $f_1(1285)$ is protected because the two-body modes are forbidden by G-parity; for the strange mesons the $K^*\pi$ and $K\rho$ channels play significant roles in mixing the 1^{++} and 1^{+-} states, while the $s\bar{s}$ state is on the borderline of the $K\bar{K}^*$ threshold.

Chiral models which focus on the hadronic colour-singlet degrees of freedom are thus the leading effect for the 0^{++} sector but subleading for the vectors. An example has been presented [101] where the N_c dependence of the coefficients of the chiral Lagrangian was studied. In the $N_c \to \infty$ limit it was found that $\Gamma(\rho) \to 0$, like

$q\bar{q}$, whereas $\Gamma(\sigma) \to \infty$, like a meson S-wave continuum. Thus there appears to be a consistency with the large N_c limit selecting out the leading S-wave components.

Conversely, the 'valence' quark model can give the leading description for the vectors or the $\Delta(1232)$ but there will be corrections that can be exposed by fine-detail data. The latter are now becoming available for the baryons from JLab; the elastic form factors of the proton and neutron show their charge and magnetic distributions to be rather subtle, and the transition to the $\Delta(1232)$ is more than simply the M1 dominance of the quark model. There are E2 and scalar multipole transitions which are absent in the leading qqq picture. The role of the πN cloud is being exposed; it is the non-leading effect in the above classification scheme. The $\Theta^+(1540)$ as a pentaquark may inspire novel insights into a potential pentaquark – or $N\pi$ cloud – component in the N and $\Delta(1232)$.

The message is to start with the best approximation – quark model or chiral – as appropriate and then seek corrections.

Bearing these thoughts in mind highlights the dangers of relying too literally on the quark model as a leading description for high-mass states unless they have high J^{PC} values for which the S-wave hadron channels may be below threshold. It also has implications for identifying glueballs and where the gluonic degrees of freedom play an explicit role and cannot simply be subsumed into the collective quasiparticle known as the constituent quark.

The lightest glueball is predicted to be scalar [78] for which the problems arising from the S-wave thresholds have already been highlighted. At least here, by exploiting the experimental strategies outlined in this overview, we are possibly going to be able to disentangle the complete picture. For the 2^{++} and 0^{-+} glueballs above 2 GeV there are copious S-wave channels open, which will obscure the deeper 'parton' structure. Little serious thinking seems to have been done here.

For the exotic hybrid nonet 1^{-+} we have a subtlety. In the flux-tube models abstracted from lattice QCD, the $q\bar{q}$ are in an effective P-wave [13,14,92], which we may describe by $|q\bar{q}g\rangle_P$. There is a leading S-wave 0^-1^+ meson pair at relatively low energies, such that

$$1^{-+} = |0^-1^+\rangle_S + |q\bar{q}g\rangle_P + \cdots.$$

The S-wave thresholds for πb_1 and πf_1 are around 1400 MeV, which is significantly below the predicted 1.8 GeV for lattice or model and tantalizingly in line with one of the claimed signals for activity in the 1^{-+} partial wave. All is not lost however; a $q\bar{q}$ or $q\bar{q}g$ nonet will have a mass pattern and decay channels into a variety of final states controlled by Clebsch–Gordan coefficients, whereas thresholds involve specific meson channels. These can, in principle, be sorted out, given enough data in a variety of production and decay channels, but it may be hard.

The essential message is that in unravelling the dynamics of hadrons, especially in the light flavour sector, no single experiment will suffice. The strategy involves a range of complementary probes and detailed assessment of their results. The electromagnetic probe, on which this book is focussed, has the special virtue of coupling in leading order to charged constituents. Electric and magnetic couplings can reveal the correlations of flavours and spins. Comparison with other data where gluonic channels are dominant can reveal patterns among hadrons which can disentangle the flavoured q, $\bar{q}$ content and expose electrically neutral gluonic degrees of freedom. Such ideas are not new but now promise to come into sharper focus than hitherto with the advent of a new generation of high-luminosity customized facilities. To this end, it is a purpose of this volume to review and sharpen the arguments and theory of the electromagnetic interactions of hadrons.

References

[1] S Eidelman *et al*, Particle Data Group, Physics Letters B592 (2004) 1
[2] R K Ellis, W J Stirling and B R Webber, *QCD and Collider Physics*, Cambridge University Press, Cambridge (1996)
[3] E Eichten *et al*, Physical Review D17 (1978) 3090
[4] E Eichten *et al*, Physical Review D21 (1980) 203
[5] T Barnes, in *Proceedings of Hadron03*, AIP Conference Proceedings 717 (2004)
[6] S Godfrey, G Karl and P J O'Donnell, Zeitschrift für Physik C31 (1986) 77
[7] F E Close, A Donnachie and Yu S Kalashnikova, Physical Review D65 (2002) 092003
[8] K Abe *et al*, BELLE Collaboration, hep-ex/0308029
[9] F E Close and P R Page, Physics Letters B578 (2004) 119
[10] N A Tornqvist, Physics Letters B590 (2004) 209
[11] G Bali *et al*, SESAM Collaboration, Nuclear Physics Proceedings Supplement 63 (1997) 209
[12] G Bali *et al*, SESAM Collaboration, Physical Review D62 (2000) 054503
[13] N Isgur and J Paton, Physics Letters 124B (1983) 247
[14] N Isgur and J Paton, Physical Review D31 (1985) 2910
[15] R L Jaffe and K Johnson, Physics Letters 60B (1976) 201
[16] T Barnes and F E Close, Physics Letters 116B (1982) 365
[17] T Barnes, F E Close and F de Viron, Nuclear Physics B224 (1983) 241
[18] M Chanowitz and S R Sharpe, Nuclear Physics B222 (1983) 211
[19] Y B Zeldovich and A D Sakharov, Soviet Journal of Nuclear Physics 4 (1967) 283
[20] A de Rujula, H Georgi and S L Glashow, Physical Review D12 (1975) 147
[21] F E Close, *Introduction to Quarks and Partons* Academic Press, London (1979)
[22] R L Jaffe, Physical Review D15 (1977) 281;
    R L Jaffe and F E Low, Physical Review D19 (1979) 2105
[23] F E Close and N Tornqvist, Journal of Physics G28 (2002) R249
[24] F E Close, A Donnachie and Yu S Kalashnikova, Physical Review D67 (2003) 074031

[25] F E Close and P R Page, Physical Review D56 (1997) 1584
[26] T Barnes, N Black and P R Page, Physical Review D68 (2003) 054014
[27] H G Blundell, S Godfrey and B Phelps, Physical Review D53 (1996) 3712
[28] H J Lipkin, Physics Letters 72B (1977) 250; Physical Review 176 (1968) 1709
[29] F E Close, Physics Letters 43B (1973) 422
[30] A Donnachie and G Shaw, Nuclear Physics 87 (1967) 556
[31] V E Krohn and C E Ringo, Physical Review D8 (1973) 1305
[32] V V Frolov et al, Physical Review Letters 82 (1998) 45
[33] M K Jones et al, Physical Review Letters 84 (2000) 1398
[34] F E Close and R H Dalitz, in *Low and Intermediate Energy KN Physics*, E Ferrari
    and G Violinin eds, Reidel, Dordrecht (1981)
[35] L Y Glozman and D Riska, Physics Reports 268 (1996) 263
[36] N Isgur, Physical Review D62 (2000) 054026
[37] N Isgur and G Karl, Physics Letters 72B (1977) 109
[38] R Hofstadter and R W McAllister, Physical Review 98 (1955) 183
[39] V L Chernyak and A R Zhitnisky, JETP Letters 25 (1977) 510
[40] N Isgur, G Karl and D W L Sprung, Physical Review D23 (1981) 163
[41] A W Thomas, Advances in Nuclear Physics 13 (1984) 1
[42] F E Close and F J Gilman, Physics Letters 38B (1972) 541
[43] F E Close, F J Gilman and I Karliner, Physical Review D6 (1972) 2533
[44] E Bloom and F J Gilman, Physical Review Letters 25 (1970) 1140; Physical Review
    D4 (1971) 2901
[45] F E Close and W Melnitchouk, Physical Review C68 (2003) 035210
[46] I Niculescu et al, Physical Review Letters 85 (2000) 1182; *ibid* 85 (2000) 1186
[47] A De Rujula, H Georgi and H D Politzer, Annals of Physics 103 (1977) 315
[48] M E Christy, in *Proceedings of the 9th International Conference on the Structure of
    Baryons*, Jefferson Laboratory, March 2002, World Scientific, Singapore (2002)
[49] V D Burkert, *AIP Conference Proceedings* 603 (2001) 3
    T Forest, in *Proceedings of the 9th International Conference on the Structure of
    Baryons*, Jefferson Laboratory, March 2002, World Scientific, Singapore (2002)
[50] F E Close, H Osborn and A M Thomson, Nuclear Physics B77 (1974) 281
[51] F E Close and F J Gilman, Physical Review D7 (1972) 2258
[52] J Kuti and V F Weisskopf, Physical Review D4 (1971) 3418
[53] F E Close and N Isgur, Physics Letters B509 (2001) 81
[54] N Isgur, S Jeschonnek, W Melnitchouk and J W Van Orden, Physical Review D64
    (2001) 054005
[55] T Nakano et al, LEPS Collaboration, Physical Review Letters 91 (2003) 012002
[56] S Stepanyan et al, CLAS Collaboration, Physical Review Letters 91 (2003) 252001
[57] V V Barmin et al, DIANA Collaboration, Physics of Atoms and Nuclei 66 (2003)
    1715
[58] J Barth et al, SAPHIR Collaboration, Physics Letters B572 (2003) 127
[59] F E Close, in *Proceedings of Hadron03*, AIP Conference Proceedings, 717 (2004)
[60] F E Close and Qiang Zhao, Journal of Physics G31 (2005) L1
[61] F E Close, in *Proceedings of ICHEP04*, H Chen, D Du, W Li and C Lu eds, World
    Scientific, Singapore (2005)
[62] Shin Jan, in *Proceedings of ICHEP04*, H Chen, D Du, W Li and C Lu eds, World
    Scientific, Singapore (2005)
[63] R L Jaffe and F Wilczek, Physical Review Letters 91 (2003) 232003
[64] M Karliner and H J Lipkin, hep-ph/0307243
[65] V Burkert, International Journal of Modern Physics A21 (2006) 1764

[66] J Weinstein and N Isgur, Physical Review Letters 48 (1982) 659; Physical Review D27 (1983) 588

[67] N A Tornqvist, Zeitschrif für Physik C68 (1995) 647

[68] M Boglione and M R Pennington, Physical Review Letters 79 (1997) 1998

[69] N N Achasov and G N Shestakov, Physical Review D56 (1997) 212

[70] N N Achasov, S A Devyanin and G N Shestakov, Physics Letters 88B (1979) 367

[71] O Krehl, R Rapp and J Speth, Physics Letters B390 (1997) 23

[72] F E Close and A Kirk, Physics Letters B483 (2000) 345

[73] F E Close and A Kirk, Physics Letters B515 (2001) 13

[74] A Aloisio *et al*, KLOE Collaboration, hep-ex/0107024

[75] R L Jaffe (private communication)

[76] J Sexton *et al*, IBM Collaboration, Physical Review Letters 75 (1995) 4563

[77] G Bali *et al*, Physics Letters B309 (1993) 378

[78] F E Close and M J Teper, *On the lightest scalar glueball* Rutherford Appleton Laboratory report RAL-96-040; Oxford University report no. OUTP-96-35P

[79] C J Morningstar and M Peardon, Physical Review D56 (1997) 4043

[80] D Weingarten, Nuclear Physics Proceedings Supplement 73 (1999) 249

[81] C Amsler and F E Close, Physics Letters B353 (1995) 385

[82] V V Anisovich, Physics-Uspekhi 41 (1998) 419

[83] C McNeile and C Michael, Physical Review D63 (2001) 114503

[84] P Minkowski and W Ochs, European Journal of Physics C9 (1999) 283

[85] F E Close, G Farrar and Z Li, Physical Review D55 (1997) 5749

[86] P Lacock, C Michael, P Boyle and P Rowland, Physics Letters B401 (1997) 308; C Bernard *et al*, Physical Review D56 (1997) 7039; P Lacock and K Schilling, Nuclear Physics Proceedings Supplement 73 (1999) 261; C McNeile, hep-lat/9904013; C Morningstar, Nuclear Physics Proceedings Supplement 90 (2000) 214

[87] C Michael, hep-ph/0009115

[88] F E Close and P R Page, Nuclear Physics B443 (1995) 233

[89] N Isgur, R Kokoski and J Paton, Physical Review Letters 54 (1985) 869

[90] F E Close and P R Page, Physical Review D52 (1995) 1706

[91] A Afanasev and P R Page, Physical Review D57 (1998) 6771

[92] F E Close and J J Dudek, Physical Review Letters 91 (2003) 142001–1

[93] N Isgur, Physical Review D 60 (1999) 114016

[94] F E Close and J J Dudek, Physical Review D70 (2004) 094015

[95] F E Close and J J Dudek, Physical Review D69 (2004) 034010

[96] A Afanasev, F E Close and J J Dudek, (unpublished) (2004)

[97] B Aubert *et al*, BaBar Collaboration, Physical Review Letters 95 (2005) 142001

[98] F E Close and P R Page, Physics Letters B628 (2005) 215

[99] L Maiani, F Piccinini, A D Polosa and V Riquer, Physical Review D72 (2005) 031502

[100] F E Close and S Godfrey, Physics Letters B574 (2003) 210

[101] J Pelaez, in *Proceedings of Hadron03*, AIP Conference Proceedings 717 (2004)

2

Elastic form factors

R Ent

A fundamental goal of nuclear and particle physics is to understand the structure and behaviour of strongly interacting matter in terms of its fundamental constituents. An excellent model of atomic nuclei is that they consist of protons and neutrons interacting by the exchange of pions. This is the view of the nucleus studied at low resolution. Hence, protons, neutrons and pions can be considered as the building blocks of the nuclei, a description valid for almost all practical purposes. Nonetheless, these building blocks are themselves composite particles, and studying their electromagnetic structure has been one of the main research thrusts of electron scattering.

The first studies of the electromagnetic structure of nucleons using energetic electron beams as probes began in the 1950s with the work by Hofstadter *et al* [1]. The experimental goal in these early measurements and in those that followed was to understand how the electromagnetic probe interacts with the charge and current distributions within nucleons. The embodiment of these interactions are the electromagnetic form factors. These quantities could be calculated if a complete theory of hadron structure existed. In the absence of such a theory, they provide an excellent meeting ground between experimental measurements and model calculations.

These nucleon form-factor measurements become increasingly difficult at high Q^2 because the cross section is found to fall as $\sim Q^{-12}$ at high Q^2 and the counting rates drop correspondingly. However, experimental techniques have progressed greatly since the early experiments and, together with the exploitation of spin degrees of freedom, have produced an impressive data set out to large Q^2.

The large range in Q^2 accessible to modern experiments, combined with the high precision of the data, allow not only a precise determination of global properties such as the charge radius of these nuclear building blocks, but also the study of

Electromagnetic Interactions and Hadronic Structure, eds Frank Close, Sandy Donnachie and Graham Shaw. Published by Cambridge University Press. © Cambridge University Press 2007

their detailed internal structure. These data have instigated considerable theoretical interest, although there is still much to be learned about the nucleon by refining both measurement and theory.

2.1 Electron scattering and form-factor measurements

Under Lorentz invariance, spatial symmetries and charge conservation, the most general form of the electromagnetic current of a nucleon can be written as

$$J^\mu = -ie\left[F_1(Q^2)\gamma^\mu + \frac{\kappa_N}{2m_N}F_2(Q^2)i\sigma^{\mu\nu}q_\nu\right], \tag{2.1}$$

where m_N is the nucleon mass, F_1 denotes the helicity-non-flip Dirac form factor, F_2 the helicity-flip Pauli form factor and $e\kappa_N/2m_N$ is the nucleon anomalous magnetic moment. The form factors at $Q^2 = 0$ are

$$F_1^p(0) = F_2^p(0) = F_2^n(0) = 1, \; F_1^n(0) = 0 \tag{2.2}$$

and $\kappa_p = 1.79, \kappa_n = -1.91$. If q^2 is the momentum transfer to the nucleon, in electron scattering q^2 is space-like, that is $q^2 < 0$ and we define $Q^2 = -q^2$. Conversely, if q^2 is time-like, that is $q^2 > 0$, as in e^+e^- annihilation, we define $Q^2 = q^2$ so that Q^2 is always positive.

Since isospin is a well-preserved symmetry in strong interactions, one can combine the proton and neutron form factors to give the isoscalar (s) and isovector (v) form factors of the nucleon:

$$F_i^s = \tfrac{1}{2}\left(F_i^p + F_i^n\right); \quad F_i^v = \tfrac{1}{2}\left(F_i^p - F_i^n\right); \quad (i = 1, 2). \tag{2.3}$$

In the space-like region ($q^2 < 0$) the form factors can be measured through electron scattering. In the time-like region ($q^2 > 0$) the form factors can be measured through the creation or annihilation reactions $e^+e^- \to N\bar{N}$ or vice-versa. The form factors are analytic in the complex q^2 plane and satisfy a dispersion relation of the form:

$$F(t) = \frac{1}{\pi}\int_{t_0}^\infty dt' \frac{\text{Im}\, F(t')}{t' - t}, \tag{2.4}$$

with $t = q^2$. G-parity requires that $t_0 = 9(4)m_\pi^2$ for the isoscalar (isovector) case, where m_π is the pion mass.

For unpolarized electron scattering with single-photon exchange, the elastic differential cross section is [2]

$$\frac{d\sigma}{d\Omega} = \frac{E'}{E}\sigma_M\left[\left(F_1^2 + \kappa^2\tau F_2^2\right) + 2\tau(F_1 + \kappa F_2)^2 \tan^2\left(\tfrac{1}{2}\theta\right)\right], \tag{2.5}$$

where $\tau = Q^2/4m_N^2$, θ is the electron scattering angle, E, E' are the incident and final electron energies, respectively, and

$$\sigma_M = \left(\frac{\alpha \cos\left(\frac{1}{2}\theta\right)}{2E \sin^2\left(\frac{1}{2}\theta\right)} \right)^2 \tag{2.6}$$

is the Mott cross section for point-like scattering [3].

Often the electric and magnetic Sachs form factors G_E and G_M are used. In the Breit frame, where the energy transfer is zero and $|\mathbf{q}| = Q$, the Fourier transforms of these Sachs form factors are directly related to the charge and magnetization distributions in the nucleon. They are defined as linear combinations of F_1 and F_2:

$$G_E = F_1 - \tau\kappa F_2, \quad G_M = F_1 + \kappa F_2. \tag{2.7}$$

Initially these form factors were separated only by using the Rosenbluth method [4], which can be understood by rewriting (2.5) in terms of the Sachs form factors:

$$\frac{d\sigma}{d\Omega} = \frac{E'}{E}\sigma_M \left[\frac{G_E^2 + \tau G_M^2}{1 + \tau} + 2\tau G_M^2 \tan^2\left(\frac{1}{2}\theta\right) \right]. \tag{2.8}$$

The polarization, ϵ, of the virtual photon is given by [2]

$$\epsilon^{-1} = 1 + 2(1 + \tau)\tan^2\left(\frac{1}{2}\theta\right), \tag{2.9}$$

which can be used to rewrite (2.8) to give the reduced cross section

$$\sigma_R \equiv \frac{d\sigma}{d\Omega}\frac{\epsilon(1+\tau)E}{\sigma_M E'} = \tau G_M^2(Q^2) + \epsilon G_E^2(Q^2). \tag{2.10}$$

The Rosenbluth method consists of making measurements at a fixed Q^2 and variable $\epsilon(\theta, E)$ and fitting the reduced cross section σ_R with a straight line of slope G_E^2 and intercept τG_M^2.

Early measurements of the form factors suggested a scaling law relating three of the four nucleon elastic form factors by a dipole formula describing their common Q^2 dependence:

$$G_E^p(Q^2) \approx \frac{G_M^p(Q^2)}{\mu_p} \approx \frac{G_M^n(Q^2)}{\mu_n} \approx G_D \equiv (1 + Q^2/0.71)^{-2}. \tag{2.11}$$

This behaviour was known as form-factor scaling. The dipole parametrization G_D corresponds, in coordinate space, to an exponentially decreasing radial density. However, high-precision polarization data show that this might not be correct: see subsection 2.2.2.

Although the neutron has no net charge, the non-zero value of its magnetic moment implies that it must have a charge distribution. The neutron electric form factor

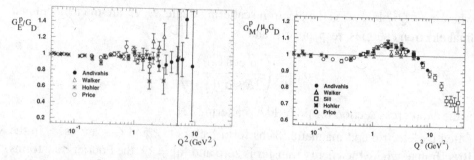

Figure 2.1. G_E^p/G_D (left) and $G_M^p/(\mu_p G_D)$ (right) versus Q^2 from the Rosenbluth method.

differs from the other three and is well parametrized [5] as

$$G_E^n = -\mu_n G_D \frac{\tau}{1 + 5.6\tau}. \tag{2.12}$$

2.2 Space-like nucleon form factors

A rich body of experiments has been devoted to determine the space-like nucleon form factors and continues today with ever-improving experimental techniques. We will first review the status of the nucleon form-factor determinations from spin-averaged measurements, and then continue with those more recent determinations using spin-dependent techniques.

2.2.1 Spin-averaged measurements

Proton form factors   The Rosenbluth method described above is problematic as it requires the measurement of absolute cross sections and at large Q^2 the cross section is insensitive to G_E. This is obvious from (2.10), recalling the definition of τ.

Figure 2.1 presents the Rosenbluth data set for the proton form factors. (See [6] for a compilation and references.) The dipole law describes the Q^2 dependence to a good approximation ($\approx 10\%$) for both form factors out to $Q^2 = 8$ GeV^2. As a result, one is inclined to believe that form-factor scaling holds well also. The latter was used to extract values of G_M^p from the forward-angle, elastic cross section measurements of SLAC experiment E136 [7], out to $Q^2 = 30$ GeV^2. Because G_M^p dominates these measurements at high Q^2, the error made due to the assumption of form-factor scaling is small.

However, the limitation of the Rosenbluth method can be seen from the large errors in G_E^p measurements for $Q^2 > 1$ GeV^2. This is even more evident in the ratio

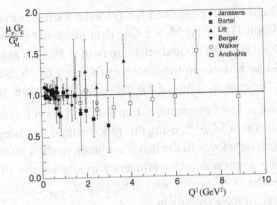

Figure 2.2. The ratio $\mu_p G_E^p / G_M^p$ as determined with the Rosenbluth method. Data are from [8–13]. The statistical and systematic errors are added in quadrature.

$\mu_p G_E^p / G_M^p$ from different data sets: data scatter by up to 50% at Q^2 values larger than 1 GeV^2 (see figure 2.2). Subsequent measurements of the proton form factors exploiting a double-polarization observable have upset the apparent form-factor scaling as we shall see in subsection 2.2.2.

Neutron form factors The lack of a free neutron target has limited the quality of the data on neutron form factors. Of all the form factors, the neutron electric form factor is the least well known. Its small intrinsic size, in combination with the dominance of G_M^n over G_E^n, makes it difficult to measure. However, the slope at $Q^2 = 0$ is known from thermal neutron scattering from electrons [14]:

$$\left. \frac{dG_E^n}{dQ^2} \right|_{Q^2=0} = 0.678 \pm 0.030. \tag{2.13}$$

The deuteron has served as an approximation to a source of a free neutron target. A good understanding of the ground- and final-state wave functions is required in order to extract reliable information about the neutron form factors. The traditional techniques (restricted to the use of unpolarized beams and targets) used to extract information about G_M^n and G_E^n have been: elastic scattering from the deuteron: d(e, e')d; inclusive quasielastic scattering: d$(e, e')X$; and scattering from the deuteron with the coincident detection of the scattered electron and recoiling neutron: d$(e, e'n)p$.

More recently, a ratio method which minimizes uncertainties in the deuteron and final state wave function has been emphasized: d$(e, e'n)/$d$(e, e'p)$. Sensitivities to nuclear binding effects and experimental fluctuations in the luminosity and detector acceptance are minimized in such a ratio measurement, the main technical difficulty being the absolute determination of the neutron detection efficiency. Such measurements have been pioneered for Q^2 values smaller than 1 GeV^2 at Mainz [15–17] and Bonn [18]. The Mainz G_M^n data are 8–10% lower than those from Bonn, at variance with the quoted uncertainty of $\sim$2%. An explanation of this discrepancy could be an error in the detector efficiency of 16–20%, but this remains unconfirmed. The current status of the data on G_M^n is shown in figure 2.3.

The lack of data at large momentum transfers will soon be remedied. A study of G_M^n at Q^2 values up to 5 GeV^2, using this ratio method, has been completed at JLab [19]. A hydrogen target was in the beam simultaneously with the deuterium target, to determine the neutron detector efficiency in situ via the $p(e, e'n\pi^+)$ reaction. Preliminary results [19] (not shown) indicate that G_M^n is within 10% of G_D over the full Q^2 range of the experiment.

Until the early 1990s the extraction of G_E^n was done most successfully through either small angle elastic e–d scattering [5,20] or by quasielastic e–d scattering [21]. In the impulse approximation, the elastic electron–deuteron cross section is the sum of the proton and neutron responses, weighted with the deuteron wave function. In the limit of small θ_e, the cross section can be written [22]

$$\frac{d\sigma}{d\Omega} \sim \left(G_E^p + G_E^n\right)^2 \int dr \left[u(r)^2 + w(r)^2\right] j_0\left(\tfrac{1}{2}qr\right), \qquad (2.14)$$

with $u(r)$ and $w(r)$ the S- and D-state wave functions of the deuteron. The coherent nature of elastic scattering gives rise here to an interference term between the neutron and proton response, which allows the smaller G_E^n contribution to be extracted. Still, the large proton contribution must be removed.

Hence, experiments have been able to achieve small statistical errors but remain sensitive to the deuteron wave function model, leaving a significant residual dependence on the nucleon–nucleon (NN) potential. The most precise data on G_E^n obtained from elastic e–d scattering are shown on the right-hand side of figure 2.3 from an experiment at Saclay, published in 1990 [20]. The curves are parametrizations of the data based on different NN potentials used in the extraction. The band they form is a natural measure of the theoretical uncertainty in this method ($\approx$50%) that reflects the knowledge of the NN potentials at that time.

2.2.2 Spin-dependent measurements

It has been known for many years that the nucleon form factors could be measured through spin-dependent elastic scattering from the nucleon, accomplished either through a measurement of the scattering asymmetry of polarized electrons from a polarized nucleon target [26,27], or equivalently by measuring the polarization transferred to the nucleon [28,29].

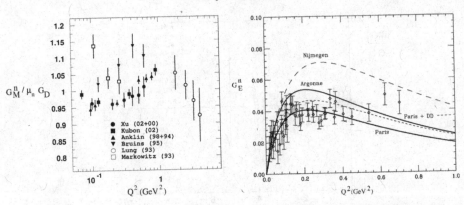

Figure 2.3. Left: G_M^n from unpolarized scattering [15–18,21,23] and polarized scattering [24,25]. Right: G_E^n from elastic e–d scattering [20]. The band defined by the lines represents the theoretical error associated with the extraction (see text).

In the scattering of polarized electrons from a polarized target, an asymmetry appears in the elastic scattering cross section when the beam helicity is reversed. In contrast, in scattering a polarized electron from an unpolarized target, the transferred polarization to the nucleon produces an azimuthal asymmetry in the secondary scattering of the nucleon (in a polarimeter) due to its dependence on its polarization. In both cases, the perpendicular asymmetry is sensitive to the product $G_E G_M$. Only since the mid 1990s have experiments exploiting these spin degrees of freedom become possible.

Like spin-averaged measurements, determination of the neutron form factors requires scattering from a nuclear target. As before, the deuteron is the target of choice but, in the case of beam–target asymmetry measurements, ^{3}He also becomes a viable target as it approximates a polarized neutron [30]. In either case, extraction of the neutron form factors is again complicated by the need to account for ground-state and final-state wave function effects. Fortunately it has been found for the deuteron that in kinematics that emphasize quasi–free neutron knockout, both the transfer polarization P_t [31] and the beam–target asymmetry A_V [32] are especially sensitive to G_E^n and relatively insensitive to the choice of NN potential and other reaction details. Calculations [30] of the beam–target asymmetry from a polarized ^{3}He target also showed only modest model dependence.

The proton form-factor ratio G_E^p/G_M^p In elastic electron–proton scattering a longitudinally polarized electron will transfer its polarization to the recoil proton. In the one-photon exchange approximation the proton can attain only polarization components in the scattering plane, parallel (P_l) and transverse (P_t) to its

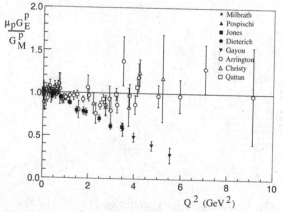

Figure 2.4. The ratio $\mu_p G_E^p/G_M^p$ from polarization-transfer experiments [33–37], compared to data using the Rosenbluth method [39,40] and the reanalysis by Arrington [38] of older SLAC data.

momentum. The ratio of the charge and magnetic form factors is directly proportional to the ratio of these polarization components:

$$\frac{G_E^p}{G_M^p} = -\frac{P_t}{P_l} \frac{E_e + E_e'}{2m_N} \tan\left(\tfrac{1}{2}\theta_e\right). \qquad (2.15)$$

The polarization-transfer technique was used for the first time by Milbrath *et al* [33] at the MIT-Bates facility. The proton form-factor ratio was measured at Q^2 values of 0.38 and 0.50 GeV^2. A further measurement at a Q^2 value of 0.4 GeV^2 was performed at the Mainz MAMI facility [34]. The greatest impact of the polarization-transfer technique, however, was made by two experiments [35,37] at JLab, extending this proton form factor ratio to a Q^2 value of 5.6 GeV^2.

Figure 2.4 shows the results for the ratio $\mu_p G_E^p/G_M^p$. The most striking feature of the data is the sharp, practically linear decline as Q^2 increases, which was parametrized [37] as:

$$\mu_p \frac{G_E^p(Q^2)}{G_M^p(Q^2)} = 1 - 0.13(Q^2 - 0.04) \qquad (2.16)$$

with Q^2 in GeV^2. Since it is known that G_M^p is well described by the dipole parametrization, it follows that G_E^p falls more rapidly with Q^2 than G_D.

This significant fall-off of the form-factor ratio is in clear disagreement with the results from the Rosenbluth method. Arrington [38] performed a careful reanalysis of earlier Rosenbluth data. He selected only experiments in which an adequate ϵ-range was covered with the same detector. The results of this reanalysis are also given in figure 2.4 and do not show the large scatter seen in figure 2.2.

Subsequently, an extensive data set on elastic electron–proton scattering collected in Hall C at JLab has been analysed [39]. The results are evidently in good agreement with Arrington's reanalysis. Lastly, a high-precision extraction of G_E^p/G_M^p has been made using the 'super-Rosenbluth' method in Hall A at JLab [40]. The latter experiment was specifically designed to reduce significantly the systematic errors compared to earlier Rosenbluth measurements. The main improvement came from detecting the recoiling protons instead of the scattered electrons, so that the proton momentum and the cross section remain practically constant when one varies ϵ at a constant Q^2 value. Results [40] of this experiment, covering Q^2 values from 2.6 to 4.1 GeV^2, are again in excellent agreement with previous Rosenbluth results. This basically rules out the possibility that the disagreement between Rosenbluth and polarization-transfer measurements of the ratio G_E^p/G_M^p is due to an underestimate of ϵ-dependent uncertainties in the Rosenbluth measurements. This is clearly illustrated in figure 2.5(left), which highlights the difference in reduced cross section measurements, as a function of ϵ, found between the Rosenbluth method and the G_E^p/G_M^p ratio from the polarization-transfer technique, at $Q^2 = 2.64$ GeV^2.

The G_E^p/G_M^p discrepancy It has been suggested [41] that two-photon exchange contributions could be responsible for the discrepancy in the G_E^p/G_M^p form-factor ratio as determined from the Rosenbluth and polarization transfer methods. In such a two-photon exchange process the nucleon undergoes a first virtual photon exchange which can lead to an intermediate excited state and then a second exchange to return to its ground state. Such corrections would provide an ϵ-dependent modification of the cross section measurements, thus directly affecting the Rosenbluth method and only to a lesser extent [42] the polarization-transfer method, which contains a direct ratio measurement of recoil polarization.

The most stringent tests of two-photon exchange effects have been carried out by measuring the ratio of electron and positron elastic scattering off a proton. Corrections due to two-photon exchange will have a different sign in these two reactions. Unfortunately, this (e^+p/e^-p) data set is quite limited [43], only extending (with poor statistics) up to a Q^2 value of ~ 5 GeV^2, while at Q^2 values larger than ~ 2 GeV^2 basically all data have been measured at ϵ values larger than ~ 0.85. Existing comparisons yield an average ratio of 1.003 ± 0.005, with $\chi^2 = 0.87$. These data have been interpreted as showing that the two-photon effects must be small. However, due to the sparsity of data at low ϵ and $Q^2 > 2$ GeV^2, it is difficult to draw definite conclusions. Arrington [43] has compiled the world's data for this ratio and show a possible ϵ-dependent slope in the (e^+p/e^-p) ratio, including all data with $Q^2 < 2$ GeV^2 (see figure 2.5(right)). This behaviour may be consistent with the two-photon exchange effects required to explain the discrepancy, if confirmed with better data, and at higher Q^2. Arrington [44] determined the proton's

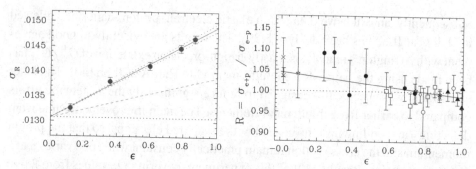

Figure 2.5. Left: the reduced cross section at $Q^2 = 2.64$ GeV^2 as determined in the JLab E01-001 experiment using the 'super-Rosenbluth' method [40]. The dashed line indicates the ϵ-dependent slope one would have measured using the polarization-transfer ratio of G_E^p/G_M^p and arbitrarily assuming that the magnetic form factor is fixed. Right: $\sigma_{e^+p}/\sigma_{e^-p}$ cross section ratio as a function of ϵ for measurements below $Q^2 = 2$ GeV^2. The solid line is a fit to the ratio, assuming a linear ϵ dependence and no Q^2 dependence. It yields a slope of $(-5.7 \pm 1.8)\%$ [43].

electric and magnetic form factors by using the Rosenbluth method data, the polarization-transfer data, and the assumption that two-photon exchange effects explain the discrepancy. This analysis is also consistent with the measured (e^+p/e^-p) ratios.

Other tests, also inconclusive, looked for non-linearities in the ϵ-dependence (the linearity of the reduced cross sections is illustrated in figure 2.5(left)) or measured the transverse (out-of-plane) polarization component of the recoiling proton, a non-zero value of which would be a direct measure of the imaginary part of the two-photon exchange amplitude.

Calculations suggest that the two-photon exchange diagram may indeed be the cause of the discrepancy. Blunden *et al* [45] resolved half of the discrepancy when calculating the elastic contribution to two-photon exchange effects, using a simple monopole Q^2 dependence of the hadronic form factors. Of course, a full calculation must include contributions where the intermediate proton is in an excited state, which was not included in [45]. Chen *et al* [46] include these contributions through two-photon exchange off partons in the proton, with emission and reabsorption of the partons described in terms of generalized parton distributions. About half of the effect needed to bring the two experimental techniques into agreement at larger Q^2 is found. Hence, it is becoming more and more likely that two-photon exchange effects are responsible for at least a major part of the measured discrepancy.

The neutron electric form factor G_E^n   Polarized targets have been used to extract G_E^n and G_M^n. The beam–target asymmetry can be written schematically (a, b, c,

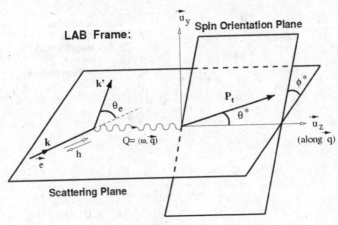

Figure 2.6. Kinematics definitions for polarized electron scattering from a polarized target.

and d are known kinematic factors) as [27]

$$A = \frac{a \cos \theta^\star G_M^2 + b \sin \theta^\star \cos \phi^\star G_E G_M}{c G_M^2 + d G_E^2},\qquad(2.17)$$

where $\theta^\star$ and $\phi^\star$ fix the target polarization axis (see figure 2.6). With the target polarization axis in the scattering plane and perpendicular to the photon momentum $\mathbf{q}$ ($\Theta^\star, \Phi^\star = 90°, 0°$), the asymmetry A_{TL} is proportional to $G_E G_M$. With the polarization axis in the scattering plane and parallel to $\mathbf{q}$ ($\Theta^\star, \Phi^\star = 0°, 0°$), measuring the asymmetry A_T allows G_M to be determined [24,25]. The latter extractions are included in figure 2.3. Here, we will show G_E^n measurements using either polarized-target (2.17) or recoil polarimetry methods (2.15).

The first measurements were carried out at MIT-Bates, both with a polarized ^{3}He target [47] and with a neutron polarimeter [48]. G_E^n was later extracted from beam–target asymmetry measurements using polarized ^{3}He targets at Mainz [49], polarized ND$_3$ targets at JLab [50] and a polarized atomic-beam target at NIKHEF [51], and from recoil polarimetry measurements using deuterium targets at Mainz [52] and JLab [53]. In all the later measurements a neutron was detected in coincidence with the scattered electron to enhance sensitivity of the measured observables to G_E^n.

At low Q^2 values, G_E^n extractions require corrections for nuclear-medium and rescattering effects, which can be sizeable: 65% for ^{2}H at 0.15 GeV^2 and 50% for ^{3}He at 0.35 GeV^2. These corrections are expected to decrease significantly with increasing Q^2, although no reliable calculations are presently available for ^{3}He above 0.5 GeV^2. There is excellent agreement among the results from the different

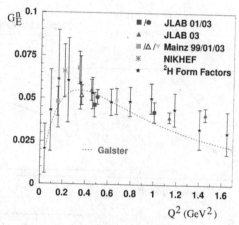

Figure 2.7. G_E^n measurements from both polarized target and recoil polarization measurements, for deuteron targets. Data are from [48,50–53]. An extraction using ^{2}H elastic form factor measurements is also shown, for comparison [54]. The dashed curve represents the Galster parametrization [5].

techniques. Moreover, nuclear-medium effects seem to become negligible at $\sim$0.7 GeV^2, even for ^{3}He [30], on the basis of a large and precise data set accumulated at Mainz [52]. Data from JLab, using either a polarimeter or a polarized target [53,50], extend up to $Q^2 \approx 1.5$ GeV^2 with an overall accuracy of $\sim$10%. For the sake of clarity, we only show G_E^n extractions using deuterium targets in figure 2.7 that have been published since 1994. A precise data set for G_E^n over a large range in Q^2: 0.2 $< Q^2 < 1.5$ GeV^2 has become available since the mid 1990s.

Lastly, Schiavilla and Sick [54] have extracted G_E^n from available data on the deuteron quadrupole form factor $F_{C2}(Q^2)$ with a much smaller sensitivity to the NN potential than from inclusive (quasi)elastic scattering. Remarkably, the Galster parametrization [5] continues to provide a reasonable description of the data.

Charge and magnetization distributions   The charge and magnetization root-mean-square radii are related to the slopes of the form factors at $Q^2 = 0$ by

$$\langle r_E^2 \rangle = 4\pi \int dr \rho(r) r^4 = -6 \frac{dG_E(Q^2)}{dQ^2}\bigg|_{Q^2=0},$$

$$\langle r_M^2 \rangle = 4\pi \int dr \mu(r) r^4 = -\frac{6}{\mu} \frac{dG_M(Q^2)}{dQ^2}\bigg|_{Q^2=0}, \qquad (2.18)$$

with $\rho(r)$ ($\mu(r)$) denoting the radial charge (magnetization) distribution. Initial results for the proton charge radius did not agree with a three-loop QED calculation [55] of the hydrogen Lamb shift. Accurately taking into account Coulomb distortion effects and higher moments of the radial distribution [56] brings the proton charge

Observable	Value $\pm$ error
$\langle (r_E^p)^2 \rangle^{1/2}$	0.895 ± 0.018 fm [56]
$\langle (r_M^p)^2 \rangle^{1/2}$	0.855 ± 0.035 fm [56]
$\langle (r_E^n)^2 \rangle$	-0.113 ± 0.005 fm^2 [14]
$\langle (r_M^n)^2 \rangle^{1/2}$	0.87 ± 0.01 fm [17]

Table 2.1. Values for the nucleon charge and magnetization radii

radius into excellent agreement. Within error bars the root-mean-square radii for the proton charge and magnetization distribution and for the neutron magnetization distribution are equal (see table 2.1).

In the Breit frame the nucleon form factors can be written as Fourier transforms of their charge and magnetization distributions (for example see [57])

$$\rho(r) = \frac{4\pi}{(2\pi)^3} \int_0^\infty dQ \, G(Q) \frac{\sin Qr}{Qr}. \tag{2.19}$$

Beyond the choice of this particular frame, the form factors are not solely determined by the internal structure of the nucleon, but will also contain dynamical effects due to relativistic boosts, complicating an intuitive interpretation in terms of spatial nucleon densities.

Nonetheless, Kelly [58] extracted spatial nucleon densities from the available form-factor data in a model in which the asymptotic behaviour of the form factors conformed to perturbative QCD scaling at large Q^2. The neutron and proton magnetization densities extracted are quite similar to each other, but narrower than the proton charge density. The neutron charge density exhibits a positive core surrounded by a negative surface charge, peaking at just below 1 fm, attributed to a negative pion cloud.

A feature common to all nucleon form factors appears to be a 'bump' or 'dip' in the ratio to the dipole form at $\sqrt{Q^2} \approx 0.5$ GeV with a width of ~ 0.2 GeV, as observed by Friedrich and Walcher [57] (see also figure 2.1). They performed a model fit to all four form factors with two ingredients: a dipole behaviour for constituent-quark form factors, and a pion cloud with an $l = 1$ harmonic oscillator behaviour, and also extracted spatial densities. Their results suggest a pion cloud peaking at a radius of ~ 1.3 fm, close to the Compton wavelength of the pion, slightly larger than the result of [58]. Note that Hammer *et al* [59] argue from general principles that this pion cloud should peak much more inside the nucleon, at ~ 0.3 fm.

Although an intuitive interpretation of form factors in terms of spatial densities alone is questionable, the general structure seen in the form factors, at $\sqrt{Q^2} \sim 0.5$ GeV, may still be valid.

2.3 Model calculations of nucleon form factors

Nucleons are composed of quarks and gluons and information about their internal structure is critical for testing quark models. For example, in a symmetric quark model, with all valence quarks having the same wave function, the charge distribution of the neutron would everywhere be zero, and thus $G_E^n = 0$. Hence, any deviation from zero exposes the details of the wave functions. Of course, detailed knowledge on the nucleon form factors is also critical for any study of nuclear structure (for example, see [29]). The accumulation of precise form-factor data, and especially the surprising polarization transfer results for G_E^p/G_M^p, have produced a flurry of theoretical studies which we summarize briefly.

Initially, the nucleon form factors were described in terms of vector-meson dominance (VMD) models [60,61], in which it is assumed that the virtual photon couples to the nucleon via intermediate light vector mesons. The coupling strengths between the virtual photon and the vector mesons and between the vector mesons and the nucleon are either constrained by other data or are fitted to the nucleon form factors. Since u and d quarks are the essential building blocks of the nucleons, all VMD models include the lowest-mass mesons, the ρ and ω. The role of the ϕ is less clear as it correlates with a possible s-quark contribution to the nucleon's ground-state wave function. The success of the simplest VMD models in describing the form factors at low and moderate Q^2 is offset by their difficulty in accommodating the Q^{-4} decrease of the form factors at high Q^2. By adding more parameters, for example by including heavier vector mesons, generalized VMD models do succeed [62] in giving an excellent description of all nucleon form-factor data, as shown in figure 2.8, but provide little predictive power.

Numerous attempts have been made to develop more prescriptive models, but these either fail quantitatively with one, or more, of the form factors or only manage to give an overall qualitative description. The VMD model has been modified by imposing [63,64] the large-Q^2 behaviour of perturbative QCD (see below), by adding a meson-cloud contribution [65] and by using dispersion relations [59] to limit the number of parameters in generalized VMD models. Other approaches include applying chiral perturbation theory [66] at low Q^2, a chiral-soliton model, in which the nucleon becomes a Skyrmion with spatial extension, plus isoscalar and isovector meson exchange [67] and an effective theory based on an SU(3) Nambu–Jona–Lasinio Lagrangian [68] that incorporates spontaneous chiral symmetry breaking and is comparable to the inclusion of vector mesons in the Skyrme model but involves fewer free parameters.

Finally, various forms of relativistic constituent-quark model have been applied [70–74], again with varying degrees of success. The three most successful results are shown in figure 2.8.

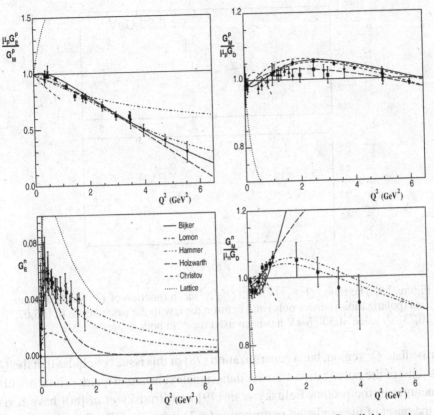

Figure 2.8. Comparison of various calculations with a sample of available nucleon form-factor data. For G_E^p only polarization-transfer data are shown. Note that not all available data are presented, especially for G_E^n and G_M^n. The calculations shown are from [59,62,65,67–69]. Where applicable, the calculations have been normalized to the calculated values of $\mu_{p,n}$.

At large Q^2, quark-dimensional scaling arguments [75] and perturbative QCD [76] can be applied. The photon couples to a single quark and each gluon exchange necessary to share the momentum among the quarks contributes a factor proportional to Q^{-2}, thereby directly obtaining $F_1 \propto 1/Q^4$, $F_2 \propto 1/Q^6$ and $F_2/F_1 \propto 1/Q^2$. The polarization-transfer data do not follow this asymptotic form-factor scaling behaviour. In the asymptotically free limit of QCD, hadron helicity is conserved. However, if a quark orbital angular momentum component is introduced [77] into the wave function of the proton, giving non-zero quark transverse momentum, then hadron helicity conservation is broken. This approach predicts a $1/Q$ behaviour for the ratio of the Dirac and Pauli form factors at intermediate values of Q^2, in excellent agreement with the polarization-transfer data for $Q^2 \geq 3$ GeV^2. It has been pointed out [65] that this $1/Q$ behaviour is accidental and only valid in an

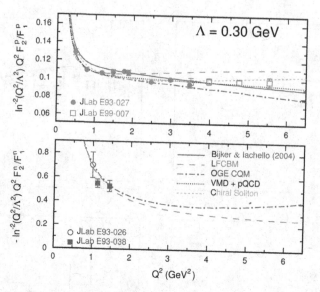

Figure 2.9. The ratio $(Q^2 F_2/F_1)/\ln^2(Q^2/\Lambda^2)$ as a function of Q^2 for the proton data (polarization-transfer only) and neutron data, with the calculations of [65,67, 70,73]. A value of 300 MeV has been used for Λ in both figures.

intermediate Q^2 region, but a generalization [78] of this issue concludes that the Q^2 behaviour of the polarization-transfer data signals substantial quark orbital angular momentum in the proton. Belitsky *et al* [79] and Brodsky *et al* [80] have revisited the onset of the predicted perturbative QCD behaviour. The former derive the following large Q^2 behaviour:

$$\frac{F_2}{F_1} \propto \frac{\ln^2 Q^2/\Lambda^2}{Q^2}, \qquad (2.20)$$

where Λ is a soft scale related to the size of the nucleon. Even though the polarization-transfer data follow this behaviour (see figure 2.9), this could very well be precocious [79], since it is not clear at what Q^2 one really reaches the perturbative QCD domain. Brodsky *et al* [80] argue that a non-zero orbital angular momentum wave function contributes to both F_1 and F_2 and thus, in the asymptotic limit, $Q^2 F_2/F_1$ should still become constant.

It has been realized that the nucleon form factors can be interpreted as the Fourier transforms of charge and current (or quark) density distributions in the plane transverse to the photon–nucleon axis. This is similar to the Feynman parton distribution, which can be interpreted in a frame of reference in which the nucleon travels with the speed of light. This is reflected in the nucleon form factors being moments of generalized parton distributions (GPDs, see chapter 9) and so providing a constraint on these distributions.

All the models that have been described are effective to some degree and highlight features of QCD. In the end, only lattice gauge theory can provide truly *ab initio* calculations, but accurate lattice QCD results are still several years away. As an example, calculations of the nucleon form factors by the QCDSF collaboration [81] are still limited to the quenched approximation, in which sea-quark contributions are neglected, a box size of 1.6 fm and a pion mass of 650 MeV are taken. Ashley *et al* [69] have extrapolated the results of these calculations to the chiral limit using chiral coefficients appropriate to full QCD. The agreement with the data is poor, a clear indication of the technology developments required before lattice QCD calculations can provide an accurate description of experimental form-factor data.

2.4 Time-like nucleon form factors

Although a great deal of effort has been put into elucidating the electromagnetic nucleon form factors in the space-like region, much less is known about the time-like form factors. In the time-like region, measurements have been made at electron–positron storage rings and, for the proton, by studying the inverse reaction of antiproton annihilation to e^+e^- on a hydrogen target.

The time-like and space-like form factors are connected by analytic continuation in the complex q^2 plane and a comprehensive model of nucleon structure should simultaneously describe the proton and neutron form factors in both regions. Hence, time-like form factors can be used to test unified descriptions of nucleon structure.

Unfortunately, data on time-like form factors are scarce, and measurements that separate the electric and magnetic form factors are only indirectly made in limited regions of Q^2 for the proton and not at all for the neutron. This is because the impact of the electric form factor on the cross section in the time-like region is kinematically reduced by a factor m_N^2/Q^2, as in the space-like region. This explains the lack of data for $G_E(Q^2)$. For the proton, Bardin *et al* [82] provided cross sections for $p\bar{p} \rightarrow e^+e^-$ at five different beam energies. Although the quality of the data does not allow for a proper separation of the proton electric and magnetic form factors, their ratio at $Q^2 \sim 4$ GeV^2 appears unity within the uncertainties.

Estimates of the proton magnetic form factor have mainly been obtained by assuming form-factor scaling, $G_E = G_M$, for which there is little justification except at the $N\bar{N}$ threshold, where the relationship $G_M(4m_N^2) = G_E(4m_N^2)$ must hold (recall (2.7), noting that $\tau = 1$ at $Q^2 = 4m_N^2$). Results range from near threshold to $Q^2 \approx 15$ GeV^2 [82–85]. Neutron magnetic form factors have been measured, with very limited statistics, by one single experiment at Frascati [86], from threshold up to $Q^2 \approx 6$ GeV^2.

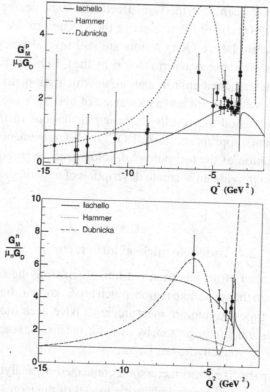

Figure 2.10. The magnetic form factors (divided by the dipole form factor G_D) in the time-like region. The proton (neutron) magnetic form factor is shown in the top (bottom) panel and is here derived using the calculated electric form factors of [92]. The data are compared to calculations of Iacchello and Wan [92], Hammer *et al* [93] and Dubnicka *et al* [94].

Very near threshold, the proton magnetic form factor exhibits a strong Q^2 dependence. A hint for a similar strong Q^2 dependence is also seen in the case of the neutron. Because the relationship $G_M(4m_N^2) = G_E(4m_N^2)$ is only valid at threshold, the strong Q^2-dependence may well originate from the G_E contribution to the cross section.

The steep Q^2-dependence of these form factors near threshold could be explained [86] by a relatively narrow structure at the boundary of the unphysical region. A similar anomaly has been reported in the total cross section $\sigma_{tot}(e^+e^- \rightarrow$ hadrons), with an apparent dip in the total multihadronic cross section [87]. This dip and the steep variation of the proton form factor near threshold may be fitted with a narrow resonance, with a mass $M \sim 1.87$ GeV and a width Γ of 10–20 MeV [86]. This narrow resonance would be consistent with the hypothesis of an $N\bar{N}$ bound state, predicted a long time ago [88–91]. However, this is not conclusive and $N\bar{N}$ experiments do not appear to support it. See chapter 4 for a further discussion.

The rather limited data for time-like magnetic form factors are shown in figure 2.10 (as mentioned above, essentially no data exist for the electric form factors). For this compilation, the magnetic form factors have been extracted from the measurements by Iachello and Wan [92] using their calculated electric form factors (these calculations aim to simultaneously describe proton and neutron, in both space- and time-like regions, in one comprehensive model based on a VMD framework). Clear deviations from the dipole form factor G_D can be seen in the time-like region, from threshold up to 6 GeV^2. This is in sharp contrast to the space-like form factors that are reasonably well described by the dipole form.

Iachello and Wan [92], Hammer *et al* [93], and Dubnicka *et al* [94] have extended their VMD calculations through analytic continuation from the space-like to the time-like domain. The model of Iachello and Wan appears to provide a consistent description of the magnetic form factors, after adding a very narrow subthreshold isoscalar resonance at $M_X = 1.870$ GeV^2 with negligible width. This explains the strong Q^2-dependence of the calculation in the threshold region, $Q^2 > 4m_N^2$, and follows the suggestion of [86].

Clearly, extensions of theoretical form-factor calculations to the time-like region are highly desirable, in combination with an extension of the present measurements to separate electric and magnetic form factors. The latter may be possible with high-luminosity e^+e^- colliders at Beijing and Frascati and the use of initial-state radiation at BELLE and BABAR. These would provide a great opportunity to unravel the structure of the nucleon with far greater detail.

2.5 The pion form factor

The pion occupies an important place in the study of the quark–gluon structure of hadrons and has been the subject of many calculations [95–104]. Due to its relatively simple $q\bar{q}$ valence structure it is expected that a perturbative QCD approach can be applied at lower energies for the pion than for the nucleon, with estimates [105] suggesting that perturbative QCD contributions start to dominate the pion form factor at $Q^2 \geq 5$ GeV^2. In addition, the asymptotic normalization of the pion wave function, in contrast to that of the nucleon, is known from pion decay. Within perturbative QCD one can derive [106]

$$\lim_{Q^2 \to \infty} F_\pi(Q^2) = \frac{8\pi\alpha_s f_\pi^2}{Q^2}, \qquad (2.21)$$

where f_π is the pion decay constant.

The charge form factor of the pion, $F_\pi(Q^2)$, is an essential element of the structure of the pion. Its behaviour at very low values of Q^2 has been determined up to $Q^2 = 0.28$ GeV^2 from scattering high-energy pions off atomic electrons [107].

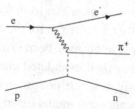

Figure 2.11. Schematic diagram representing quasielastic electron scattering off a pion in the proton.

Such data may be used to extract the charge radius of the pion, which leads to a root-mean-squares radius of 0.663 ± 0.006 fm. For the determination of the pion form factor at higher values of Q^2 one has to resort to high-energy electroproduction of pions on a nucleon, that is employ the $p(e, e'\pi^+)n$ reaction. At sufficiently small momentum transfer t to the nucleon, the pion-exchange contribution can be determined. This may be considered as quasielastic scattering of the electron from a virtual pion in the proton, and gives access to F_π (see figure 2.11). The cross section for this process can be written as

$$\frac{d^3\sigma}{dE'd\Omega_{e'}d\Omega_\pi} = \Gamma_V \frac{d^2\sigma}{dt d\phi}, \tag{2.22}$$

where Γ_V is the virtual-photon flux factor, ϕ is the azimuthal angle of the outgoing pion with respect to the electron scattering plane and t is the Mandelstam variable $t = (p_\pi - q)^2$. The two-fold differential cross section can be written as

$$2\pi \frac{d^2\sigma}{dt d\phi} = \epsilon \frac{d\sigma_L}{dt} + \frac{d\sigma_T}{dt} + \sqrt{2\epsilon(\epsilon+1)} \frac{d\sigma_{LT}}{dt} \cos\phi$$
$$+ \epsilon \frac{d\sigma_{TT}}{dt} \cos 2\phi, \tag{2.23}$$

where ϵ is the virtual-photon polarization parameter. The cross sections $\sigma_X \equiv d\sigma_X/dt$ depend on W, Q^2 and t. In the pole approximation to pion exchange, the longitudinal cross section σ_L is proportional to the square of the pion form factor:

$$\sigma_L \propto \frac{-t\,Q^2}{\left(t - m_\pi^2\right)^2} F_\pi^2. \tag{2.24}$$

The ϕ acceptance of the experiments should be large enough for the interference terms σ_{LT} and σ_{TT} to be determined. Then, by taking data at two energies at every Q^2, σ_L can be separated from σ_T by means of a longitudinal/transverse (LT) separation.

The pion form factor was studied in the 1970s for Q^2 between 0.4 and 9.8 GeV^2 at CEA/Cornell [108] and at $Q^2 = 0.7$ GeV^2 at DESY [109]. In the DESY experiment

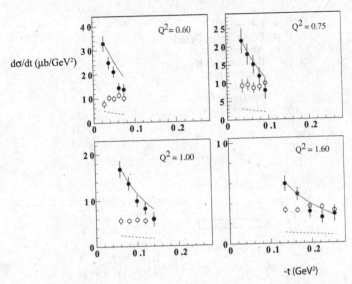

Figure 2.12. Separated cross sections σ_L and σ_T (full and open symbols, respectively) compared to the Regge model (full curve for L, dashed curve for T) [110]. The Q^2 values are in units of GeV^2.

an LT separation was performed by taking data at two values of the electron energy. In the experiments done at CEA/Cornell, however, this was done in a few cases only, and even then the resulting uncertainties in σ_L were so large that the LT-separated data were not used. Instead (2.24) was used to determine the pion form factor from σ_L, which was estimated by subtracting σ_T, assumed to be proportional to the virtual-photon cross section, from the measured (differential) cross section. No uncertainty in σ_T was included in this subtraction. Hence values of F_π above $Q^2 = 0.7$ GeV^2 published in the 1970s were not based on LT-separated cross sections.

Additional LT-separated data for the pion form factor have been published for Q^2 up to 1.6 GeV^2 [110]. Using the High Momentum Spectrometer and the Short Orbit Spectrometer at JLab, and electron energies between 2.4 and 4.0 GeV, data for the reaction $p(e, e'\pi^+)n$ were taken for central values of Q^2 of 0.6, 0.75, 1.0 and 1.6 GeV^2, at a central value of the invariant mass W of 1.95 GeV. Because of the excellent properties of the electron beam and experimental setup, LT-separated cross sections could be determined with high accuracy. The extracted cross sections are displayed in figure 2.12. The error bars represent the combined statistical and systematic uncertainties. As a result of the LT separation technique the total error bars on σ_L are enlarged considerably, typically to about 10%.

The pion-pole contribution sits on a large background from other exchanges. An attempt to model the $\gamma^* p \to \pi^+ n$ reaction was made by Vanderhaeghen *et al* [111] in terms of Regge theory (see chapter 5) with the pion and ρ-meson trajectories.

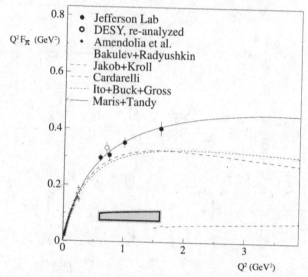

Figure 2.13. The JLab and reanalysed DESY values for F_π in comparison to the results of several calculations [96,98,99,100,104]. The model uncertainty is represented by the grey area. The (model-independent) pion-atomic electron scattering data from [107] are also shown. A monopole behaviour of the form factor obeying the measured charge radius is almost identical to the Maris and Tandy curve.

A comparison with the separated σ_L and σ_T data at $W = 1.95$ GeV is given in figure 2.12 from which it can be seen that the description of the σ_L data is only qualitative and the σ_T data are underestimated.

To determine F_π, the experimental data of [110] were compared to the results of a Regge theory fit [111], although the centre-of-mass energy is rather low for this approach. The pion form factor and the $\pi\rho$ transition form factor were treated as free parameters, together with t-independent backgrounds at each Q^2. The values of F_π, extracted this way, are shown in figure 2.13. For consistency F_π was determined in [110] in the same way from the published DESY cross sections [109] at $Q^2 = 0.7$ GeV^2 and $W = 2.19$ GeV. The resulting best value for F_π, also shown in figure 2.13, is larger by 12% than the original result, which was obtained using the Born-term model of Gutbrod and Kramer [112].

The application of pion-pole dominance for σ_L at small $|t|$, as prescribed by (2.24), was checked by Volmer *et al* [110] by studying the reactions d$(e, e'\pi^+)nn$ and d$(e, e'\pi^-)pp$, which gave within the uncertainties a ratio of unity for the longitudinal cross sections.

The data for F_π in the region of Q^2 up to 1.6 GeV^2, as shown in figure 2.13, globally follow a monopole form obeying the pion charge radius [107]. This can be understood phenomenologically in a VMD picture, with the pion radius well described by $\sqrt{6}/M_\rho \sim 0.63$ fm. In this picture, the pion form factor is fitted well by a simple monopole form dominated by the $\rho(770)$.

The F_π data are also compared to a sample of quark-based calculations. The model by Maris and Tandy [104] provides a good description of the data. It is based on the Bethe–Salpeter equation with dressed quark and gluon propagators, and includes parameters that were determined without the use of these F_π data. The data are also well described by the QCD sum rule plus the hard-scattering estimate of [95,96]. Other models [100] were fitted to the older F_π data and therefore underestimate the present data. A perturbative QCD calculation [102], extending the basic dependence given by (2.21) to next-to-leading order, and including transverse momenta of the quarks, Sudakov factors, and a way to regularize the infrared divergence, renders an approximately constant value of $Q^2 F_\pi \approx 0.18$ over the whole range of Q^2 shown. Other perturbative QCD calculations yield similar results, but with a lower value of $Q^2 F_\pi$ [99]. Hence it is clear that in the region below $Q^2 \approx 2\,\text{GeV}^2$, where accurate data now exist, soft contributions are still larger than perturbative QCD ones. It is of great interest to extend the data to higher values of $Q^2 \approx 5\,\text{GeV}^2$, where as mentioned estimates [105] suggest that perturbative QCD contributions may start to dominate the pion form factor.

The success of the single-pole approximation to the pion form factor with an effective mass $m_{eff} \approx m_\rho$

$$F_\pi(q^2) = \frac{1}{1 - q^2/m_{eff}^2} \qquad (2.25)$$

for space-like momenta q^2 can be understood through a simple dispersion-relation calculation [113]. The simplest version, a superconvergent dispersion relation, has the form

$$F_\pi(t) = \int_{t_0}^{\infty} dt' \, \frac{\mathrm{Im} F_\pi(t')}{t' - t}, \qquad (2.26)$$

provided that $F_\pi(t) \to 0$ sufficiently rapidly as $t \to \infty$ to ensure the convergence of the integral in (2.26). Here $t_0 = 4m_\pi^2$ is the physical π–π threshold.

At large space-like momenta, perturbative QCD gives the rigorous prediction [114] for the asymptotic behaviour of the form factor

$$F_\pi(q^2) \sim \frac{8\pi f_\pi^2 \alpha_s(-q^2)}{-q^2}, \qquad (2.27)$$

where α_s is the QCD coupling parameter and $f_\pi = 130.7 \pm 0.4$ MeV is the pion decay constant. In leading order, (2.27) becomes

$$F_\pi(q^2) \sim \frac{32\pi^2 f_\pi^2}{-\beta_0 q^2 \log(|q^2/\Lambda^2|)}, \qquad (2.28)$$

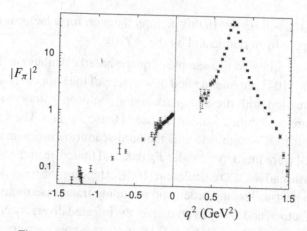

Figure 2.14. The squared modulus of the pion form factor.

where $\beta_0 = 11 - 2n_f/3$ is the usual β-function of QCD. To satisfy (2.28), at large positive q^2 the absorptive part of the pion form factor should have the asymptotic behaviour

$$\text{Im}F_\pi(t) \to \frac{32\pi^3 f_\pi^2}{q^2\beta_0 \log^2(q^2/\Lambda^2)}. \tag{2.29}$$

This is more than sufficient to ensure the convergence of the integral in (2.26) and indeed to ensure that the absorptive part of the pion form factor satisfies the superconvergence condition

$$\int_{s_0}^{\infty} dt' \, \text{Im}F_\pi(t') = 0. \tag{2.30}$$

The square of the modulus of the pion form factor in the time-like region is related to the cross section for $e^+e^- \to \pi^+\pi^-$ by

$$\sigma_{e^+e^-\to\pi\pi} = \frac{\pi\alpha^2}{3q^2}\left(1 - \frac{4m_\pi^2}{q^2}\right)^{\frac{3}{2}}|F_\pi(q^2)|^2. \tag{2.31}$$

However, what is required for the dispersion integral (2.26) is $\text{Im}F_\pi(q^2)$, not $|F_\pi(q^2)|^2$ and this requires a model-dependent fit. This is discussed in chapter 4 and although there are uncertainties at $\pi-\pi$ masses above 1 GeV the contribution from this region is small as the integral is dominated by the $\rho(770)$. This can be seen in figure 2.14, which shows the squared modulus of the space-like and time-like pion form factor. As the P-wave $\pi\pi$ phase shift is known and the amplitude is elastic until the $\pi\omega$ threshold, the phase of $F_\pi(q^2)$ is given by Watson's theorem [115] (see chapter 3) over this mass range, so there is no model dependence in the dominant mass region.

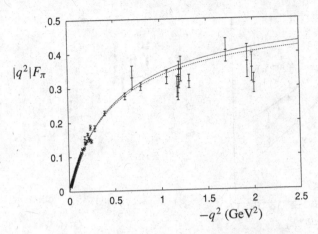

Figure 2.15. $|q^2|F_\pi(q^2)$ in the space-like region from the dispersion-relation cal-culation. The curves are defined in the text.

To obtain $\mathrm{Im}F_\pi$ for the dispersion integral (2.26), a fit is made to the $e^+e^- \to \pi\pi$ cross section assuming that the low-momentum region is given by a coherent sum over a few vector mesons and that the amplitude matches the QCD prediction asymptotically [113]. Evaluating the dispersion integral is then straightforward. Because of the dominance of the contribution from the $\rho(770)$ to the integral, the uncertainty in the resulting space-like form factor is small, well within the errors of the present data. The result [113] for the space-like form factor is given in figure 2.15, the two curves showing the range of model dependence. In the single-pole approximation, the upper curve corresponds to $m_{eff}^2 \approx 0.525$ GeV^2 and the lower curve to $m_{eff}^2 \approx 0.505$ GeV^2.

2.6 The axial form factor

In electroweak interactions we also have to consider matrix elements of the axial-vector current operator as well as the vector current. For weak charged currents, the most general expression for the axial current of the nucleon is

$$A^\mu = -ie\left[G_A(Q^2)\gamma^\mu\gamma_5 + G_P(Q^2)\frac{1}{2m_N}q^\mu\gamma_5\right], \qquad (2.32)$$

with G_A the axial and G_P the induced pseudoscalar form factor, respectively. In principle, in the most general form also a scalar form factor and a tensor form factor are present, which disappear in the absence of 'second class currents' [116]. Such second class currents would not be invariant under G-parity, and are experimentally excluded to a high precision.

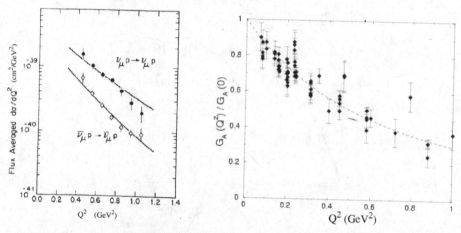

Figure 2.16. Left: cross sections obtained from neutrino and antineutrino scattering [118]. The solid lines are from the best dipole fit to the combined data ($m_A = 1.06$ GeV). Right: experimental data for the normalized axial form factor extracted from pion electroproduction experiments in the threshold region. Note that these G_A results were extracted using various theoretical models. The dashed curve shows a dipole fit with an arbitrary axial mass $m_A = 1.1$ GeV to guide the eye.

Consider first the axial form factor. For low and moderate momentum transfer, $Q^2 < 1$ GeV^2, where the bulk of the experimental data exists, the axial form factor can be represented by a dipole fit

$$G_A(Q^2) = \frac{g_A}{\left(1 + Q^2/m_A^2\right)^2} \qquad (2.33)$$

in terms of one adjustable parameter, m_A, the axial mass, and the axial-vector coupling constant $g_A = 1.2670 \pm 0.035$ [117]. This axial-vector coupling constant is measured in (polarized) neutron β decay. The axial root-mean-square radius is given by

$$\langle r_A^2 \rangle = -\frac{6}{g_A} \frac{dG_A(Q^2)}{dQ^2}\bigg|_{Q^2=0} = \frac{12}{m_A^2} \qquad (2.34)$$

in analogy with the corresponding relations (2.18) for the electromagnetic root-mean-square radii. Two methods have been used to determine the axial form factor of the nucleon. The most direct method is to use (quasi)elastic (anti)neutrino scattering off nucleons, typically $\nu_\mu + n \rightarrow \mu^- + p$ [118]. Data are shown in figure 2.16(left). With new and precise results for the nucleon electromagnetic form factors now available, and the latest value for the axial-vector coupling constant, a new global fit to the neutrino world data has been carried out [119]. The improved fit gives $m_A = 1.001 \pm 0.020$ GeV. This corresponds to an axial radius of $\langle r_A^2 \rangle^{1/2} = (0.683 \pm 0.014)$ fm.

The second method to determine the axial form factor is based on the analysis of charged pion electroproduction data off protons slightly above pion production threshold. First, the transverse component (see also (2.23)) of the near-threshold electroproduction cross section is determined. Then a multipole expansion is used to determine the electric-dipole transition amplitude $E_{0+}^{(-)}$. The axial form factor is then linked to this transition amplitude through the low-energy theorem of Nambu, Lurié and Shrauner [120], valid for massless pions. Model-dependent corrections to allow for finite pion mass were subsequently developed. The resulting axial-vector form factor is shown in figure 2.16(right). In a review, Bernard *et al* [121] used chiral perturbation theory to compute a finite mass correction to the axial mass extraction that is appreciable, correcting the value extracted from a Mainz Microtron experiment, $m_A = 1.068 \pm 0.015$ GeV [122], to $m_A = 1.013 \pm 0.015$ GeV. This brings the value in agreement with that extracted from the neutrino scattering method. Therefore, it appears m_A, and thus the axial radius, is reasonably well determined.

In contrast, the induced pseudoscalar form factor $G_P(Q^2)$ is essentially unknown. Pion-pole dominance arguments relate the induced pseudoscalar form factor to the axial form factor. Under the assumptions of axial-current conservation and massless pions (partially-conserved axial current approximation or PCAC) one can write [123]:

$$G_P(Q^2) = 2m_N \frac{G_A(Q^2)}{m_\pi^2 - Q^2}. \qquad (2.35)$$

Hence, the pseudoscalar coupling constant g_P equals $2m_N g_A / m_\pi^2$.

The most direct measurement of G_P is from muon capture on the proton, $\mu^- + p \rightarrow \nu_\mu + n$. Such a measurement is constrained to $Q^2 = 0.88 m_\mu^2$, the momentum transfer for muon capture by the proton at rest. The results of such measurements often disagree, with atomic physics corrections complicating matters at such low Q^2 [121]. The weighted world average for g_P from such measurements amounts to 8.79 ± 1.92, in good agreement with expectations from PCAC, although within large uncertainties.

In principle, $G_P(Q^2)$ can also be measured from the longitudinal cross section determined in pion electroproduction. The leading dependence due to the pion pole (and its chiral corrections) is unique and can be tested. However, such experiments are challenging as the pion momentum has to be small in the centre-of-mass system to justify the chiral corrections. The only data published at larger Q^2 are from a Saclay experiment [124], where only the lowest pion centre-of-mass momentum could be used to extract G_P reliably.

The existing world data, both the weighted world average from the muon capture experiments and the determination from pion electroproduction, are shown in

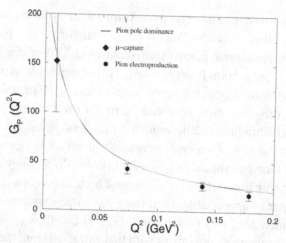

Figure 2.17. The 'world data' for the induced pseudoscalar form factor $G_P(Q^2)$. The pion electroproduction data are from [124]. The datum at $Q^2 = 0.88 m_\mu^2$ represents the world average of muon capture. The solid line indicates an expectation from pion-pole dominance.

figure 2.17. A curve indicating the pion-pole prediction has been added, which illustrates well the usual interpretation of G_P in terms of a nucleon pion-cloud picture. It is clear that more precise data, from both processes, would be very welcome – there is obvious room for improvement in data constraining the G_P form factor. More precise data could, for example, discriminate between the pion-pole prediction and corrections within chiral perturbation theory [121].

2.7 Nucleon strange form factors

The electromagnetic vector and the axial nucleon currents, as defined in (2.1) and (2.32), can be related to the currents of the elementary quarks in the nucleon, assuming that the quarks are point-like Dirac particles [125]. Hence, the nucleon form factors can also be expressed as linear combinations of the currents of the different flavours of quarks. This allows a relatively clean determination of the contribution of u, d and s quarks to the vector currents of the nucleon by means of comparison of neutral weak and electromagnetic elastic scattering measurements off the nucleons, as shown by Kaplan and Manohar in 1988 [126], assuming that the proton and neutron obey charge symmetry [127].

The lowest-order Feynman diagrams contributing to electron–nucleon scattering are given in figure 2.18. The Z-boson has both vector and axial-vector coupling that gives rise to parity violation in electron scattering.

Figure 2.18. Lowest-order Feynman diagrams contributing to electron–nucleon scattering. The electroweak currents of the nucleon are indicated by solid blobs.

For electroweak electron–neutron scattering the electromagnetic charges, e_f^γ, of the quarks are related to the neutral weak vector charges, e_f^Z, by

$$e_f^Z = 2T_{3,f} - 4e_f^\gamma \sin^2(\theta_W), \tag{2.36}$$

where f denotes the quark flavour (u, d, or s) and $T_{3,f}$ is the weak isospin. Factoring out the quark charges, the electromagnetic and neutral weak vector form factors can be written as:

$$
\begin{aligned}
G_{E,M}^\gamma &= \tfrac{2}{3} G_{E,M}^u - \tfrac{1}{3}\left(G_{E,M}^d + G_{E,M}^s\right), \\
G_{E,M}^Z &= \left(1 - \tfrac{8}{3}\sin^2\theta_W\right) G_{E,M}^u + \left(-1 + \tfrac{4}{3}\sin^2\theta_W\right)\left(G_{E,M}^d + G_{E,M}^s\right),
\end{aligned}
\tag{2.37}
$$

with θ_W the weak mixing angle [117]. Similarly, the neutral weak axial currents of the quarks can be identified in the overall axial current as

$$G_A = G_A^u - \left(G_A^d + G_A^s\right). \tag{2.38}$$

The key point here is that the electromagnetic and neutral weak vector form factors (currents) represent different linear combinations of the same matrix elements of contributions from the different flavours of quarks, in the end allowing for a decomposition. Because they have charges of the opposite sign, quarks and antiquarks contribute to the matrix elements $G_{E,M,A}^f$ with opposite signs. As a result, if the spatial distributions of s and $\bar{s}$ quarks were the same, their charges would cancel everywhere and G_E^s would vanish.

It is straightforward to solve for the contributions of the three flavours in the case where one more observable can be found that depends on a different linear combination of these matrix elements. This is what the parity-violating electron–nucleon scattering programme hinges on [125,126].

Because the weak interaction violates parity (that is the weak current contains both vector and axial-vector contributions), the interference of the electromagnetic and weak currents violates parity. Observation of this small effect requires comparison of an experiment and its mirror image. In parity-violating electron scattering, the mirror measurement is made by reversing the beam helicity. The asymmetry in cross sections formed with respect to this beam helicity is

$$A^{PV} \equiv \frac{\sigma_+ - \sigma_-}{\sigma_+ + \sigma_-},\qquad\qquad(2.39)$$

with its overall scale set by the ratio of the neutral weak and electromagnetic propagators, that is

$$\frac{\mathcal{M}^Z}{\mathcal{M}^\gamma} \sim \frac{Q^2}{M_Z^2}\qquad\qquad(2.40)$$

or about 10^{-4} at $Q^2 = 1\ \mathrm{GeV}^2$.

Of course, a simple three-quark picture of the nucleon yields zero for strange matrix elements. However, this is clearly too simplistic. A non-zero nucleon's strangeness content was historically viewed as a meson-cloud effect – that is the feature that a ('dressed') nucleon can, for a short time consistent with the uncertainty principle, transform into a ('bare') nucleon plus a multi-pion state – within which description one can intuitively understand the origin of the nucleon's charge structure. In the case of the strangeness matrix elements, of course, it is not the pion cloud which is responsible for the effect but rather the transformation into states containing strange quarks – $K\Lambda$, $K\Sigma$, ηN – which yields a non-zero effect. A more contemporary way to describe the origin of non-zero strangeness current matrix elements is in terms of the strange quark sea, which can be represented by the fragmentation of gluons into $s\bar{s}$ pairs.

The parity-violating interaction of electrons with nucleons also involves an axial-vector coupling to the nucleon, G_A^e. This axial-vector coupling can slightly differ from the corresponding quantity in neutrino scattering, G_A, as described in section 2.6. Explicitly, it contains additional contributions from the strange-quark helicity content of the nucleon, Δs, the anapole form factor, that is an additional term to (2.32), including the possibility that parity is not strictly conserved, and radiative corrections. The axial form factor G_A^e, or at least its isovector piece $G_A^{e(T=1)}$, can be determined from the parity-violating asymmetry in quasielastic electron scattering from deuterium, where the strange-quark effects in the neutron and proton tend to cancel. Hence, this is an experimentally observable quantity that can be used to correct strange form-factor measurements.

The SAMPLE experiment at the Bates Linear Accelerator Center was the first to study strange form factors and the anapole contribution. The elastic asymmetry from the proton was measured, in addition to the quasielastic asymmetry from the deuteron, for backward angles at $Q^2 = 0.1\ \mathrm{GeV}^2$. In [128], analysis of the SAMPLE results is presented which yields a value of the strange quark contribution to the

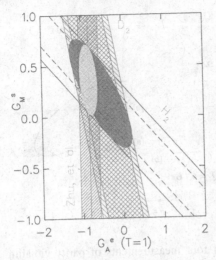

Figure 2.19. Uncertainty bands of G_M^s vs. $G_A^{e(T=1)}$ at $Q^2 = 0.1$ GeV^2 resulting from the two 200 MeV data sets of the SAMPLE experiment [128]. Also shown is the uncertainty in the theoretical expectation (vertical band) for $G_A^{e(T=1)}$ as computed by [129] extrapolated to the same Q^2. The smaller ellipse corresponds to the 1σ overlap of the hydrogen data and the theoretical prediction, the larger ellipse to the 1σ overlap of the two data sets.

magnetic form factor

$$G_M^s(Q^2 = 0.1) = +0.37 \pm 0.20 \pm 0.26 \pm 0.07, \qquad (2.41)$$

where the uncertainties are due to statistics, systematics, and radiative corrections, respectively. The value for G_A^e is consistent with the theoretical expectation [129], as illustrated in figure 2.19. A similar consistency in $G_A^{e(T=1)}$ is found at $Q^2 = 0.03$ GeV^2 [130].

For forward angle experiments in particular, it is convenient to express the deviation due to strange quark contributions in terms of

$$G_E^s + \eta G_M^s \sim (A^{PV} - A^{NVS}), \qquad (2.42)$$

where $\eta = \tau G_M^p / \epsilon G_E^p$ and A^{NVS} is the no-vector-strange (NVS) asymmetry hypothesis ($G_E^s = G_M^s = 0$).

The HAPPEX experiments utilized the two spectrometers in Hall A at JLab to measure parity violation in elastic scattering at very forward angles. In the original HAPPEX measurement at $Q^2 = 0.477$ GeV^2 the result was close to that expected with the NVS hypothesis [131]. The A4 experiment at the Mainz Microtron used an array of PbF$_2$ detectors for their forward angle measurements, at $Q^2 = 0.23$ GeV^2 [132] and subsequently at $Q^2 = 0.108$ GeV^2 [133]. While no measurement

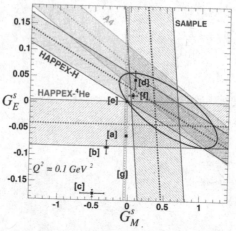

Figure 2.20. Results of four measurements of parity-violating asymmetries in elastic electron–proton scattering at $Q^2 \sim 0.1$ GeV^2. Shaded bands represent the 1-sigma combined statistical and systematic uncertainty. Also shown is the 95% confidence level ellipse from all four measurements. The black squares and narrow vertical band represent various theoretical calculations (with the labels [a]–[g] corresponding to [136,93,137–141], respectively).

independently indicates a significant strange form-factor contribution, the A4 measurement at $Q^2 = 0.108$ GeV^2 is 2σ away from the theoretical expectation neglecting strange quarks. HAPPEX measurements have been performed at $Q^2 = 0.1$ GeV^2 on both hydrogen [134] and ^{4}He [135] targets, and a summary of the results at $Q^2 = 0.1$ GeV^2 is presented in figure 2.20. Note that the elastic scattering from the spin-0 ^{4}He nucleus does not allow for magnetic or axial-vector current contributions.

Although each independent measurement does not show a statistically significant non-zero $G_E^s + \eta G_M^s$, the data at $Q^2 = 0.1$ GeV^2 do indeed suggest a positive contribution consistent with $G_M^s \sim +0.5$ and a small, possibly negative, G_E^s at $Q^2 = 0.1$ GeV^2. At this Q^2, improvements by a factor of 2–3 in precision are expected from a future HAPPEX experiment. Such precision has the potential to dramatically impact our understanding of the role of strange quarks in the nucleon. The role of strange quarks in parity-violating elastic electron–proton scattering has been mapped out over a large range in Q^2, $0.12 \leq Q^2 \leq 1.0$ GeV^2, by the G0 experiment at JLab [142]. This experiment used a dedicated toroidal spectrometer to detect the recoiling protons following elastic scattering. The results indicate non-zero, Q^2-dependent, strange-quark contributions, as shown in figure 2.21, which also includes the earlier HAPPEX and A4 measurements. The agreement between the experiments is excellent. The results show a systematic and intriguing Q^2-dependence. The G0 measurements at $Q^2 \sim 0.12$ GeV^2 are consistent with the hypothesis mentioned earlier that $G_M^s \sim +0.5$ at $Q^2 = 0.1$ GeV^2. Because η

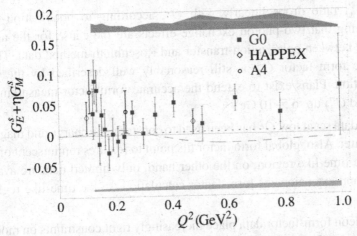

Figure 2.21. The combination $G_E^s + \eta G_M^s$ for the G0 measurements. The grey bands indicate systematic uncertainties (to be added in quadrature); the lines indicate the sensitivity to different electromagnetic form-factor parametrizations [58,57,38]. For comparison, the HAPPEX and A4 results [131–134] are also shown.

increases linearly throughout, the apparent decline of the data in the intermediate region up to $Q^2 \sim 0.3$ may hint at a negative value for G_E^s in this range. Lastly, there is a trend to positive values of $G_E^s + \eta G_M^s$ at higher Q^2. Backward angle measurements planned by G0 at JLab and A4 at Mainz will provide precise separations of G_E^s and G_M^s over a range of Q^2 to address this issue.

The planned backward angle G0 experiments include measurements of the parity-violating asymmetry in quasielastic scattering from deuterium, up to $Q^2 \sim 0.8$ GeV^2. Hence, future parity-violating experiments will not only allow precise determination of G_E^s and G_M^s from the hydrogen data, but also the first experimental information on (the isovector piece of) the axial form factor $G_A^{e(T=1)}$ away from the static limit [143]. In combination with neutrino scattering data, this should also allow extraction of the axial strange form factor G_A^s [144].

2.8 Outlook and conclusions

Advances in polarized electron sources, polarized nucleon targets and nucleon recoil polarimeters have enabled accurate measurements of the spin-dependent elastic electron–nucleon cross section. New data on nucleon electromagnetic form factors with unprecedented precision have (and will continue to) become available in an ever increasing Q^2-domain. The two magnetic form factors G_M^p and G_M^n approximately follow the simple dipole form factor G_D, although a clear deviation appears for G_M^p for $Q^2 > 8$ GeV^2, increasing with Q^2.

The G_E^p/G_M^p ratio drops linearly with Q^2, according to polarization-transfer data, assuming that two-photon exchange effects are the cause for the apparent discrepancy between polarization-transfer and Rosenbluth-method data. The neutron electric form factor G_E^n is still reasonably well described by the Galster parametrization. Plans exist to extend the accurate form-factor measurements of G_E^p, G_E^n, and G_M^n up to 5–10 GeV^2.

The precise data set at low Q^2 has been used to constrain the charge and magnetization radii better. Also, global form factor fits point to features reminiscent of a pion cloud. In the time-like region, on the other hand, only limited data are available. Better data and extensions of form-factor calculations to the time-like region are required.

The full nucleon form-factor data place increasingly tight constraints on models of nucleon structure. So far, all available theories are at least to some extent effective, although only a few adequately describe all four form factors. Only lattice gauge theory will in the end provide truly *ab initio* calculations, but accurate lattice QCD results for the form factors will not be available for some time. An excellent model description of the present data on the pion form factor does already exist. Here, it will be very interesting to extend the Q^2 range of these measurements to shed more light on the onset of the anticipated asymptotic scaling behaviour. Hadron form factors can also be connected directly to the quark structure of matter using the generalized parton distribution formalism, discussed in chapter 9.

When extended to electroweak lepton–nucleon scattering, knowledge can be gained on the axial and pseudoscalar form factors, or, through a program of parity-violating elastic electron–nucleon scattering experiments, on the strange quark contributions to the nucleon form factors. Some outstanding discrepancies between neutrino scattering and pion electroproduction methods have been solved, with the axial mass m_A now well established at 1.001 ± 0.020 GeV. Data on the pseudoscalar form factor remain scarce.

References

[1] R Hofstadter, Annual Review of Nuclear Science 7 (1957) 231
[2] F E Close, *An Introduction to Quarks and Partons*, Academic Press, London (1979)
[3] N F Mott, Proceedings Royal Society A124 (1929) 425
[4] M Rosenbluth, Physical Review 79 (1950) 615
[5] S Galster *et al*, Nuclear Physics B32 (1971) 221
[6] P Bosted, Physical Review C51 (1995) 409
[7] A F Sill *et al*, Physical Review D48 (1993) 29
[8] T Janssens *et al*, Physical Review 142 (1966) 922

[9] W Bartel *et al*, Physical Review Letters 17 (1966) 608

[10] J Litt *et al*, Physics Letters 31B (1970) 40

[11] C Berger *et al*, Physics Letters 35B (1971) 87

[12] R C Walker *et al*, Physical Review D49 (1994) 5671

[13] L Andivahis *et al*, Physical Review D50 (1994) 5491

[14] S Kopecky *et al*, Physical Review C56 (1997) 2229

[15] H Anklin *et al*, Physics Letters B336 (1994) 313

[16] H Anklin *et al*, Physics Letters B428 (1998) 248

[17] G Kubon *et al*, Physics Letters B524 (2002) 26

[18] E E W Bruins *et al*, Physical Review Letters 75 (1995) 21

[19] W Brooks, private communications; JLab experiment E94-017, W Brooks and M Vineyard, spokespersons

[20] S Platchkov *et al*, Nuclear Physics A510 (1990) 740

[21] A Lung *et al*, Physical Review Letters 70 (1993) 718

[22] E L Lomon, Annals of Physics 125 (1980) 309

[23] P Markowitz *et al*, Physical Review C48 (1993) R5

[24] W Xu *et al*, Physical Review Letters 85 (2000) 2900

[25] W Xu *et al*, Physical Review C67 (2003) 012201

[26] N Dombey, Reviews of Modern Physics 41 (1969) 236

[27] T W Donnelly and A S Raskin, Annals of Physics 169 (1986) 247; *ibid* 191 (1989) 81

[28] A I Akhiezer, L N Rozentsweig and I M Snushkevich, Soviet Physics JETP 6 (1958) 588

[29] R G Arnold, C Carlson and F Gross, Physical Review C23 (1981) 363

[30] J Golak *et al*, Physical Review C63 (2001) 034006

[31] H Arenhövel, Physics Letters B199 (1987) 13

[32] H Arenhövel, W Leidemann, E L Tomusiak, Zeitschrift für Physik A331 (1988) 123; *ibid* A334 (1989) 363(E); Physical Review C46 (1992) 455

[33] B Milbrath *et al*, Physical Review Letters 80 (1998) 452; *erratum* 82 (1999) 2221

[34] T Pospischil *et al*, European Physics Journal 12 (2001) 125

[35] M K Jones *et al*, Physical Review Letters 84 (2000) 1398

[36] S Dieterich *et al*, Physics Letters B500 (2001) 47

[37] O Gayou *et al*, Physical Review Letters 88 (2002) 092301

[38] J Arrington, Physical Review C68 (2003) 034325

[39] M E Christy *et al*, Physical Review C70 (2004) 015206

[40] I A Quattan *et al*, Physical Review Letters 94 (2005) 142301

[41] P A M Guichon and M Vanderhaeghen, Physical Review Letters 91 (2003) 142303

[42] A V Afanasev, I Akushevich, A Ilyichev and N P Merenkov, Physics Letters B514 (2001) 269

[43] J Arrington, Physical Review C69 (2004) 032201R

[44] J Arrington, Physical Review C69 (2004) 022201R

[45] P G Blunden, W Melnitchouk and J A Tjon, Physical Review Letters 91 (2003) 142304

[46] Y-C Chen *et al*, Physical Review Letters 93 (2004) 122301

[47] C E Jones-Woodward *et al*, Physical Review C44 (1991) 571R

[48] T Eden *et al*, Physical Review C50 (1994) 1749R

[49] J Bermuth *et al*, Physics Letters B564 (2003) 199; D Rohe *et al*, Physical Review Letters 83 (1999) 4257

[50] G Warren *et al*, Physical Review Letters 92 (2004) 042301; H Zhu *et al*, Physical Review Letters 87 (2001) 081801

[51] I Passchier *et al*, Physical Review Letters 82 (1999) 4988
[52] D I Glazier *et al*, European Physical Journal A24 (2005) 101;
  M Ostrick *et al*, Physical Review Letters 83 (1999) 276;
  C Herberg *et al*, European Physics Journal A5 (1999) 131
[53] R Madey *et al*, Physical Review Letters 91 (2003) 122002
[54] R Schiavilla and I Sick, Physical Review C64 (2001) 041002
[55] K Melnikov and T van Ritbergen, Physical Review Letters 84 (2000) 1673
[56] I Sick, Physics Letters B576 (2003) 62
[57] J Friedrich and Th Walcher, European Physics Journal 17 (2003) 607
[58] J J Kelly, Physical Review C66 (2002) 065203
[59] H W Hammer, D Drechsel and U G Meissner, Physics Letters B586 (2004) 291
[60] F Iachello, A Jackson and A Lande, Physics Letters B43 (1973) 191
[61] G Höhler *et al*, Nuclear Physics B114 (1976) 505
[62] E L Lomon, Physical Review C64 (2001) 035204; ibid C66 (2002) 045501
[63] M Gari and W Krümpelmann, Zeitschrift für Physik A322 (1985) 689; Physics Letters B173 (1986) 10
[64] M Gari and W Krümpelmann, Physics Letters B274 (1992) 159
[65] R Bijker and F Iachello, Physical Review C69 (2004) 068201
[66] T Fuchs, J Gegelia, and S Scherer, European Physics Journal 19 (Suppl. 1) (2004) 35
[67] G Holzwarth, Zeitschrift für Physik A356 (1996) 339; hep-ph/0201138 (2002)
[68] C V Christov *et al*, Nuclear Physics A592 (1995) 513;
  H C Kim *et al*, Physical Review D53 (1996) 4013
[69] J D Ashley *et al*, European Physics Journal 19 (Suppl. 1) (2004) 9
[70] G A Miller, Physical Review C66 (2002) 032001R
[71] F Cardarelli and S Simula, Physical Review C62 (2000) 065201
[72] R F Wagenbrunn *et al*, Physics Letters B511 (2001) 33;
  S Boffi *et al*, European Physics Journal 14 (2002) 17
[73] M Giannini, E Santopinto and A Vassallo, Progress in Particle and Nuclear Physics 50 (2003) 263
[74] D Merten *et al*, European Physics Journal 14 (2002) 477
[75] S Brodsky and G Farrar, Physical Review Letters 31 (1973) 1153; Physical Review D11 (1975) 1309
[76] S J Brodsky and G P Lepage, Physical Review D22 (1980) 2157; *ibid* D24 (1981) 2848
[77] G A Miller, Physical Review C65 (2002) 065205
[78] J P Ralston and P Jain, Physical Review D69 (2004) 053008
[79] A V Belitsky, X Ji and F Yuan, Physical Review Letters 91 (2003) 092003
[80] S J Brodsky *et al*, Physical Review D69 (2004) 076001
[81] M Göckeler *et al*, Physical Review D71 (2005) 034508
[82] G Bardin *et al*, Nuclear Physics B411 (1994) 3
[83] A Antonelli *et al*, Physics Letters B334 (1994) 431; and references therein.
[84] T A Armstrong *et al*, Physical Review Letters 70 (1993) 1212
[85] M Ambrogiani *et al*, Physics Review D60 (1999) 032002
[86] A Antonelli *et al*, Nuclear Physics B517 (1998) 3
[87] A Antonelli *et al*, Physics Letters B365 (1996) 427
[88] F Myhrer and A W Thomas, Physics Letters B64 (1976) 59
[89] I S Shapiro, Physics Reports 35 (1978) 129
[90] R L Jaffe, Physical Review D17 (1978) 1444
[91] C B Dover and J M Richard, Annals of Physics 121 (1979) 70

[92] F Iachello and Q Wan, Physical Review C69 (2004) 055204

[93] H W Hammer, U-G Meissner, and D Drechsel, Physics Letters B385 (1996) 343

[94] S Dubnicka, A Z Dubnickova and P Weisenpacher, Journal of Physics G29 (2003) 405

[95] V A Nesterenko and A V Radyushkin, Physics Letters B115 (1982) 410

[96] A P Bakulev and A V Radyushkin, Physics Letters B271 (1991) 223; A V Radyushkin, Nuclear Physics A532 (1991) 141c

[97] H-N Li and G Sterman, Nuclear Physics B381 (1992) 129

[98] H Ito, W W Buck and F Gross, Physical Review C45 (1992) 1918

[99] R Jakob and P Kroll, Physics Letters B315 (1993) 463

[100] F Cardarelli *et al*, Physics Letters B332 (1994) 1; Physics Letters B357 (1995) 267

[101] P Maris and C D Roberts, Physical Review C58 (1998) 3659

[102] N G Stefanis, W Schroers and H-Ch Kim, Physics Letters B449 (1999) 299; European Physical Journal C18 (2000) 137

[103] V M Braun, A Khodjamirian and M Maul, Physical Review D61 (2000) 073004

[104] P Maris and P C Tandy, Physical Review C62 (2000) 055204

[105] W Schweiger, Nuclear Physics (Proc. Supp.) 108 (2002) 242

[106] G R Farrar and D R Jackson, Physical Review Letters 43 (1979) 246

[107] S R Amendolia *et al*, Nuclear Physics B277 (1986) 168

[108] C J Bebek *et al*, Physical Review D17 (1978) 1693

[109] P Brauel *et al*, Zeitschrift für Physik C3 (1979) 101

[110] J Volmer *et al*, Physical Review Letters 86 (2001) 1713

[111] M Vanderhaeghen, M Guidal and J-M Laget, Physical Review C57 (1998) 1454; Nuclear Physics A627 (1997) 645

[112] F Gutbrod and G Kramer, Nuclear Physics B49 (1972) 461

[113] K Watanabe, H Ishikawa and M Nakagawa, hep-ph/0111168

[114] A V Efremov and A V Radyushkin, JETP Letters 25 (1977) 210 S Brodsky and G P Lepage, Physics Letters 87B (1979) 359

[115] K M Watson, Physical Review 95 (1954) 226

[116] S Weinberg, Physical Review 112 (1958) 1375

[117] K Hagiwara *et al*, Physical Review D66 (2002) 010001

[118] L Ahrens *et al*, Physical Review D35 (1987) 785

[119] H Budd, A Bodek and J Arrington, Nuclear Physics (Proc. Supp.) B139 (2005) 90

[120] Y Nambu and D Lurié, Physical Review 125 (1962) 1429; Y Nambu and E Shrauner, Physical Review 128 (1962) 862

[121] V Bernard, L Elouadhriri and U-G Meissner, Journal of Physics G28 (2002) R1

[122] A Liesenfeld *et al*, Physics Letters B468 (1999) 20

[123] S L Adler *et al*, Physical Review D11 (1975) 3309

[124] S Choi *et al*, Physical Review Letters 71 (1993) 3927

[125] D H Beck and B R Holstein, International Journal of Modern Physics E10 (2001) 1

[126] D Kaplan and A Manohar, Nuclear Physics B310 (1988) 527

[127] G A Miller, Physical Review C57 (1997) 1492

[128] D T Spayde *et al*, Physics Letters B583 (2004) 79

[129] S-L Zhu, S J Puglia, B R Holstein and M J Ramsey-Musolf, Physical Review D62 (2000) 033008

[130] T M Ito *et al*, Physical Review Letters 92 (2004) 102003

[131] K A Aniol *et al*, Physical Review Letters 82 (1999) 1096; Physical Review C69 (2004) 065501

[132] F E Maas *et al*, Physical Review Letters 93 (2004) 022002
[133] F E Maas *et al*, Physical Review Letters 94 (2005) 152001
[134] K A Aniol *et al*, HAPPEX Collaboration, Physics Letters B635 (2006) 275
[135] K A Aniol *et al*, HAPPEX Collaboration, Physical Review Letters 96 (2006) 022503
[136] N W Park and H Weigel, Nuclear Physics A451 (1992) 453
[137] H W Hammer and M J Ramsey-Musolf, Physical Review C60 (1999) 045204
[138] A Silva *et al*, Physical Review D65 (2001) 014015
[139] V Lyubovitskij *et al*, Physical Review C66 (2002) 055204
[140] R Lewis *et al*, Physical Review D67 (2003) 013003
[141] D B Leinweber *et al*, Physical Review Letters 94 (2005) 212001
[142] D S Armstrong *et al*, G0 Collaboration, Physical Review Letters 95 (2005) 092001
[143] E J Beise, European Physical Journal A24S2 (2005) 43
[144] S Pate, Physical Review Letters 92 (2004) 082002

3

Electromagnetic excitations of nucleon resonances

V Burkert and T-S H Lee

It has long been recognized that the study of nucleon resonances (N^*) is one of the important steps in the development of a fundamental understanding of strong interactions. While the existing data on the nucleon resonances are consistent with the well-studied $SU(6) \otimes O(3)$ constituent-quark-model classification, many open questions remain. On the fundamental level, there exists only very limited understanding of the relationship between QCD [1], the fundamental theory of strong interactions, and the constituent-quark model or alternative hadron models. Experimentally, we still do not have sufficiently complete data that can be used to uncover unambiguously the structure of the nucleon and its excited states. For instance, precise and consistent data on the simplest nucleon form factors and the nucleon–$\Delta(1232)$ resonance (N–Δ) transition form factors up to sufficiently high momentum transfer are becoming available. Thus the study of N^* structure remains an important task in hadron physics, despite its long history.

With the development since the 1990s of various facilities with electron and photon beams, extensive data on electromagnetic production of mesons have now been accumulated for the study of N^* physics. These facilities include JLab, LEGS at Brookhaven National Laboratory and MIT-Bates in the USA; MAMI at Mainz and ELSA at Bonn in Germany; GRAAL at Grenoble in France and LEPS at Spring-8 in Japan. The details of these facilities are summarized in [2] and will not be included here. The main purpose of this chapter is to review the theoretical models used in analysing these data, and to highlight the results obtained.

3.1 Issues

It is useful first to review briefly the status of theoretical efforts in understanding the structure of the nucleon and the N^* states. The most fundamental approach is

Electromagnetic Interactions and Hadronic Structure, eds Frank Close, Sandy Donnachie and Graham Shaw.
Published by Cambridge University Press. © Cambridge University Press 2007

to develop accurate numerical simulation of QCD on the lattice [3] (lattice QCD). Significant progress has been made. Some basic properties of baryons, such as masses of ground states and low-lying excited states, have been reproduced by lattice QCD [4–6]. The first calculation of the electromagnetic transition form factors from the ground state proton to the $\Delta(1232)$ within unquenched QCD has been attempted [7]. However, reliable lattice QCD calculations for electromagnetic meson-production reactions seem to be in the distant future, and models of hadron structure and reactions will likely continue to play an important role and provide theoretical guidance for experimenters.

The development of hadron models for the nucleon and nucleon resonances has a long history. Since the 1970s, the constituent-quark model has been greatly refined to account for residual quark–quark interactions due to one-gluon exchange [8–10] and/or Goldstone boson exchange [11,12]. Efforts are underway to reformulate the model within relativistic quantum mechanics [13–15]. Conceptually completely different models have also been developed, such as bag models [16], chiral bag models [17,18], algebraic models [19], soliton models [20], colour-dielectric models [21], Skyrme models [22], and covariant models based on Dyson–Schwinger equations [23]. With suitable phenomenological procedures, most of these models are comparable in reproducing the low-lying N^* spectrum as determined by the amplitude analyses of elastic πN scattering. However, they have rather different predictions on the number and the ordering of the highly excited N^* states. They also differ significantly in predicting some dynamical quantities such as the electromagnetic and mesonic N–N^* transition form factors. Clearly, accurate experimental information for these N^* observables is needed to distinguish these models. This information can be extracted from the data on electromagnetic meson-production reactions. With the very intense experimental efforts in the past few years, such data have now been accumulated with high precision at the various facilities mentioned above. We are now in a very good position to make progress.

We next discuss how the new data can be used to address some of the long-standing problems in the study of N^* physics. The first one is the so-called *missing-resonance problem*. This problem originated from the observation that some of the N^* states predicted by the constituent-quark model are not seen in the baryon spectrum that is determined mainly from the amplitude analyses of πN elastic scattering. There are two possible solutions to this problem. First, it is possible that the constituent-quark model has the wrong effective degrees of freedom of QCD in describing the highly-excited baryon states. Other models with fewer degrees of freedom, such as quark–diquark models, or models based on alternative symmetry schemes [24], could be more accurate in reproducing the baryon spectrum. The second possibility is that these missing resonances do not couple strongly to the πN channel and can only be observed in other processes. Data from experiments measuring as many meson–baryon channels as possible are needed to resolve the missing-resonance problem.

The second outstanding problem in the study of N^* physics is that the partial decay widths of baryon resonances compiled and published periodically by the Particle Data Group (PDG) have very large uncertainties in most cases [25]. For some decay channels, such as ηN, $K\Sigma$ and ωN, the large uncertainties are mainly due to insufficient data. However, the discrepancies between the results from different methods of amplitude analysis are also sources of uncertainties. This problem can be resolved only with a sufficiently large database that allows much stronger constraints on amplitude analyses, and a strong reduction in the model dependence of the extracted N^* masses and partial decay widths. This requires that the data must be precise and must cover very large kinematic regions in scattering angles, energies, and momentum transfers. The data on polarization observables must also be as extensive as possible.

These two experimental challenges are now being met. The modern accelerator facilities are well equipped with sophisticated detectors for measuring not only the dominant single-pion channel but also kaon, vector-meson, and two-pion channels.

The third long-standing problem is in the theoretical interpretation of the N^* parameters. Most of the model predictions on $N^* \to \gamma N$ helicity amplitudes are only in qualitative agreement with the experimental results. In some cases they disagree even in sign. One could attribute this to the large experimental uncertainties, as discussed above. However, the well-determined empirical values of the simplest and most unambiguous $\Delta(1232) \to \gamma N$ helicity amplitudes are about 40% larger than the predictions from practically all of the hadron models. This raises the question about how the hadron models, as well as the lattice QCD calculations, are related to the N^* parameters extracted from empirical amplitude analyses. We need to evaluate critically their relationships from the point of view of fundamental reaction theory. The discrepancies in the $\Delta(1232)$ region must be understood before meaningful comparisons between theoretical predictions and empirical values can be made. Much progress has been achieved. The results, as will be detailed in section 3.4.1, strongly indicate that it is necessary to apply an appropriate reaction theory in making meaningful comparisons of the empirical values from amplitude analyses and the predictions from hadron models and lattice QCD.

The above discussions clearly indicate that a close collaboration between experimental and theoretical efforts is needed to make progress in the study of N^* physics. A possible interplay between them is illustrated in figure 3.1. On the theoretical side, we need to use lattice QCD calculations and/or hadron structure models to predict properties of nucleon resonances, such as the $N–N^*$ transition form factors indicated in figure 3.1. On the experimental side, sufficiently extensive and precise

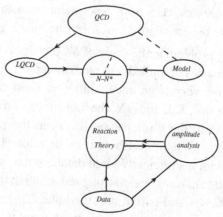

Figure 3.1. Scheme for N^* study.

data of meson-production reactions must be accumulated. We then must develop reaction models for interpreting the data in terms of hadron structure calculations. The development of empirical amplitude analyses of the data is an important part of this task.

In this chapter, we will focus on the lower parts of figure 3.1. The N^* structure calculations can be found in textbooks, for example [26], or review articles [27]. In section 3.2, the general formulation for calculating cross sections for electromagnetic meson production is presented. Section 3.3 is devoted to reviewing and assessing most of the models being used to analyse the data. Results from the analyses of electromagnetic meson-production data are highlighted in section 3.4. In section 3.5 we discuss future directions.

3.2 General formalism

Most of the experiments for N^* study involve reactions with a single meson and baryon in the final state. We therefore only present the formulation for analysing the data for such reactions. The generalization of the formulation to the cases that the final states are three-body states, such as $\gamma^* N \to \pi \pi N$, is straightforward.

The $N(e, e'M)B$ reaction is illustrated in figure 3.2. The final meson–baryon $(M–B)$ states can be composed of two-body states, such as πN, ηN, $K \Lambda$, ωN and ϕN, or of quasi-two-body states, such as $\pi \Delta$, σN and ρN. Within relativistic quantum field theory, the Hamiltonian density for describing this process can be written concisely as

$$H_{em}(x) = e A^{\mu}(x)[j_{\mu}(x) + J_{\mu}(x)],$$

(3.1)

where A^{μ} is the photon field,

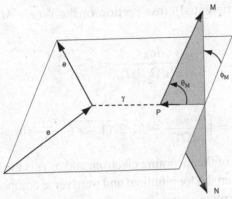

Figure 3.2. Kinematics of the meson electroproduction reaction. M is a meson, B represents a nucleon or any baryon (Δ, Λ, ...), ϕ_M is the angle between the e–e' plane and the M–B plane and θ_M is the angle between the outgoing meson and the virtual photon.

$$j^\mu(x) = \bar{\psi}_e(x)\gamma^\mu\psi_e(x) \tag{3.2}$$

is the lepton current, and the electromagnetic interactions involving hadrons are induced by the hadron current J^μ.

With the convention of Bjorken and Drell [28], the Hamiltonian density (3.1) leads to

$$\left\langle kp' \left| \int dx\, A^\mu(x)J_\mu(x) \right| qp \right\rangle = (2\pi)^4\delta^4(p+q-k-p')$$
$$\times \langle kp' \mid \epsilon_\mu(q)J^\mu(0) \mid qp \rangle, \tag{3.3}$$

where q, p, k, and p' are the four-momenta for the initial photon, initial nucleon, final meson and final nucleon respectively and $\epsilon_\mu(q)$ is the photon polarization vector. It is convenient to write

$$\langle kp' \mid \epsilon_\mu(q)J^\mu(0) \mid qp \rangle = \frac{1}{(2\pi)^6}\sqrt{\frac{m_N}{E_N(p')}}\frac{1}{\sqrt{2E_\pi(k)}}\epsilon_\mu(q)J^\mu(k,p';q,p)$$
$$\times \sqrt{\frac{m_N}{E_N(p)}}\frac{1}{\sqrt{2\omega}}, \tag{3.4}$$

where $E_a(p) = \sqrt{\mathbf{p}^2 + m_a^2}$, with m_a denoting the mass of particle a, and $\omega = q^0$ is the photon energy. Throughout this chapter, we will suppress the spin and isospin indices unless they are needed for detailed explanations. The formula for calculating electromagnetic meson-production cross sections can be expressed in terms of $J^\mu(k,p';q,p)$.

For evaluating electroproduction cross sections, it is common and convenient to choose a coordinate system in which the virtual photon is in the quantization z-direction and the angle between the e–e' plane and M–B plane is ϕ_M, as illustrated in figure 3.2. The differential cross section of the $N(e, e'M)B$ reaction can be written as [29]

$$\frac{d\sigma_h}{dE'_e d\Omega'_e d\Omega^*_M} = \Gamma \frac{d\sigma_h}{d\Omega^*_M}$$

with

$$\frac{d\sigma_h}{d\Omega^*_M} = \left[\frac{d\sigma_{unpol}}{d\Omega^*_M} + h\sqrt{2\epsilon(1-\epsilon)} \frac{d\sigma_{LT'}}{d\Omega^*_M} \sin\phi_M \right], \tag{3.5}$$

where h is the helicity of the incoming electron, and $\sigma_{LT'}$ is the cross section due to the interference between the longitudinal and transverse components of the current matrix elements. An explicit expression for the unpolarized cross section σ_{unpol} will be given later. All cross sections with scattering angle Ω^* in (3.5) are defined in the centre-of-mass frame of the final M–B system. The kinematic factors associated with the incoming and outgoing electrons are only contained in the following two variables:

$$\epsilon = \left\{ 1 + \frac{2 |\mathbf{q}|^2}{Q^2} \tan^2(\tfrac{1}{2}\theta_e) \right\}^{-1},$$

$$\Gamma = \frac{\alpha K_H}{2\pi^2 Q^2} \frac{E'_e}{E_e} \frac{1}{1-\epsilon}, \tag{3.6}$$

where $K_H = \omega - Q^2/2m_N$ is the virtual photon flux, $\alpha = e^2/(4\pi) = 1/137$ is the electromagnetic coupling strength, $q^\mu = (\omega, \mathbf{q})$ is the four-momentum of the photon in the laboratory frame, and $Q^2 = -q^2 = |\mathbf{q}|^2 - \omega^2$. The incident and outgoing electron energies E_e and E'_e are related to q^μ by

$$\omega = E_e - E'_e, \tag{3.7}$$

$$Q^2 = 4E_e E'_e \sin^2(\tfrac{1}{2}\theta_e), \tag{3.8}$$

where θ_e is the angle between the incident and outgoing electrons.

We next present formulae for calculating the centre-of-mass differential cross sections on the right-hand side of (3.5). The unpolarized cross section is given by

$$\frac{d\sigma_{unpol}}{d\Omega^*_M} = \frac{d\sigma_T}{d\Omega^*_M} + \epsilon \frac{d\sigma_L}{d\Omega^*_M} + \epsilon \frac{d\sigma_{TT}}{d\Omega^*_M} \cos 2\phi_M + \sqrt{2\epsilon(1+\epsilon)} \frac{d\sigma_{LT}}{d\Omega^*_M} \cos\phi_M,$$

$$\tag{3.9}$$

where σ_T, σ_L, σ_{TT}, and σ_{LT} are called the transverse, longitudinal, polarization, and interference cross sections. These four cross sections and $\sigma_{LT'}$ in (3.5) can be written as

$$\frac{d\sigma_\alpha}{d\Omega^*_M} = \frac{|\mathbf{k}_c|}{q_c^\gamma} M_\alpha(k_c, p'_c; q_c, p_c). \tag{3.10}$$

Here $q_c^\gamma = (W^2 - m_N^2)/(2m_N) = K_H$ is the magnitude of the effective photon centre-of-mass momentum, $W = \sqrt{s}$ is the centre-of-mass energy, given by $W^2 = (p + q)^2 = (k + p')^2$ and $\alpha = T, L, TT, LT$ and LT'. The centre-of-mass momenta k_c, p_c', q_c, and p_c in (3.10) can be calculated from the corresponding momenta in the laboratory frame by a Lorentz boost with $\vec{\beta} = \hat{z} \mid \vec{q} \mid /(\omega + m_N)$. Obviously $q_c^\gamma = \mid \mathbf{q_c} \mid$ at the real photon point $Q^2 = 0$.

The meson-production dynamics is contained in the M_α of (3.10). They are calculated from various combinations of current matrix elements evaluated on the $\phi_M = 0$ plane (see figure 3.2):

$$M_T(k_c, p_c'; q_c, p_c) = \tfrac{1}{4} F \sum_{spins} [\mid J^x(k_c, p_c'; q_c, p_c) \mid^2 + \mid J^y(k_c, p_c'; q_c, p_c) \mid^2],$$

$$M_L(k_c, p_c'; q_c, p_c) = \tfrac{1}{2} F \sum_{spins} \frac{Q^2}{\omega^2} [\mid J^z(k_c, p_c'; q_c, p_c) \mid^2],$$

$$M_{TT}(k_c, p_c'; q_c, p_c) = \tfrac{1}{4} F \sum_{spins} [\mid J^x(k_c, p_c'; q_c, p_c) \mid^2 - \mid J^y(k_c, p_c'; q_c, p_c) \mid^2],$$

$$M_{LT}(k_c, p_c'; q_c, p_c) = -\tfrac{1}{2} F \sum_{spins} \sqrt{\frac{Q^2}{\omega^2}} \text{Re}\{J^z(k_c, p_c'; q_c, p_c) J^{x*}(k_c, p_c'; q_c, p_c)\},$$

$$M_{LT'}(k_c, p_c'; q_c, p_c) = \tfrac{1}{2} F \sum_{spins} \sqrt{\frac{Q^2}{\omega^2}} \text{Im}\{J^z(k_c, p_c'; q_c, p_c) J^{x*}(k_c, p_c'; q_c, p_c)\},$$

$$\text{(3.11)}$$

with

$$F(k_c, pc'; q_c, p_c) = \frac{e^2}{(2\pi)^2} \frac{1}{2E_M(k_c)} \frac{m_N}{E_B(p_c')} \frac{m_N}{E_N(p_c)} \frac{E_M(k_c)E_B(p_c')}{2W}. \quad \text{(3.12)}$$

This formulation can be readily used to calculate various polarization observables with a polarized initial nucleon. For observables with a polarized final baryon, the situation is more complicated. The relevant formulae have been explicitly derived for pseudo-scalar meson production [30,31] and those for analysing spin observables in vector-meson production were developed in [32].

We also note that the unpolarized photoproduction cross section is given by $d\sigma_T/d\Omega_M^*$ evaluated at $Q^2 = 0$ and $q_c^\gamma \to \mid \mathbf{q_c} \mid$. Calculations of pion photoproduction polarization observables are given, for example, in Appendix C of [68].

For investigating nucleon resonances, it is necessary to express the meson-production cross sections in terms of multipole amplitudes corresponding to definite angular momentum, parity and isospin. This is well known when the final hadron state consists of only a pseudo-scalar and a spin-$\tfrac{1}{2}$ baryon, such as πN, KY and

ηN. Equation (3.4) is written in terms of invariants in the centre-of-mass frame of the final M–B system [33]

$$\epsilon_\mu J^\mu(k_c, p_c'; q_c, p_c) = \sum_{i=1,6} F_i(s, t, Q^2)\bar{u}(\mathbf{p_c'})O_i u(\mathbf{p_c}), \qquad (3.13)$$

where $s = (k_c + p_c')^2$ and $t = (k_c - q_c)^2$ are the usual Mandelstam variables, $F_i(s, t, Q^2)$ are the Lorentz-invariant amplitudes and O_i are operators defined in the baryon spin space

$$O_1 = i\sigma \cdot \mathbf{b},$$
$$O_2 = \sigma \cdot \hat{\mathbf{k}}_c \sigma \cdot (\hat{\mathbf{q}} \times \mathbf{b}),$$
$$O_3 = i\sigma \cdot \hat{\mathbf{q}}_c \hat{\mathbf{k}}_c \cdot \mathbf{b},$$
$$O_4 = i\sigma \cdot \hat{\mathbf{k}}_c \hat{\mathbf{k}}_c \cdot \mathbf{b},$$
$$O_5 = -i\sigma \cdot \hat{\mathbf{k}}_c b_0,$$
$$O_6 = -i\sigma \cdot \hat{\mathbf{q}}_c b_0, \qquad (3.14)$$

with

$$b^\mu = \epsilon^\mu(q_c) - \frac{\hat{\varepsilon} \cdot \hat{\mathbf{q}}_c}{|\mathbf{q}_c|} q_c^\mu. \qquad (3.15)$$

Obviously we have $\mathbf{b} \cdot \hat{\mathbf{q}}_c = 0$.

The invariant amplitudes $F_i(s, t, Q^2)$ can be expanded in terms of multipole amplitudes characterized by the angular momentum and quantum numbers of the initial $\gamma^* N$ and the final M–B system:

$$F_1 = \sum_\ell [P_{\ell+1}'(x)E_{\ell+} + P_{\ell-1}'(x)E_{\ell-} + P_{\ell+1}'(x)M_{\ell+} + (\ell+1)P_{\ell-1}'(x)M_{\ell-}],$$

$$F_2 = \sum_\ell [(\ell+1)P_\ell'(x)M_{\ell+} + \ell P_\ell'(x)M_\ell],$$

$$F_3 = \sum_\ell [P_{\ell+1}''(x)E_{\ell+} + P_{\ell-1}''(x)E_{\ell-} - P_{\ell+1}''(x)M_{\ell+} + P_{\ell-1}''(x)M_{\ell-}],$$

$$F_4 = \sum_\ell [-P_\ell''(x)E_{\ell+} - P_\ell''(x)E_{\ell-} + P_\ell''(x)M_{\ell+} - P_\ell''(x)M_{\ell-}],$$

$$F_5 = \sum_\ell [-(\ell+1)P_\ell'(x)S_{\ell+} + \ell P_\ell'(x)S_{\ell-}],$$

$$F_6 = \sum_\ell [(\ell+1)P_{\ell+1}'(x)S_{\ell+} - \ell P_{\ell-1}'(x)S_{\ell-}]. \qquad (3.16)$$

In the above equations, the multipole amplitudes $E_{\ell\pm}$, $M_{\ell\pm}$ and $S_{\ell\pm}$ are functions of W and Q^2 only, with the subindices $\pm$ denoting the total angular momentum $J = \ell \pm \frac{1}{2}$ of the final M–B state. They describe the transitions which can be classified according to the transverse or scalar (S) character of the photon. The transverse photon states can either be electric (E) with parity $(-1)^{J_\gamma}$, or magnetic

(M) with parity $(-1)^{J_\gamma+1}$, where J_γ is the angular momentum of the initial photon. The relation between the longitudinal multipoles and the scalar multipoles is $L_{\ell\pm} = (\omega/|\mathbf{q}|)S_{\ell\pm}$.

The application of the standard isospin formalism to electromagnetic processes [34] rests on the assumption that the electromagnetic current of the hadrons, like the electric charge, transforms as an isoscalar plus the third component of an isovector. In first-order processes, this is equivalent to treating the photon as a linear combination of isoscalar and isovector particles, both with $I_3 = 0$, and assuming isospin conservation. For any multipole $M_\ell \equiv E_{\ell\pm}, M_{\ell\pm}, S_{\ell\pm}$, this leads to three isospin amplitudes M_ℓ^I with $I = 0, \frac{1}{2}, \frac{3}{2}$. The superscript 0 indicates an isoscalar-photon amplitude leading to an $I = \frac{1}{2}$ final state, while the superscripts $\frac{1}{2}$ and $\frac{3}{2}$ indicate isovector-photon amplitudes leading to $I = \frac{1}{2}$ and $I = \frac{3}{2}$ final states respectively.

In the energy region where pion–nucleon scattering is purely elastic, the multipole amplitudes for $\gamma N \to \pi N$ satisfy Watson's theorem [35]

$$\mathrm{Im}M_\ell^I(W) = \pm|M_\ell^I(W)|e^{-i\delta_\ell^I(W)} \tag{3.17}$$

or equivalently

$$\mathrm{Im}M_\ell^I(W) = M_\ell^I(W)e^{-i\delta_\ell^I(W)}\sin\delta_\ell^I(W), \tag{3.18}$$

where $\delta_\ell^I(W)$ is the phase shift for pion–nucleon scattering in the partial wave with the same quantum numbers (ℓ, I). This result follows from unitarity and time-reversal invariance and implies that the phases of the multipole amplitudes are given by the corresponding elastic scattering phase shifts.

3.3 Theoretical models

A theoretical framework for investigating pion photoproduction reactions was first developed by Watson [34,35] and by Chew, Goldberger, Low and Nambu [36] (CGLN) using dispersion relations. With the advent of accurate experimental pion–nucleon phase shifts, the application of dispersion relations [37–42] led rapidly to the elucidation of all the main features of pion photoproduction in the $\Delta(1232)$ resonance region, including detailed values of both resonant excitation amplitudes [38]. Subsequently the method was extended to discuss both the higher-mass N^* resonances [43,44] and pion electroproduction [45–47]. It was further exploited [48,49] to give the only significant test, to date, of the isospin properties of the electromagnetic current discussed above. The dispersion-relation approach has been revived [50,51] to analyse new pion photoproduction and electroproduction data and has been extended [52,53] to the analysis of η photoproduction.

Isobar models [54,55] were developed to extract the parameters of the higher-mass nucleon resonances and to incorporate final states other than the pion, for example two pions and other mesons such as η, K, ω, and ϕ. Later, the K-matrix effective-Lagrangian models [56,57] were developed to study the $\Delta(1232)$ excitation. The K-matrix method and isobar parametrization have been used subsequently as tools for performing amplitude analyses of the data and determining resonance parameters. Examples are the dial-in codes SAID [58] and MAID [59]. In SAID, the dispersion relations are also used in their analyses of πN elastic scattering. Progress has also been made in extracting resonance parameters using the multi-channel K-matrix method [60–62] and the unitary coupled-channel isobar model [63–65].

More recently, a rather different theoretical point of view has been taken to develop dynamical models of meson production reactions [66–81]. These models account for off-shell scattering effects and can therefore provide a much more direct way to interpret the resonance parameters in terms of existing models of hadron structure. So far, the dynamical reaction model has been able to interpret the resonance parameters, in particular of the $\Delta(1232)$ resonance, in terms of constituent-quark models. Its connection with the results from quenched and unquenched lattice QCD calculations remains to be established. In the first part of this section, we derive schematically the formalism for the models which have often been used to analyse the data of electromagnetic meson production reactions. This enables us to see the differences between these models and to assess the results from applying them in analysing the data. We then give some detailed formulae for the dynamical models that are needed for discussing the results in section 3.4. The approach based on dispersion relations will be described in the last part of this section.

3.3.1 Hamiltonian formulation

To give a general derivation of most of the existing models of meson production in electromagnetic interactions, it is convenient to consider a Hamiltonian formulation. Such a formulation can be derived from relativistic quantum field theory by using various approximations such as the unitary transformation method developed in [70,78]. The Hamiltonian formulation is most commonly used in the study of nucleon–nucleon and pion–nucleon scattering. It is equally valid in the study of reactions involving photons as we will now describe in this subsection.

Within this framework, the Hamiltonian describing the meson–baryon reactions can be cast into the form

$$H = H_0 + V, \tag{3.19}$$

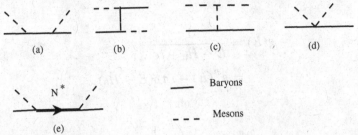

(a) (b) (c) (d)

——— Baryons

N*

--- Mesons

(e)

Figure 3.3. Tree diagrams for meson–baryon reactions. N^* represents an excited baryon state.

where the free Hamiltonian H_0 is the sum of kinetic energy operators of the photon and of the mesons and baryons in the reaction. The interaction term is defined as

$$V = v^{bg} + v^R(E),\qquad(3.20)$$

where v^{bg} is the non-resonant (background) term due to mechanisms such as the tree diagrams illustrated in figure 3.3(a)–(d), and v^R describes the N^* excitation, figure 3.3(e). Schematically, the resonant term of (3.20) can be written as

$$v^R(E) = \sum_{a,b}\sum_{N_i^*}\frac{\Gamma_{i,a}^\dagger\Gamma_{i,b}}{E - M_i^0}.\qquad(3.21)$$

Here E is the total energy of the system, $\Gamma_{i,a}$ defines the decay of the ith N^* state into a meson–baryon state $a : \gamma N, \pi N, \eta N, \pi\Delta, \rho N, \sigma N, \cdots$. The mass parameter M_i^0 is related to the resonance position. For example, M_Δ^0 is a bare mass of the $\Delta(1232)$ state which will be dressed by the non-resonant interaction v^{bg} to generate a physical mass $M_\Delta \sim (1232 + i\,120)$ MeV at $E = 1232$ MeV resonance position.

With the Hamiltonian (3.19), the S-matrix of the reaction (a,b), where (a,b) denote specific channels, is defined by

$$S_{a,b}(E) = \delta_{a,b} - 2\pi i\delta(E - H_0)T_{a,b}(E),\qquad(3.22)$$

and the scattering T-matrix is given by

$$T_{a,b}(E) = V_{a,b} + \sum_c V_{a,c}g_c(E)T_{c,b}(E).\qquad(3.23)$$

Here the propagator of channel c is

$$g_c(E) = \langle c\,|\,g(E)\,|\,c\rangle$$

with

$$g(E) = \frac{1}{E - H_0 + i\epsilon}$$

$$= g^P(E) - i\pi\delta(E - H_0), \tag{3.24}$$

where

$$g^P(E) \equiv P\left(\frac{1}{E - H_0}\right). \tag{3.25}$$

Here P denotes taking the principal-value part of any integration over the propagator. We can also define the K-matrix as

$$K_{a,b}(E) = V_{a,b} + \sum_c V_{a,c} g_c^P(E) K_{c,b}(E). \tag{3.26}$$

Equations (3.23) and (3.26) then define the following relation between the K-matrix and T-matrix:

$$T_{a,b}(E) = K_{a,b}(E) - \sum_c T_{a,c}(E)[i\pi\delta(E - H_0)]_c K_{c,b}(E). \tag{3.27}$$

By using the two-potential formulation [82], one can cast (3.23) into the form

$$T_{a,b}(E) = t_{a,b}^{bg}(E) + t_{a,b}^R(E), \tag{3.28}$$

where the first term

$$t_{a,b}^{bg}(E) = v_{a,b}^{bg} + \sum_c v_{a,c}^{bg} g_c(E) t_{c,b}^{bg}(E) \tag{3.29}$$

is determined only by the non-resonant interaction. The resonant amplitude

$$t_{a,b}^R(E) = \sum_{i,j} \bar{\Gamma}_{N_i^*,a}^\dagger(E)[G(E)]_{i,j} \bar{\Gamma}_{N_j^*,b}(E) \tag{3.30}$$

is determined by the dressed vertex

$$\bar{\Gamma}_{N^*,a}(E) = \Gamma_{N_i^*,a} + \sum_b \Gamma_{N_i^*,b} g_b(E) t_{b,a}^{bg}(E) \tag{3.31}$$

and the dressed propagator

$$[G(E)^{-1}]_{i,j}(E) = (E - M_i^0)\delta_{i,j} - \Sigma_{i,j}(E). \tag{3.32}$$

Here M_i^0 is the bare mass of the ith N^* resonance and the self-energy is

$$\Sigma_{i,j}(E) = \sum_a \Gamma_{N_i^*,a}^\dagger g_a(E) \bar{\Gamma}_{N_j^*,a}(E). \tag{3.33}$$

Note that the propagator $g_a(E)$ for channels including an unstable particle, such as $\pi\Delta$, ρN and σN, must be modified to include a width. In the Hamiltonian formulation this amounts to the replacement

$$g_a(E) \to \left\langle a \left| \frac{1}{E - H_0 - \sigma_v(E)} \right| a \right\rangle, \tag{3.34}$$

where the energy shift $\sigma_v(E)$ is due to the decay of an unstable particle in the presence of a spectator. Schematically it can be written as

$$\sigma_v(E) = g_v^\dagger \frac{P_{\pi\pi N}}{E - H_0 + i\epsilon} g_v, \tag{3.35}$$

where g_v describes the decays of ρ, σ or Δ and $P_{\pi\pi N}$ is the projection operator for the $\pi\pi N$ subspace.

Equations (3.23) and (3.27)–(3.35) are the starting points of our derivations. From now on, we consider the formulation in the partial-wave representation. The channel labels, (a, b, c), will also include the usual angular momentum and isospin quantum numbers.

3.3.2 Tree-diagram models

The tree-diagram models are based on the simplification that $T \approx V = v^{bg} + v^R$. The resonant effect is included by modifying the mass parameter of v^R, defined in (3.21), to include a width, such as $M_i^0 = M_i - \frac{1}{2}i\Gamma_i^{tot}(E)$. Equation (3.23) then simplifies to

$$T_{a,b}(tree) = v_{a,b}^{bg} + \sum_{N_i^*} \frac{\Gamma_{i,a}^\dagger \Gamma_{i,b}}{E - M_i + \frac{1}{2}i\Gamma_i^{tot}(E)}, \tag{3.36}$$

where v^{bg} is calculated from the tree diagrams of a chosen Lagrangian (see figures 3.3(a)–(d)) and Γ_i^{tot} is the total decay width of the ith N^*.

Tree-diagram models have been applied mainly to the photoproduction and electroproduction of K mesons [83–90], vector mesons (ω, ϕ) [91–93] and two pions [94]. At high energies, the t- and u-channel amplitudes (figures 3.3(b)–(c)) are replaced in some tree-diagram models by a Regge parametrization [95,96]. The validity of using tree-diagram models to investigate nucleon resonances is obviously very questionable, as discussed in the studies of ω photoproduction [93] and kaon photoproduction [76]. For example, it is found that the coupling with the πN channel can change the total cross section of $\gamma p \to K^+\Lambda$ by about 20% [76] and can drastically change the photon asymmetry of ω photoproduction reactions [93].

3.3.3 *Unitary isobar models*

The unitary isobar model (UIM) developed [59] by the Mainz group (MAID) is based on the on-shell relation (3.27). By including only one hadron channel, πN, (3.27) leads to

$$T_{\pi N, \gamma N} = \frac{1}{1 + i K_{\pi N, \pi N}} K_{\pi N, \gamma N}. \qquad (3.37)$$

In the energy region below the two-pion production threshold, we can use the relation $K_{\pi N, \pi N} = -\tan \delta_{\pi N}$ with $\delta_{\pi N}$ being the pion–nucleon scattering phase shift. One then can write the right-hand side of (3.37) as $\sim e^{i\delta_{\pi N}} K_{\pi N, \gamma}$, which is consistent with Watson's theorem [35]. This relation with the πN phase shift is simply extended in MAID to higher energies where the phase shifts become complex. By further assuming that $K = V = v^{bg} + v^R$, one can cast the above equation into the form

$$T_{\pi N, \gamma N}(UIM) = \eta_{\pi N} e^{i\delta_{\pi N}} \left[v_{\pi N, \gamma N}^{bg} \right] + \sum_{N_i^*} T_{\pi N, \gamma N}^{N_i^*}(E), \qquad (3.38)$$

where $\eta_{\pi N}$ is the inelasticity parameter of the πN elastic scattering. Clearly, the non-resonant multi-channel effects, such as $\gamma N \to (\rho N, \pi \Delta) \to \pi N$ which could be important in the second and third resonance regions, are neglected in MAID. In addition, the non-resonant amplitude $v_{\pi N, \gamma N}^{bg}$ is evaluated using an energy-dependent mixture of pseudo-vector (PV) and pseudo-scalar (PS) $\pi N N$ coupling

$$L_{\pi N N} = \frac{\Lambda_m^2}{\Lambda_m^2 + q_0^2} L_{\pi N N}^{PV} + \frac{q_0^2}{\Lambda_m^2 + q_0^2} L_{\pi N N}^{PS}, \qquad (3.39)$$

where q_0 is the on-shell photon energy. With a cutoff $\Lambda_m = 450$ MeV, one then gets PV coupling at low energies and PS coupling at high energies.

For the resonant terms in (3.38), MAID uses the parametrization due to Walker [54]:

$$T_{\pi N, \gamma N}^{N_i^*}(E) = f_{\pi N}^i(E) \frac{\Gamma_{tot} M_i e^{i\Phi}}{M_i^2 - E^2 - i M_i \Gamma^{tot}} f_{\gamma N}^i(E) \bar{A}^i, \qquad (3.40)$$

where $f_{\pi N}^i(E)$ and $f_{\gamma N}^i(E)$ are the form factors describing the decays of N^*, Γ_{tot} is the total decay width and $\bar{A}^i$ is the $\gamma N \to N^*$ excitation strength. The phase Φ is determined by the unitarity condition and the assumption that the phase ψ of the total amplitude is related to the πN phase shift $\delta_{\pi N}$ and inelasticity $\eta_{\pi N}$ by

$$\psi(E) = \tan^{-1} \left[\frac{1 - \eta_{\pi N}(E) \cos 2\delta_{\pi N}(E)}{\eta_{\pi N}(E) \sin 2\delta_{\pi N}(E)} \right]. \qquad (3.41)$$

At energies below the two-pion production threshold, where $\eta_{\pi N} = 1$, then $\psi = \delta_{\pi N}$ so that Watson's theorem is recovered. This is the motivation for choosing this prescription. It is, however, not clear whether the prescription (3.40)–(3.41) can be derived within a Hamiltonian formulation. In addition to the assumption (3.39) on the πNN coupling, this phenomenological aspect of MAID must be clarified before the N^* parameters extracted from the analyses using MAID can be interpreted correctly.

The UIM developed by the JLab/Yerevan group (JANR) [53] is similar to that of MAID, but it implements a Regge parametrization in calculating the amplitudes at high energies. MAID and JANR-UIM have been used extensively to analyse the data on π and η production. Information for a few N^* resonances has been extracted from data and will be discussed in some detail in section 3.4.

3.3.4 Multi-channel K-matrix models

The SAID model   The model employed in the SAID collaboration [58] is based on the on-shell relation (3.27) with three channels: γN, πN, and $\pi \Delta$ that represents all other open channels. The solution of the resulting 3×3 matrix equation can be written as

$$T_{\gamma N, \pi N}(SAID) = A_I(1 + iT_{\pi N, \pi N}) + A_R T_{\pi N, \pi N}, \tag{3.42}$$

where

$$A_I = K_{\gamma N, \pi N} - \frac{K_{\gamma N, \pi \Delta} K_{\pi N, \pi N}}{K_{\pi N, \pi \Delta}}, \tag{3.43}$$

$$A_R = \frac{K_{\gamma N, \pi \Delta}}{K_{\pi N, \pi \Delta}}. \tag{3.44}$$

In the analysis, A_I and A_R are simply parametrized as

$$A_I = \left[v_{\gamma N, \pi N}^{bg} \right] + \sum_{n=0}^{M} \bar{p}_n z Q_{l_\alpha + n}(z), \tag{3.45}$$

$$A_R = \frac{m_\pi}{k_0} \left(\frac{q_0}{k_0} \right)^{l_\alpha} \sum_{n=0}^{N} p_n \left(\frac{E_\pi}{m_\pi} \right)^n, \tag{3.46}$$

where k_0 and q_0 are the on-shell momenta for the pion and photon respectively, $z = \sqrt{k_0^2 + 4m_\pi^2}/k_0$, $Q_L(z)$ is the Legendre function of the second kind, $E_\pi = E_\gamma - m_\pi(1 + m_\pi/(2m_N))$, and p_n and $\bar{p}_n$ are free parameters. SAID calculates $v_{\gamma N, \pi N}^{bg}$ of (3.45) from the standard PS Born term and ρ and ω exchanges. The empirical πN amplitude $T_{\pi N, \pi N}$ needed to evaluate (3.42) is also included in SAID.

Once the parameters $\bar{p}_n$ and p_n in (3.45) and (3.46) are determined, the N^* parameters are then extracted by fitting the resulting amplitude $T_{\gamma N, \pi N}$ at energies near the resonance position to a Breit–Wigner parametrization similar to (3.40). Extensive data on pion photoproduction have been analysed by SAID and the method extended to pion electroproduction.

While the the parametrizations (3.45) and (3.46) are convenient for fitting the data, it is not clear whether they are sound theoretically. If the data were extensive enough to cover most of the possible observables, perhaps the extracted N^* parameters would not depend too much on the parametrizations of the amplitudes. On the other hand, this is unlikely over to be the case. A theoretical understanding of the prescription (3.45) and (3.46) used in SAID is needed to make progress.

The Giessen model   The coupled-channel model developed by the Giessen group [62] can be obtained from (3.27) by making the approximation $K = V$; that is by neglecting all multiple-scattering effects for the K-matrix included in (3.26). This leads to a matrix equation involving only the on-shell matrix elements of V:

$$T_{a,b}(Giessen) = \sum_c [(1 + iV(E))^{-1}]_{a,c} V_{c,b}(E). \qquad (3.47)$$

The interaction $V = v^{bg} + v^R$ is evaluated from tree diagrams corresponding to various effective Lagrangians. The form factors, coupling constants, and resonance parameters are adjusted to fit both the πN and γN reaction data. They include up to five channels in some fits, and have claimed to identify several new N^* states. However, further confirmation is needed to establish their findings conclusively. In particular, it remains to investigate how the off-shell multiple scattering effect, neglected in this model, influences the resonant amplitudes.

3.3.5 Dynamical models

The development of dynamical models has been aimed at separating the reaction mechanisms from the hadron structure in interpreting the data of electromagnetic meson-production reactions. The importance of this theoretical effort can be understood by recalling the experience in the development of nuclear physics. For example, the information about the deformation of ^{12}C can be extracted from the inelastic scattering reaction $^{12}C(p, p')^{12}C(2^+, 4.44 \text{ MeV})$ only when a reliable theory [97], such as the distorted-wave impulse approximation and the coupled-channels method, is used to calculate the initial and final proton–^{12}C interactions including off-shell scattering effects. Accordingly, one expects that the N^* structure can be determined only when the interactions in its decay channels $\pi N, \pi\pi N, \ldots$ can be calculated from a reliable reaction theory. This has been pursued by developing

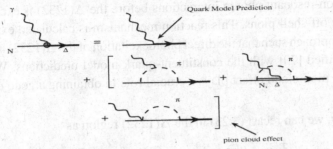

Figure 3.4. Graphic representation of the dressed $\gamma N \rightarrow \Delta(1232)$ vertex within the SL model [70,72].

dynamical models which are guided to a very large extent by the early work on meson-exchange models of nucleon–nucleon and pion–nucleon scattering.

We will give a detailed account of the dynamical models in the $\Delta(1232)$ region. Dynamical models for higher-mass N^* states are still being developed and will only be briefly described.

The $\Delta(1232)$ region Keeping only the $\Delta(1232)$ and the two channels $a, b = \pi N, \gamma N$, (3.28)–(3.33) reduce to the Sato–Lee (SL) model [70,72].

Explicitly, we have

$$T_{\pi N, \pi N}(E) = t^{bg}_{\pi N, \pi N}(E) + \frac{\bar{\Gamma}^{\dagger}_{\Delta, \pi N}(E) \bar{\Gamma}_{\Delta, \pi N}(E)}{E - M^0_{\Delta} - \Sigma_{\Delta}(E)}, \tag{3.48}$$

$$T_{\gamma N, \pi N}(E) = t^{bg}_{\gamma N, \pi N}(E) + \frac{\bar{\Gamma}^{\dagger}_{\Delta, \gamma N}(E) \bar{\Gamma}_{\Delta, \pi N}(E)}{E - M^0_{\Delta} - \Sigma_{\Delta}(E)}, \tag{3.49}$$

with

$$\bar{\Gamma}_{\Delta, \gamma N}(E) = \Gamma_{\Delta, \gamma N} + \Gamma_{\Delta, \pi N} G_{\pi N}(E) t^{bg}_{\pi N, \gamma N}(E), \tag{3.50}$$

$$\bar{\Gamma}_{\Delta, \pi N}(E) = \Gamma_{\Delta, \pi N} + \Gamma_{\Delta, \pi N} G_{\pi N}(E) t^{bg}_{\pi N, \pi N}(E), \tag{3.51}$$

$$t^{bg}_{\pi N, \gamma N}(E) = v^{bg}_{\pi N, \gamma N} + t^{bg}_{\pi N, \pi N}(E) G_{\pi N}(E) v^{bg}_{\pi N, \gamma N}, \tag{3.52}$$

$$t^{bg}_{\pi N, \pi N}(E) = v^{bg}_{\pi N, \pi N} + v^{bg}_{\pi N, \pi N} G_{\pi N}(E) t^{bg}_{\pi N, \pi N}(E) \tag{3.53}$$

and

$$\Sigma_{\Delta}(E) = \Gamma^{\dagger}_{\Delta, \pi N}(E) G_{\pi N}(E) \bar{\Gamma}_{\Delta, \pi N}(E). \tag{3.54}$$

These equations indicate clearly how the non-resonant interactions modify the resonant amplitude. Specifically, (3.50) is illustrated in figure 3.4 for the dressed $\gamma N \rightarrow \Delta(1232)$ transition in the SL model. We see that the pion-cloud effect is due to the non-resonant photon interactions before the $\Delta(1232)$ is excited by the intermediate off-shell pions. This reaction mechanism is calculated explicitly in a dynamical approach such that the direct photoexcitation of the $\Delta(1232)$ in figure 3.4 can be identified [70] with the constituent-quark-model predictions. We will see later that the pion-cloud effect plays a crucial role in obtaining a good description of the data.

Alternatively, we can recast (3.23) in the $\Delta(1232)$ region as

$$T_{\gamma N, \pi N}(E) = t^{B}_{\gamma N, \pi N}(E) + t^{R}_{\gamma N, \pi N}(E), \tag{3.55}$$

with

$$t^{B}_{\gamma N, \pi N}(E) = v^{bg}_{\gamma N, \pi N} + v^{bg}_{\gamma N, \pi N} G_{\pi N}(E) T_{\pi N, \pi N}(E),$$

$$t^{R}_{\gamma N, \pi N}(E) = v^{R}_{\gamma N, \pi N} + v^{R}_{\gamma N, \pi N} G_{\pi N}(E) T_{\pi N, \pi N}(E). \tag{3.56}$$

These latter equations are used by the Dubna–Mainz–Taiwan (DMT) model [73,75], except that it departs from a consistent Hamiltonian formulation and replaces the term t^{R} by Walker's parametrization, (3.40):

$$t^{R}_{\gamma N, \pi N}(E) = f_{\pi N}(E) \frac{\Gamma_{tot} M_{\Delta} e^{i\Phi}}{M_{\Delta}^{2} - E^{2} - i M_{\Delta} \Gamma^{tot}} f_{\gamma N}(E) \bar{A}_{\gamma N}. \tag{3.57}$$

Other differences between the SL model and the DMT model are in the πN potential employed and how the non-resonant $\gamma N \to \pi N$ amplitudes are regularized. The πN potentials in both models are constrained by the fits to πN scattering phase shifts, but they could generate different off-shell effects because of the differences in the approximations used in deriving the meson-exchange mechanisms from relativistic quantum field theory. In the DMT model, the non-resonant $\gamma N \to \pi N$ amplitudes are calculated by using MAID's mixture (3.39) of PS and PV couplings, while their πN potential is from a model [98] using PV coupling. In the SL model, the standard PV coupling is used in a consistent derivation of both the πN potential and $\gamma N \to \pi N$ transition interaction using a unitary transformation method.

We now give the formulae required for the discussions in section 3.4.1 on the $\Delta(1232)$ resonance. The $\Delta(1232)$ excitation is parametrized in terms of a Rarita–Schwinger field. In the Δ rest frame, where $m_{\Delta} = q_0 + E_N(\mathbf{q})$, the resulting $\gamma N \to \Delta(1232)$ vertex function can be written in the more transparent form

$$\langle \Delta \mid \Gamma_{\gamma N \to \Delta} \mid q \rangle = -\frac{e}{(2\pi)^{3/2}} \sqrt{\frac{E_N(\mathbf{q}) + m_N}{2 E_N(\vec{q})}} \frac{1}{\sqrt{2\omega}} \frac{3(m_\Delta + m_N)}{4 m_N (E_N(\mathbf{q}) + m_N)} T_3$$

$$\times [i G_M(q^2) \mathbf{S} \times \mathbf{q} \cdot \epsilon + G_E(q^2)(\mathbf{S} \cdot \epsilon \sigma \cdot \mathbf{q} + \mathbf{S} \cdot \mathbf{q} \sigma \cdot \epsilon)$$

$$+ \frac{G_C(q^2)}{m_\Delta} \mathbf{S} \cdot \mathbf{q} \sigma \cdot \mathbf{q} \epsilon_0], \tag{3.58}$$

where $q = (\omega, \mathbf{q})$ is the photon four-momentum, and $\epsilon = (\epsilon_0, \epsilon)$ is the photon polarization vector. The transition operators $\mathbf{S}$ and $\mathbf{T}$ are normalized by the reduced matrix elements $\langle \Delta || \vec{S} || N \rangle = \langle \Delta || \vec{T} || N \rangle = 2$. The convention

$$\langle J'M' \mid T_{kq} \mid JM \rangle = (-1)^{2k} \langle J'M' \mid JkMq \rangle / \langle J' \mid\mid T \mid\mid J \rangle / \sqrt{2J'+1}$$

has been used.

By using (3.58) and the standard definitions [99,100] for multipole amplitudes, it is straightforward to evaluate the magnetic M1, electric E2 and longitudinal ('Coulomb' in nuclear-physics convention) C2 amplitudes of the $\gamma N \to \Delta(1232)$ transition. With the commonly-used convention $Q^2 = -q^2 > 0$, it is found [72] that

$$G_M(Q^2) = \frac{1}{N}[\Gamma_{\gamma N \to \Delta}]_{M1}, \tag{3.59}$$

$$G_E(Q^2) = -\frac{1}{N}[\Gamma_{\gamma N \to \Delta}]_{E2}, \tag{3.60}$$

$$G_C(Q^2) = \frac{2m_\Delta}{qN}[\Gamma_{\gamma N \to \Delta}]_{C2}, \tag{3.61}$$

with

$$N = \frac{e}{2m_N} \sqrt{\frac{m_\Delta q}{m_N}} \frac{1}{[1 + Q^2/(m_N + m_\Delta)^2]^{1/2}}.$$

At $Q^2 = 0$, the above relations agree with that given in Appendix A of [68]. Equations (3.59)–(3.61) can also be defined to relate the dressed vertex $\bar{\Gamma}_{\gamma N \to \Delta}$, defined by (3.50), to the corresponding dressed transition form factors:

$$G_M^*(Q^2) = \frac{1}{N}[\bar{\Gamma}_{\gamma N \to \Delta}^K]_{M1}, \tag{3.62}$$

$$G_E^*(Q^2) = \frac{-1}{N}[\bar{\Gamma}_{\gamma N \to \Delta}^K]_{E2}, \tag{3.63}$$

$$G_C^*(Q^2) = \frac{2m_\Delta}{qN}[\bar{\Gamma}_{\gamma N \to \Delta}^K]_{C2}. \tag{3.64}$$

Note that in these equations, the upper index K in $\Gamma_{\Delta, \gamma N}^K$ means taking only the principal-value integration in evaluating the second term of (3.50). Details are discussed in [70].

At the $Q^2 = 0$ real-photon point, we will also compare the theoretical predictions with the helicity amplitudes listed by the PDG [25]. These amplitudes are related to the multipole amplitudes by

$$A_{3/2} = \frac{\sqrt{3}}{2}[[\bar{\Gamma}_{\gamma N \to \Delta}^K]_{E2} - [\bar{\Gamma}_{\gamma N \to \Delta}^K]_{M1}], \tag{3.65}$$

$$A_{1/2} = -\frac{1}{2}[3[\bar{\Gamma}_{\gamma N \to \Delta}^K]_{E2} + [\bar{\Gamma}_{\gamma N \to \Delta}^K]_{M1}]. \tag{3.66}$$

At the $\Delta(1232)$ resonance position $E = M_R = 1232$ MeV, the πN phase shift in the P_{33} channel passes through 90 degrees. As derived in detail in [70], this leads to a relation between the multipole components of the dressed vertex $\bar{\Gamma}_{\gamma N \to \Delta(1232)}$ and the imaginary parts of the $\gamma N \to \pi N$ multipole amplitudes in the πN P_{33} channel:

$$G_M^*(Q^2) = \frac{1}{N} \sqrt{\frac{8\pi m_\Delta k \Gamma_\Delta}{3 m_N q}} \times \mathrm{Im}\left\{ M_{1+}^{3/2} \right\}, \tag{3.67}$$

$$G_E^*(Q^2) = \frac{1}{N} \sqrt{\frac{8\pi m_\Delta k \Gamma_\Delta}{3 m_N q}} \times \mathrm{Im}\left\{ E_{1+}^{3/2} \right\}, \tag{3.68}$$

$$\frac{q}{2 m_\Delta} G_C^*(Q^2) = \frac{1}{N} \sqrt{\frac{8\pi m_\Delta k \Gamma_\Delta}{3 m_N q}} \times \mathrm{Im}\left\{ S_{1+}^{3/2} \right\}, \tag{3.69}$$

where Γ_Δ is the $\Delta(1232)$ width, k and q are respectively the magnitudes of the three-momenta of the pion and photon in the rest frame of the Δ. We can then obtain the useful relations that the $E2/M1$ ratio, R_{EM}, and the $C2/M1$ ratio, R_{SM}, of the dressed $\gamma N \to \Delta(1232)$ transition at $W = 1232$ MeV can be evaluated directly by using the $\gamma N \to \pi N$ multipole amplitudes:

$$R_{EM} = \frac{[\bar{\Gamma}_{\gamma N \to \Delta}^K]_{E2}}{[\bar{\Gamma}_{\gamma N \to \Delta}^K]_{M1}} = \frac{\mathrm{Im}\left\{ E_{1+}^{3/2} \right\}}{\mathrm{Im}\left\{ M_{1+}^{3/2} \right\}}, \tag{3.70}$$

$$R_{SM} = \frac{[\bar{\Gamma}_{\gamma N \to \Delta}^K]_{C2}}{[\bar{\Gamma}_{\gamma N \to \Delta}^K]_{M1}} = \frac{\mathrm{Im}\left\{ S_{1+}^{3/2} \right\}}{\mathrm{Im}\left\{ M_{1+}^{3/2} \right\}}. \tag{3.71}$$

Equations (3.70) and (3.71) can be used in empirical amplitude analyses to extract the transition form factors and the $E2/M1$ and $C2/M1$ ratios of the $\gamma N \to \Delta(1232)$ transition. The extractions of the bare vertices, which can be compared with the predictions from some hadron-structure calculations, can only be achieved by using dynamical models through (3.50). This indicates why an appropriate reaction theory is needed in the N^* study, as illustrated in figure 3.1 and the results of [70,72].

The second and third resonance regions   In these regions we need to include more than the πN channel to solve equations (3.28)–(3.35). In addition, these formulae must be extended [78] to account explicitly for the $\pi\pi N$ channel, instead of using the quasi-two-particle channels $\pi\Delta$, ρN, and σN to simulate the $\pi\pi N$ continuum. This, however, has not been fully developed. Studies in the second and third resonance regions are within the formulation defined by (3.28)–(3.35). These equations are the basis of examining N^* effects [92] and coupled-channel effects

[93] on ω meson photoproduction [92,93] and K photoproduction [76]. Both effects are found to be important in determining the differential cross sections and various polarization observables. In these studies, all N^* resonances listed by PDG [25] or predicted by constituent-quark models are considered, but only the coupling with πN channel is emphasized. Clearly, their results should be further examined by considering the effects due to other channels.

A coupled-channel study of both πN scattering and $\gamma N \to \pi N$ in the S_{11} partial wave [74] included the πN, ηN and γN channels. In their $\gamma N \to \pi N$ calculation, they neglected the $\gamma N \to \eta N \to \pi N$ coupled-channel effect, and followed the procedure of the DMT model to evaluate the resonant contribution in terms of Walker's parametrization (3.40). They found that four N^* resonances are required to fit the empirical amplitudes in the S_{11} partial wave up to $W = 2$ GeV. Obviously their results must be re-examined because of the lack of a complete treatment of coupled-channel effects, in particular those due to two-pion channels.

There are also some coupled-channel calculations [77,69] of pion photoproduction up to the second resonance region, $W = 1.5$ GeV, although the treatment is not complete. While they can give reasonable fits to the pion photoproduction data, their results on the N^* parameters cannot be interpreted until further investigations of coupled-channel effects are made.

The dynamical study of electromagnetic meson-production reactions in the higher mass N^* region are still being developed. It is an outstanding challenge to develop a complete dynamical coupled-channel calculation including all N^* resonances listed by the PDG [25] or predicted by the constituent-quark model to fit all meson-production data up to $W = 2.5$ GeV. Such a complex task requires close collaboration between theoretical and experimental efforts.

3.3.6 Dispersion-relation approaches

The dispersion-relation approach is based on the unitarity of the S-matrix and on the analyticity and crossing properties of the scattering amplitudes, which can be proven with some degree of rigour [101]. This approach played a key role in the early study of pion photoproduction, as outlined at the beginning of this section. Here we summarize briefly later work that has been used in analysing the data discussed in this chapter, referring to the many review articles, for example [44,102,100], for a more complete discussion.

To discuss models based on dispersion relations, it is necessary first to write the hadronic current matrix element defined by (3.3) as

$$\langle kp' \mid \epsilon_\mu J^\mu(0) \mid qp \rangle = \sum_{i=1,6} \bar{u}(\mathbf{p}')[A_i(s, t, u)M_i]u(\mathbf{p}), \tag{3.72}$$

where $u(\mathbf{p})$ is the Dirac spinor, $A_i(s, t, u)$ are Lorentz-invariant functions of Mandelstam variables s, t and u. The M_i are independent invariants formed from γ^μ, γ_5 and momenta. The expressions for M_i can be found, for example, in [44]. For pion production, the amplitudes can be further classified by isospin quantum numbers. These are $A^{(0)}$ for the isoscalar photon, and for the isovector the two amplitudes $A^{(1/2)}$ and $A^{(3/2)}$ for the final πN system with total isospin $I = \frac{1}{2}$ and $I = \frac{3}{2}$ respectively. Each invariant amplitude in (3.72) can be expanded as

$$A_i = \tfrac{1}{2} A_i^{(-)}[\tau_\alpha, \tau_3] + A_i^{(+)}\delta_{\alpha,3} + A_i^{(0)}\tau_\alpha, \tag{3.73}$$

where τ is the isospin Pauli operator, and α is the isospin quantum number associated with the produced pion. Equation (3.73) then leads to $A_i^{(1/2)} = A_i^{(+)} + 2A_i^{(-)}$ and $A_i^{(3/2)} = A_i^{(+)} - A_i^{(-)}$.

For π and η production, the starting point is the fixed-t dispersion relation [37] for the invariant amplitudes A_i^I:

$$\text{Re}[A_i^I(s, t)] = A_i^{I, pole} + \frac{1}{\pi}P\int_{S_{thr}}^\infty ds'\left[\frac{1}{s' - s} + \frac{\epsilon^I \xi_i}{s' - u}\right]\text{Im}[A_i^I(s', t)], \tag{3.74}$$

where $I = 0, +, -$ denotes the isospin component and $\xi_1 = \xi_2 = -\xi_3 = -\xi_4 = 1$, $\epsilon^+ = \epsilon^0 = -\epsilon^- = 1$ are defined such that the crossing symmetry relation $A_i^I(s, t, u) = \xi_i \epsilon^I A_i^I(u, t, s)$ is satisfied. With the relations (3.13) and (3.72) and the multipole expansion defined by (3.17), the fixed-t dispersion relation (3.74) leads to the following set of coupled equations relating the real part and imaginary parts of multipole amplitudes:

$$\text{Re}[M_\ell^I(W)] = M_\ell^{I, pole}(W) + \frac{1}{\pi}P\int_{W_{thr}}^\infty dW'\frac{\text{Im}[M_\ell^I(W)]}{W' - W}$$

$$+ \frac{1}{\pi}\int_{W_{thr}}^\infty dW'\sum_{\ell'} K_{\ell\ell'}^I(W, W')\text{Im}[M_{\ell'}^I(W)], \tag{3.75}$$

where M_ℓ^I is the multipole amplitude, $M_\ell^{I, pole}(W)$ is calculated from the PS Born term and the kernel $K_{\ell, \ell'}^I$ contains various kinematic factors. These equations are valid up to a centre-of-mass energy of $W \approx 1.3$ GeV and should form the basis of a reasonable approximation at somewhat higher energies [39]. In the work of [50], the procedures of [42] are used to solve the above equations by using the method of Omnes [103]. It assumes that the multipole amplitude can be written as

$$M_\ell^I(W) = \frac{e^{i\phi_\ell(W)}}{r_{\ell I}}\bar{M}_\ell^I(W), \tag{3.76}$$

where $\bar{M}_\ell^I$ is a real function and $r_{\ell I}$ is some kinematic factor. Hence

$$\text{Im}[M_\ell^I(W)] = h_\ell^{I*}(W)M_\ell^I(W), \tag{3.77}$$

with $h_\ell^I(w) = \sin(\phi_\ell^I) \exp(i\phi_\ell^I(W))$. In the elastic region, the phase $\phi_\ell^I = \delta_\ell^I$ by Watson's theorem [35]: see (3.18). At higher energies it is assumed to be

$$\phi_\ell^I(W) = \arctan\left[\frac{1 - \eta_\ell^I(W)\cos 2\delta_\ell^I(W)}{\eta_\ell^I(W)\sin 2\delta_\ell^I(W)}\right], \tag{3.78}$$

where δ_ℓ^I and η_ℓ^I are the phase and inelasticity of πN scattering in the partial wave with quantum numbers (ℓ, I).

The next approximation is to limit the sum over ℓ' in the right-hand-side of (3.75) to a cutoff ℓ_{max}. For investigating production below $E_\gamma = 500$ MeV, $\ell_{max} = 1$ is taken. Another approximation is needed to handle the integration over W in (3.75). In [50], the integration is cutoff at $W = \Lambda = 2$ GeV such that the required phases Φ_ℓ can be determined by the empirical πN phase shifts. The contribution neglected from $W > 2$ GeV is then accounted for by adding vector-meson exchange terms, $M_\ell^{I,V}(W)$. Equation (3.75) then takes the form

$$\bar{M}_\ell^I(W) = \bar{M}_\ell^{I,pole}(W) + \frac{1}{\pi}\int_{W_{thr}}^{\Lambda} dW' \frac{h_\ell^{I*}(W')\bar{M}_\ell^I(W')}{W' - W - i\epsilon}$$

$$+ \frac{1}{\pi}\sum_{\ell',I'}\int_{W_{thr}}^{\Lambda} dW' \bar{K}_{\ell\ell'}^{I,I'}(WW')h_{\ell'}^{I'*}(W')\bar{M}_\ell^I(W')$$

$$+ M_\ell^{I,V}(W). \tag{3.79}$$

The method of solving (3.79) is given in [50]. With the above procedures, the model contains ten adjustable parameters. An excellent fit to all $\gamma N \to \pi N$ data up to $E_\gamma = 500$ MeV has been obtained in [50]. The resonance parameters of the $\Delta(1232)$ resonance have been extracted.

The calculations in [51] follow the same approach with the additional simplification that the couplings among different multipoles and the contribution from $W > \Lambda$ to the integration are neglected; setting $\bar{K}_{\ell,\ell'}^{I,I'} = 0$ and $M_\ell^{I,V} = 0$ in solving (3.79). These simplifications are justified in calculating the dominant $\Delta(1232)$ excitation amplitude $M_{1+}^{(3/2)}$. But it is questionable if they can be applied for calculating weaker amplitudes. Thus no attempt was made in [51,52] to fit the data directly using dispersion relations. Rather, the emphasis was in the interpretation of the empirical amplitudes $M_{1+}^{(3/2)}$, $E_{1+}^{(3/2)}$ in terms of rescattering effects and constituent-quark model predictions. By assuming that the multipole expansion is also valid in electroproduction, the Q^2 dependence of these $\Delta(1232)$ excitation amplitudes is then predicted.

The dispersion-relation approach is also used in [52] to analyse the pion photoproduction and electroproduction data in the second and third resonance regions. It is assumed that the imaginary parts of the amplitudes for $M_N + m_\pi < W < 2$ GeV are from the resonant amplitudes parametrized as Walker's Breit–Wigner form (3.40) and above $W = 2.5\,\text{GeV}$ are from a Regge-pole model. The imaginary parts of the amplitudes in $2\,\text{GeV} < W < 2.5\,\text{GeV}$ are obtained by interpolation. The real parts of the amplitude in the second and third resonance regions are then calculated from the dispersion relations. The empirical amplitudes are then fitted by adjusting the resonant parameters. It turns out that the resulting parameters are close to those obtained using the UIM described in subsection 3.3.3.

With appropriate modification, the dispersion-relation approach can be applied to investigate the production of other PS mesons. This has been achieved in [53] in analysing the η photo- and electroproduction data.

3.4 Data and results

A large volume of data is needed to extract the fundamental physics on resonance parameters or discover new baryon states from the electromagnetic meson-production reactions. Efforts in this direction in the 1970s and 1980s at various laboratories were hampered by the low-duty-cycle synchrotrons that were available and by the limitations of magnetic spectrometers with relatively small apertures. For a discussion on these results see the review by Foster and Hughes [104]. The construction of continuous-wave (CW) electron accelerators, and advances in detector technologies, making it possible to use detector systems with nearly 4π solid angle coverage and the ability to operate at high luminosity, have revolutionized the subject. These experimental advances have been reviewed in [2]. In this section we focus on the main results from the analyses of the new data.

3.4.1 Single-pion production

Single-pion photoproduction and electroproduction have been the main processes in the study of the electromagnetic transition amplitudes of the lower-mass nucleon resonances. A large amount of data on pion photoproduction now exists [105–108], including results from measuring beam asymmetries and beam–target double-polarization observables [109,110]. Similar advances have also been made for π^0 and π^+ electroproduction from protons [111–124], covering a large range in W and Q^2 and the full polar and azimuthal angles. For the first time, high-precision data from measurements using polarized electron beams and polarized nucleon targets and detection of recoil nucleon polarization have been obtained. A summary of the new electroproduction data is presented in table 3.2 (see below). In addition to the unpolarized differential cross section ($d\sigma_{unpol}/d\Omega_M^*$), various polarization observables have also been measured.

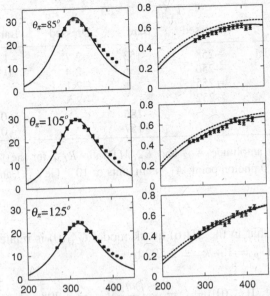

Figure 3.5. The differential cross section $d\sigma/d\Omega$ (left) and the photon asymmetry Σ (right) of the $p(\gamma, \pi^0)p$ reaction calculated from the SL model [70] compared with the MAMI data [106] for photons in the energy range 200 to 450 MeV. The Σ photoproduction data from LEGS [107] agree well with the MAMI data and hence are not shown. The dashed curves are obtained by setting $R_{EM} = 0$. The dashed curves for the differential cross section (left panels) are indistinguishable from the solid curves.

One of the main outcomes from analyses of these single-pion production data is a more detailed understanding of the $\Delta(1232)$ resonance. The focus has been on the determination of the magnetic M1, electric E2, and Coulomb C2 form factors of the $\gamma N \rightarrow \Delta(1232)$ transition. This development is discussed in detail in this subsection. The analyses of the data for the single-pion production in the second and third resonance regions are discussed in sections 3.4.2 and 3.4.3.

Pion photoproduction High-statistics data on the photon asymmetry of the $\gamma N \rightarrow \pi N$ reaction are essential in extracting the small E_{1+} amplitude which can be used in (3.68) to determine the electric E2 strength of the $\gamma N \rightarrow \Delta(1232)$ transition. Figure 3.5 shows the $\gamma p \rightarrow p\pi^0$ data from MAMI [106]. The results from the SL model [70] are also displayed. When the E_{1+} amplitude is turned off in the dynamical calculation, the predicted photon asymmetries (dotted curves) deviate from the data. Amplitude analyses of these new data have been performed by several groups and we now have a world-averaged value of the R_{EM} ratio, defined by (3.70), $R_{EM}(0) = (-2.38 \pm 0.27)\%$ [125]. The magnetic M1 transition strength has been determined as $G_M^*(0) = 3.18 \pm 0.04$ (unitless as defined by (3.58)). It is

	$A_{3/2}$		R_{EM}		
	Dressed	Bare	Dressed	Bare	
Dynamical model	−258	−153	−2.7	−1.3	[70]
	−256	−136	−2.4	0.25	[73]
K-matrix	−255		−2.1		[57]
Dispersion	−252		−2.5		[50]
Quark model		−186		~ 0	[126]
		−157		~ 0	[19]

Table 3.1. Helicity amplitude $A_{3/2}$ and $E2/M1$ ratio R_{EM} for the $\gamma N \to \Delta(1232)$ transition at $Q^2 = 0$ photon point. $A_{3/2}$ is in units of 10^{-3} GeV$^{-1/2}$ and R_{EM} is given as a percentage.

instructive to note that in the SU(6) quark model $G_M^*(0)$ is related to the proton magnetic moment $G_p = 1 + \kappa_p = 2.793$ by

$$[G_M^*(0)]_{SU(6)} = \frac{4}{3}\sqrt{\frac{m_N}{2m_\Delta}}G_p = 2.298. \tag{3.80}$$

This relation can be obtained by following the derivations given in Appendix A of [68] and using the relationship between the decay width $\Gamma_{\Delta \to \gamma N}$ and the multipole amplitudes M_{1+} and E_{1+}:

$$\Gamma_{\Delta \to \gamma N} = \frac{q^2 m_N}{2\pi m_\Delta}[|M_{1+}|^2 + 3|E_{1+}|^2]. \tag{3.81}$$

The SU(6) value 2.298 given in (3.80) clearly seriously underestimates the empirical value 3.18 ± 0.04. This has long been noticed in many quark-model calculations.

Using the dynamical model of section 3.3.5, the bare transition strengths can be extracted by separating the pion-cloud effects from the full (dressed) transition strengths, as defined in (3.50) and illustrated in figure 3.4. The pion-cloud effect can be more clearly seen in table 3.1. The dressed values of the helicity amplitude $A_{3/2}$, calculated from two dynamical models [70,73], agree well with the empirical values from amplitude analyses based on the K-matrix method and dispersion relations. Within the dynamical models, the dressed values are about 40% larger in magnitude than the bare strengths, indicating the importance of the pion cloud. These bare values are within the range predicted by two constituent-quark models [19,126]. This suggests that the bare parameters of the dynamical models can be identified with some hadron structure calculations. From table 3.1 it can also be seen that the difference between the dressed and bare values of R_{EM} is even larger. The bare values from the two dynamical-model analyses are quite different, indicating some significant differences in their formulations as discussed in section 3.3.5.

Reaction	W range (GeV)	Q^2 range (GeV^2)	Lab/experiment	
$ep \rightarrow ep\pi^0$	< 1.8	0.4–1.8	JLab-CLAS	[113]
	$\Delta(1232)$	0.1–0.9	ELSA-Elan	[119]
	$\Delta(1232)$	2.8, 4.0	JLab-HallC	[111]
	< 2	2–6	JLab-CLAS	[120]
	$\Delta(1232)$	7.5	JLab-HallC	[121]
	< 2.0	1.0	JLab-HallA	[115]
$ep \rightarrow en\pi^+$	< 1.6	0.3–0.65	JLab-CLAS	[112]
	$\Delta(1232)$	0.1–0.9	ELSA-Elan	[119]
	< 2	2–6	JLab-CLAS	[120]
$\vec{e}p \rightarrow ep\pi^0$	$\Delta(1232)$	0.2	MAMI-A1	[117]
	$\Delta(1232)$	0.3–0.65	JLab-CLAS	[114]
	$\Delta(1232)$	0.12	Bates-OOPS	[118]
$\vec{e}p \rightarrow en\pi^+$	< 1.6	0.3–0.65	JLab-CLAS	[116]
$\vec{e}p \rightarrow e\vec{p}\pi^0$	$\Delta(1232)$	1.0	JLab-HallA	[122]
$\vec{e}\vec{p} \rightarrow ep\pi^0$	$\Delta(1232)$	0.5–1.5	JLab-CLAS	[124]
$\vec{e}\vec{p} \rightarrow en\pi^+$	< 1.85	0.4, 0.65, 1.1	JLab-CLAS	[123]

Table 3.2. Summary of data of single-pion electroproduction reactions. The arrows in the reaction indicate if the electron beam or the target nucleon is polarized.

Pion electroproduction　High-precision pion electroproduction data started to become available in 1999 with the publication [111] of $p(e, e'\pi^0)p$ data at $Q^2 = 2.8$ and 4 GeV^2 from JLab. These data allowed the determination of the parametrization of the $\gamma^*N \rightarrow \Delta(1232)$ transition form factors within the SL [72], MAID [59], and DMT [73] models. For example, the resulting bare magnetic M1 form factor $G_M(Q^2)$ of the SL model takes the form

$$G_M(Q^2) = G_M(0)[(1 + aQ^2)\exp(-bQ^2)]G_D(Q^2), \qquad (3.82)$$

where $G_M(0) = 1.9 \pm 0.05$, $a = 0.154$ GeV^{-2}, $b = 0.166$ GeV^{-2} and $G_D(Q^2) = 1/(1 + Q^2/0.71)^2$ is the dipole form factor of the proton. With this phenomenological step, these three models have since been most commonly used to make predictions for the data listed in table 3.2. Sample comparisons with the data from JLab are shown in figures 3.6 and 3.7. In general, the predictions from the SL, MAID, and DMT models are in good agreement with all of the available data in the $\Delta(1232)$ region. However, work is still needed to remove the remaining discrepancies so that the $\gamma^*N \rightarrow \Delta(1232)$ form factors can be more precisely determined. In particular, the electric E2 form factor $G_E(Q^2)$ and Coulomb C2 form factor $G_C(Q^2)$ within each model must be further examined and refined before they can be used for testing the predictions from hadron models or lattice QCD calculations.

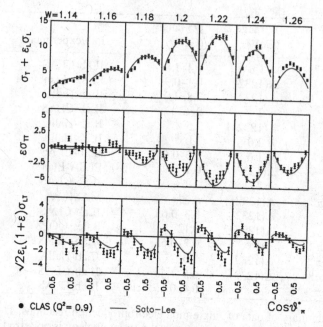

Figure 3.6. $p(e, e'\pi^0)p$ cross section data [113] from CLAS at JLab compared with the predictions from the SL Model [72].

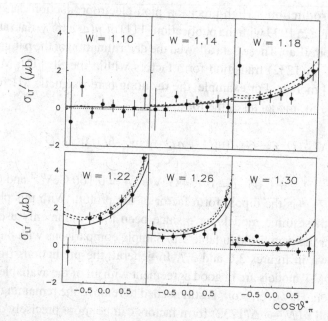

Figure 3.7. CLAS data [114] on $\sigma_{LT'}$ of the $p(e, e'\pi^+)n$ reaction in the $\Delta(1232)$ region are compared with predictions from SL [72] (solid), MAID [59] (dashed) and DMT [73] (dash-dotted) models.

The data from JLab have extensive coverage in angle and energy and hence allow nearly model-independent determination of the $\gamma^* N \to \Delta(1232)$ transition form factors. Analyses are based on the consideration that, at the $\Delta(1232)$ peak, the dominant amplitude is M_{1+} and the small quadrupole E_{1+} and S_{1+} are accessible through their interference with it. Thus analysis can be started by using a truncation in which only terms involving M_{1+} are retained. With the partial-wave decompositions of (3.17), the differential cross sections in (3.9) can be expanded in terms of Legendre polynomials:

$$\frac{d\sigma_T}{d\Omega} + \epsilon \frac{d\sigma_L}{d\Omega} = \sum_{l=0}^{\infty} A_l P_l(\cos\theta), \tag{3.83}$$

$$\sqrt{2\epsilon(\epsilon+1)} \frac{d\sigma_{LT}}{d\Omega} = \sum_{l=1}^{\infty} B_l P_l'(\cos\theta), \tag{3.84}$$

$$\epsilon \frac{d\sigma_{TT}}{d\Omega} = \sum_{l=2}^{\infty} C_l P_l''(\cos\theta), \tag{3.85}$$

where $P_\ell'(x) = dP_\ell(x)/dx$ and $P_\ell''(x) = d^2 P_\ell(x)/d^2 x$.

In the approximation that we retain only terms that contain the dominant multipole M_{1+}, the coefficients in these equations are related to $|M_{1+}|^2$ and its projections on to the other S- and P-wave multipoles E_{1+}, S_{1+}, M_{1-}, E_{0+} and S_{0+}:

$$|M_{1+}|^2 = \tfrac{1}{2} A_0, \tag{3.86}$$

$$\mathrm{Re}(E_{1+} M_{1+}^*) = \tfrac{1}{8}(A_2 - \tfrac{2}{3} C_2), \tag{3.87}$$

$$\mathrm{Re}(M_{1-} M_{1+}^*) = -\tfrac{1}{8}(A_2 + 2(A_0 + \tfrac{1}{3} C_2)), \tag{3.88}$$

$$\mathrm{Re}(E_{0+} M_{1+}^*) = \tfrac{1}{2} A_1, \tag{3.89}$$

$$\mathrm{Re}(S_{0+} M_{1+}^*) = B_1, \tag{3.90}$$

$$\mathrm{Re}(S_{1+} M_{1+}^*) = \tfrac{1}{6} B_2. \tag{3.91}$$

The coefficients A_ℓ, B_ℓ and C_ℓ of (3.86)–(3.91) are determined by fitting (3.83)–(3.85) to the data. As the pion–nucleon P_{33} phase shift is known, (3.86) can be used to give both the real and imaginary parts of the M_{1+} amplitude and, through the interference terms (3.87) and (3.91), the E_{1+} and S_{1+} amplitudes. Finally, the $\gamma^* N \to \Delta(1232)$ transition form factors are determined through (3.67)–(3.69). Although this procedure is largely model-independent, the corrections from the truncation of multipoles in fitting the data can be estimated by using, for example, the UIM of the JLab/Yerevan group (JANR) described in section 3.3. The resulting values (solid and open squares) of G_M^* of the $\gamma N \to \Delta(1232)$ transition up to $Q^2 = 6 \ \mathrm{GeV}^2$ are displayed in figure 3.8, along with the values given by earlier

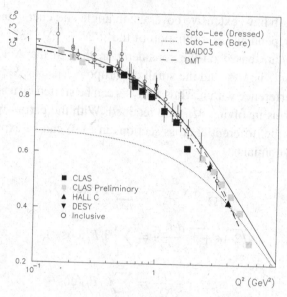

Figure 3.8. Data on the magnetic transition form factor G_M^* for the $\gamma^*N \rightarrow$ $\Delta(1232)$ transition are compared with the predictions from SL [72] (Sato–Lee (dressed)), MAID [59] of 2003 (MAID03) and DMT [73] (DMT) models. The dotted curve (Sato–Lee (bare)) is obtained from the SL model by turning off the pion-cloud effect. $G_D(Q^2) = 1/(1 + Q^2/0.71)^2$ is the usual dipole form factor of the proton. Data extracted from old single-arm electron scattering experiments are labelled 'inclusive'. All other data have been obtained from multipole fits to the data of exclusive π^0 production from protons. The data are from a compilation in [2].

work. They are compared with the results calculated from the models of SL (Sato–Lee (dressed)), MAID of 2003 (MAID03) and DMT. For the SL model, we also show the bare form factor (Sato–Lee (bare), dotted curve) that is obtained by turning off the pion-cloud effect and using (3.82). As mentioned above, the parameters of these three models have been determined mainly by using the data [111] at $Q^2 = 2.8$ and 4 GeV^2. The results at other low Q^2 and in the $Q^2 > 4$ GeV^2 region are the predictions. The agreement between these three model predictions with the new empirical values from JLab clearly indicates that the magnetic M1 form factors in the three models considered are consistent with the new data.

Comparing the solid and dotted curves in figure 3.8, it is seen that the pion-cloud effect is about 40% at $Q^2 = 0$, but becomes much smaller at high Q^2. This suggests that results at $Q^2 \geq 6$ (GeV)^2 will essentially probe the bare form factor which describes the direct excitation of the quark core as illustrated in figure 3.4. It remains to be verified that the bare form factor (3.82) of the SL model can be identified with the constituent-quark-model predictions, as is the case for the $Q^2 = 0$

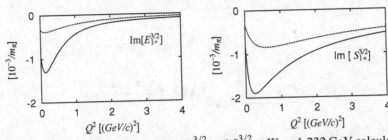

Figure 3.9. The imaginary parts of the $E_{1+}^{3/2}$ and $S_{1+}^{3/2}$ at $W = 1.232$ GeV calculated from the SL model [72]. The dotted curves are obtained from setting the pion-cloud effect to zero.

photoproduction amplitudes given in table 3.1. How the results shown in figure 3.8 can be related to lattice QCD calculations [7] is still an open question mainly because of the difficulties involved in handling the meson–baryon continuum states on the lattice.

The pion-cloud effect is even more dramatic in the E_{1+} and S_{1+} amplitudes. This is illustrated in figure 3.9. We see that the pion-cloud effect greatly enhances these two amplitudes at low Q^2 and yields non-trivial Q^2 dependence. The ratios R_{EM} and R_{SM}, which are defined by these two amplitudes in (3.70) and (3.71), calculated from the SL and DMT models are compared with the empirical values in figure 3.10. The models agree well with each other and with the empirical values of R_{EM}. However, they differ greatly in R_{SM} at low Q^2, indicating significant differences in their formulations as discussed in section 3.3. The results from the SL model appear to be consistent with the JLab results. On the other hand, efforts are still being made to improve the empirical values at low Q^2 for a more detailed verification of the pion-cloud effects illustrated in figure 3.9.

3.4.2 η-meson production

The photo- and electroproduction of the η meson on protons is another example of a successful single-channel analysis and the photocoupling amplitude of the $S_{11}(1535)$ resonance has now been extracted with confidence.

As η is an isoscalar meson it can only couple with nucleons to form isospin $I = \frac{1}{2}$ states. This makes the production of η from nucleon targets an ideal tool to separate isospin-$\frac{1}{2}$ resonances from isospin-$\frac{3}{2}$ resonances. The total photoproduction cross section, shown in figure 3.11(left), exhibits a rapid rise just above the $N\eta$ threshold, indicative of a strong S-wave contribution. This behaviour is known to be due to the first negative-parity nucleon resonance, the $S_{11}(1535)$, which couples to the $N\eta$ channel with a branching fraction of approximately 55% [127]. The next

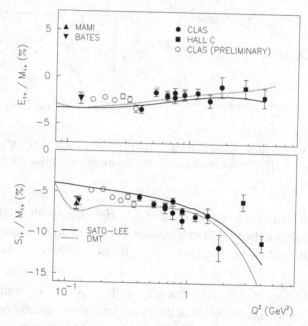

Figure 3.10. Data on ratios R_{EM} (denoted as E_{1+}/M_{1+}) and R_{SM} (denoted as S_{1+}/M_{1+}) for the $\gamma N \rightarrow \Delta(1232)$ transition compared with predictions from the SL [72] and DMT [73] models. These two ratios are related to the $E_{1+}^{3/2}$, $S_{1+}^{3/2}$, and $M_{1+}^{3/2}$ multipole amplitudes of $\gamma^* N \rightarrow \pi N$, by (3.70) and (3.71). Preliminary data from CLAS at low Q^2 are also included (open circles). The data are from a compilation in [2].

higher-mass nucleon resonance with a significant $N\eta$ coupling is the $P_{11}(1710)$. This makes the production of η from nucleon targets the reaction of choice for detailed studies of the electromagnetic excitation of the $S_{11}(1535)$ resonance. As can be seen in figure 3.11(right) the S_{11} remains dominant up to rather high values of Q^2.

While the various data sets agree in their general behaviour, there are apparent discrepancies between two of them near 1070 MeV photon energy. The discrepancy may be traced to the differences in the way the extrapolations to $\theta_\eta^* = 0°$ and $\theta_\eta^* = 180°$ have been made in the two experiments. In order to obtain a total cross section, extrapolations are necessary if the experiments do not cover the full angular range, and models are used to guide the extrapolation into the unmeasured regions.

Table 3.3 gives an overview of the kinematics covered in η-production measurements [128–138].

η photoproduction from protons   The η-photoproduction data on differential cross sections [128,129,133,135] cover the $p\eta$ invariant mass range from production threshold up to $W = 2.3$ GeV, and have allowed a detailed analysis of the $S_{11}(1535)$ resonance.

Reaction	Observables	W range (GeV)	Q^2 range (GeV^2)	Laboratory	
$\gamma p \to p\eta$	$d\sigma/d\Omega$	< 2.0		JLab-CLAS	[128]
	$d\sigma/d\Omega$	< 1.7		GRAAL	[130,135]
	$d\sigma/d\Omega$	< 2.3		ELSA-CB	[129]
	$d\sigma/d\Omega$	< 1.53		MAMI-TAPS	[133]
$\gamma(n/p) \to (n/p)\eta$	$d\sigma_n/d\sigma_p$	< 2.3		GRAAL	[131]
$\vec\gamma p \to p\eta$	Σ	< 2.3		GRAAL	[136]
$\gamma\vec p \to p\eta$	T	< 2.3		ELSA	[138]
$\vec\gamma\vec p \to p\eta$	Σ	< 1.53		MAMI-A2	[132]
$ep \to ep\eta$	$\sigma_{LT}, \sigma_{TT},$ $\sigma_T + \epsilon\sigma_L$	< 2.2	2.8–4.0	JLab-Hall-C	[127]
	$\sigma_{LT}, \sigma_{TT},$ $\sigma_T + \epsilon\sigma_L$	< 2.2	0.3–4.0	JLab-CLAS	[134,137]

Table 3.3. Summary of η photoproduction data. Σ is the linearly-polarized photon asymmetry and T is the polarized-target asymmetry.

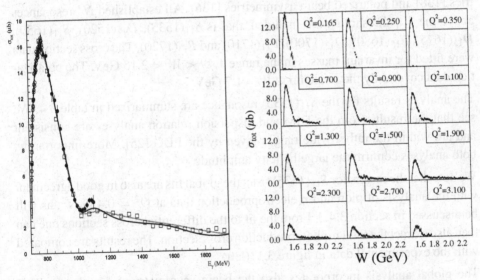

Figure 3.11. Left: total cross section for $\gamma p \to p\eta$. The open circles are data from MAMI [132,133]. The solid circles are from GRAAL [130], and open squares are from JLab [128]. Right: the Q^2 dependence of the total cross sections for $\gamma^* p \to \eta p$ from JLab [134,137].

The $S_{11}(1535)$ has been treated as a three-quark state in the quark model. However, its nature as a three-quark state has been questioned. For example, in [139] the state is characterized as a dynamically-generated $\Sigma \bar K$ state. On the other hand, a three-quark state with $J^P = \frac{1}{2}^-$ is clearly seen in quenched lattice QCD

Resonance	Mass (MeV)	Γ (MeV)	$A^p_{1/2}$	Model
$S_{11}(1535)$	1527	142	96	UIM
	1542	195	119	Dispersion relations
	1520–1555	100–200	60–120	PDG

Table 3.4. $S_{11}(1535)$ photocoupling in units (10^{-3} GeV$^{-1/2}$) from global fit [53] using the UIM and dispersion relations compared with the PDG values [25].

calculations. The photocoupling amplitudes and their Q^2 dependence could be powerful tools in determining the internal resonance structure.

Several analyses have been performed on η photoproduction [53,140,141]. We describe the results from a global analysis [53] that uses both the UIM, as explained in section 3.3.3, and a dispersion-relation approach, as also described in section 3.3.6, to assess the model dependence of the resulting amplitudes. In addition to the differential cross sections this global analysis includes polarized target asymmetries [138] and polarized beam asymmetries [136]. All established N^* resonances above the $N\eta$ threshold were included, that is $S_{11}(1535)$, $D_{13}(1520)$, $S_{11}(1650)$, $D_{15}(1675)$, $F_{15}(1680)$, $D_{13}(1700)$, $P_{11}(1710)$ and $P_{13}(1720)$. The cross section data were fitted for invariant masses in the range $1.49 < W < 2.15$ GeV. The polarization data cover only the range up to $W = 1.7$ GeV.

The analysis results for the $S_{11}(1535)$ resonance are summarized in table 3.4. We see that the results from the UIM and dispersion relation analyses are consistent and are within the rather wide ranges given by the PDG [25]. More importantly, both analyses confirm the large helicity amplitude $A^p_{1/2}$.

We note that the results for $S_{11}(1535)$ from the global fits are also in good agreement with the analysis of pion and η electroproduction data at $Q^2 = 0.4$ GeV^2, as will be discussed in section 3.4.3. From the fit to the differential cross sections one can then also extract the total η-photoproduction cross section. The results are compared with the experimental data in figure 3.11(left).

The global analysis incorporates also the beam asymmetry, Σ, in the fit. This observable is highly sensitive to the interference of the dominant E_{0+} multipole with the E_{2-} and M_{2-} multipoles of the neighbouring $D_{13}(1520)$ state. The sensitivity of Σ to the contributions from the $D_{13}(1520)$ can be seen if we express Σ in the approximation that only S-waves, P-waves, and D-waves with spin $J \leq \frac{3}{2}$ contribute, and only terms containing the dominant E^η_{0+} multipole are retained [140]:

$$\Sigma = \frac{3\sin^2\theta \operatorname{Re}[E^{*\eta}_{0+}(E^\eta_{2-} + M^\eta_{2-})]}{|E^\eta_{0+}|^2}. \tag{3.92}$$

Resonance	Mass (MeV)	Γ (MeV)	$\beta_{\eta N}$ (%)	$\beta_{\pi N}$ (%)
$D_{13}(1520)$	1520	120	0.05 ± 0.02	50–60
$F_{15}(1680)$	1675	130	0.15 ± 0.03	60–70

Table 3.5. The mass, width (Γ), and branching fractions $\beta_{\eta N}$ and $\beta_{\pi N}$ for the $D_{13}(1520)$ and $F_{15}(1680)$ states extracted from the global fits [53] of η photoproduction data.

This expression can be fitted to the measured beam asymmetry Σ. Using E_{0+}^{η} from fits to the cross section data, the multipoles $E_{2-}^{\eta} + M_{2-}^{\eta}$ for the $D_{13}(1520)$ can then be determined. Using the known pion multipoles $E_{2-}^{\pi} + M_{2-}^{\pi}$ and the branching fraction $\beta_{\pi N}$, the branching fraction $\beta_{\eta N}$ can then be extracted.

In the mass range of the $F_{15}(1680)$, the beam asymmetry is also sensitive to the branching ratio $\beta_{\eta N}$ for that state. The results for the $D_{13}(1520)$ and $F_{15}(1680)$ are summarized in table 3.5. Both results represent significantly- improved values for the branching fractions.

η **electroproduction from protons** η electroproduction experiments have focussed on the Q^2 evolution of the $S_{11}(1535)$ transverse photocoupling amplitude $A_{1/2}(Q^2)$. Experiments in the 1970s [142–145] indicated a very slow falloff with Q^2. More recent experiments [127,134,137] have studied this behaviour in detail with high statistics over a wide kinematic range. Applying the Legendre polynomial expansions (3.83)–(3.85) with a limit of $l = 2$ to fit the differential cross section data, the coefficients A_0, A_1, A_2, B_1, B_2 and C_2 can be extracted. The resulting values of these coefficients are shown in figure 3.12(left). We note that A_0 is mostly due to the $S_{11}(1535)$ resonance, and is by far the largest amplitude. The longitudinal and transverse amplitudes cannot be separated in this analysis. However, the global analysis that includes single-pion channels, discussed below, finds a rather small longitudinal contribution to A_0. Assuming $\sigma_L = 0$, $|A_{1/2}|$ can then be computed from a Breit–Wigner resonance fit to A_0. The figure 3.12(right) shows a compilation of results for $A_{1/2}(Q^2)$. The striking feature is the slow falloff with Q^2, indicating a hard transition form factor throughout the entire Q^2 range. It should be noted that while the Q^2 evolution is well determined, the absolute normalization of $A_{1/2}(Q^2)$ is uncertain to the extent that the branching fraction $\beta_{N\eta}(S_{11}) = 0.55$ and a total width of 150 MeV have been used in extracting $A_{1/2}$. The PDG [25] allows a large range of 0.30–0.55 for this branching fraction. However, the analysis of Armstrong *et al* [127], gives a most-probable value of $\beta_{N\eta}(S_{11}) \approx 0.55$. The use of this value is consistent with the values $\beta_{N\eta}(S_{11}) = 0.55$ and $\beta_{N\pi}(S_{11}) = 0.4$ used in the combined analysis [146] of π and η electroproduction, which is the subject of the next section.

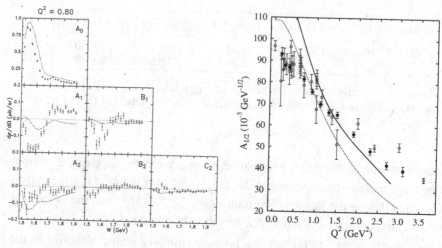

Figure 3.12. Left: the fitted values of the partial-wave coefficients (defined in (3.83) to (3.85)) of η production are compared to the predictions of ηMAID [30]. Right: the Q^2 evolution of the photocoupling helicity amplitudes ($A_{1/2}$) determined from Breit–Wigner resonance fits to the parameter $A_0(W)$ at fixed W. The full circles are the most recent CLAS data [134,137]. The open diamonds are JLab data [127] and the open circles are from the compilation of older data in [134]. Solid and dashed lines are quark-model predictions from [153] and [150] respectively.

3.4.3 Combined analysis of $N\pi$ and $N\eta$ electroproduction

Most previous results on the Q^2 evolution of the $S_{11}(1535)$ transition amplitude were obtained from Breit–Wigner resonance parametrizations to fit the η-production data. This has been justified because of the dominant contribution of the $S_{11}(1535)$ to the $p\eta$ channel. However, this is not a fully satisfactory solution since the higher-mass states that couple to $N\eta$ may also contribute to the lower-mass region. Furthermore, there are also non-resonant contributions that must be taken into account. A much more constrained approach is to carry out a global analysis that takes higher-mass resonances as well as non-resonant terms into account and fits both the π and η production data.

A large-acceptance detector allows simultaneous measurements of cross sections and polarization observables for several channels, for example $p\pi^0$, $n\pi^+$ and $p\eta$. Also, the large acceptance provides complete angular distributions, including the full azimuthal dependence. As can be seen in (3.9) this allows the separation of three response functions in the unpolarized cross section. Furthermore, the use of a highly-polarized electron beam provides data on the helicity-dependent response function $\sigma_{LT'}$. The full set of these data taken with a hydrogen target has been analysed using both the UIM [52] and the dispersion-relation approach [53]. The results for the $S_{11}(1535)$ are shown in table 3.6.

Resonance	Mass (MeV)	Width (MeV)	Q^2 (GeV^2)	$A_{1/2} \times 10^3$ (GeV$^{-1/2}$)	$S_{1/2} \times 10^3$ (GeV$^{-1/2}$)
$S_{11}(1535)$	1530	150	0.375	92 ± 2	-12 ± 3
			0.75	91 ± 1	-13 ± 3
			0.40	91 ± 4	-21 ± 4
			0.65	95 ± 3	-15 ± 3

Table 3.6. Results for $S_{11}(1535)$ from the combined analysis [146] of cross section and polarized beam asymmetry data of electroproduction with $n\pi^+$, $p\pi^0$, and $p\eta$ final states. $A_{1/2}$ and $S_{1/2}$ are the transverse and scalar helicity amplitudes.

In table 3.6, we see that the transverse helicity amplitude $A_{1/2}(Q^2)$ at low Q^2 shows little Q^2 dependence. These values of $A_{1/2}(Q^2)$ are slightly higher than those obtained in the single-resonance analyses shown in figure 3.12(right). We also note that the longitudinal amplitude $S_{1/2}$ for the $S_{11}(1535)$ is rather small. The uncertainties are the differences between the results obtained from the two conceptually very different approaches, and thus indicate the model dependence of the analysis. Clearly, the two approaches are remarkably consistent in extracting the $A_{1/2}$ and $S_{1/2}$ amplitudes. As can be seen in figure 3.12, the slow falloff with Q^2 is only qualitatively reproduced by constituent quark model calculations [150,153].

The results for the $P_{11}(1440)$ and $D_{13}(1520)$ are compared with models in figures 3.13 and 3.14. For the $P_{11}(1440)$, the magnitude of $A_{1/2}(Q^2)$ drops rapidly with Q^2 and changes sign near $Q^2 = 0.5$ GeV^2. The $P_{11}(1440)$ also shows a strong longitudinal coupling. The comparison with a variety of models shows that none of them describes the data, for either the transverse or longitudinal amplitudes. The data need to establish whether there is a sign change in $A_{1/2}^p$ as models [147–153] differ as to whether this is to be expected.

The results of the global analysis for the $D_{13}(1520)$ are shown in figure 3.14. These new results are compared with the previous data in the top panels of figure 3.16. We see from figure 3.14 that the $A_{1/2}$ and $A_{3/2}$ amplitudes have a strikingly-different Q^2 dependence. While $A_{3/2}$ is large at the real photon point and drops rapidly with Q^2, $A_{1/2}$ is small at $Q^2 = 0$ and then rises in magnitude with increasing Q^2 before falling slowly at high Q^2. One can discuss this in terms of the helicity asymmetry A_1, defined as:

$$A_1 = \frac{A_{1/2}^2 - A_{3/2}^2}{A_{1/2}^2 + A_{3/2}^2}. \tag{3.93}$$

The data imply that $A_1 \approx -1$ at $Q^2 = 0$, and crosses zero near $Q^2 = 0.5$ GeV^2, where $A_{1/2} \approx A_{3/2}$. The ratio A_1 approaches $+1$ at high Q^2. Such a 'helicity switch'

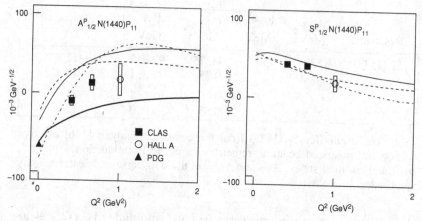

Figure 3.13. JLab results for the Q^2 evolution of the $P_{11}(1440)$ photocoupling helicity amplitudes $A_{1/2}(Q^2)$ (left), and $S_{1/2}(Q^2)$ (right) from a combined analysis [146] of πN and ηN data. Curves are the predictions of the hybrid model of [147] (bold solid), the light-cone model of [149] (thin solid), the light-cone model of [152] (dashed) and the meson-cloud model of [148] (dashed-dot).

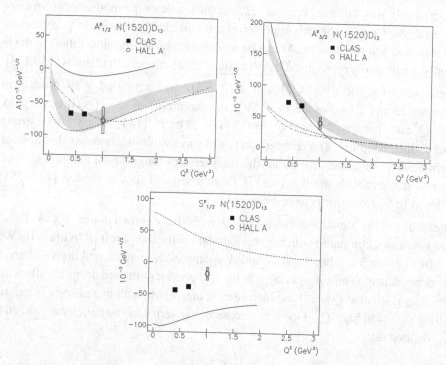

Figure 3.14. JLab results for the Q^2 evolution of the $D_{13}(1520)$ photocoupling amplitudes $A_{1/2}$, $A_{3/2}$ and $S_{1/2}$ from a combined analysis [146] of πN and ηN data. The shaded bands indicate the uncertainties seen in previous analyses using mostly $p\pi^0$ cross section data. Solid, dot, and dot-dashed lines show predictions of theoretical calculations [149,152,151].

was indeed predicted [154,155] within the non-relativistic quark model. In such a model [156], the ratio of helicity amplitudes for $\gamma p \to D_{13}^+(1520)$ is given by:

$$\frac{A_{1/2}^{D13}}{A_{3/2}^{D13}} = \frac{1}{\sqrt{3}}\left(\frac{\mathbf{q}^2}{\beta^2} - 1\right), \tag{3.94}$$

where $\mathbf{q}^2$ is the square of the three-momentum transfer to the $D_{13}(1520)$ at the real photon point, evaluated in the laboratory system, and β is the strength of the oscillator string constant in the simple-harmonic-oscillator model. The q^2-dependent term corresponds to the spin-flip term B in the single-quark transition model (SQTM), to be explained in section 3.4.4, while the constant term is the quark-orbit flip amplitude A. At the real photon point, the ratio $A_1 \approx -1$ in the model as $\beta^2 \approx \mathbf{q}^2$ for the photoproduction of $D_{13}(1520)$. This agrees remarkably with the data for $\gamma p \to D_{13}^+(1520)$. For electroproduction, where $Q^2 \neq 0$, the momentum transfer $\mathbf{q}^2$ increases causing the ratio A_1 to increase and $A_{1/2}$ becomes dominant at high Q^2. Modern quark models include other terms in addition to the quark spin-flip and orbit-flip, but predict qualitatively the same behaviour [126,149,150]. A similar behaviour is predicted by quark models [126,153,155,156] for the transition to the $F_{15}(1680)$. Here the data quality is not sufficient to allow stringent tests of the prediction.

In summary, the combined analysis of complete data sets with high statistics in $p\eta$, $n\pi^+$ and $p\pi^0$ cross sections and the beam-spin response functions $\sigma_{LT'}$, has produced a nearly model-independent result for the electrocoupling amplitudes of the $P_{11}(1440)$, $S_{11}(1535)$ and $D_{13}(1520)$ resonances. The Q^2 evolution of the $A_{1/2}$ and $S_{1/2}$ amplitudes for the $P_{11}(1440)$ is qualitatively consistent with the predictions of a model including a meson cloud. The latter result confirms what was found for the $\gamma^* N \to \Delta(1232)$ transition. As discussed in section 3.4.1, the meson-cloud effects can be sizeable and must be taken into account in modelling the $\gamma N \to N^*$ transition amplitudes.

3.4.4 *Electromagnetic transitions to the* $[70, 1^-]$ *supermultiplet*

Analyses of the data on π and η production have established the electromagnetic transition amplitudes for the $S_{11}(1535)$, $P_{11}(1440)$ and $D_{13}(1520)$ resonances. In this section, we discuss how this information can be used to test the constituent-quark model.

The existing data for nucleon resonance properties, such as mass, spin-parity and flavour, fit well into the representations of the SU(6) $\otimes$ O(3) symmetry group. This symmetry group leads to supermultiplets of baryon states with the same orbital angular momentum **L**. Within an SU(6) $\otimes$ O(3) supermultiplet the quark spins are aligned to form a total quark spin **S**, with $s = \frac{1}{2}, \frac{3}{2}$, which combines with **L** to form the total angular momentum $\mathbf{J} = \mathbf{L} + \mathbf{S}$. A large number of explicit dynamical quark models have been developed to predict the electromagnetic transitions between the nucleon ground state and its excited states [157,148–150]. Many of these calculations belong to the SQTM within which it is assumed that only a single quark is involved in the electromagnetic transition. The fundamentals of the SQTM are described in [158,159,154], where the symmetry properties for the transition from the ground state nucleon [56, 0$^+$] to the [70, 1$^-$] and the [56, 2$^+$] supermultiplet are discussed. Within the SQTM, algebraic relations between resonance excitations can be derived from symmetry properties. The parameters of these algebraic equations can be determined [160] from the experimental information on just a few resonances. Predictions for other resonances belonging to the same SU(6) $\otimes$ O(3) supermultiplet can then be made.

The [70, 1$^-$] supermultiplet contains $S_{11}(1535)$ and $D_{13}(1520)$ which have been studied with a large amount of data of π and η production, as discussed in the previous subsections. Thus we can make use of this new information to predict electromagnetic transitions to the [70, 1$^-$] supermultiplet. For the transition to the [70, 1$^-$], the quark transverse current can be written within the SQTM as a sum of three terms [158,159,154]:

$$J^+ = AL^+ + B\sigma^+ + C\sigma_z L^+, \tag{3.95}$$

where σ is the quark Pauli-spin operator, L_z is the projection of the quark orbital angular momentum **L** onto the photon direction (z-axis), and $L^+ = -\frac{1}{\sqrt{2}}(L_x + iL_y)$ is a raising operator for **L**. Clearly, the term A in (3.95) corresponds to a quark orbit-flip with $\Delta L_z = +1$, and the term B to a quark spin-flip with $\Delta L_z = 0$. The term C corresponds to both a quark spin-flip and an orbit-flip with $\Delta L_z = +1$. The relationships between the A, B and C amplitudes and the usual helicity photocoupling amplitudes $A_{1/2}$ and $A_{3/2}$ are listed in table 3.7.

The coefficients A, B and C of (3.95) can be determined for the $\gamma + [56, 0^+] \rightarrow$ [70, 1$^-$] transitions [160], using table 3.7 and the data on the photocoupling amplitudes of the $S_{11}(1535)$ and $D_{13}(1520)$ resonances. The results are shown in figure 3.15. The shaded bands are the ranges of these three coefficients allowed by the errors on the photocoupling amplitudes of the $S_{11}(1535)$ and $D_{13}(1520)$ resonances. Knowledge of these three coefficients and of two mixing angles for the transition to the [70, 1$^-$] allows predictions for 16 amplitudes of states belonging to the same supermultiplet. The first mixing angle θ is between two $J^P = \frac{1}{2}^-$ states resulting in physical $S_{11}(1535)$ and $S_{11}(1650)$ states. There is also a small mixing angle of 6° for the two $\frac{3}{2}^-$ states resulting in the physical states $D_{13}(1520)$ and $D_{13}(1700)$.

State	Proton target	Neutron target
$S_{11}(1535)$	$A^+_{1/2} = \frac{1}{6}(A + B - C)\cos\theta$	$A^0_{1/2} = -\frac{1}{6}(A + \frac{1}{6}B - \frac{1}{3}C)$
$D_{13}(1520)$	$A^+_{1/2} = \frac{1}{6\sqrt{2}}(A - 2B - C)$	$A^0_{1/2} = -\frac{1}{18\sqrt{2}}(3A - 2B - C)$
	$A^+_{3/2} = \frac{1}{2\sqrt{6}}(A + C)$	$A^0_{3/2} = \frac{1}{6\sqrt{6}}(3A - C)$
$S_{11}(1650)$	$A^+_{1/2} = \frac{1}{6}(A + B - C)\sin\theta$	$A^0_{1/2} = \frac{1}{18}(B - C)$
$D_{13}(1700)$	$A^+_{1/2} = 0$	$A^0_{1/2} = \frac{1}{18\sqrt{5}}(B - 4C)$
	$A^+_{3/2} = 0$	$A^0_{3/2} = \frac{1}{6\sqrt{15}}(3B - 2C)$
$D_{15}(1675)$	$A^+_{1/2} = 0$	$A^0_{1/2} = -\frac{1}{6\sqrt{5}}(B + C)$
	$A^+_{3/2} = 0$	$A^0_{3/2} = -\frac{1}{6}\sqrt{\frac{2}{5}}(B + C)$
$D_{33}(1700)$	$A^+_{1/2} = \frac{1}{6\sqrt{2}}(A - 2B - C)$	Same
	$A^+_{3/2} = \frac{1}{2\sqrt{6}}(A + C)$	Same
$S_{31}(1620)$	$A^+_{1/2} = \frac{1}{18}(3A - B + C)$	Same

Table 3.7. Helicity amplitudes for the electromagnetic transition from the ground state $[56, 0^+]$ to the $[70, 1^-]$ multiplet expressed in terms of the SQTM amplitudes. θ is the mixing angle relating two $J^P = \frac{1}{2}^-$ states resulting in the physical states $S_{11}(1535)$ and $S_{11}(1650)$. The mixing angle used in the calculations is $\theta = 31°$ [9,158].

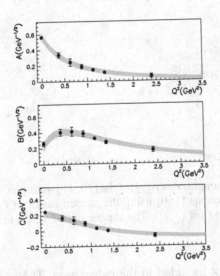

Figure 3.15. The single-quark transition amplitudes A, B, C (defined in (3.95)) extracted [160] using world data on the $S_{11}(1535)$ and on the $D_{13}(1520)$. The shaded bands are the ranges of these three coefficients allowed by the errors on the photocoupling amplitudes of the $S_{11}(1535)$ and $D_{13}(1520)$ resonances.

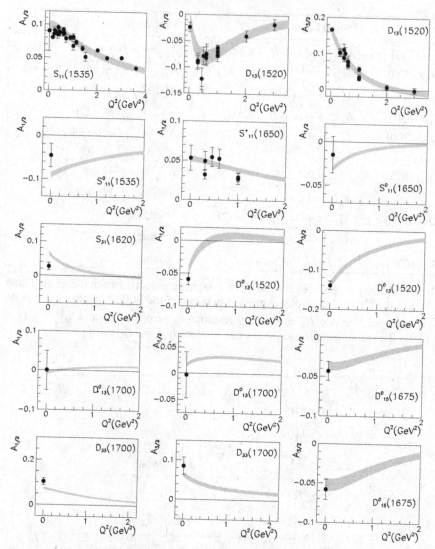

Figure 3.16. Helicity amplitudes $A_{1/2}$ for the $[70, 1^-]$ multiplet. The shaded bands are the SQTM predictions [160] using the shaded bands of figure 3.15 for the coefficients A, B, and C of (3.95). The superscript '0' refers to amplitude for the neutron target.

This mixing angle is not included in the calculation. The predictions, calculated from the shaded bands in figure 3.15, are shown in figure 3.16 and compared with the available helicity amplitude data. Note that the data points for $S_{11}(1535)$ and $D_{13}(1520)$ (three figures on the top row of figure 3.16) are used to determine the coefficients A, B and C of (3.95) shown in figure 3.15. The rest of the figures 3.16 are the SQTM predictions.

In the second to fifth rows of figure 3.16 we see that the SQTM predictions agree very well with the data at the $Q^2 = 0$ real-photon point. This suggests that the single-quark transition may be the dominant contribution to the excitation of the states in $[70, 1^-]$. More electroproduction data for extracting transition amplitudes at $Q^2 > 0$ and for other members of the $[70, 1^-]$ multiplet are needed for a conclusive test of the SQTM. Most of the states in the $[70, 1^-]$ with masses near 1700 MeV couple strongly to $N\pi\pi$ channels and require more involved analysis techniques. Progress in this area is discussed in the next section. There are similar SQTM relations for the transition from the nucleon ground state to the members of the $[56, 2^+]$ supermultiplet. In this case four SQTM amplitudes can contribute. The only state for which the two transverse photocoupling amplitudes have been measured in electroproduction is the $F_{15}(1680)$, which is insufficient to extract the four SQTM amplitudes.

3.4.5 Two-pion production

Two-pion channels dominate the electromagnetic meson-production cross sections in the second and third resonance regions where we hope to resolve the missing resonance problem [157] and ultimately determine what basic symmetry group [24] is underlying the baryon spectrum. Thus a detailed understanding of two-pion production is very important in the N^* study, and has been pursued very actively. Extensive two-pion production data have now been accumulated at JLab, Mainz and Bonn, but have not been fully analysed and understood theoretically. Here we will mainly report on the status of the data and describe some very preliminary attempts to identify N^* states.

The study of $\pi\pi N$ channels requires the use of detectors with nearly 4π solid angle coverage for charged or neutral particle detection. Several such detectors have been in operation for a number of years, and have generated large data sets for the reactions

$$\gamma p \rightarrow p\pi^0\pi^0, \tag{3.96}$$

$$\gamma p \rightarrow p\pi^+\pi^-, \tag{3.97}$$

$$ep \rightarrow ep\pi^+\pi^-. \tag{3.98}$$

These processes can be projected onto various isobar channels which are useful in identifying the nucleon resonances from the data. The $p\pi^+\pi^-$ final state contains the $\Delta^{++}\pi^-$ and $\rho^0 p$ isobar components, which could have large sensitivity to resonance decays. But it also has very strong contributions from non-resonant mechanisms which complicate the analysis of the data. The $p\pi^0\pi^0$ final state has the advantage of high sensitivity to resonances because it has fewer non-resonant

Reaction	Observables	W range (GeV)	Q^2 range (GeV^2)	Lab.
$\gamma^* p \to p\pi^+\pi^-$	σ_{tot}, $\dfrac{d\sigma}{dM_{p\pi^+}}$ $\dfrac{d\sigma}{dM_{\pi^+\pi^-}}$, $\dfrac{d\sigma}{d\cos\theta_{\pi^-}}$	< 2.1	0.65–1.3	CLAS [162]
$\gamma p \to p\pi^+\pi^-$	σ_{tot}, $\dfrac{d\sigma}{dM_{p\pi^+}}$ $\dfrac{d\sigma}{dM_{\pi^+\pi^-}}$, $\dfrac{d\sigma}{d\cos\theta_{\pi^-}}$	< 2.0	0	CLAS [163]
$\gamma p \to p\pi^0\pi^0$	σ_{tot}, $\dfrac{d\sigma}{dM_{p\pi^0}}$ $\dfrac{d\sigma}{dM_{\pi^0\pi^0}}$	< 1.9	0	GRAAL [165]
$\gamma p \to p\pi^0\pi^0$	σ_{tot}, Σ, $\dfrac{d\sigma}{dM_{p\pi^0}}$ $\dfrac{d\sigma}{dM_{\pi^0\pi^0}}$	< 1.55	0	MAMI [164,166]

Table 3.8. Summary of $\gamma p \to p\pi\pi$ reaction data. $M_{N\pi}$ denotes the invariant mass of the produced $N\pi$ subsystem. Σ is the photon asymmetry.

contributions. It also does not couple to the $p\rho^0$ isobar state. Table 3.8 gives a summary of two-pion production data obtained at various laboratories.

Analysis of $p\pi^+\pi^-$ final state   The $p\pi^+\pi^-$ channel has been studied in both photoproduction [161,163] and electroproduction [162] reactions. Two distinctly different approaches, based on isobar models, have been applied to analyse the data from JLab [163,167] and CB-ELSA [161]. The first approach is to adjust the parameters of an isobar model [167] to fit the fully-extracted cross section and polarization-asymmetry data [167,169]. The second one is to fit directly the unbinned data [161].

The first approach makes use of knowledge obtained from production by pion beams. The energy dependence of non-resonant processes is parametrized and resonance photocouplings and hadronic couplings are fixed if known, for example from single-pion processes. Resonances in specific partial waves can be introduced to search for undiscovered states. Model parameters are then obtained by fitting to the one-dimensional projections of the multi-dimensional differential cross section. Such a model can lead to a qualitatively good description of the projected data. The method has been used in the analysis of CLAS electroproduction data [162]. As displayed in figure 3.17(left), a significant discrepancy was

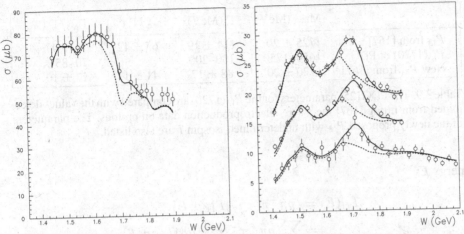

Figure 3.17. Total cross sections for photoproduction (left) and electroproduction (right) of $\pi^+\pi^-$ on protons at $Q^2 = 0.65, 0.95$ and $1.30\,\text{GeV}^2$ (from the top). Data sets are from CLAS [162,163]. The curves are explained in the text.

found near $W = 1.7$ GeV between the data and the predictions (dotted curves) from the isobar model employed. This discrepancy was attributed either to inaccurate hadronic couplings for the well-known $P_{13}(1720)$ resonance, determined from the analysis of hadronic experiments, or to an additional resonance with $J^P = \frac{3}{2}^+$ with either $I = \frac{1}{2}$ or $I = \frac{3}{2}$. The discrepancy is best visible in the total cross sections for electroproduction, shown in figure 3.17(right). The dotted line shows the model predictions using resonance parameters from single-pion electroproduction and from the analysis of $\pi N \to N\pi\pi$ data [25,60,65]. The solid line represents the fit when the hadronic couplings of the $P_{13}(1720)$ to $\Delta\pi$ and to $N\rho$ are allowed to vary well beyond the ranges established in the analysis of hadronic data. Alternatively, a new state was introduced with hadronic couplings extracted from the data. Table 3.9 summarizes the results of the analysis using a single P_{13} with modified hadronic couplings, and a new P_{13} state with undetermined isospin while keeping the PDG $P_{13}(1720)$ hadronic couplings unchanged.

The total photoproduction cross section in figure 3.17(left) shows a dependence on W that is very different from the electroproduction data in the right panel. In particular, the photoproduction data have a much higher non-resonant contribution largely due to increased non-resonant ρ^0 production. Both data sets are consistent with a strong resonance near $W = 1.72$ GeV in the P_{13} partial wave [170].

The drawback of this approach is that when fitting one-dimensional projections of cross sections, correlations between data sets are lost. The second approach [163] is based on a partial-wave formalism starting from the T matrix at a given photon

	Mass (MeV)	Γ (MeV)	$\Gamma_{\pi\Delta}/\Gamma$ (%)	$\Gamma_{N\rho}/\Gamma$ (%)
P_{13} from [167]	1725 ± 20	114 ± 19	63 ± 12	19 ± 9
$P_{13}(1720)$ of PDG	1650–1750	100–200		70–85
New P_{13} from [167]	1720 ± 20	88 ± 17	41 ± 13	17 ± 10

Table 3.9. The PDG [25] parameters for the $P_{13}(1720)$ are compared with the values determined from fitting [167] the $\pi^+\pi^-$ electroproduction data on protons. The parameters of the newly proposed P_{13} with undetermined isospin I are also listed.

energy E:

$$
\begin{aligned}
T_{fi}(E) &= \langle p\pi^+\pi^-; \tau_f | T | \gamma p; E \rangle \\
&= \sum_\alpha \langle p\pi^+\pi^-; \tau_f | \alpha \rangle \langle \alpha | T | \gamma p; E \rangle \\
&= \sum_\alpha \psi^\alpha(\tau_f) V^\alpha(E),
\end{aligned}
\tag{3.99}
$$

where α denotes all intermediate states, and τ_f characterizes the final state kinematics. The decay amplitude $\psi^\alpha(\tau_f) = \langle p\pi^+\pi^-; \tau_f | \alpha \rangle$ is calculated using an isobar model for specific decay channels, for example $\Delta^{++}\pi^-$, $\Delta^0\pi^+$ or $p\rho^0$. The production amplitude $V^\alpha = \langle \alpha | T | \gamma p; E \rangle$ is then determined by using an unbinned maximum-likelihood procedure to fit the data event by event. This method takes into account all correlations between the variables.

In this analysis, 35 partial waves were included in addition to t-channel processes with adjustable parameters. Figure 3.18 shows intensity distributions in different isobar channels, for the $J = \frac{5}{2}(m = \frac{1}{2})$ and $J = \frac{3}{2}(m = \frac{1}{2})$ partial waves. Clear signals of the well-known $F_{15}(1680)$ and the poorly-known $P_{33}(1600)$ are seen. In the final analysis the energy dependence is fitted to a Breit–Wigner form to determine masses and widths of resonant states. This method is closer to a model-independent approach and can directly 'discover' new resonances in specific partial waves.

Analysis of the $p\pi^0\pi^0$ final state   The CB-ELSA collaboration has analysed the $p\pi^0\pi^0$ final state using a more model-dependent version of the partial-wave analysis. Here s-channel Breit–Wigner distributions are fitted to the data on an event-by-event basis, therefore retaining the correlations in the data. However, the fit is constrained by the parametrized energy dependence of the Breit–Wigner function. Neglecting t-channel processes may be suitable for the $p\pi^0\pi^0$ channel, but is insufficient for the $p\pi^+\pi^-$ channel. The method has been applied in the mass range up to 1800 MeV [161]. The decay channels $\Delta(1232)\pi$, $N(\pi\pi)_s$, and $P_{11}(1440)\pi$ are included. The data can be fitted with partial waves of known states,

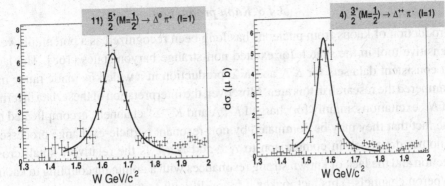

Figure 3.18. Preliminary results [163] of a partial-wave analysis of the $\gamma p \rightarrow p\pi^+\pi^-$ reaction showing the well-known $F_{15}(1680)$ (left) and evidence for the poorly-known $P_{33}(1600)$ (right).

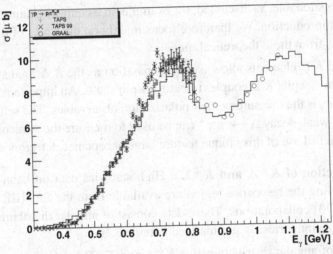

Figure 3.19. Preliminary results [161] of a partial-wave analysis for the reaction $\gamma p \rightarrow p\pi^0\pi^0$, shown by the histogram, compared with the total cross section measured by TAPS [164] and GRAAL [165].

and do not require additional new states. At the present time the solutions are not unique, and equally good fits are obtained with different partial waves. This points to a need for the inclusion of polarization observables or other final states into the analysis. In figure 3.19 the total cross section for $\gamma p \rightarrow \pi^0\pi^0 p$ is shown as extracted from the integral over all partial-wave contributions and compared to previous data from TABS [164] and GRAAL [165].

3.4.6 Kaon production

Production of kaons from nucleons has long been recognized as a potentially very sensitive tool in the search for excited non-strange baryon states [168]. The lack of consistent data sets for $K\Lambda$ and $K\Sigma$ production in a wide kinematic range has hampered the research in this area. Moreover, the interpretation of these data in terms of N^* excitations, mainly for charged $K^+\Lambda$ and $K^+\Sigma^0$ channels, is complicated by the fact that they may be dominated by non-resonant particle-exchange processes. Another drawback in comparison to $N\pi$ and $N\pi\pi$ is the relatively small cross section and the lack of known strong resonances with a dominant coupling to kaon–hyperon channels. This fact makes it more difficult to use strangeness production as a tool in the study of excited baryons, and specifically in the search for new resonances.

Most of the theoretical models [83–90] for kaon production are based on the tree-diagrams approach, as described in section 3.3.2. The validity of these tree-diagram models is questionable, as discussed, for example, in a coupled-channel study [76] of kaon photoproduction. We therefore focus mainly on the status of the data, not on the results from these theoretical models.

The $K\Lambda$ and $K\Sigma$ channels allow isospin separation as the $K\Lambda$ final state selects isospin-$\frac{1}{2}$ states, while $K\Sigma$ couples to both isospin states. An important tool in resonance studies is the measurement of polarization observables. The self-analysing power of the weak decay $\Lambda \to p\pi^-$ can be used to measure the Λ recoil polarization. To make full use of this unique feature large acceptance detectors are needed.

Photoproduction of $K^+\Lambda$ and $K^+\Sigma$ High-statistics data on kaon photoproduction covering the resonance region are available from the SAPHIR [172] and the CLAS [173] collaborations. These data consist of angular distributions and Λ polarization asymmetries, as summarized in table 3.10.

Typical data of angular distributions for $K^+\Lambda$ and $K^+\Sigma$ production are shown in figure 3.20. We see that the $K^+\Lambda$ data (left panel) show a strong forward peaking for photon energies greater than 1 GeV, indicating large t-channel contributions. For the $K^+\Sigma^0$ channel (right panel) the angular distributions are more symmetric or 'resonance-like' at low energies, but become somewhat more forward-peaked at energies above 1.3 GeV.

The high statistics of these data allows, for the first time, the identification of the structures in the differential cross section that hint at interference between resonances and the non-resonant background. The presence of s-channel resonances is particularly evident in the W dependence shown in figure 3.21. At the most forward angles (upper panel), two resonance-like structures are visible at $W \approx 1.7$ GeV,

Reaction	Observables	W range (GeV)	$\cos\theta_K^*$ range	Q^2 (GeV^2)	Laboratory
$\gamma p \to \Lambda K$	$d\sigma/d\Omega$	< 2.15	-0.95 to 0.95		SAPHIR [171]
	$d\sigma/d\Omega$	< 2.6	-0.95 to 0.95		SAPHIR [172]
	$d\sigma/d\Omega$	< 2.3	-0.85 to 0.85		CLAS [173]
$\gamma p \to \Sigma K$	$d\sigma/d\Omega$	< 2.15	-0.95 to 0.95		SAPHIR [171]
	$d\sigma/d\Omega$	< 2.6	-0.95 to 0.95		SAPHIR [172]
	$d\sigma/d\Omega$	< 2.3	-0.85 to 0.85		CLAS [173]
$\gamma p \to K^+\vec{\Lambda}, \vec{\Sigma}$	P	< 2.3	-0.85 to 0.85		CLAS [173]
	p.r.f.	< 2.6	-0.95 to 0.95		SAPHIR [172]
$ep \to eK^+\vec{\Lambda}$	$\sigma_{LT}, \sigma_{TT},$ $\sigma_T + \epsilon\sigma_L$	< 2.2	-1.0 to 1.0	< 3	CLAS [174]
$ep \to eK^+\vec{\Sigma}$	$\sigma_T, \sigma_{TT},$ $\sigma_T + \epsilon\sigma_L$	< 2.2	-1.0 to 1.0	< 3	CLAS [174]
$\vec{e}p \to eK^+\Lambda$	P_x', P_z'	< 2.2	-1.0 to 1.0	< 3	CLAS [174]

Table 3.10. Summary of hyperon photo- and electroproduction data. The transfer polarization components P_x' and P_z' are defined in (3.100) and p.r.f. is the polarization response function.

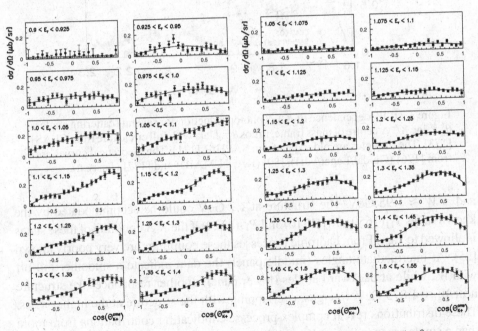

Figure 3.20. Angular distributions of $K^+\Lambda$ photoproduction (left) and $K^+\Sigma^0$ photoproduction (right). Both data sets are from SAPHIR [172]. The curves represent Legendre polynomial fits to the data.

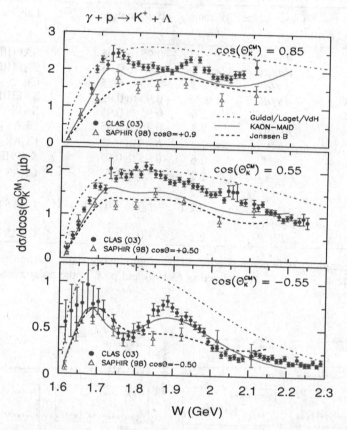

Figure 3.21. W-dependence of $K^+\Lambda$ photoproduction for several scattering angles θ_K^{CM} inc $K^+\Lambda$ centre of mass frame. in $\cos\theta_K^*$. Data with full circles are from CLAS [173]. The triangles are older data from SAPHIR [171]. The theoretical curves are from [176] (Guidal/Laget/vdH), [89] (KAON-MAID) and [177] (Janssen).

and at $W \approx 1.95$ GeV. The structure at 1.7 GeV could be accommodated by the known states $S_{11}(1650)$, $P_{11}(1710)$ and $P_{13}(1720)$ if the $K\Lambda$ coupling of these states is allowed to vary. From hadronic processes these couplings are very poorly known [25]. At intermediate angles (middle panel) the data indicate a smoother fall-off with W, while at backward angles (lower panel) another resonance-like structure near $W \approx 1.875$ GeV emerges, overlapping with the structure at the higher mass. These distributions reveal complex processes, indicating contributions from more than a single resonance near $W = 1.9$ GeV.

Electroproduction of $K^+\Lambda$ and $K^+\Sigma$ Kaon electroproduction is another tool in the study of non-strange nucleon resonances. While the $K^+\Lambda$ and $K^+\Sigma^0$ photoproduction cross sections exhibit complex structures of resonant and non-resonant

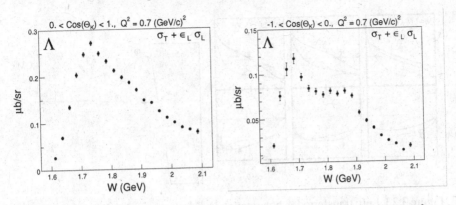

Figure 3.22. Cross section from CLAS [174] for $K^+\Lambda$ electroproduction at $Q^2 = 0.7$ GeV^2 integrated over to the forward hemisphere (left panel) and backward hemisphere (right panel) in the centre of mass angle θ_K. Here $\epsilon_L \equiv \epsilon$ of (3.6).

contributions that are difficult to disentangle, some of the resonance contributions in electroproduction may be enhanced at higher Q^2 due to their slower form factor falloff compared to other resonances and compared to the background amplitudes. A significant amount of data is available from CLAS [174,175]. In these experiments the electron beam is polarized, and hence the virtual photon also has a net circular polarization.

Figure 3.22 shows samples of the $K^+\Lambda$ production cross sections integrated over either the forward hemisphere (left panel) and backward hemisphere (right panel) at fixed Q^2. Clearly, the angular dependence of electroproduction data at fixed W also show strong forward peaking, indicating significant t-channel mechanisms. This is similar to what has been observed in photoproduction data. The results shown in figure 3.22(right) reveal resonance-like behaviour near $W = 1.7$ GeV and 1.87 GeV at large angles, while at the forward angles (left panel) the resonant structures are masked by the large non-resonant contributions. The enhancements in the cross section appear in the same range of W as in photoproduction and are likely to be due to the same resonances contributions.

The data on Λ recoil polarization have been obtained in measurements with polarized electron beams. The total Λ recoil polarization can be written as

$$\mathbf{P}_\Lambda = \mathbf{P}^0 \pm P_e \mathbf{P}', \qquad (3.100)$$

where P_e is the electron beam polarization, $\mathbf{P}^0$ is the induced polarization, which is present without beam polarization, and $\mathbf{P}'$ is the transferred polarization. Figure 3.23 displays the data of the transferred Λ polarization integrated over all Q^2 for three bins in W. The quantities $P'_{x'}$ and $P'_{z'}$ are the projections of the polarization vector $\mathbf{P}'$ on to the x'- and z'-axes, which are also defined in figure 3.23. The data show that the

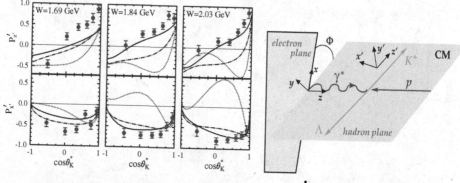

Figure 3.23. Left: transferred Λ polarization in K^+ electroproduction. Right: coordinate system defining the Λ polarization projections. Data, integrated over all ϕ angles, are from CLAS [175]. The curves are predictions from models of [85] (dotted), [90] (solid) and [172] (dot-dashed).

z' polarization is large and is rising with $\cos\theta_K^*$, indicating a t-channel mechanism. On the other hand, the x' polarization is large and remains negative throughout the angular range. None of the theoretical results from tree-diagram models [83–90] or a Regge model [176] can give an adequate description of the data.

To summarize, production of $K^+\Lambda$ and $K^+\Sigma^0$ from protons exhibits evidence of s-channel nucleon resonance contributions in the mass range where no N or $\Delta(1232)$ resonances have been well established. However, resonances are masked by large t-channel processes. In order to extract reliable information on contributing resonances a better understanding of non-resonant processes is needed. Currently, the most important task is to continue experimentally to establish a broad and solid base of consistent data in the strangeness sector, including extensive differential cross sections, beam and target polarization asymmetries and polarization transfer measurements. A 'complete' measurement of all observables, which is needed to extract unambiguously all helicity amplitudes, can be achieved [178,179]. This requires use of a polarized photon beam and of a polarized target and the measurement of the hyperon recoil polarization. Experimental effort in this direction will continue with a series of new measurements planned at JLab. On the theoretical side, a dynamical coupled-channels approach, such as that described in section 3.3.5, must be developed to interpret the resonance parameters extracted from the data.

3.4.7 Photoproduction and electroproduction of vector mesons

Early investigations of photoproduction and electroproduction of vector mesons were mainly in the high-energy region where the data can be explained largely by diffractive pomeron exchange. In the low-energy region, the meson-exchange

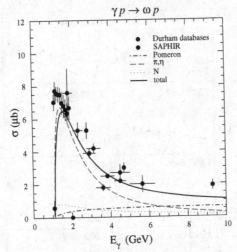

$\gamma p \rightarrow \omega p$

Figure 3.24. Data [180,181] on the $\gamma p \rightarrow \omega p$ total cross section compared with the theoretical results of [92]. The contributions from the exchanges of the pomeron (dot-dashed), π and η (dashed), and nucleon (dotted) are also displayed.

mechanisms play an important role. This is illustrated in figure 3.24, which is taken from the calculation of [92]. We see that the diffractive pomeron-exchange (dash-dotted curve) becomes negligible at energies near the ω photoproduction threshold. The s- and u-channel nucleon terms, and π and η exchanges, account for the main part of the total cross section.

Here we describe only the status of ω photoproduction in the resonance region $W <$ 2.5 GeV, where the excitations of nucleon resonances can be studied. High-quality data in this region have been obtained at ELSA, JLab and GRAAL. As an example, the data from ELSA are shown in figure 3.25. Photoproduction of the ρ will not be discussed since the ρ is broad and the coupling of the ρN channel to N^* states can be meaningfully defined only in an analysis involving two-pion production channels as discussed in section 3.4.5. Photoproduction of the ϕ will also not be covered here since the ϕ meson has little, if any, contribution from s-channel resonances, as the predominantly $s\bar{s}$ quark structure of the ϕ makes $N^* \rightarrow N\phi$ an OZI suppressed decay[182].

Quark models that also couple to hadronic channels predict that ω photoproduction off protons is a promising tool in the search for undiscovered N^* states [126,183]. As in the case of $N\eta$ and $K^+\Lambda$, the $p\omega$ final state, due to the isoscalar nature of the ω, is only sensitive to isospin-$\frac{1}{2}$ N^* resonances. Experimentally, ω production has been measured in both magnetic detectors and neutral-particle detectors. In magnetic detectors the $p\omega$ channel is usually identified through the $\omega \rightarrow \pi^+\pi^-\pi^0$ decay. This channel has an 89% branching ratio. Detectors with large acceptance

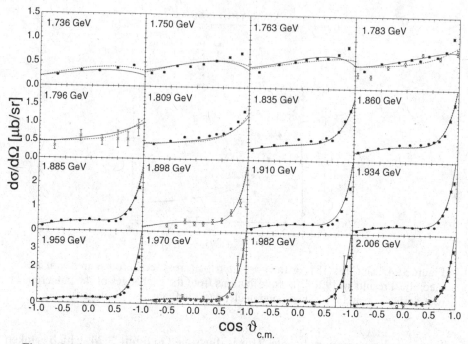

Figure 3.25. Data of differential cross section for $\gamma p \to p\omega$ from ELSA [181]. The curves are from empirical fits to the data.

for the detection of photons allow the $\omega \to \pi^0 \gamma$ channel with an 8.5% branching ratio to be measured.

Theoretical models for investigating low-energy ω production are still being developed. Most of the calculations, such as that displayed in figure 3.24, are based on tree diagrams. It has been recognized that coupled-channels effects must be accounted for before the models can be used reliably to extract resonance parameters from the data. The importance of coupled-channels effects on ω photoproduction has been demonstrated in a one-loop calculation [93] based on the dynamical coupled-channel formulation, and also in a K-matrix coupled-channels model of the Giessen group [62]. More effort is clearly needed to improve these theoretical approaches.

3.5 Concluding remarks and outlook

Very significant progress is being made in the study of N^* physics. We now have fairly extensive data for π, η, K, ω, ϕ and $\pi\pi$ production. Theoretical models for interpreting these new data and/or extracting the N^* parameters have also been developed.

From the analyses of the single-pion data in the $\Delta(1232)$ region, quantitative information about the $\gamma N \to \Delta(1232)$ transition form factors has been obtained. With the development of dynamical reaction models, the role of pion-cloud effects in determining the $\Delta(1232)$ excitation has been identified as the source of the long-standing discrepancy between the data and the constituent-quark-model predictions. Moreover, the Q^2-dependence of the $\gamma N \to \Delta(1232)$ form factors has also been determined up to about $Q^2 \sim 6\,\mathrm{GeV}^2$. The $M1$, $E2$ and $C2\,\gamma^*N \to \Delta(1232)$ transition form factors should be considered along with the proton and neutron form factors as benchmark data for testing various hadron models as well as lattice QCD calculations.

The combined analysis of the π and η production data has led to a quantitative determination of several N^* parameters in the second resonance region. However, a correct interpretation of the N^* parameters in terms of current hadron-model predictions requires a rigorous investigation of the dynamical coupled-channels effects which are not included in amplitude analyses based on either the K-matrix isobar model or the dispersion relation approach.

Analyses of the data for $K\Lambda$, $K\Sigma$, $N\pi\pi$ and $p\omega$ channels are still being developed. So far, most of the analyses are based on tree-diagram models with isobar parametrizations for the resonances. Final state interactions, as required by the unitarity condition, are either neglected completely or calculated perturbatively using effective Lagrangians. The coupled-channels K-matrix effective Lagrangian model, pioneered by the Giessen Group, looks very promising for extracting the resonance parameters from a combined analysis of the data for all channels. However, much work is needed to reduce the uncertainties in their non-resonant parameters and to account for the $\pi\pi N$ unitarity condition. There is also a need for better tracking of systematics and model-dependent uncertainties in such complex fits. For a rigorous interpretation of the resonance parameters in terms of predictions from hadron models or lattice QCD calculations, the analyses based on the dynamical coupled-channels model, as given in section 3.3.5, are indispensable.

Progress made since the 1990s has resulted from close collaboration between experimentalists and theorists. With much more complex data to be analysed and interpreted, such collaborations must be continued and extended in order to bring the study of N^* physics to a complete success.

References

[1] J F Donoghue, E Golowich and B R Holstein, *Dynamics of the Standard Model*, Cambridge University Press, Cambridge (1992)
[2] V D Burkert and T-S H Lee, International Journal of Modern Physics E13 (2004) 1035

[3] H J Rothe, *Lattice Gauge Theories: An Introduction*, World Scientific, Singapore (1997)

[4] D B Leinweber, A W Thomas, K Tsushima and S V Wright, Physical Review D61 (2000) 074502

[5] S Aoki *et al*, Physical Review D60 (1999) 114508

[6] D B Leinweber, D H Lu and A W Thomas, Physical Review D60 (1999) 034014

[7] C Alexandrou *et al*, Physical Review D69 (2004) 114506 ; Physical Review Letters 94 (2005) 021601

[8] A De Rujula, H Georgi, and S L Glashow, Physical Review D12 (1975) 147

[9] N Isgur and G Karl, Physics Letters B72 (1977) 109; Physical Review D19 (1979) 2653; *ibid* D23 (1981) 817

[10] S Capstick and N Isgur, Physical Review D34 (1986) 2809

[11] L Ya Glozman and D O Riska, Physics Reports 268 (1996) 263

[12] L Ya Glozman, Z Rapp and W Plessas, Physics Letters B381 (1996) 311
B F Wagenbrunn *et al*, Physics Letters B511 (2001) 33

[13] A Amghar, B Desplanques and L Theussl, Nuclear Physics A714 (2003) 213

[14] F Coester and D O Riska, Nuclear Physics A728 (2003) 439
B Julia-Diaz, D O Riska and F Coester, Physical Review C69 (2004) 035212

[15] S Boffi *et al*, European Physical Journal A14 (2002) 17

[16] A Chodos *et al*, Physical Review D9 (1974) 3471; *ibid* D10 (1974) 2599

[17] G E Brown and M Rho, Physics Letters 82B (1979) 177
G E Brown, M Rho and V Vento, Physics Letters 94B (1979) 383

[18] S Theberge, A W Thomas and G A Miller, Physical Review D22 (1980) 2838; *ibid* D24 (1981) 216

[19] R Bijker, F Iachello and A Leviatan, Annals of Physics (N.Y.) 236 (1994) 69

[20] R Friedberg and T D Lee, Physical Review D15 (1977) 1694; *ibid* D16 (1977) 1096

[21] H B Neilsen and A Patkos, Nuclear Physics B195 (1982) 137
H J Pirner, Progress in Particle and Nuclear Physics 29 (1992) 33

[22] G Eckart and B Schwesinger, Nuclear Physics A458 (1980) 620

[23] P Maris and C D Roberts, International Journal of Modern Physics E12 (2003) 297

[24] M Kirchbach, Modern Physics Letters A12 (1997) 3177; International Journal of Modern Physics A15 (2000) 1435
M Kirchbach, M Moshinsky and Yu F Smirnov, Physical Review D64 (2001) 114005

[25] Particle Data Group, Physics Letters B592 (2004) 1

[26] A W Thomas and W Weise, *The Structure of the Nucleon*, Wiley 2001

[27] S Capstick and W Roberts, Progress in Particle and Nuclear Physics 45 (2000) S241

[28] J D Bjorken and S D Drell, *Relativistic Quantum Mechanics*, McGraw-Hill, New York (1964)

[29] S Nozawa and T-S H Lee, Nuclear Physics A513 (1990) 511

[30] W-T Chiang, S N Yang, L Tiator and D Drechsel, Nuclear Physics A700 (2002) 429

[31] G Knochlein, D Drechsel and L Tiator, Zeischrift für Physik A352 (1995) 327

[32] M Pichowsky, C Savakli and F Tabakin, Physical Review C53 (1996) 593

[33] P Dennery, Physical Review 124 (1961) 2000

[34] K M Watson, Physical Review 85 (1952) 852

[35] K M Watson, Physical Review 95 (1954) 226

[36] F Chew, M Goldberger, F E Low and Y Nambu, Physical Review 106 (1957) 1345

[37] J S Ball, Physical Review 124 (1961) 2014

[38] A Donnachie and G Shaw, Annals of Physics (N.Y.) 37 (1966) 333; Nuclear Physics 87 (1967) 556

[39] G Shaw, Il Nuovo Cimento 44 (1966) 1276

[40] D Schwela, H Rollnik, R Weizel and W Korth, Zeitschrift für Physik 202 (1967) 452

[41] F A Berends, A Donnachie and D L Weaver, Nuclear Physics B4 (1967) 1, 54, 103

[42] D Schwela and R Weizel, Zeitschrift für Physik 221 (1969) 71

[43] F A Berends and A Donnachie, Physics Letters 25B (1967) 278

[44] A Donnachie, in *High Energy Physics*, E H S Burhop ed, Academic Press, New York (1972)

[45] G von Gehlen, Nuclear Physics B9 (1969) 17; *ibid* B20 (1970) 102

[46] R L Crawford, Nuclear Physics B28 (1971) 573

[47] R C E Devenish and D H Lyth, Physical Review D5 (1972) 47; Nuclear Physics B43 (1972) 228; *ibid* B97 (1975) 109

[48] A I Sanda and G Shaw, Physical Review Letters 24 (1970) 1310

[49] A Donnachie and G Shaw, Physical Review D5 (1972) 1117

[50] O Hanstein, D Drechsel and L Tiator, Nuclear Physics A632 (1998) 561

[51] I G Aznauryan, Physical Review C57 (1998) 2727

[52] I G Aznauryan, Physical Review C67 (2003) 015209

[53] I G Aznauryan, Physical Review C68 (2003) 065204

[54] R L Walker, Physical Review 182 (1969) 1729

[55] R C E Devenish, and D H Lyth, Nuclear Physics B94 (1975) 109
R G Moorhouse in *Electromagnetic Interactions of Hadrons*, volume 1, A Donnachie and G Shaw eds, Plenum Press, New York (1978)

[56] M G Olsson and T Osypowski, Nuclear Physics B87 (1975) 399; Physical Review D17 (1978) 174

[57] R M Davidson and N C Mukhopadhyay, Physical Review D42 (1990) 20
R M Davidson, N C Mukhopadhyay and R S Wittman, Physical Review D43 (1991) 71

[58] R A Arndt, I I Strakovsky, R L Workman and M M Pavan, Physical Review C52 (1995) 2120
R A Arndt, I I Strakovsky and R L Workman, Physical Review 53 (1996) 430; International Journal of Modern Physics A18 (2003) 449

[59] D Drechsel, O Hanstein, S S Kamalov and L Tiator, Nuclear Physics A645 (1999) 145

[60] D M Manley, International Journal of Modern Physics A18 (2003) 441

[61] D M Manley and E M Saleski, Physical Review D45 (1992) 4002

[62] U Feuster and U Mosel, Physical Review C58 (1998) 457; *ibid* C59 (1999) 460
G Penner and U Mosel, Physical Review C66 (2002) 055211; *ibid* C66 (2002) 055212

[63] R E Cutkosky, C P Forsyth, R E Hendrick and R L Kelly, Physical Review D20 (1979) 2839

[64] M Batinic, I Slaus and A Svarc, Physical Review C51 (1995) 2310

[65] T P Vrana, S A Dytman and T-S H Lee, Physics Reports 328 (2000) 181

[66] H Tanabe and K Ohta, Physical Review C31 (1985) 1876

[67] S N Yang, Journal of Physics G11 (1985) L200

[68] S Nozawa, B Blankenleider and T-S H Lee, Nuclear Physics A513 (1990) 459

[69] F Gross and Y Surya, Physical Review C47 (1993) 703
Y Surya and F Gross, Physical Review C53 (1996) 2422

[70] T Sato and T-S H Lee, Physical Review C54 (1996) 2660

[71] T Yoshimoto, T Sato, M Arima and T-S H Lee, Physical Review C61 (2000) 065203

[72] T Sato and T-S H Lee, Physical Review C63 (2001) 055201

[73] S S Kamalov and S N Yang, Physical Review Letters 83 (1999) 4494

[74] G-Y Chen *et al*, Nuclear Physics A723 (2003) 447

[75] S S Kamalov *et al*, Physical Review C64 (2001) 032201(R)

[76] W-T Chiang, F Tabakin, T-S H Lee and B Saghai, Physics Letters B517 (2001) 101

[77] M G Fuda and H Alharbi, Physical Review C68 (2003) 064002

[78] T-S H Lee, A Matsuyama and T Sato, in *Proceedings of the Workshop on Physics of Excited Nucleons*, J-P Bocquet, V Kuznetsov and D Rebreyend eds, World Scientific, Singapore (2004)

[79] N Kaiser, P B Siegel and W Weise, Nuclear Physics A594 (1995) 325
N Kaiser, T Waas, and W Weise, Nuclear Physics A612 (1997) 297

[80] E Oset and A Ramos, Nuclear Physics A635 (1998) 99

[81] M F M Lutz and E E Kolomeitsev, Nuclear Physics A700 (2002) 193

[82] M L Goldberger and K M Watson, *Collision Theory*, Robert E. Krieger Publishing Company (1975)

[83] R A Adelseck, C Bennhold and L E Wright, Physical Review C32 (1985) 1681

[84] R A Adeleck and B Saghai, Physical Review C42 (1990) 108

[85] R A Williams, C R Ji and S R Cotanch, Physical Review C46 (1992) 1667

[86] J C David, C Fayard, G H Lamot and B Saghai, Physical Review C53 (1996) 2613

[87] Z Li and B Saghai, Nuclear Physics A644 (1998) 345

[88] H Haberzettl, C Bennhold and T Mart, Physical Review C58 (1998) 40

[89] T Mart and C Bennhold, Physical Review C61 (2000) 12201

[90] S Janssen, J Ryckebusch, D Debruyne and T V Cauteren, Physical Review C65 (2001) 015201

[91] Q Zhao, Z Li, and C Bennhold, Physical Review C58 (1998) 2393
Q Zhao, Physical Review C63 (2001) 025203

[92] Y Oh, A Titov and T-S H Lee, Physical Review C63 (2001) 025201

[93] Y Oh and T-S H Lee, Physical Review C66 (2002) 045201

[94] J C Nacher and E Oset, Nuclear Physics A674 (2000) 205

[95] M Guidal, J M Laget and M Vanderhaeghen, Nuclear Physics A627 (1997) 645

[96] W-T Chiang *et al*, Physical Review C68 (2003) 045202

[97] H Feshbach, *Theoretical Nuclear Physics: Nuclear Reactions*, Wiley, New York (1992)

[98] C T Hung, S N Yang and T-S H Lee, Physical Review C64 (2001) 034309

[99] T deForest and J D Walecka, Advances in Phyics 15 (1966) 1

[100] E Amaldi, S Fubini and G Furlan, *Springer Tracts in Modern Physics* 83 (1979)

[101] R J Eden, P V Landshoff, D I Olive and J C Polkinghorne, *The Analytic S-Matrix* Cambridge University Press, Cambridge (1966)

[102] A Donnachie and G Shaw in *Electromagnetic Interactions of Hadrons*, volume 1, A Donnachie and G Shaw eds, Plenum Press, New York 1978

[103] R Omnes, Il Nuovo Cimento 8 (1958) 316

[104] F Foster and G Hughes, Reports on Progess in Physics 46 (1983) 1445

[105] A D'Angelo *et al*, GRAAL collaboration, in *Proceedings International Conference on the Structure of Baryons*, C Carlson and B Mecking eds, World Scientific, Singapore (2002)

[106] R Beck *et al*, Physical Review Letters 78 (1997) 606
R Beck *et al*, Physical Review C61 (2000) 035204

[107] G Blanpied *et al*, Physical Review C61 (2000) 024604

[108] J Ajaka *et al*, Physics Letters B475 (2000) 372
O Bartalini *et al*, Physics Letters B544 (2002) 113

[109] J Ahrens *et al*, Physical Review Letters 84 (2000) 5950

[110] J Ahrens *et al*, Physical Review Letters 88 (2002) 232002

[111] V Frolov *et al*, Physical Review Letters 82 (1999) 45

[112] H Egiyan *et al*, CLAS Collaboration, *Proceedings Workshop on the Physics of Excited Nucleons*, S A Dytman and E S Swanson eds, World Scientific, Singapore (2002)

[113] K Joo *et al*, CLAS collaboration, Physical Review Letters 88 (2002) 122001

[114] K Joo *et al*, CLAS collaboration, Physical Review C68 (2003) 032201

[115] G Laveissière *et al*, Physical Review C69 (2004) 045203

[116] K Joo *et al*, CLAS collaboration, *Proceedings International Conference on the Structure of Baryons*, C Carlson and B Mecking eds, World Scientific, Singapore (2002)

[117] P Bartsch *et al*, Physical Review Letters 88 (2002) 142001

[118] C Kunz *et al*, OOPS Collaboration, Physics Letters B564 (2003) 21

[119] R Gothe, *Proceedings Workshop on the Physics of Excited Nucleons*, S A Dytman and E S Swanson eds, World Scientific, Singapore (2002)

[120] P Stoler *et al*, JLab experiment E99-107

[121] P Stoler *et al*, JLab experiment E01-002

[122] S Frullani *et al*, JLab experiment E99-011

[123] R De Vita *et al*, Physical Review Letters 88 (2002) 082001-1

[124] A Biselli *et al*, Physical Review C68 (2003) 035202

[125] A Arndt *et al*, *Proceedings of the Workshop on the Physics of Excited Nucleons*, D Drechsel and L Tiator eds, World Scientific, Singapore (2001)

[126] S Capstick, Physical Review C46 (1992) 2864

[127] C S Armstrong *et al*, Physical Review D60 (1999) 052004

[128] M Dugger *et al*, CLAS collaboration, Physical Review Letters 89 (2002) 222002

[129] V Crede *et al*, CB-ELSA collaboration, Physical Review Letters 94 (2005) 012004

[130] D Rebreyend, *Proceedings of NSTAR 2002*, S A Dytman and E S Swanson eds, World Scientific, Singapore (2002)

[131] V Kuznetsov, private communication.

[132] J Ahrens *et al*, A2 collaboration, European Physical Journal A17 (2003) 241

[133] B Krusche *et al*, Physical Review Letters 74 (1995) 3736

[134] R Thompson *et al*, CLAS collaboration, Physical Review Letters 86 (2001) 1702

[135] F Renard *et al*, Physics Letters B258 (2002) 215

[136] J Ajaka *et al*, Physical Review Letters 81 (1998) 1797

[137] H Denizli *et al*, *Proceedings International Conference on the Structure of Baryons*, C Carlson and B Mecking eds, World Scientific, Singapore (2002)

[138] A Bock *et al*, Physical Review Letters 81 (1998) 534
      J Price *et al*, Physical Review C51 (1995) 2283

[139] N Kaiser, P B Siegel and W Weise, Physics Letters B362 (1995) 23

[140] W-T Chiang *et al*, Physical Review C68 (2003) 045202

[141] B Saghai and Z P Li, European Physical Journal A11 (2001) 217

[142] J-C Alder *et al*, Nuclear Physics B91 (1975) 386

[143] W Brasse *et al*, Nuclear Physics B139 (1978) 37; Zeitschrift für Physik C22 (1984) 33

[144] U Beck *et al*, Physics Letters 51B (1974) 103

[145] H Breuker *et al*, Physics Letters 74B (1978) 409

[146] I G Aznauryan, *et al*, Physical Review C71 (2005) 015201

[147] Z P Li, V Burkert and Zh Li, Physical Review D46 (1992) 70

[148] M Warns, H Schröder, W Pfeil and H Rollnik, Zeitschrift für Physik C45 (1990) 627
[149] S Capstick and B D Keister, Physical Review D51 (1995) 3598
[150] M Giannini, E Santopinto and A Vassallo, Progress in Particle and Nuclear Physics 50 (2003) 263
[151] F Cano and P Gonzalez, Physics Letters B431 (1998) 270
[152] F Cardarelli and S Simula, Physical Review Letters 62 (2000) 06520
[153] F E Close and Z P Li, Physical Review D42 (1990) 2194
[154] F E Close, *An Introduction to Quarks and Partons*, Academic Press, London (1979)
[155] F E Close and F J Gilman, Physics Letters 38B (1972) 541
[156] L A Copley, G Karl and E Obryk, Physics Letters 29B (1969) 117
[157] P Koniuk and N Isgur, Physical Review D21 (1980) 1868
[158] A J G Hey and J Weyers, Physics Letters 48B (1974) 69
[159] W N Cottingham and I H Dunbar, Zeitschrift für Physik C2 (1979) 41
[160] V Burkert *et al*, Physical Review C67 (2003) 035204
[161] U Thoma, *Proceedings Workshop on the Physics of Excited Nucleons*, S A Dytman and E S Swanson eds, World Scientific, Singapore (2002)
[162] M Ripani *et al*, Physical Review Letters 91 (2003) 022002
[163] M. Bellis *et al*, CLAS collaboration, *Proceedings of NSTAR 2004*, D Rebreyend, J-P Bocquet and V Kuznetsov eds, World Scientific, Singapore (2004)
[164] M Wolf *et al*, European Physics Journal 9 (2000) 5
[165] Y Assafiri *et al*, Physical Review Letters 90 (2003) 222001
[166] F Harter *et al*, Physics Letters B401 (1997) 229
[167] V Mokeev *et al*, Physics of Atomic Nuclei 64 (2001) 1292
[168] S Capstick and W Roberts, Physical Review D58 (1998) 074011
[169] V Burkert *et al*, Physics of Atomic Nuclei 66 (2003) 2149
[170] V I Mokeev *et al*, *Proceedings of NSTAR 2004*, D Rebreyend, J-P Bocquet and V Kuznetsov eds, World Scientific, Singapore (2004)
[171] M Q Tran *et al*, Physics Letters B445 (1998) 20
[172] K H Glander *et al*, European Physical Journal A19 (2004) 251
[173] J W C McNabb *et al*, CLAS Collaboration, Physical Review C69 2004) 042201
[174] R Feuerbach *et al*, *Proceedings International Conference on the Structure of Baryons*, C Carlson and B Mecking eds, World Scientific, Singapore (2002); M D Mestayer *et al*, *Proceedings PANIC'99*, G Faeldt, B Höistad and S Kullander eds, North-Holland, Amsterdam (1999)
[175] D S Carman *et al*, Physical Review Letters 90 (2003) 131804
[176] M Guidal, J M Laget and M Vanderhaeghen, Physical Review C68 (2003) 058201
[177] D G Ireland, S Janssen and J Ryckebusch, Physics Letters B562 (2003) 51
[178] I S Barker, A Donnachie and J K Storrow, Nuclear Physics B79 (1974) 431
[179] I S Barker, A Donnachie and J K Storrow, Nuclear Physics B95 (1975) 347
[180] J Ballam *et al*, Physical Review D7 (1973) 3150; W Struczinski *et al*, Nuclear Physics B108 (1976) 45; D P Barber *et al*, Zeitschrift für Physik. C 26 (1984) 343
[181] F J Klein *et al*, πN Newsletter 14 (1998) 141
[182] S Okobo, Physics Letters 5 (1966) 1173
G Zweig, CERN report TH-412, Geneva (1964)
G Zweig in Developments in the Quark Theory of Hadrons volume 1, DB Lichtenberg and S P Rosen eds, Hadronic Press, Nomantum (1980)
J Iizuka, Progress in Theoretical Physics Supplement 37/38 (1966) 21
[183] S Capstick and W Roberts, Physical Review C49 (1994) 4570; *ibid* D57 (1998) 4301

4

Meson radiative decays

F E Close, A Donnachie and Yu S Kalashnikova

Three aspects of meson radiative decays are discussed in this chapter. The first, electron–positron annihilation to hadrons, is aimed primarily at the study of the light-quark vector mesons, both isovector and isoscalar, to determine their spectrum, decay modes and nature. An important byproduct is the evaluation of the contribution from hadronic vacuum polarization to the muon magnetic moment. Directly related to the isovector states, through the assumption of conserved vector current, are the semileptonic decays of the τ and hadronic decays of B mesons. The second topic deals with single-photon transitions between two mesons $A \to \gamma B$ that are a much cleaner probe of the meson wave functions than are hadronic decays. The direct coupling of the photon to the charges and spins of constituents makes it possible for single-photon decays to discriminate among different models for mesons. In particular, they can be very relevant for distinguishing hybrids and glueballs from conventional $q\bar{q}$ excitations. Two-photon decays of mesons, discussed in the third section, provide an even more powerful probe of meson structure and a qualitative distinction between glueballs and $q\bar{q}$ excitations, although restricted to a more limited set of states than are single-photon decays.

4.1 Electron–positron annihilation and τ decay

There are two principal fields of interest in e^+e^- annihilation and the corresponding channels in τ decay. These are the spectroscopy and decays of the vector-mesons and the contribution from the total e^+e^- annihilation cross section to vacuum polarization integrals relevant for Standard Model calculations of the muon anomalous magnetic moment.

To the extent that the conserved vector current (CVC) hypothesis is correct, the isovector part of $\sigma(e^+e^- \to \text{hadrons})$ is related to the corresponding τ decay by an

Electromagnetic Interactions and Hadronic Structure, eds Frank Close, Sandy Donnachie and Graham Shaw.
Published by Cambridge University Press. © Cambridge University Press 2007

isospin rotation [1]. The e^+e^- annihilation cross section to a final hadronic state X may be written in terms of a spectral function $v_0^X(s)$ as

$$\sigma(e^+e^- \to X) = \frac{4\pi\alpha^2}{s} v_0^X(s), \tag{4.1}$$

where $s \equiv M_X^2$ is the square of the e^+e^- centre-of-mass energy. For the decay $\tau^- \to X^-\nu_\tau$,

$$\frac{d\Gamma}{ds} = \frac{G_F^2|V_{ud}|^2 S_{EW}}{32\pi^2 M_\tau^3}(M_\tau^2 - s)^2(M_\tau^2 + 2s)v_-^X(s), \tag{4.2}$$

where G_F is the Fermi constant, $|V_{ud}|$ is the Cabibbo–Kobayashi–Maskawa (CKM) weak mixing-matrix element, S_{EW} accounts for electroweak radiative corrections [2] and M_τ is the mass of the τ. The assumption of CVC then implies

$$v_0^X = v_-^X, \tag{4.3}$$

where v_-^X is the τ-decay spectral function with an appropriate combination of final states. For example,

$$\sigma(e^+e^- \to \pi^+\pi^-) = \frac{4\pi\alpha^2}{s} v^{\pi^-\pi^0}, \tag{4.4}$$

$$\sigma(e^+e^- \to \pi^+\pi^-\pi^+\pi^-) = \frac{4\pi\alpha^2}{s} v^{\pi^-3\pi^0}, \tag{4.5}$$

$$\sigma(e^+e^- \to \pi^+\pi^-\pi^0\pi^0) = \frac{4\pi\alpha^2}{s}(v^{2\pi^-\pi^+\pi^0} - v^{\pi^-3\pi^0}). \tag{4.6}$$

The reaction $e^+e^- \to \gamma + X$, where the photon emission is caused by initial-state radiation (ISR) and X is a hadronic final state, can be used [3–5] to measure the cross section for $e^+e^- \to X$. The ISR cross section for a particular hadronic final state X is related to the corresponding e^+e^- annihilation cross section by

$$\frac{d\sigma_{e^+e^- \to \gamma X}(s, x)}{dx} = W(s, x)\sigma_{e^+e^- \to X}(s(1 - x)), \tag{4.7}$$

where $x = 2E_\gamma/\sqrt{s}$, E_γ is the energy of the ISR photon in the e^+e^- centre-of-mass frame, $\sqrt{s}$ is the e^+e^- centre-of-mass energy and $\sqrt{s(1 - x)}$ is the mass of the final state X. The function $W(s, x)$ describes the energy dependence of the ISR photons and is given by [5]

$$W(s, x) = \beta[(1 + \delta)x^{(\beta-1)} - 1 + \tfrac{1}{2}x], \tag{4.8}$$

where

$$\beta = \frac{2\alpha}{\pi}[2\log(\sqrt{s}/m_e) - 1] \tag{4.9}$$

and δ takes into account vertex and self-energy corrections.

An advantage of ISR data is that an entire range of centre-of-mass energies is scanned in one experiment, avoiding relative normalization uncertainties. A disadvantage is that the mass resolution is less good than that obtained in direct annihilation.

The vector-meson spectrum can also be accessed via appropriate hadronic B-meson decays, $\bar{B} \to D^{(*)}X$, where X is a light-quark hadronic system. The particular case of $X = \omega\pi^-$ has been observed [6]. Conversely, knowledge of the vector-meson spectrum and decay modes can be used to study the mechanism of factorization in B decays [7].

4.1.1 The ρ, ω and ϕ

For a given final state X, the e^+e^- cross section at a centre-of-mass energy $\sqrt{s}$ produced via a single Breit–Wigner resonance of mass m_V is

$$\sigma_{e^+e^- \to X}(s) = 4\pi\alpha^2 \frac{4\pi}{\gamma_V^2} \frac{m_V^2}{s} \frac{m_V \Gamma_X(s)}{(s - m_V^2)^2 + m_V^2 \Gamma_{tot}^2(s)}, \tag{4.10}$$

where $\Gamma_X(s)$ is the partial width for the final state X, $\Gamma_{tot}(s)$ is the total width and $4\pi/\gamma_V^2$ is a measure of the strength of the radiative decay width

$$\Gamma_{V \to e^+e^-} = \frac{m_V \alpha^2}{3} \frac{4\pi}{\gamma_V^2}. \tag{4.11}$$

Hence (4.10) may be rewritten as

$$\sigma_{e^+e^- \to X}(s) = \frac{12\pi}{s} \frac{(m_V \Gamma_{e^+e^-})(m_V \Gamma_X(s))}{(s - m_V^2)^2 + m_V^2 \Gamma_{tot}^2(s)}. \tag{4.12}$$

The radiative decay width is determined by the wave function of the $q\bar{q}$ bound state at the origin, $\psi(0, m_V)$, by

$$\Gamma_{V \to e^+e^-} = \frac{16\pi\alpha^2}{m_V^2} C_V^2 |\psi(0, m_V)|^2, \tag{4.13}$$

where C_V is the mean electric charge of the valence quarks inside the vector meson: $C_\rho^2 = \frac{1}{2}$, $C_\omega^2 = \frac{1}{18}$, $C_\phi^2 = \frac{1}{9}$, $C_{J/\psi}^2 = \frac{4}{9}$, $C_\Upsilon^2 = \frac{1}{9}$. If the bound state is given by a Coulomb-like non-relativistic potential, then $|\psi(0, m_V)|^2 \propto m_V^3$ as, in general, $|\psi(0, m_V)|^2 \propto m_V^{3/(2+n)}$ for $V(r) \propto r^n$. Then (4.13) does not give the correct ratios of widths which are [8]

$$\Gamma_{\rho \to e^+e^-} : \Gamma_{\omega \to e^+e^-} : \Gamma_{\phi \to e^+e^-} : \Gamma_{J/\psi \to e^+e^-} : \Gamma_{\Upsilon \to e^+e^-} =$$
$$10.68 \pm 0.29 : 0.913 \pm 0.036 : 1.93 \pm 0.074 : 8.22 \pm 0.31 : 2. \tag{4.14}$$

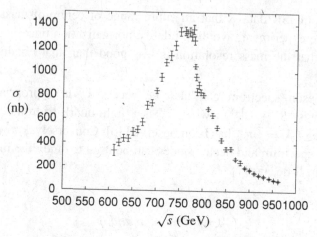

Figure 4.1. Cross section for $e^+e^- \rightarrow \pi^+\pi^-$. The data are from [9].

To obtain the ratios (4.14) it is necessary to assume that $|\psi(0, m_V)|^2 \propto m_V^2$ in which case the ratios of widths becomes 9:1:2:8:2 in the absence of mixing, in good agreement with (4.14).

The cross section for the reaction $e^+e^- \rightarrow \pi^+\pi^-$ is dominated by the ρ, although the line-shape does not correspond to a relativistic Breit–Wigner with a conventional P-wave width. An additional shape parameter is required and this, together with the large width of the ρ, creates difficulties in determining the resonance parameters. Another effect on the cross section comes from interference with the isospin-violating $\pi^+\pi^-$ decay of the ω. Although the $\pi^+\pi^-$ branching fraction of the ω is small, less than 2% [8,9], the interference results in a significant distortion of the line shape, as can be seen in figure 4.1. The sharp drop in the cross section at the ω mass is clearly visible. Conversely there must be an isospin-violating $\pi^+\pi^-\pi^0$ decay of the ρ, although this is much more difficult to measure as, because of the large ρ width, the interference term appears as a small background under the ω peak, giving a slight distortion of the ω line-shape. The branching fraction has been determined [10] as $B_{\rho \rightarrow 3\pi} = (1.0^{+0.54}_{-0.36} \pm 0.34)10^{-4}$.

A third topic of interest is the $\pi^+\pi^-\pi^0$ decay of the ϕ. Because of the Okuba–Zweig–Iizuka (OZI) rule [11], which implies suppression of this decay mode, the conventional view is that it is due to ω–ϕ mixing. That is

$$|\phi\rangle \approx |\phi^0\rangle + \epsilon_{\phi\omega}|\omega^0\rangle, \quad |\phi^0\rangle = s\bar{s}, \quad |\omega^0\rangle = \frac{1}{\sqrt{2}}(u\bar{u} + d\bar{d}), \qquad (4.15)$$

where $|\phi^0\rangle$ and $|\omega^0\rangle$ are the unmixed states and $\epsilon_{\phi\omega}$ is the ω–ϕ mixing parameter. An alternative to ω–ϕ mixing is direct decay. A particularly sensitive test is provided [12] by the ratio $\Gamma(\phi \rightarrow e^+e^-)/\Gamma(\omega \rightarrow e^+e^-)$. If the 3π decay of the ϕ is due entirely to mixing, then $\epsilon_{\phi\omega} \approx 0.05$. However, the preferred value appears to be $\epsilon_{\phi\omega} \approx 0.015$ [10], indicating that the direct transition is the principal mechanism.

4.1.2 The continuum: vector-meson spectroscopy

The existence of the isovectors $\rho(1450)$ and $\rho(1700)$ and their isoscalar counterparts $\omega(1420)$ and $\omega(1600)$ appears to be well established [8]. Following the original suggestion [13] from a theoretical analysis on the consistency of the 2π and 4π electromagnetic form factors and the $\pi\pi$ scattering length, a full analysis [14] of the 2π and 4π channels in e^+e^- annihilation and photoproduction reactions confirmed the conjecture that there are two isovector vector states in this mass region and determined their masses. The existence of the $\rho(1450)$ was supported by the analysis [15] of the $\eta\rho^0$ mass spectra obtained in photoproduction and e^+e^- annihilation. Numerous subsequent analyses, summarized in [8], of the 2π, 4π and $\eta\rho^0$ channels in e^+e^- annihilation and τ decay, confirm the earlier conclusions with a caveat on the mass of the $\rho(1450)$ that we discuss below. The evidence for the $\rho(1450)$ and the $\rho(1700)$ comes primarily from the $\pi^+\pi^-$ and $\omega\pi^0$ channels in e^+e^- annihilation and the corresponding charged channels in τ decay. The data on $e^+e^- \to \pi^+\pi^-\pi^+\pi^-$ and $e^+e^- \to \pi^+\pi^-\pi^0\pi^0$ (excluding $\omega\pi^0$) and the corresponding charged channels in τ decay are compatible with the two-resonance interpretation [16–18] but do not provide such good discrimination, despite these being the two major decay channels in the continuum. The reason is that in $\pi\pi$ and $\omega\pi$ there is strong interference with the tail of the ρ which is absent in the other two channels. The $\pi\pi$ data show the interference quite unambiguously, as can be seen in figure 4.2. There is a clear shoulder below 1.5 GeV from the interference of the $\rho(1450)$ with the (primarily real) high-mass tail of the ρ followed by destructive interference with the $\rho(1700)$ and then possibly constructive interference with yet another state at higher mass. In contrast the 4π data, an example of which is given in figure 4.3, are rather featureless with no detailed structure.

The emphasis has been on the isovector states for several reasons. The two-body $\pi\pi$ and $\omega\pi$ channels have no counterpart in the isoscalar sector, the cross sections for the excitation of the $\rho(1450)$ and the $\rho(1700)$ are an order of magnitude larger than those for the $\omega(1420)$ and $\omega(1650)$ and τ decay provides an additional data source with comparable statistics to e^+e^- annihilation but very different systematics. The data available for the study of the $\omega(1420)$ and $\omega(1650)$ are $e^+e^- \to \pi^+\pi^-\pi^0$ (which is dominated by $\rho\pi$) and $\omega\pi^+\pi^-$. The latter cross section shows a clear peak that is apparently dominated by the $\omega(1650)$. The e^+e^- data for the former cross section show little structure, but are appreciably larger than estimated from the tails of the ω and ϕ, indicating that an additional contribution was required.

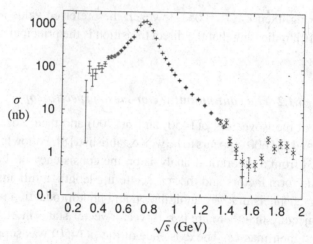

Figure 4.2. Cross section for $e^+e^- \to \pi^+\pi^-$. The data are from [19] (crosses) and [20] (bars).

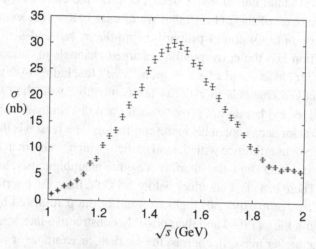

Figure 4.3. Cross section for $e^+e^- \to \pi^+\pi^-\pi^+\pi^-$. The data are from [18].

A best fit was obtained with two states [16], although a fit with only the $\omega(1650)$ could not be completely excluded. However ISR data [21] are quite unambiguous with two clear resonance peaks, see figure 4.4.

There is an interesting structure in $e^+e^- \to 6\pi$ near 1.9 GeV, as shown in the combination $\sigma(e^+e^- \to 2\pi^+2\pi^-2\pi^0) + \sigma(e^+e^- \to 3\pi^+3\pi^-)$ in figure 4.5. A good fit to these data is obtained [24] with two resonances, one of mass about 1.78 GeV and one of mass about 2.11 GeV. The latter is a candidate for another resonance, but the former could be the known $\rho(1700)$ with the resonance peak distorted

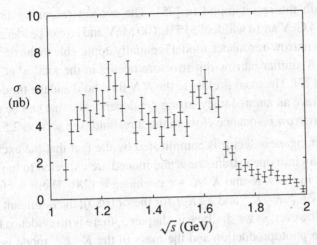

Figure 4.4. Cross section for $e^+e^- \to \pi^+\pi^-\pi^0$. The data are from [21].

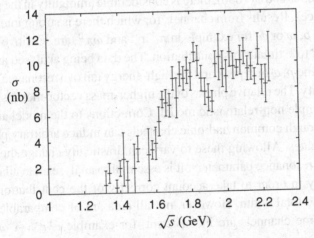

Figure 4.5. Combined cross sections $\sigma(e^+e^- \to 2\pi^+2\pi^-2\pi^0) + \sigma(e^+e^- \to 3\pi^+3\pi^-)$. The data are from [22] and [23], quoted in [24].

by threshold effects, causing an apparent increase in the mass. Similar structure is found [25,26] in diffractive photoproduction of 6π states, although the interpretation put on the data is different. A satisfactory fit can be obtained assuming that the dip is due to the interference of a narrow resonance either with a continuum [25] or with a continuum plus a second resonance [26]. In the latter case the narrow resonance has a mass of 1.91 ± 0.01 GeV and a width of 37 ± 13 MeV, consistent with the result found in [25]. The second resonance has a mass of $1.730 \pm 0.034 \, \text{GeV}$ and a width of $315 \pm 100 \, \text{MeV}$ and so can be identified with the $\rho(1700)$. This narrow-resonance model is equally applicable to the 6π data in e^+e^- annihilation. A similar narrow-dip structure occurs in the total e^+e^- annihilation cross section [27]. The mass is close to the $N\bar{N}$ threshold and the time-like nucleon form factors have an anomalously strong Q^2 dependence that can be explained by postulating a narrow resonance close to the threshold: see section 2.5.

The hidden-strangeness sector is complicated by the fact that the excited ρ and ω states can decay into open strangeness and indeed are expected to have significant partial widths in the $K\bar{K}$ and $K^*\bar{K} + $ c.c. channels [28]. There is clear evidence in $e^+e^- \rightarrow K_L^0 K_S^0$, K^+K^- and $K_S^0 K\pi$ for the $\phi(1680)$, the dominant decay mode being KK^*. However, as we shall see in chapter 5, there is no evidence for this latter decay mode in photoproduction and the mass in the K^+K^- mode is appreciably higher, at 1750 MeV [29].

Although there is general consensus on the existence of the $\rho(1450)$, $\rho(1700)$, $\omega(1420)$, $\omega(1650)$ and $\phi(1680)$, there is considerable ambiguity in the parameters of these resonances. Results from channels for which there is strong interference with the tail of the ρ, ω or ϕ, for example in $\pi^+\pi^-$ and $\omega\pi^0$, are sensitive to the choice of model used to estimate this contribution. The data being analysed are many half-widths above the ρ, ω or ϕ peak and the high-energy tail of a resonance is a rather ill-defined quantity. The relative phases of the higher-mass vector-mesons are unknown except in a simple non-relativistic model. Corrections to the model and mixing of bare states through common hadronic channels can induce arbitrary phases among the physical states. Allowing these to vary can drastically change the interference and hence the resonance parameters. It is essential to analyse all available channels simultaneously in order to take account correctly of the contribution of opening channels to the total width. However, not all data are of comparable quality and the data in some channels are inconsistent, for example $e^+e^- \rightarrow \omega\pi$ shown in figure 4.6.

Independent evidence for both the $\rho(1450)$ and $\rho(1700)$ in their 2π and 4π decay modes has come from the study of the reactions $\bar{p}n \rightarrow \pi^-\pi^0\pi^0$ [33] and $\bar{p}n \rightarrow \pi^-4\pi^0$, $\bar{p}n \rightarrow 2\pi^-2\pi^-\pi^+$ [34]. However, in a reanalysis [35] of the CERN–Munich data [36] for the reaction $\pi^-p \rightarrow \pi^-\pi^+n$ at 17.2 GeV beam energy, no significant evidence of the $\rho(1450)$ was found, but a definite contribution was required from the $\rho(1700)$. There is also clear evidence [37] for the $\omega(1650)$ in the reaction $\pi^-p \rightarrow \omega\eta n$ at a beam energy of 18 GeV.

The initial explanation of these vector-meson states was that they are the first radial, 2^3S_1, and first orbital, 1^3D_1, excitations of the ρ and ω and the first radial excitation of the ϕ as the masses are close to those predicted by the quark model [38]. However,

Channel	$\pi\pi$	$\pi\omega$	$\rho\eta$	πh_1	πa_1	$\rho\rho$	$\rho(\pi\pi)_S$	Other	Total
ρ_{2S}	68	115	18	1	3	10	1	80	295
ρ_{1D}	27	23	13	104	105	6	0	137	415

Table 4.1. The $^3P_0$ partial widths for ρ_{2S} and ρ_{1D}.

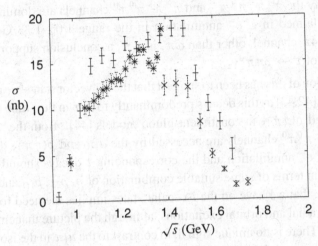

Figure 4.6. Cross section for $e^+e^- \rightarrow \omega\pi^0$. The data are from [30] (crosses), [31] (stars) and [32] (bars).

the data on the 4π channels do not appear to be compatible with what is expected for the isovector radial and orbital excitations of the ρ. Of course, this is a model-dependent statement as it assumes that the hadronic decays of the vector $q\bar{q}$ can be predicted. The standard decay model, the $^3P_0$ model [28,39], does appear to allow this with some accuracy. A systematic study [28] of known light-$q\bar{q}$ decays shows that a $^3P_0$-type amplitude dominates and calculated widths agree with data to 25–40%. Assuming that their masses are 1.45 and 1.7 GeV, the $^3P_0$ partial widths for ρ_{2S} and ρ_{1D} are given in table 4.1, where 'other' includes $K\bar{K}$, $K^*\bar{K}$+ c.c. and 6π channels, and the σ is the broad S-wave $\pi\pi$ enhancement. Altogether 16 channels have been incorporated in the calculation [28,40].

From table 4.1 one can see that the 4π decays of the ρ_{2S}, other than $\omega\pi$, are negligible and so the ρ_{2S} effectively makes no contribution to the 4π channel. In contrast, the 4π decays of the ρ_{1D} are large and the two dominant ones, $a_1\pi$ and $h_1\pi$, are comparable. However, in the non-relativistic limit the e^+e^- width of the 1^3D_1 state vanishes. Some non-zero width will be created by relativistic corrections [38], but this is expected to be small, so again one does not anticipate a large contribution to the 4π channel. This is in direct contrast to the data, as

Channel	$\rho\pi$	$\omega\eta$	$b_1\pi$	$\omega(\pi\pi_S)$	Other	Total
ω_{2S}	328	12	1	8	36	385
ω_{1D}	101	13	371	0	53	561

Table 4.2. The $^3P_0$ widths for the ω_{2S} and ω_{1D}.

experimentally the $\pi^+\pi^-\pi^+\pi^-$ and $\pi^+\pi^-\pi^0\pi^0$ channels are dominant. Further, it has been claimed in e^+e^- annihilation in the range 1.05–1.38 GeV [41] that the principal 4π channel, other than $\omega\pi$, is $a_1\pi$, a conclusion supported from the analysis [32] of $\tau \to 3\pi\pi^0\nu_\tau$.

One explanation of this has been to suggest that the $q\bar{q}$ vector states are mixed with a hybrid vector [42,43] as this decays predominantly to $a_1\pi$ in flux-tube models [43], and to $a_1\pi$ and $\rho(\pi\pi)_S$ in constituent-gluon models [44]. Both the $\pi^+\pi^-\pi^+\pi^-$ and the $\pi^+\pi^-\pi^0\pi^0$ channels are accessed by the $a_1\pi$ and $\rho(\pi\pi)_S$ decays so, in either case, e^+e^- annihilation and the corresponding τ decays should in principle be explicable in terms of some suitable combination of ρ, ρ_{2S}, ρ_{1D} and hybrid ρ_H. Such evidence that we have on the isoscalar states implies the need for more than the ω plus its radial and orbital excitations, although the picture that emerges is not entirely clear. There is no unique signal, in contrast to the $a_1\pi$ in the isovector case. The ω_{2S}, the ω_{1D} and the hybrid ω_H are all expected to have $\rho\pi$, which is known to dominate the $\pi^+\pi^-\pi^0$ channel, as a strong decay mode. The $^3P_0$ predictions for the decays of the ω_{2S} and ω_{1D} are shown in table 4.2. It is known [45,46] that the 5π channel is dominated by $\omega\pi^+\pi^-$, so is consistent with $b_1\pi$. In the flux-tube model the width of the hybrid ω_H is predicted to be small, ~ 20 MeV, and is essentially all $\rho\pi$. The ω_H width is predicted to be appreciably larger in constituent-gluon models [44] with $\rho\pi$ dominant, although some $\omega(\pi\pi)_S$ is allowed.

In [40] a simple mixing scheme was proposed for the isovector channels assuming that there is no direct mixing between the 2^3S_1 and 1^3D_1 $q\bar{q}$ states and that the bare hybrid has no direct e^+e^- coupling. The results were reasonable and qualitatively consistent with observation. Two of the three physical states were identified with the $\rho(1450)$ and the $\rho(1700)$ and the third state was put 'off stage' at a higher mass, provisionally identified with the possible isovector state in the vicinity of 2.0 GeV.

However, there is another possibility, namely that the $\rho(1450)$ is not the lightest of the isovector states. Identifying the $\rho(1450)$ and $\rho(1700)$ states as the first radial and first orbital excitations of the ρ is suspect as the masses of the corresponding $J^P = 1^-$ strange mesons are less than the predictions, particularly for the 2^3S_1 at 1414 ± 15 MeV [8] compared to the predicted 1580 MeV [38]. Quite apart from comparing predicted and observed masses, one would expect the $n\bar{n}$ mesons to be 100–150 MeV lighter than their strange counterparts, putting the $2^3 S_1$ at less than 1300 MeV and the $1^3 D_1$ at less than 1600 MeV. Analysis of $\tau \to \nu_\tau \pi^+ \pi^-$ [20] gives a mass in the range 1300–1350 MeV, depending on the model used for the tail of the ρ, and there are numerous indications in other reactions supporting this.

Regge theory (see chapter 5) requires the existence of an infinite set of trajectories, daughter trajectories, successively one unit lower in angular momentum. This concept was developed at a time when little was known about meson spectroscopy. The evidence for daughter trajectories is now very convincing [47]. This then requires isoscalar and isovector $J^P = 1^-$ mesons approximately degenerate with the $f_2(1270)$ and $a_2(1320)$. Over the years there have been numerous indications of an excited ρ in the region of 1250–1300 MeV. A vector state at about this mass, decaying predominantly to $\pi \omega$ and approximately decoupling from $\pi \pi$, was predicted long ago from an analysis of form-factor data [48]. More generally, it emerged naturally in applications of generalized vector dominance (GVD) [49] and corresponded to that given by the Veneziano spectrum [50] with a universal slope: see chapter 5.

At the time photoproduction data on $\gamma p \to (\omega \pi) p$, with the $\omega \pi$ peaking at around 1240 MeV, were believed to provide direct experimental evidence for the $\rho(1250)$ [51]. However, the conclusion of spin-parity analyses of later data [52,53] was that the enhancement is consistent with predominant $J^P = 1^+ b_1(1235)$ production, with $\sim 20\%$ $J^P = 1^-$ background, if only these two spin-parity states are included in the analysis. However, if a $J^P = 0^-$ state is added, then $J^P = 1^-$ becomes the largest partial wave. It has been suggested [54,55] that there is strong evidence for a $\rho(1200)$ in the $\omega \pi$ channel in the reaction $\bar{p}n \to \omega \pi^- \pi^0$. Intriguingly, many years ago evidence was presented [56] for two $J^P = 1^-$ states with masses (1097^{+16}_{-19}) and (1266 ± 5) MeV in the reaction $\gamma p \to e^+ e^- p$. The evidence for these two states was obtained from the interference between the Bethe–Heitler amplitude and the real part of the hadronic photoproduction amplitude. There is further evidence from the analysis [57] of the $\pi^+ \pi^-$ channel in the reaction $K^- p \to \pi^+ \pi^- \Lambda$ for a $J^P = 1^-$ state with mass (1302^{+28}_{-25}) MeV. Finally, a spin-parity analysis [58] of the $\omega \pi^-$ system in the $D \omega \pi^-$ and $D^* \omega \pi^-$ decays of the $\bar{B}$ shows preference for a $J^P = 1^-$ resonance with mass $(1349 \pm 25^{+10}_{-5})$ MeV and width $(547 \pm 86^{+46}_{-45})$ MeV.

If there is a $\rho(1250)$, then the picture proposed in [40] is incorrect. A more reasonable interpretation would be that the $\rho(1250)$ is the $2^3 S_1$ radial excitation of the ρ, with the $\rho(1450)$ and $\rho(1700)$ being mixed states of the hybrid and the $1^3 D_1$ orbital excitation of the ρ. In some ways this is natural as the hybrid and the $1^3 D_1$

are believed to have a common channel, namely $a_1\pi$, so mixing between these two states can be expected.

An alternative explanation for the strong $a_1\pi$ component in the 4π channel is direct non-resonant production as the $a_1-\rho-\pi$ coupling is strong. A current algebra calculation of the $e^+e^- \to 2\pi^+2\pi^-$ cross section using the axial-current matrix element obtained from $\tau \to \rho\pi\nu$ indicated that this dominates the low-mass part of the $2\pi^+2\pi^-$ spectrum [59]. This would explain the observed strong $a_1\pi$ component in the 4π channels without the need to invoke a hybrid meson. In this scenario, the $\rho(1250)$, if it exists, would be identified with the 2^3S_1 radial excitation of the ρ and the $\rho(1700)$ identified with the 1^3D_1 orbital excitation or this mixed with a higher-mass hybrid meson. A second effect of direct $a_1\pi$ production is that the e^+e^- coupling of the $\rho(1700)$ is greatly reduced, making its interpretation as the orbital excitation of the ρ more plausible.

It is clear that neither the number nor the nature of the vector states is known with certainty. Detailed knowledge of the specific hadronic channels in their decays is lacking and is an essential requisite to make progress. We shall see in chapter 5 that photoproduction data add further complexity to this already confused situation. High-statistics e^+e^- experiments using ISR and photoproduction experiments at JLab have the potential to change this situation.

4.1.3 The contribution to the muon magnetic moment

The Standard Model prediction for the anomalous magnetic moment of the muon can be separated into three basic contributions, the purely electromagnetic [60,61], the weak [62] and the hadronic. The hadronic component, a_μ^{had}, can itself be divided into three:

$$a_\mu^{had} = a_\mu^{had,LO} + a_\mu^{had,HO} + a_\mu^{had,LBL}, \tag{4.16}$$

where $a_\mu^{had,LO}$ is the lowest-order contribution from hadronic vacuum polarization [63–65], $a_\mu^{had,HO}$ is the corresponding higher-order part [66] and $a_\mu^{had,LBL}$ is a small contribution from light-by-light scattering [67,68] that includes, among other terms, a pion-pole contribution.

The contribution from $a_\mu^{had,LO}$ is calculated from the dispersion relation [69]

$$a_\mu^{had,LO} = \frac{\alpha^2}{3\pi^2} \int_{4m_\pi^2}^{\infty} ds \, \frac{K(s)}{s} R(s), \tag{4.17}$$

where $R(s)$ is the ratio of the 'bare' cross section for e^+e^- annihilation into hadrons to the point-like muon-pair cross section and $K(s)$ is the QED kernel [70]

$$K(s) = x^2(1 + \tfrac{1}{2}x^2) + (1 + x)^2 \left(1 + \frac{1}{x^2}\right)$$

$$\times (\log(1 + x) - x + \tfrac{1}{2}x^2) + \frac{1+x}{1-x}x^2 \log x \qquad (4.18)$$

with $x = (1 - \beta_\mu)/(1 + \beta_\mu)$ and $\beta_\mu = \sqrt{(1 - 4m_\mu^2/s)}$. The reason for using the 'bare' cross section, defined as the measured cross section corrected for initial-state radiation, electron-vertex loop and photon vacuum polarization contributions, is to avoid possible double-counting of higher-order contributions which are already part of a_μ. As the kernel $K(s)$ decreases with increasing s, the integrand in (4.17) is weighted to low energies, emphasizing the contribution from the ρ, which is about 73%. About 93% comes from values of $\sqrt{s}$ below 1.8 GeV [65].

As the directly-measured total cross section for e^+e^- annihilation into hadrons is not known with sufficient accuracy above $\sqrt{s} \approx 1$ GeV, it is customary to use the individual exclusive cross sections. Both e^+e^- and τ-decay are available, but there is disagreement between the 2π and 4π spectral functions from these two reactions. This poses a serious problem as these are the two dominant channels. Evaluating all exclusive cross sections [65] gives the lowest-order hadronic vacuum polarization contribution to a_μ as $(684.7 \pm 6.0 \pm 3.6)10^{-10}$, based on e^+e^- data, and $(709 \pm 5.1 \pm 1.2 \pm 2.8)10^{-10}$, based on τ-decay data. These lead respectively to 3σ and 0.9σ deviations of the Standard Model predictions for a_μ from the measured value [71] of $(11\,659\,204 \pm 7 \pm 5)10^{-10}$.

It has been suggested [72] that the difference between the 2π spectral functions in e^+e^- annihilation and τ decay may be explained by the isospin breaking due to a difference between the masses and widths of the charged and neutral ρ mesons. The τ data are about 10% larger than the e^+e^- data in the tail above the ρ and this can be reduced significantly, to 2 or 3%, by a mass difference of a few MeV between ρ^0 and $\rho^\pm$, the latter being the heavier, and a consequential change in the widths, the $\rho^\pm$ width being the larger by several MeV. This then leads to the lowest-order hadronic vacuum polarization contribution to a_μ being $(694.8 \pm 8.6)10^{-10}$, a Standard Model prediction for a_μ of $(11\,659\,179.4 \pm 9.3)10^{-10}$ and a discrepancy of $(23.6 \pm 12.3)10^{-10}$, corresponding to a deviation of 1.9σ.

4.1.4 Testing factorization in hadronic B decay

An understanding of non-leptonic B decays is important for the study of CP violation. In certain kinematic situations it has been argued that factorization may be applied, that is matrix elements of four-quark operators may be written as the product of pairs of matrix elements of two-quark operators. Two different

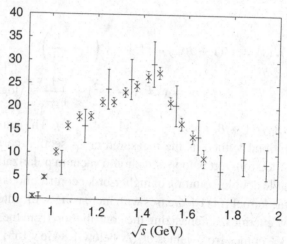

Figure 4.7. Comparison of the $\omega\pi$ mass distribution in τ decay [32] (crosses) and in B decay [6] (bars).

approaches have been used to support this assumption, perturbative QCD [73] and the large-N_c limit of QCD [74]. Reactions such as $B \to D^{(*)}X$, where X is a hadronic state of low invariant mass and $D^{(*)}$ retains the light quark from the B meson, provide specific tests of the factorization hypothesis. The perturbative QCD arguments for factorization depend on the light quarks being produced in an almost collinear state and corrections are expected to grow with m_X/E_X, where m_X is the invariant mass of the final state X and E_X is its energy in the B rest frame. However, if factorization works primarily due to the large N_c limit, then its accuracy is not expected to decrease as m_X increases. Thus multi-body final states provide more information than do two-body final states.

Factorization has been shown to be consistent with experiment in two-body decays, such as $B \to D^{(*)}\pi$ and $D^{(*)}\pi\pi$ [75], but these do not allow the study of corrections to factorization as a function of m_X. This is obviously so for $X \equiv \pi$ and, as the ρ dominates for $X \equiv \pi\pi$, $d\Gamma(B \to D^{(*)}\pi\pi)/dm_X$ is a steeply-falling function of m_X. The case of $X \equiv \omega\pi^-$, which has been measured [6], does permit factorization to be studied over a range of m_X. The factorization prediction for the decay $B \to D^{(*)}X$ is [7]

$$\frac{d\Gamma(B \to D^{(*)}X)/dm_X^2}{d\Gamma(B \to D^{(*)}l\bar{\nu})/dm_X^2} = 3\pi \left(c_1(m_B) + \tfrac{1}{3}c_2(m_B)\right)^2 v_X \left(m_X^2\right), \qquad (4.19)$$

where c_1 and c_2 are known functions that occur in the weak Hamiltonian, and v_X is the τ-decay spectral function defined in (4.2). A comparison of τ-decay and β-decay data is made in figure 4.7, from which it can be deduced that factorization is a reasonable approximation that does not appear to worsen as m_X increases. However, it should be recalled, see figure 4.6, that there are considerable discrepancies among the different $\omega\pi$ data sets in τ decay and e^+e^- annihilation.

4.2 Single-photon transitions between mesons

As the electromagnetic interaction is much better known than the strong one, radiative decays are a cleaner probe of wave functions than hadronic decays. The photon couples directly to the charges and spins of constituents, and it is possible to discriminate among different models for mesons. This can be particularly relevant in distinguishing gluonic excitations (hybrids and glueballs) from conventional excitations with the same J^{PC}.

The simplest single-photon transition process is that between two mesons $A \to \gamma B$. For the initial meson A with spin-parity $J_A^{P_A}$ and final meson B with spin-parity $J_B^{P_B}$, conservation of angular momentum and parity gives the selection rules

$$|J_A - J_B| \leq J \leq J_A + J_B, \tag{4.20}$$

where J is the photon angular momentum,

$$P_A P_B = (-1)^J \tag{4.21}$$

for electric-multipole EJ photon emission and

$$P_A P_B = (-1)^{J+1} \tag{4.22}$$

for magnetic-multipole MJ photon emission.

Electromagnetic gauge invariance requires the matrix element M_μ of the radiative decay current to be transverse,

$$k_\mu M_\mu = 0, \tag{4.23}$$

where k_μ is the photon four-momentum. For the transition $A \to \gamma B$ it means that the matrix elements of the current vanish in the limit $k \to 0$.

4.2.1 Quark model description of $A \to \gamma B$

The transition amplitude is readily calculated in the framework of the non-relativistic quark model, where mesonic spectra and wave functions are defined from the Schrödinger equation with the Hamiltonian

$$H = \frac{p^2}{2\mu} + V(r), \tag{4.24}$$

where $\mu = (m_q m_{\bar{q}})/(m_q + m_{\bar{q}})$ is the reduced mass in the quark–antiquark system, and r is the interquark distance. A popular choice for the potential is 'linear plus Coulomb', motivated by lattice Wilson-loop calculations [76].

As soon as the model for the effective potential V is defined and mesonic wave functions are found, the amplitude for the radiative transition $A \rightarrow \gamma B$ is given by well-known quantum-mechanical expressions. Let the initial meson A, with mass m_A, decay at rest to the final meson B, with mass m_B, and a photon with three-momentum $\mathbf{k}$. The transition amplitude has the form

$$\mathbf{M}_{A \rightarrow B} = e_q \left\langle B \left| \frac{1}{2m_q} (\mathbf{p}_q e^{-i\mathbf{k}\mathbf{r}_q} + e^{-i\mathbf{k}\mathbf{r}_q} \mathbf{p}_q) - \frac{i}{2m_q} \sigma \times \mathbf{k} e^{-i\mathbf{k}\mathbf{r}_q} \right| A \right\rangle$$
$$+ (q \leftrightarrow \bar{q}), \qquad (4.25)$$

where m_q is the constituent-quark mass, $\mathbf{r}_q$ is the quark coordinate and $\mathbf{p}_q$ is the quark three-momentum.

The differential decay rate is given by

$$\frac{d\Gamma}{d\cos\theta} = k \frac{E_B}{m_A} \alpha \sum |M_{A \rightarrow B}|^2, \qquad (4.26)$$

where the sum is over final-state polarizations, and E_B is the centre-of-mass energy of the final meson.

Expanding (4.25) in powers of kR, where R is the meson radius, one obtains standard multipole expansion formulae. The lowest multipoles are electric dipole, $E1$, and magnetic dipole, $M1$, transitions. Those most readily accessible experimentally are $E1$ transitions between S and P levels and $M1$ transitions between S levels.

If the quarks are treated non-relativistically, the long-wavelength approximation $kR \ll 1$ should be applicable. $M1$ transitions proceed via the spin-flip part of the amplitude (the second term in (4.25)), so that the amplitude is proportional to the overlap of wave functions,

$$M \sim \int d^3 r \, \psi_B^*(\mathbf{r}) \psi_A(\mathbf{r}). \qquad (4.27)$$

This is either 1 or 0, so in this limit $M1$ transitions do not require knowledge of exact wave functions.

The $E1$ transitions are not so simple even in the long-wavelength approximation. Not only is knowledge of wave functions required, the amplitude must be proportional to the photon momentum. While this is obviously the case for the spin-flip part of the amplitude, the convective term (the first term in (4.25)) responsible for $E1$ transitions apparently does not exhibit such behaviour. To reproduce low-energy theorems in electromagnetic interactions one should invoke the Schrödinger equation for the wave functions together with the commutation relation

$$p_i = \frac{i}{2}[p^2, r_i].$$ (4.28)

As the strong Hamiltonian takes the form (4.24), one has

$$\left\langle B \left| \frac{p_k}{2\mu} \right| A \right\rangle = -\frac{i}{2}\omega\langle B \mid r_k \mid A \rangle,$$ (4.29)

where $\omega = m_A - m_B$ is the photon energy in this approximation. In this way the well-known dipole formula is established:

$$M_k \sim \omega \int d^3 r \, \psi_B^*(\mathbf{r}) r_k \psi_A(\mathbf{r}),$$ (4.30)

where $\psi_A(\mathbf{r})$ and $\psi_B(\mathbf{r})$ are the wave functions of initial and final mesons in the coordinate representation.

The result (4.30) is of a more general nature. The amplitude should vanish in the $k_\mu \to 0$ limit. This property follows from the QCD Lagrangian and manifests itself as electromagnetic-current conservation (4.23). At the level of the Lagrangian, properties like vector-current conservation are obvious, but in order to retain such properties for physical matrix elements at the mesonic level the initial and final states must be proper solutions of the bound-state problem.

$E1$ rates are sensitive to the potential, especially if the mesons involved are radial excitations with nodes in their wave functions. However, the main uncertainty comes from the relation (4.30), which holds true only if the masses of the initial and final states are the 'correct' eigenvalues of the quark-model Hamiltonian (4.24), and not the physical masses. In other words, the naive quark model (4.24) does not take into account the fine and hyperfine corrections to the spectra.

Such corrections are relativistic, $O(1/m^2)$, and come from the Fermi–Breit reduction of a Dirac-type equation with vector and scalar interaction (for example see [77]). To be self-consistent, one should take into account not only relativistic corrections to the spectra and the wave functions, but also the corrections of the same order to the radiative decay amplitude. The latter are not exhausted by retaining the subleading $O((kR)^2)$ corrections in (4.25), but add extra terms as was shown in [77–79].

A comprehensive analysis of these extra terms, including the spin–orbit term in radiative transitions, is given in [80], where the expression for the radiative decay amplitude is derived using a variety of methods. The result for the transition amplitude with next-to-leading relativistic corrections is

$$M_k = \sum_{q(\bar{q})} e_q \left\langle B \left| -\frac{1}{2m_q}(p_k e^{-i\mathbf{kr}} + e^{-i\mathbf{kr}} p_k) + \frac{i}{2m_q}\sigma_n \epsilon_{nik} k_i e^{-i\mathbf{kr}} \right. \right.$$

$$\left. \left. + \frac{1}{2m_q^3}k^2 k_k - \frac{1}{2m_q^2}\epsilon_{kin}\sigma_n \frac{\partial V_v(r)}{\partial r_i} + \frac{1}{2m_q^2}[p_k, V_s(r)]_+ \right| A \right\rangle. \quad (4.31)$$

Here V_v and V_s are vector and scalar confining potentials in the Dirac bound-state equation. Hence the answer for the radiative transition amplitude depends not only on the form of the binding potential, but also on the Lorentz nature of the confining force. In particular, the scalar piece of the confining potential does not contribute to the spin–orbit term in (4.31).

The states A, B are eigenfunctions of the full Fermi–Breit Hamiltonian. Using the explicit form of this Hamiltonian and the commutation relation

$$p_i e^{-i\mathbf{kr}} = \frac{i}{2}\left([p^2, r_i e^{-i\mathbf{kr}}] + r_i[e^{-i\mathbf{kr}}, p^2]\right), \quad (4.32)$$

one can obtain the modified dipole formula

$$M_k = \sum_{q(\bar{q})} e_q \left\langle B \left| -i\omega r_k e^{-i\mathbf{kr}} + \frac{i}{2m_q}\sigma_n \epsilon_{njk} P_j e^{-i\mathbf{kr}} - \frac{i\omega}{4m_q^2}\sigma_n \epsilon_{njk} k_j \right| A \right\rangle, \quad (4.33)$$

where ω is the difference of eigenvalues of the Fermi–Breit Hamiltonian. This dipole formula contains the same model dependence on the Lorentz nature of confinement as the equivalent formula (4.31), as this dependence is hidden in (4.33) in the initial and final meson wave functions.

Various relativistic models based on Bethe–Salpeter equations with instantaneous [81] or fully-covariant [82] interaction kernels not only reproduce the formulae (4.31) and (4.33) for heavy quarks, but also justify some phenomenological concepts employed in the description of radiative decay amplitudes. For example, quarks are dressed by the strong interaction that leads, in particular, to a modification of the quark–photon vertex [83]. In a theory with confinement such a modification should give rise to a vector-meson-dominant quark form factor, thus confirming the vector-meson-dominance hypothesis [84]. On the other hand, it can be viewed as a mechanism generating in a natural way an anomalous magnetic moment for the quark, a quantity introduced phenomenologically to fit the data on radiative mesonic transitions (see, for example, [85]). Another point concerns the backward-in-time motion of the quark pair in a meson. In the radiative transition amplitude these negative-energy components of the mesonic wave function lead to exchange-current contributions [86,87]. One should have in mind, however, that the results of relativistic quark models are very model-dependent.

4.2.2 Applications: heavy flavours

In practice, the calculations of radiative decay amplitudes often include lowest $E1$ and $M1$ amplitudes and subleading $M2$ transitions only, neglecting the relativistic corrections described. The most notable successes of this approach have been in reproducing qualitatively the magnitudes and relative phases of over 100 helicity amplitudes for photoexcitation of the proton and neutron [88–90]. These give a clear indication of which amplitudes are large or small, and of their relative sizes and signs. This suggests that although corrections may be individually significant, their collective effect is small.

For the case $m_q = m_{\bar{q}} = m$ the total width for the $M1$ transition between the ground state $^3S_1$ and excited $^1S_0$ levels in the dipole approximation is

$$\Gamma(1S \to 1S) = \frac{4}{3}\alpha\frac{E_B}{m_A}\langle e_q^2\rangle\frac{k^3}{m}, \tag{4.34}$$

where $\langle e_q^2\rangle$ is the charge factor:

$$\langle e_q^2\rangle = \tfrac{1}{36} \quad \text{for } n\bar{n} \to n\bar{n} \text{ with the same isospin,}$$

$$\langle e_q^2\rangle = \tfrac{1}{4} \quad \text{for } n\bar{n} \to n\bar{n} \text{ with different isospin,}$$

$$\langle e_q^2\rangle = \tfrac{1}{9} \quad \text{for } s\bar{s} \to s\bar{s},$$

$$\langle e_q^2\rangle = \tfrac{4}{9} \quad \text{for } c\bar{c} \to c\bar{c}. \tag{4.35}$$

The first successes in calculations of meson radiative decay rates were in describing these transitions. The rate for the decay $\omega \to \pi^0\gamma$, calculated with (4.34) and the standard value of 0.33 GeV for the constituent light-quark mass, is about 600 keV, which is not far from the empirical value 717 ± 43 keV [8]. This agreement should not be taken too seriously as, in spite of the phenomenological successes of the naive quark model (4.24), the pion cannot be described in such a simple way as its Goldstone-boson nature is completely lost. Nonetheless, the relations (4.35) give the value $\tfrac{1}{9}$ for the ratio $\Gamma(\rho^0 \to \pi^0\gamma)/\Gamma(\omega \to \pi^0\gamma)$, in satisfactory agreement with data: 0.09 ± 0.02 [8].

The quark model gives a natural explanation for the ratios of radiative transition rates of charged and neutral $^3S_1$ heavy–light mesons, such as $D^* \to D\gamma$. Indeed, from (4.25) one has

$$\frac{M(D^{*0} \to D^0\gamma)}{M(D^{*+} \to D^+\gamma)} = \frac{2/3m_c + 2/3m_u}{2/3m_c - 1/3m_d}. \tag{4.36}$$

Substituting the values of constituent-quark masses one immediately sees that $BR(D^{*0} \to D^0\gamma) \gg BR(D^{*+} \to D^+\gamma)$.

Transition	Calculated width (keV)	Measured width (keV)
$\chi_{c0} \to \gamma J/\psi$	120	119 ± 16
$\chi_{c1} \to \gamma J/\psi$	242	287 ± 24
$\chi_{c2} \to \gamma J/\psi$	315	426 ± 38
$\psi' \to \gamma \chi_{c0}$	46	24 ± 2
$\psi' \to \gamma \chi_{c1}$	41	24 ± 2
$\psi' \to \gamma \chi_{c2}$	29	18 ± 2

Table 4.3. $E1$ transition rates calculated in the Cornell model [93] for charmonium.

One expects that the most reliable calculations are those of transitions between heavy quarkonia levels, as this situation is non-relativistic. Taking the mass of charmed quarks as 1.8 GeV, the naive calculation (4.34) gives

$$\Gamma(J/\psi \to \gamma \eta_c) = 1.94 \text{ keV}, \tag{4.37}$$

while the experimental number is 1.13 ± 0.41 keV [8]. More elaborate calculations in the framework of the Cornell model [91,92] for the binding potential give essentially the same number, 1.92 keV [93].

The transition $\psi' \to \gamma \eta_c$ is forbidden in the dipole approximation due to the orthogonality of the wave functions (see (4.27)), and one must expand the factors $e^{-i\mathbf{kr}}$ in (4.25) further. To calculate this knowledge of wave functions is required and, due to the node in the wave function of the radial excitation, these calculations are rather unstable. Nevertheless, the Cornell model describes this decay quite satisfactorily: the calculated width is 0.91 keV, while data give 0.78 ± 0.25 keV.

The S–P radiative transitions in charmonium are known and can be used to check the formalism. The leading $E1$ transition rate in the dipole approximation is

$$\Gamma = \frac{4(2J+1)}{27} \alpha \langle e_q^2 \rangle k^3 r_{AB}^2 \tag{4.38}$$

for $^3S_1 \to {}^3P_J \gamma$ decays, and

$$\Gamma = \tfrac{4}{9} \alpha \langle e_q^2 \rangle k^3 r_{AB}^2 \tag{4.39}$$

for $^3P_J \to {}^3S_1 \gamma$ decays, where r_{AB} is the radial part of the dipole matrix element (4.30). Up to differences in phase space, (4.38) gives

$$\Gamma(\psi' \to \gamma \chi_{c0}) : \Gamma(\psi' \to \gamma \chi_{c1}) : \Gamma(\psi' \to \gamma \chi_{c2}) = 1 : 3 : 5 \tag{4.40}$$

and the rates $\chi_{cJ} \to \gamma J/\psi$ should be equal. The results of the full calculations with the Cornell model are summarized in table 4.3.

The agreement is not very satisfactory, though the trends are more or less reproduced. It is argued in [87] that including exchange-current contributions brings the calculated widths into closer agreement with data. Note, however, that as was discussed above, accounting for exchange currents also brings uncertainties due to the strong model dependence.

Other corrections discussed in this regard are those due to coupled channels effects [92]. A well-known example is the position of the $^3D_1$ charmonium level. Potential models place it at 3815 MeV, while it is the $\psi(3770)$ that is usually identified as $^3D_1$ charmonium. On the other hand, the relevant open-charm thresholds are $D^0\bar{D}^0$ at 3729 MeV and D^+D^- at 3739 MeV. It is usually assumed that the coupling to $D\bar{D}$ pulls the $^3D_1$ charmonium down to the observed mass of 3770 MeV.

For the transitions $^3D_1 \to \gamma\,^3P_J$ in the dipole approximation one has the ratio

$$\Gamma(^3D_1 \to \gamma\chi_{c0}) : \Gamma(^3D_1 \to \gamma\chi_{c1}) : \Gamma(^3D_1 \to \gamma\chi_{c2}) = 20 : 15 : 1, \qquad (4.41)$$

a trend quite opposite to (4.40). Full Cornell-model calculations [94] yield the rates

$$\Gamma(\psi(3770) \to \gamma\chi_{c0}) = 225 \ \text{keV} \ (254 \ \text{keV}),$$
$$\Gamma(\psi(3770) \to \gamma\chi_{c1}) = 59 \ \text{keV} \ (183 \ \text{keV}),$$
$$\Gamma(\psi(3770) \to \gamma\chi_{c2}) = 3.2 \ \text{keV} \ (3.9 \ \text{keV}), \qquad (4.42)$$

where the numbers in parentheses are calculated without coupled-channel corrections. There are no experimental data with which to compare, but the results (4.42), as well as other results of the Cornell model, demonstrate that coupled-channel corrections can be substantial for higher quarkonia.

4.2.3 Angular distributions

Further information can be obtained from angular distributions [95], which is most naturally done in terms of helicity amplitudes. For example, consider the process

$$e^+e^- \to 1^{--} \to J^{P+}\gamma, \qquad (4.43)$$

where the only allowed values of the projection of the vector-meson spin are ± 1, with the z-axis taken along the electron beam direction. Then, for $J = 0$ there is only one helicity amplitude $\mathcal{A}$, which leads to

$$\frac{d\sigma}{d\cos\theta} \sim |\mathcal{A}|^2(1 + \cos^2\theta). \qquad (4.44)$$

For $J = 1$ there are two helicity amplitudes, $\mathcal{B}_0$ and $\mathcal{B}_1$, and

$$\frac{d\sigma}{d\cos\theta} \sim (|\mathcal{B}_0|^2 + 2|\mathcal{B}_1|^2)\left(1 + \frac{|\mathcal{B}_0|^2 - 2|\mathcal{B}_1|^2}{|\mathcal{B}_0|^2 + 2|\mathcal{B}_1|^2}\cos^2\theta\right). \qquad (4.45)$$

For $J = 2$ there are three helicity amplitudes, $\mathcal{C}_0$, $\mathcal{C}_1$ and $\mathcal{C}_2$, and

$$\frac{d\sigma}{d\cos\theta} \sim (|\mathcal{C}_0|^2 + |\mathcal{C}_2|^2 + 2|\mathcal{C}_1|^2)\left(1 + \frac{|\mathcal{C}_0|^2 + |\mathcal{C}_2|^2 - 2|\mathcal{C}_1|^2}{|\mathcal{C}_0|^2 + |\mathcal{C}_2|^2 + 2|\mathcal{C}_1|^2}\cos^2\theta\right). \qquad (4.46)$$

Quark-model predictions constrain the relations between various helicity amplitudes. For example, the structure of the current responsible for the radiative transition with a transversely-polarized photon, between $q\bar{q}$ S and P states is [84,96,97]

$$J_{\pm 1} \equiv A\langle L_\pm\rangle + B\langle S_\pm\rangle + C\langle S_z L_\pm\rangle. \qquad (4.47)$$

Here $\langle L_\pm\rangle$, $\langle S_\pm\rangle$ and $\langle S_z L_\pm\rangle$ are the relevant Clebsch–Gordan coefficients for coupling $\mathbf{L} \times \mathbf{S} \to \mathbf{J}$, and the z-axis is directed along the photon momentum. In terms of unknown quantities A, B and C the helicity amplitudes can be written as

$$\mathcal{A} = \sqrt{2}(A - B - C), \qquad (4.48)$$

$$\mathcal{B}_0 = \sqrt{3}(A - C), \quad \mathcal{B}_1 = \sqrt{3}(A - B), \qquad (4.49)$$

$$\mathcal{C}_0 = (A + 2B - C), \quad \mathcal{C}_1 = \sqrt{3}(A + B), \quad \mathcal{C}_2 = \sqrt{6}(A + C). \qquad (4.50)$$

As the electric and magnetic multipole amplitudes are linear combinations of A, B and C, the amplitudes may be written alternatively in terms of multipoles, as in [98], giving

$$A \equiv E_1; \quad B \equiv (M - E_R); \quad C \equiv -(M + E_R). \qquad (4.51)$$

Here E_1 is the leading electric-dipole term, E_R is the 'extra' electric-dipole term and M is the magnetic-quadrupole term. Note that electric octupole contributions, while allowed in general for radiative transitions between tensor and vector-mesons, will vanish if the vector is a pure $^3S_1$ state, and the tensor is a pure $^3P_2$ state [96].

In accordance with (4.44), (4.45) and (4.46) the angular distributions can measure the ratios of helicity amplitudes, which, in turn, allows possible deviations from $q\bar{q}$ structure to be studied.

4.2.4 Light flavours and a flavour filter

Calculations in the light-quark sector are more ambiguous. Nevertheless, one may hope that quark-model calculations can provide a reliable estimate of which radiative decay rates are large or small and which ones are accessible experimentally.

Decay	Width (keV)	Final state
$\rho^0(1450) \rightarrow f_2(1270)\gamma$	712	$\pi\pi\gamma$
$\phi(1690) \rightarrow f_2(1525)\gamma$	148	$K\bar{K}\gamma$
$\phi(1900) \rightarrow f_1(1420)\gamma$	408	$K\bar{K}\pi\gamma$

Table 4.4. Rates and final states for some radiative decays calculated in [99].

The calculations [99] in the model given by strong Hamiltonian (4.24) and transition current (4.25) indeed demonstrate that some radiative transitions in the light-quark sector should be measurable at high-intensity facilities. Results for some interesting radiative decays are listed in table 4.4 together with the corresponding final states. The results quoted in table 4.4 were obtained assuming $\rho(1450)$ and $\phi(1690)$ to be 2^3S_1 $n\bar{n}$ and $s\bar{s}$ states respectively, while the (unobserved) $\phi(1900)$ was taken to be the 1^3D_1 $s\bar{s}$ state.

The advantages of $\rho^0(1450) \rightarrow f_2(1270)\gamma$ and $\phi(1690) \rightarrow f_2(1525)\gamma$ are that the final-state mesons are comparatively narrow, their decays are two-body and there are no neutrals in the final state other than the photon.

These two decays are unique identifiers of the $\rho(1450)$ and the $\phi(1690)$ as quarkonia, and discriminate between quarkonia and hybrid assignments for these states. In a hybrid vector the quark–antiquark pair is in a spin-singlet, while in quarkonia it is in a spin-triplet. Thus for hybrids the transition to ^3P_J $q\bar{q}$ states is necessarily magnetic spin-flip and is suppressed.

The decay $\phi(1900) \rightarrow f_1(1420)\gamma$ discriminates between the $f_1(1420)$ and the $\eta(1440)$, as the width of $\phi(1900) \rightarrow f_1(1420)\gamma$ appears necessarily to be much larger than that of $\phi(1900) \rightarrow \eta(1440)\gamma$. The nearness of the masses and widths of the $f_1(1420)$ and the $\eta(1440)$, and several common hadronic decay modes, have hitherto been sources of confusion.

Radiative decays to and from scalar mesons offer the possibility to determine the flavour content of scalar mesons. Moreover, as lattice calculations [100] identify the lightest glueball to be a scalar 0^{++} with mass around 1.5 GeV, it should mix with scalar $q\bar{q}$ mesons in the 1.3–1.7 GeV range. Radiative decays may disentangle the role of the glueball in these scalar mesons.

In the absence of glueball mixing one has in the model [97]

$$\Gamma(f_0(1370) \rightarrow \gamma\rho) \sim 2300 \text{ keV} \tag{4.52}$$

and

$$\Gamma(f_0(1710) \rightarrow \gamma\phi) \sim 870 \text{ keV}, \tag{4.53}$$

assuming $f_0(1370)$ and $f_0(1710)$ to be $n\bar{n}$ and $s\bar{s}$ scalar states.

	$\rho(770)$			$\phi(1020)$		
	L	M	H	L	M	H
$f_0(1370)$	443	1121	1540	8	9	32
$f_0(1500)$	2519	1458	476	9	60	454
$f_0(1710)$	42	94	705	800	718	78

Table 4.5. Effect of mixing in the scalar sector of the 1^3P_0 nonet for radiative decays to ρ and ϕ. The radiative widths, in keV, are given for three different mixing scenarios as described in the text: light glueball (L), medium-weight glueball (M) and heavy glueball (H).

The width of the decay $f_1(1285) \to \gamma\rho$ is measured and provides a check on the model. The calculated width is 1400 keV, which compares well with the data [8,101,102] of 1320 ± 312 keV. The width for $f_2 \to \gamma\rho$ appears to be rather smaller, 644 keV. Experimentally this width is small as neither the MARKIII [101] nor the WA102 [102] experiments has any evidence for it, so it is reasonable to suppose that the results for the f_0 radiative decays are valid as well.

The widths (4.52) and (4.53) can be changed substantially when glueball mixing is included, the degree of modification depending on the mass of the bare glueball. Three different mixing scenarios have been proposed: the bare glueball is lighter than the bare $n\bar{n}$ state [103]; its mass lies between the bare $n\bar{n}$ state and the bare $s\bar{s}$ state [103]; or it is heavier than the bare $s\bar{s}$ state [104]. We denote these three possibilities by L, M and H respectively. The predicted widths for the decays of $f_0(1370)$, $f_0(1500)$ and $f_0(1710)$ to $\gamma\rho$ and $\gamma\phi$ for each of these possibilities are given in table 4.5.

It is clear that the discrimination between different mixing scenarios is strong. Other potentially powerful ways of determining glueball mixing in the scalar mesons were identified in [99] through the radiative transitions $\rho_D \to \gamma f_0(1370)$ and $\gamma f_0(1500)$. There is also similar sensitivity in the decays $\phi_D \to \gamma f_0(1500)$ and $\phi_D \to \gamma f_0(1710)$.

4.2.5 Photoexcitation of hybrid mesons

Theory [105] has provided compelling arguments from QCD that confinement occurs via the formation of a flux tube: a relativistic object with an infinite number of degrees of freedom. Conventional mesons arise when this flux tube is unexcited, acting as an effective potential, which underpins the application of potential models such as those described elsewhere in this chapter. Excitations of the flux tube give new states, known as hybrids.

A significant plank in the proposed upgrade of JLab is its assumed ability to expose the predicted hybrid mesons in photo- and electroproduction. Direct calculations [106,107] of such electromagnetic transitions predict that the $E1$ transition amplitudes may be large and accessible in forthcoming experiments.

In [105,108,109] the flux tube was discretized into $N + 1$ cells, and then $N \to \infty$. Up to N modes may be excited. We shall focus on the first excited state, with excitation energy $\omega = \pi/r$.

The flux tube is dynamic, with degrees of freedom in the two dimensions transverse to the $Q\bar{Q}$-axis. The state of the flux tube can be written in terms of a complete set of transverse eigenstates $|\vec{y}_1 \ldots \vec{y}_n \ldots \vec{y}_N\rangle$ and the Fourier mode for the first excited state is

$$\vec{y}_n = \sqrt{\frac{2}{N+1}} \vec{a}_1 \sin \frac{\pi n}{N+1}.$$

In the small-oscillation approximation the system becomes harmonic in $\vec{y}$ ($\vec{a}$). The states of the flux tube are then described by Gaussians (see equations (11), (12) and (13) in [108]). For a pedagogic illustration, consider the tube to be modelled by a single bead, mass br.

If the transverse displacement is $\vec{y}$, then conservation of the position of the centre of mass and of orbital angular momentum about the centre of mass leads to a mean transverse displacement of the Q and $\bar{Q}$. If these have masses m_Q, then relative to the centre of mass, the position vector of the quark has components in the longitudinal $\vec{r}$ and transverse $\vec{y}$ directions:

$$\vec{r}_Q = \left[\tfrac{1}{2}\vec{r} \; ; \; \left(\frac{br}{2m_Q} \right) \vec{y} \right].$$

The dependence of $\vec{r}_Q$ on $\vec{y}$ enables a quark–current interaction at r_Q to excite transitions in the $\vec{y}$ oscillator, leading to excitation of the flux tube.

This is the essential physics behind the excitation of hybrid modes by current interactions with the quark or antiquark. Extending to N beads leads to more mathematical detail, but the underlying principles are the same. The position vector of the quark becomes [108]

$$\vec{r}_{Q;\bar{Q}} = \vec{R} \pm \frac{1}{2}\vec{r} + \frac{br}{\pi m_Q} \sqrt{\frac{2}{N+1}} \vec{a}_1$$

with $\vec{R}$ the position of the centre of mass of the $q\bar{q}$-tube system.

It has been argued that this dependence $\vec{r}_Q = f(\vec{r}, \vec{y})$ gives significant contributions to static properties of hadrons, such as charge radii, $\langle r_\pi^2 \rangle$, and to the slope of the Isgur–Wise function $\rho(v \cdot v')$, which seem to be required experimentally [108]. Isgur showed that these 'transverse excursions' give huge $\sim 51\%$ corrections in light-quark systems where $m_Q = m_d$, and $\sim 13\%$ corrections in heavy–light $Q\bar{q}$ systems. Furthermore, the $\sum_1^\infty (1/p^3)$ is $\sim 80\%$ saturated by its $p = 1$ term. Together, these suggest that the transition amplitudes to the lowest hybrids ($p = 1$ phonon modes) could be substantial. In [106] it was demonstrated that this can be so, at least for certain quantum numbers.

The respective amplitudes for conventional $E1$ transitions and the hybrid excitation come from expanding the incoming plane wave to leading order in the momentum transfer, thereby enabling the linear terms in $\vec{q} \cdot \vec{r}_Q$ to break the orthogonality of initial and final wave functions and cause the transition.

By combining with the tensor decomposition of the current–quark interaction, we may calculate excitation amplitudes to hybrids, and compare with those for conventional mesons in various multipoles. Full details are in [107].

A general feature of operators required to excite the lowest hybrid states (the first flux-tube mode) is the presence of the transverse position vector $\vec{y}$ to break the orthogonality between the lowest $Q\bar{Q}$ state and the '$\vec{y}$-excited' hybrid states. Hence in photoproduction one accesses $E1$ or (orbitally excited) $M1$ transitions in leading order. These are $\Delta S = 0$, for example $0_Q^{-+} \to 1_H^{\pm\pm}$ or $1_Q^{--} \to (0, 1, 2)_H^{\mp\pm}$. (Note that states with the 'wrong' charge conjugation will only be accessible for flavoured mesons, for example in $\gamma p \to H^+ n$, and hence will have no analogue for $c\bar{c}$ and other $I = 0$ states.)

Transitions involving spin-flip, $\Delta S = 1$, will need a $\vec{\sigma}$ spin operator as well as the above. Such terms arise as finite-size corrections to the $\vec{\sigma} \cdot \vec{B}$ magnetic interaction and also in the spin–orbit interaction $\vec{\sigma} \cdot \vec{p}_Q \times \vec{E}$, in J_{em}. These are normally non-leading effects at $O(v/c)^2$ in amplitude and hence much suppressed for heavy flavours.

The familiar $E1$ amplitude between $Q_1\bar{Q}_2$ conventional states (e.g $\gamma\pi \leftrightarrow b_1$) may be summarized by

$$\tilde{\mathcal{M}}(\gamma\pi \leftrightarrow b_1) = \left(\frac{e_1}{m_1} - \frac{e_2}{m_2}\right)_b \langle r \rangle_\pi |\vec{q}| \frac{\mu}{\sqrt{3}}, \qquad (4.54)$$

where $_b\langle r \rangle_\pi$ is the radial wave function moment $\int_0^\infty r^2 dr\, R_b(r) r R_\pi(r)$, and μ is the reduced mass of the $Q\bar{Q}$ system.

Following [105] we denote the number of positive or negative helicity phonon modes transverse to the body vector $\vec{r}$ by $\{n_+, n_-\}$, which for our present purposes will be $\{1, 0\}$ or $\{0, 1\}$. The analogous amplitude for exciting the $\vec{y}$ oscillator between spin-singlet states leads to $\tilde{\mathcal{M}} \equiv \mathcal{M}(\delta_+ - \delta_-)$, where

$$\mathcal{M}(\gamma\pi \to a_{1H}) = \left(\frac{e_1}{m_1} + \frac{e_2}{m_2}\right) \cdot _H\langle r\rangle_\pi |\vec{q}| \sqrt{\frac{b}{3\pi^3}}\, \delta_{m,+1} \qquad (4.55)$$

(where the factors $\delta_{+,-}$ correspond to $\{1, 0\}$ and $\{0, 1\}$ respectively, while the $\delta_{m,\pm1}$ refers to the hybrid polarization in the fixed axes x, y, z [105]). The transition $\gamma\pi \leftrightarrow a_{1H}$ is seen to vanish when $m_1 \equiv m_2$ and $e_1 = -e_2$ in accord with the constraints of charge conjugation. The above formula can be immediately taken over to flavoured states where $m_1 \neq m_2$.

The parity eigenstates in the flux tube are given in [105]. Parity eigenstates $\pm$ are then the linear superpositions $\frac{1}{\sqrt{2}}(|\{1, 0\}\rangle \mp |\{0, 1\}\rangle)$ such that for $\pi\gamma$ $E1$ transitions we have

$$\langle P = -|\pi\gamma\rangle = 0; \quad \langle P = +|\pi\gamma\rangle = \sqrt{2}\mathcal{M}.$$

This applies immediately to the excitation of the hybrid $a_{1H}^\pm$ in $\gamma\pi^\pm \to a_{1H}^\pm$, where there is no spin-flip between the spin-singlet π and a_{1H}. In general we can write the radiative width $\Gamma(A \to B\gamma)$ as

$$4\frac{E_B}{m_A}\frac{|\vec{q}|}{(2J_A + 1)} \sum_{m_J^A} |\tilde{\mathcal{M}}(m_J^A, m_J^B = m_J^A + 1)|^2,$$

where the sum is over all possible helicities of the initial meson. The ratio of widths $\Gamma_{E1}(a_{1H}^+ \to \pi^+\gamma)/\Gamma_{E1}(b_1^+ \to \pi^+\gamma)$ is then

$$\frac{72}{\pi^3}\frac{b}{m_n^2}\left|\frac{_H\langle r\rangle_\pi}{_b\langle r\rangle_\pi}\right|^2 \left[\frac{|\vec{q}_H|^3 \exp\left(-|\vec{q}_H|^2/8\bar{\beta}_H^2\right)}{|\vec{q}_b|^3 \exp\left(-|\vec{q}_b|^2/8\bar{\beta}_b^2\right)}\right], \qquad (4.56)$$

where $m_m = m_q m_{\bar{q}}/(m_q + m_{\bar{q}})$ and the factor in square brackets includes the q^3 phase-space and a 'typical' form factor taken from the case of harmonic-oscillator binding [99].

The adiabatic model of [105], with a variational harmonic-oscillator solution, gives $|_H\langle r\rangle_\pi/_b\langle r\rangle_\pi|^2 \approx 1.0$, so the radial moments do not suppress hybrids [107]. The main uncertainty is the computed size of the π [99]. Assuming that this hybrid has mass $\sim$1.9 GeV [105,109], and using the measured width $\Gamma(b_1^+ \to \pi^+\gamma) = 230 \pm 60$ keV [8] we predict using (4.56) that

$$\Gamma(a_{1H}^+ \to \pi^+\gamma) = 2.1 \pm 0.9 \text{ MeV},$$

where the error allows for the uncertainty in β_π.

The equivalent $E1$ process for spin-triplet $Q\bar{Q}$ states is $(0, 1, 2)_H^{+-} \leftrightarrow \rho\gamma$, where the only difference from the $S = 0$ case is the addition of L, S Clebsch–Gordan factors coupling the $Q\bar{Q}$ spin and flux-tube angular momentum to the total J of the hybrid meson in question. The matrix element is analogous to (4.55) multiplied by the Clebsch–Cordan $\langle 1+1; 1m_\rho | J m_J \rangle$. We find (for $J = 0, 1, 2$ in this $E1$ limit)

$$\Gamma(b^+_{JH} \to \rho^+ \gamma) = 2.3 \pm 0.8 \text{ MeV},$$

where the error reflects the uncertainties in the conventional $E1$ strength and β_{f_1} and where we have taken $m_H = 1.9$ GeV.

The transition to the exotic 2^{+-} may be significant from either ρ or π exchange [107,110]. This state may also be excited diffractively by photons, the gluon exchanges (pomeron) exciting the flux tube. Hence the exotic 2^{+-} is expected to be photoproduced at least as copiously as the 1^{-+} and should be sought at JLab via meson exchange and at higher energy diffractively.

4.2.6 Radiative transitions $A \to \gamma M_1 M_2$

The process $A \to \gamma B$ can be viewed as the $O((1/N_C)^0)$ single-photon emission process in the $1/N_C$ expansion. More complicated transitions $A \to \gamma M_1 M_2$ are of order $1/\sqrt{N_C}$ at most, and are described with decay chains

$$A \to \gamma B, \quad B \to M_1 M_2,$$
$$A \to M_1 B, \quad B \to \gamma M_2. \tag{4.57}$$

Consider, for example, the radiative decay of a neutral vector-meson V^0 into a pair of charged pseudoscalars $P^+ P^-$. It can go via scalar intermediate state S,

$$V^0 \to \gamma S, \quad S \to P^+ P^-, \tag{4.58}$$

that is an $E1$ radiative transition, and, as was discussed above, the amplitude is proportional to the photon energy ω. It can also go via a vector intermediate state V',

$$V^0 \to V'^{\pm} P^{\mp}, \quad V'^{\pm} \to \gamma P^{\pm} \tag{4.59}$$

($M1$ radiative transition), and the amplitude is again proportional to ω.

However, the photon couples to charged constituents in all possible ways, including emission from the charged external lines, which corresponds to the decay chain

$$V^0 \to P^+ P^-, \quad P^{\pm} \to \gamma P^{\pm}. \tag{4.60}$$

In the soft photon limit $\omega \to 0$ the amplitude is

$$M_\mu = eg_V \left(((q_+ \cdot \epsilon) - (q_- \cdot \epsilon) + (k \cdot \epsilon)) \frac{q_{+\mu}}{(q_+ \cdot k)} \right.$$
$$\left. + ((q_- \cdot \epsilon) - (q_+ \cdot \epsilon) + (k \cdot \epsilon)) \frac{q_{-\mu}}{(q_- \cdot k)} - 2\epsilon_\mu \right), \qquad (4.61)$$

where g_V is the VPP coupling constant, $q_\pm$ are the pseudoscalar four-momenta, k is the photon four-momentum, and ϵ is the polarization four-vector of the vector meson. The first terms in (4.61) correspond to the emission from the external pseudoscalar lines, while the last term is the so-called contact term required by electromagnetic gauge invariance.

Equation (4.61) is, actually, the Low theorem [111] for a given process. In the soft-photon limit there are the terms $O(1/\omega)$ and $O(\omega^0)$ in the radiative transition amplitude; these terms are proportional to the elastic $V \to P^+P^-$ amplitude g_V, and can be obtained in a model-independent form (4.61). As the amplitudes (4.58) and (4.59) are proportional to ω, they are next-to-next-to-leading in the limit $\omega \to 0$.

There are no such terms if the final-state mesons are neutral, so the mechanisms (4.58) and (4.59) compete, depending on the flavour content of the mesons involved. Nevertheless, the existence of the decay chain (4.60) suggests that the process $V \to \gamma P^0 P^0$ can proceed also via an intermediate loop of charged pseudoscalars:

$$V^0 \to \gamma P^+P^-, \quad P^+P^- \to P^0P^0. \qquad (4.62)$$

The latter amplitude is formally $O(1/N_C^{3/2})$, but may be dominant in some kinematical circumstances due to threshold effects.

Special cases are the decays $\phi \to \gamma\pi^0\pi^0$ and $\phi \to \gamma\pi^0\eta$. In the upper part of the final-meson invariant mass spectra narrow peaks are observed [112], which are the manifestation of scalar $f_0(980)$ and $a_0(980)$ resonances.

The ϕ-meson is predominantly $s\bar{s}$. In the quarkonium picture the isovector $a_0(980)$ is made of light quarks. Thus the amplitude (4.25) cannot be responsible for the decay $\phi \to \gamma\pi^0\eta$ in the a_0 resonance region. The same is true for the decay $\phi \to \pi^0\pi^0\gamma$, if the isoscalar $f_0(980)$ is made of light quarks. Conversely, if $f_0(980)$ is an $s\bar{s}$ state, then its decay to $\pi^0\pi^0$ final state is suppressed by the OZI rule [11]. On the other hand, both $a_0(980)$ and $f_0(980)$ are known to couple strongly to the $K\bar{K}$ channel, so independently of the model for these scalars, the decay mechanism via a charged kaon loop is operative, as was suggested in [113] and [114], and this amplitude is enhanced due to the nearby $K\bar{K}$ threshold.

In the kaon loop model the transition current describing the decay $\phi \to \gamma S$, where S is a_0 or f_0, is written as

$$M_\mu = e\frac{g_\phi g_S}{2\pi^2 i m_K^2}I(a, b)(\epsilon_\mu(p \cdot k) - p_\mu(k \cdot \epsilon)), \qquad (4.63)$$

where p and k are the ϕ-meson and photon momenta, m_K is kaon mass, g_ϕ and g_S are $\phi K^+ K^-$ and $SK^+ K^-$ coupling constants, $a = m_\phi^2/m_K^2$, $b = m_S^2/m_K^2$, and $I(a, b)$ is the loop integral function:

$$I(a, b) = \frac{1}{2(a - b)} - \frac{2}{(a - b)^2}\left(f(b^{-1}) - f(a^{-1})\right) + \frac{a}{(a - b)^2}\left(g(b^{-1}) - g(a^{-1})\right),$$
$$(4.64)$$

where

$$f(x) = \begin{cases} -(\arcsin(1/2\sqrt{x}))^2 & x > \frac{1}{4} \\ \frac{1}{4}(\log(\eta_+/\eta_-) - i\pi) & x < \frac{1}{4} \end{cases},$$

$$g(x) = \begin{cases} (4x - 1)^{1/2}\arcsin(1/2\sqrt{x}) & x > \frac{1}{4} \\ \frac{1}{2}(1 - 4x)^{1/2}(\log(\eta_+/\eta_-) - i\pi) & x < \frac{1}{4} \end{cases},$$

$$\eta_\pm = \frac{1}{2x}\left(1 \pm (1 - 4x)^{1/2}\right).$$

Calculation of the loop integral (4.64) is rather instructive, as it is governed to a large extent by electromagnetic gauge invariance. As already mentioned, one should add the contact term in (4.61) to make the $\phi\gamma K\bar{K}$ interaction gauge invariant. Substituting the vertex (4.61) into the loop integral one immediately notes that both the contact graph and the one describing the emission from the kaon line diverge, and only the sum of these graphs is finite. Moreover, the correct finite part of the loop integral is extracted by appealing to gauge invariance. This can be done either by imposing the condition $M_\mu k_\mu = 0$ [114], or by calculating the imaginary part of the amplitude and reconstructing the real part by means of a subtracted dispersion relation, with the subtraction constrained by gauge invariance [113]. In such a way, the integral $I(A, b)$ remains finite in the limit $a \to b$, and the amplitude (4.63) is proportional to the photon momentum.

The data [112] on $\phi \to \gamma\pi^0\eta$ and $\phi \to \gamma\pi^0\pi^0$ are compatible with the kaon loop model [115].

4.3 Two-photon decays of mesons

4.3.1 General motivation

As photons couple to charged particles, $\gamma\gamma$ decays are a useful tool for studying the structure of mesons. The cross sections also reveal the dynamics rather directly. In $\gamma\gamma \to \pi\pi$ at low energies the photon sees the pion's overall charge, whereby $\sigma(\gamma\gamma \to \pi^+\pi^-)$ is much larger near threshold than $\sigma(\gamma\gamma \to \pi^0\pi^0)$. At higher energies the photon begins to resolve the π, coupling to its constituent $q\bar{q}$ and causing them to resonate. The $f_2(1270)$ and various $f_0(980; 400{-}1200; 1500;1710)$ scalar resonances then dominate the cross sections of both charged and neutral $\pi\pi$.

If one had full information on the polarization of the incident photons and angular correlations between the initial and final directions and spins, it would be possible to extract the couplings of states with a definite set of J^{PC}. In practice, experiments have limited angular coverage and initial polarizations are not measured. This restricted information affects determination of the resonance parameters [116] and progress has come from the imposition of constraints.

Low-energy theorems for Compton scattering $\gamma\pi \to \gamma\pi$ at threshold [117,118] absolutely normalize the threshold cross sections for $\gamma\gamma \to \pi\pi$ [119].

At threshold the amplitude is proportional to the squared charge of the π and hence $\sigma(\gamma\gamma \to \pi^+\pi^-) >> \sigma(\gamma\gamma \to \pi^0\pi^0)$. The pion pole that determines the Born amplitude is so near the $\gamma\gamma \to \pi\pi$ physical region that it dominates the behaviour of the $\gamma\gamma \to \pi\pi$ amplitude in the low-energy region [116,120,119].

Above about 500 MeV unitarity adds constraints where each $\gamma\gamma \to h \to \pi\pi$ partial wave amplitude is related to the corresponding hadronic process $h \to \pi\pi$ [121]. At energies below 1 GeV, where $h \equiv \pi\pi$, the constraints are very restrictive but above this energy the $K\bar{K}$ threshold is crossed and coupled channel unitarity requires inputs from $\pi\pi \to K\bar{K}$. The $\eta\eta$ channel is relatively weak but above 1.4 GeV $\gamma\gamma \to 4\pi$ becomes important and analysis has not yet been made. Thus, at the present, highly constrained analyses of $\gamma\gamma \to \pi\pi$ exist only up to ~ 1.4 GeV.

The $\gamma\gamma$ couplings of resonances determined in such amplitude analyses are calculated by two different methods [120]. One involves analytic continuation to the pole position of the extracted amplitudes found in the complex s-plane and is formally correct; the other is a more naive approach based on the height of the Breit–Wigner-like peak. In the case of prominent isolated resonances, such as the $f_2(1270)$, the two methods give nearly identical results. However, for the $f_0(980)$, which overlaps both the $K\bar{K}$ threshold and the very broad S-wave enhancement described as $f_0(400-1200)$ [121], only the pole method is applicable. Conversely, for the $f_0(400-1200)$ itself, the pole is far from the real axis and only the peak-height provides a sensible measure of its $\gamma\gamma$ width via

$$\Gamma_{\gamma\gamma}^R(peak) = \frac{\sigma_{\gamma\gamma}^{peak} m_R^2 \Gamma_{tot}}{8\pi(\hbar c)^2(2J+1)BR}, \tag{4.65}$$

where BR is the hadronic branching ratio to the particular final state.

The resulting widths are found to be [121]

$$\Gamma(f_0(980) \to \gamma\gamma) = (0.28^{+0.09}_{-0.13}) \text{ keV},$$
$$\Gamma(f_0(400-1200) \to \gamma\gamma) = (3.8 \pm 1.5) \text{ keV}. \tag{4.66}$$

The implications of these for models are discussed later.

The $\gamma\gamma$ widths of the prominent tensor resonances are abstracted from data on $\gamma\gamma \to \pi\eta(a_2)$ and $\gamma\gamma \to K\bar{K}(f_2(1525))$ relatively directly. The current values in keV are [8]

$$\Gamma_{\gamma\gamma}(a_2(1320) : f_2(1270) : f_2(1525)) =$$
$$1.00 \pm 0.08 : 2.61 \pm 0.30 : 0.093 \pm 0.015. \tag{4.67}$$

These electromagnetic couplings of the 2^{++} states confirm their tendency towards ideal flavour states. Independently we know that the $f_2(1270)$ is to good approximation the $n\bar{n}$ member of an ideal flavour nonet. Here the decays to $\gamma\gamma$ of the neutral members have amplitudes proportional to the sum of the **squares** of the electric charges of the quarks weighted by their relative phases. Thus for the relative squared amplitudes

$$a_2(u\bar{u} - d\bar{d})/\sqrt{2} : f_2(u\bar{u} + d\bar{d})/\sqrt{2} : f_2(s\bar{s}) = 9 : 25 : 2, \tag{4.68}$$

which are very close to the experimental values in (4.67) (the shortfall for the $s\bar{s}$ state being consistent with the suppression of magnetic couplings $\sim (m_n/m_s)^2$).

The predictions for the $\gamma\gamma$ widths depend not only on the flavour content but also on the probability for the constituents to overlap in order to annihilate. Within a $q\bar{q}$ multiplet we expect these to be approximately the same, as confirmed above. If this is true throughout the supermultiplet of $q\bar{q}$ states $^3P_{2,1,0}$, one can relate the tensor widths to those of the scalars, with some model dependence. We now review this and then consider the implications for the light scalar mesons.

4.3.2 Non-relativistic approach to $Q\bar{Q} \to \gamma\gamma$

Two-photon widths have been estimated by modelling mesons as non-relativistic $Q\bar{Q}$ states. In lowest-order perturbation theory, begin with the free-quark scattering amplitude $\mathcal{M}_{scat}$ and use the prescription

$$\mathcal{M}_{BS} = \int d^3 p \, \Phi(\mathbf{p}) \mathcal{M}_{scat} \frac{(2\pi)^{3/2}(2M)^{1/2}}{[(2\pi)^{3/2}(2m)^{1/2}]^2}, \qquad (4.69)$$

where $\mathcal{M}_{BS}$ is the invariant amplitude for the $\gamma\gamma$ transition from the bound state to the final state, $\Phi(\mathbf{p})$ is the momentum-space wave function, M is the bound-state mass and $m = \frac{1}{2}M$ is the quark mass. Then if $R(r)$ is the radial part of the wave function normalized to $\int dr \, r^2 |R(r)|^2 = 1$, including a factor of three for colour, and $\langle e_q^2 \rangle^2$ is the square of the effective quark charge for the process, one has [122–124]

$$\Gamma_{\gamma\gamma}(^1S_0) = 12\alpha^2 \langle e_q^2 \rangle^2 \frac{|R(0)|^2}{M^2}, \qquad (4.70)$$

$$\Gamma_{\gamma\gamma}(^3P_0) = \frac{15}{4}\Gamma_{\gamma\gamma}(^3P_2) = 432\alpha^2 \langle e_q^2 \rangle^2 \frac{|R'(0)|^2}{M^4} \qquad (4.71)$$

and

$$\Gamma_{\gamma\gamma}(^1D_2) = 192\alpha^2 \langle e_q^2 \rangle^2 \frac{|R''(0)|^2}{M^6}. \qquad (4.72)$$

These results have been generalized in [125]. Denoting $R^{(l)}(0)$ as the lth derivative of the wave function, the non-relativistic $\gamma\gamma$ widths of positronium are of the form

$$\Gamma[^{2S+1}L_J(e^+e^-) \to \gamma\gamma] = \Sigma_{\lambda=0,2}\Gamma_\lambda \frac{\alpha^2}{m_e^{2L+2}}|R^{(l)}(0)|^2. \qquad (4.73)$$

The allowed values of the $\gamma\gamma$ helicity λ are 0 and 2 for $(S = 1, J = L \pm 1)$, and $\lambda = 2$ only for the 'middle of the multiplet' triplet states. For the $q\bar{q}$ case again, replace $m_e \to m_q$, and multiply by a flavour factor $|\langle e_q^2/e^2 \rangle|^2$ and a colour factor of three. The authors argue (see next subsection) that M in the above formulae should be understood as $2m$, where m is the constituent mass. This distinction becomes important for high partial waves.

For the spin-singlet case, $J^{PC} = \text{even}^{-+}$, the reduced partial width is

$$\Gamma_{\lambda=0}(^1L_{J=L}) = 1 \qquad (4.74)$$

for all L [126]. This reproduces (4.70) and (4.72) as special cases. For spin-triplet states, $S = 1$ and $L = \text{odd}$, the coefficients Γ_λ have non-trivial J, L, λ dependences. The general results are [125]

$$\Gamma_{\lambda=2}(^3L_{J=L+1}) = \frac{(L+2)(L+3)}{L(2L+3)}, \qquad (4.75)$$

$$\Gamma_{\lambda=0}(^3L_{J=L+1}) = 0, \qquad (4.76)$$

$$\Gamma_{\lambda=2}(^3L_{J=L}) = \frac{(L-1)(L+2)}{L^2}, \tag{4.77}$$

$$\Gamma_{\lambda=2}(^3L_{J=L-1}) = \frac{(L-2)(L-1)(L+1)}{L^2(2L-1)}(1+\chi)^2, \tag{4.78}$$

$$\Gamma_{\lambda=0}(^3L_{J=L-1}) = \frac{(2L+1)^2}{L(2L-1)}, \tag{4.79}$$

where χ is

$$\chi = 0(L=1); = \frac{L(2L+1)}{2(L-2)(L-1)}(L \geq 3). \tag{4.80}$$

Spin-triplet decays of immediate interest are those with $L = 1, 3$. For $L = 1$ the above results recover the well-known *relative* widths of

$$[\Gamma_{\lambda=2}(^3P_2) : \Gamma_{\lambda=0}(^3P_2)] : \Gamma_\lambda(^3P_1) : \Gamma_{\lambda=0}(^3P_0) = [1:0] : 0 : \tfrac{15}{4}. \tag{4.81}$$

The light-quark $L = 3$ $q\bar{q}$ states will probably be the first $L > 2$ states to be detected in $\gamma\gamma$. The theoretical partial widths and helicity couplings implicit in the above relations are potentially useful as experimental signatures. For the $L = 3$ multiplet, the relative $\gamma\gamma$ widths, summed over helicities, are

$$\Gamma(^3F_4) : \Gamma(^3F_3) : \Gamma(^3F_2) = 1 : 1 : \tfrac{919}{100}, \tag{4.82}$$

which imply that the 2^{++} $^3F_2$ state should be the easiest to observe by an order of magnitude in the partial width (but somewhat less in the cross section due to $(2J + 1)$ factors). We expect the 3^{++} and 4^{++} ^3F_J states to have similar $\gamma\gamma$ widths; this is in marked contrast to the analogous $L = 1$ states for which the $J = l$; 1^{++} state is forbidden to decay to on-shell photons by the Landau–Yang symmetry theorem. As the 4^{++} state $f_4(2050)$ is well established and has known branching ratios to $\pi\pi$ and $\omega\omega$ [8], a search for $\gamma\gamma$ production of this state could be carried out using existing data sets. Both the helicity couplings of $^3F_2$ $q\bar{q}$ states are present with comparable amplitudes

$$\frac{\Gamma_{\lambda=0}(^3F_2)}{\Gamma_{\lambda=2}(^3F_2)} = \frac{294}{625}, \tag{4.83}$$

so it should be possible to distinguish this state from a radially excited $^3P_2$ for which the $\lambda = 2$ amplitude is dominant. Close and Li [127] found that $\lambda = 2$ dominance is characteristic of the lightest 2^{++} hybrids as well; hence a large $\lambda = 0$ width for a 2^{++} state would be a clear signature for $^3F_2$.

The matrix element for $Q\bar{Q} \rightarrow \gamma\gamma$ is the coherent sum of e_q^2 contributions from the individual quark flavours. For $I = 1$, $\frac{1}{\sqrt{2}}(u\bar{u} - d\bar{d})$ this is $\langle e_q^2 \rangle^2 = \frac{1}{18}$. For $I = 0$ the mixing between $n\bar{n}$ and $s\bar{s}$, or equivalently between $SU(3)_F$ **8** and **1**, can be deduced by comparison of the $\gamma\gamma$ widths. As a function of the mixing angle (where $\theta = 0$ is ideal mixing) the two $\langle e_q^2 \rangle^2$ are

$$\langle e_q^2 \rangle_a^2 = \left[\frac{5}{9\sqrt{2}} \cos\theta - \frac{1}{9} \sin\theta \right]^2; \quad \langle e_q^2 \rangle_b^2 = \left[\frac{5}{9\sqrt{2}} \sin\theta + \frac{1}{9} \cos\theta \right]^2. \quad (4.84)$$

This implies that at least one of the two isoscalars must have a significantly larger $\gamma\gamma$ width than the $I = 1$ partner. Furthermore, the correlation of $d\bar{d}$ (lighter) and $s\bar{s}$ (heavier) masses and the above couplings implies that one expects the lighter of the two isoscalars to have the larger $\gamma\gamma$ partial width. As we have seen above, the tensor mesons agree with these generalities with $\theta \sim 0$.

The above relations are among different flavoured states of the same J^{PC}. Complementary to this we have relations among common flavour states of different J^{PC}, typified by the relations (4.70)–(4.72).

These formulae can be applied to the hadronic decays of heavy flavour $Q\bar{Q}$ where in perturbative QCD these are driven by $Q\bar{Q} \rightarrow gg$ annihilation. At leading order the relation is

$$\Gamma_{\gamma\gamma}(R) = \frac{9\langle e_q^2 \rangle^2}{2} \left(\frac{\alpha}{\alpha_s} \right)^2 \Gamma_{gg}(R), \quad (4.85)$$

though there are considerable QCD corrections, in particular the possibility of $c\bar{c}g$ components in the wave function such that the (colour octet) $c\bar{c}$ can annihilate into a gluon in S-wave. These may be responsible for the excessive relative strength in the 0^{++} [8], where for $c\bar{c}$ the ratio of $0^{++} : 2^{++}$ hadronic widths can be compared with (4.71):

$$\Gamma_T(0^{++}) : \Gamma_T(2^{++}) = 8 \pm 1 \quad (4.86)$$

and for the $\gamma\gamma$ widths

$$\Gamma_{\gamma\gamma}(0^{++}) : \Gamma_{\gamma\gamma}(2^{++}) = 7 \pm 2. \quad (4.87)$$

Although the magnitudes of the ratio are not precise, the clear message is that the scalar width to $\gamma\gamma$ is significantly larger than that of the tensor, as predicted in the non-relativistic picture.

4.3.3 Phenomenology

The absolute scale of $\gamma\gamma$ widths of light $q\bar{q}$ states is a rather sensitive quantity and one can easily be misled into believing that predictions of $\gamma\gamma$ widths are stable by models that reproduce the $0^{-+}, 2^{++}$ widths to $\sim 30\%$. The $^1D_2$ is an example. Ackleh and Barnes [126] stressed that the calculations involve m_q more directly than $\frac{1}{2}M$. The resulting width for the $^1D_2$ is proportional to m_q^{-6} and so is highly sensitive. With $m_q \sim 330$ MeV, Ackleh and Barnes [126] found $\Gamma_{\gamma\gamma} \sim 0.3 - 2$ keV, whereas with $M = 1670$ MeV the width is predicted to be only 1–10 eV! Comparison with data [8] is unclear. Signals have been claimed corresponding to $\Gamma_{\gamma\gamma}(\pi_2(1670)) \sim 1$ keV, whereas other experiments report only upper limits of less than 72 eV. The test of this procedure is to apply it in searching for the $I = 0$ partner, $\eta_2(\sim 1700)$ for which $\Gamma_{\gamma\gamma} \sim (25/9)\Gamma_{\gamma\gamma}(\pi_2)$.

Ackleh and Barnes [126] derived the exact relativistic result for the singlet $q\bar{q}$:

$$\Gamma^J(\gamma\gamma) = \frac{6\alpha^2}{\pi m^2}\left\langle\frac{e_q^2}{e^2}\right\rangle^2 \mid \int_0^\infty dp\, p^2\phi(p)\left(\frac{1-\beta^2}{\beta}\right) Q_J(\beta^{-1})\mid^2, \tag{4.88}$$

where $Q_J(x)$ is the Legendre function of the second kind. As $\beta \to 0$ this is

$$\Gamma_{NR}(^1D_2 \to \gamma\gamma) = \frac{3\alpha^2}{m^6}\left\langle\frac{e_q^2}{e^2}\right\rangle^2 |R''(0)|^2, \tag{4.89}$$

which agrees with Anderson *et al* [124] so long as their $M \to 2m_q$.

The limitations of such calculations are exposed for the $\gamma\gamma$ decay widths of π, η, η' that are approximately proportional to M_R^3. This can be motivated by an effective Lagrangian

$$\mathcal{L} = \frac{1}{2}g\phi F_{\mu\nu}F^*_{\mu\nu} \to \Gamma_{\gamma\gamma} = \frac{1}{64\pi}g^2 M_R^3. \tag{4.90}$$

The problem with the quark-model calculation is that it does not incorporate the physical mass of the initial state, which must be restored as a physical input. In effect the quark-model calculation determines g and then imposes the M_R^3 dependence. Similar arguments lead to M_R^3 dependence for the $L = 1, 2$ $q\bar{q}$ state also.

The relativistic calculations of [126] find no significant suppression with increasing L for $\Gamma(q\bar{q} \to \gamma\gamma)$ for light flavours. This would be consistent if the signals claimed for π_2 [8] are confirmed, as they are of similar strength to that of the $a_2 \to \gamma\gamma$. If this is the case, then one might plausibly expect to observe higher-L states such as $L = 3 (2, 3, 4)^{++}$.

This is in marked contrast to the predictions for heavy flavours. The prediction for $c\bar{c}$ is successful

$$\Gamma_{\gamma\gamma}(\eta_c) = 4.8 \text{ keV (Theory, [126])}; \quad 7.5 \pm 1.8 \text{ keV (Data [8])}, \tag{4.91}$$

though the predictions are rather sensitive to the choice of m_c [128]. There is strong orbital suppression, $0^{-+} \rightarrow 2^{-+}$ falling by $\sim 10^2$ for $c\bar{c}$ and by $\sim 10^3$ for $b\bar{b}$. Hence it is unrealistic to expect to see $L \geq 2$ $c\bar{c}$ and $b\bar{b}$ coupling to $\gamma\gamma$.

4.3.4 Scalar mesons in $\gamma\gamma$

The non-relativistic calculations that give rise to the above also imply that the tensor is produced polarized, $A(\gamma\gamma \rightarrow 2^{++})_{J_z=0} = 0$. This result is predicted to be robust even for light relativistic $q\bar{q}$ [129] and appears to be good empirically for the $f_2(1270)$. Thus it may be reasonable to apply these ideas to light flavours. Including M^3 factors as discussed above

$$\Gamma_{\gamma\gamma}({}^3P_0) = \frac{15}{4} \left(\frac{m_0}{m_2}\right)^3 \Gamma_{\gamma\gamma}({}^3P_2). \tag{4.92}$$

The relativistic corrections to this have been calculated [129]. For $m_q \rightarrow 0$ the ratio $\rightarrow 0$ as a result of chiral symmetry. For constituent masses of ~ 300 MeV the ratio $\Gamma_{\gamma\gamma}({}^3P_0)/\Gamma_{\gamma\gamma}({}^3P_2)$ is reduced by ~ 2; a larger $\gamma\gamma$ width for 0^{++} relative to 2^{++} is still expected. The width for $f_0(400 - 1200)$ in (4.66) is thus consistent with ${}^3P_0$ $n\bar{n}$, whereas that for $f_0(980)$ is not, though ${}^3P_0$ $s\bar{s}$ is possible.

There is much discussion that the $f_0(980)$ may be a $K\bar{K}$ molecule [130] or a $qq\bar{q}\bar{q}$ state [131] (for detailed discussion of this issue and further references see [132]).

For a $K\bar{K}$ molecule, the fourth power of the constituent charges is much larger than for an $s\bar{s}$ state but the state is more diffuse and so the annihilation probability will be much smaller. The actual magnitude of the $\gamma\gamma$ width is thus model-dependent. In the non-relativistic model of [130], the $\gamma\gamma$ width has been calculated [133] to be $\Gamma(f_0(K\bar{K}) \rightarrow \gamma\gamma) = 0.6$ keV, which is somewhat larger but not inconsistent with the data in (4.66). Within the compact $qq\bar{q}\bar{q}$ picture the predictions are approximately 0.3 keV for both f_0 and a_0 [134,113]. Experimentally [8]

$$\Gamma_{\gamma\gamma}(f_0(980)) = 0.39^{+0.10}_{-0.13}\text{keV}; \quad \Gamma_{\gamma\gamma}(a_0(980)) = 0.24^{+0.08}_{-0.07}\text{keV}, \tag{4.93}$$

where for the a_0 it is assumed that the branching ratio to $\eta\pi$ is 100%. Thus technically its $\gamma\gamma$ width is larger than this value.

There is also discussion as to whether these states are compact $qq\bar{q}\bar{q}$ states and what implications this has for their $\gamma\gamma$ widths [113]. Rather than focussing on absolute magnitudes there may be lessons to learn from the ratio of the $f_0/a_0 \rightarrow \gamma\gamma$ widths.

From (4.93) this is in the region of 1–2. This is somewhat below the $\frac{25}{9}$ expected if these were pure $s\bar{s}$ and $n\bar{n}$ states, as could be the case given their mass degeneracy [135]. It is consistent with unity, as expected in the simplest form of $K\bar{K}$ molecule, where only the charged kaons contribute and isospin is assumed to be exact. It is possible that, due to the closeness of $K\bar{K}$ thresholds, and the mass difference between charged and neutral kaons, there could be significant breaking of G-parity in these states. The effect on the $\gamma\gamma$ ratio has not been definitively discussed.

It is generally agreed that the $f_0(980)$ and $a_0(980)$ are not simply $q\bar{q}$ but have significant $qq\bar{q}\bar{q}$ or $K\bar{K}$ components in their wave functions. It is not clear how $\gamma\gamma$ probes the deep structure of their wave functions. We know that for 2^{++} the $\gamma\gamma$ reads the compact $q\bar{q}$ flavours; there is no two-body S-wave competition in the imaginary part as $\rho\rho$ and other allowed hadronic channels are all too heavy. One would expect that for the 0^{++} $K\bar{K}$ will dominate if there is a long-range $K\bar{K}$ component in the wave function. At the other extreme, were the state a pure compact four-quark, then the higher intermediate states – KK, KK^*, KK^{**}, $\cdots$ – would all be present but the *ratios* of the $\gamma\gamma$ couplings to f_0/a_0 would probably be sensitive and more reliable.

The production by highly virtual $\gamma^*\gamma^*$ in $e^+e^- \to e^+e^- f_0/a_0$ could probe the spatial dependence of their wave functions. It would be especially instructive were the ratio to be strongly Q^2 dependent.

Above 1 GeV there are other scalars, notably the $f_0(1370, 1500, 1710)$ and an isovector $a_0(1450)$. The $\gamma\gamma$ widths of these states can be important in elucidating their structure. This is particularly interesting in view of the expectation that there is a scalar glueball in this region and may mix with the $I = 0$ states [136–138]. The effects of such mixings on the pattern of $\gamma\gamma$ widths can be dramatic [139]. Only rough estimates of these widths are available and a detailed analysis is awaited. ALEPH have quoted [140] $\Gamma(f_0(1500) \to \gamma\gamma) < 0.17$ keV. This small value could point towards significant $s\bar{s}$ content (though this seems at odds with hadronic decays of this state [137,138]) or that it is a glueball or, as is certainly plausible, that the $\gamma\gamma$ width is not yet well determined.

To extend amplitude analysis of $\gamma\gamma$ beyond 1.4 GeV will require studies of $\gamma\gamma \to 2\pi, 4\pi$ and $K\bar{K}$. Only when these channels are analysed simultaneously can reliable scalar signals be extracted from beneath the dominant tensor effects. The width $f_0(1710) \to \gamma\gamma$ in particular is important to measure in order to complete the analysis and isolate the role of a scalar glueball [139].

4.3.5 Glueballs and $\gamma\gamma$ constraints

If a state is a glueball it will occur in $\psi \to \gamma R_J$ as a singleton and be strongly suppresssed in $R_J \to \gamma\gamma$. By contrast, if R_J is an $I = 0$ member of a $q\bar{q}$ nonet, there will be two orthogonal states in the **1-8** flavour basis for production in both $\psi \to \gamma R_J$ and $R_J \to \gamma\gamma$. Flavour mixing angles may suppress one or the other of the pair in either $R_J \to \gamma\gamma$ or $\psi \to \gamma R_J$ but there are strong correlations between the two processes, so that a comparison between them can help to distinguish glueball from $q\bar{q}$. In particular, if a $q\bar{q}$ state is flavour 'favoured' in $\psi \to \gamma R_J$, so that it is prominent and superficially 'glueball-like', it will also be flavour favoured in $\gamma\gamma \to R(q\bar{q})$ in dramatic contrast to a glueball. The detailed arguments are presented in [139].

Such discrimination between a glueball G and a $q\bar{q}$ M can be rather powerful. Chanowitz [141] suggests a quantitative measure for glueballs via 'stickiness', which after phase space is removed is given by

$$S_R = \frac{\Gamma(\psi \to \gamma R)}{\Gamma(R \to \gamma\gamma)}, \tag{4.94}$$

such that for G and M with the same J^{PC}, $S_G/S_M \sim 1/\alpha_s^4 \gg 1$.

The essence of stickiness is that $G \to \gamma\gamma$ proceeds via a $q\bar{q}$ loop whereby

$$\frac{\Gamma(G \to \gamma\gamma)}{\Gamma(M \to \gamma\gamma)} \sim \left(\frac{\alpha_s}{\pi}\right)^2, \tag{4.95}$$

while in perturbation theory

$$\frac{\Gamma(\psi \to \gamma G)}{\Gamma(\psi \to \gamma M)} \sim \left(\frac{1}{\alpha_s}\right)^2. \tag{4.96}$$

This provides a qualitative distinction between G and $q\bar{q}$ with some empirical success, for example that $S(\eta(1440)) \sim 10S(\eta(550))$. However, the absolute normalization of S was not defined until [139].

The detailed analysis in section 5 of [139] showed how the relationship has to be applied to the set of $I = 0$ states in a nonet and that for the $q\bar{q}$ states $f_2(1270)$ and $f_2(1520)$ the $\gamma\gamma$ and $\psi \to \gamma R$ data give consistent results if this pair are orthogonal $q\bar{q}$ states. For the scalar mesons above 1 GeV the analysis involves both $q\bar{q}$ and G states mixed together. It is data, such as $\gamma\gamma$ couplings, that are required to determine these mixings.

Within a tightly-constrained mixing analysis based on the known hadronic decays of these states, two possible solutions were found. The details are in [139] and [138]. They can, in principle, be distinguished once $\gamma\gamma$ widths are established. Both schemes imply that $\Gamma(f_0(1710) \to \gamma\gamma) = 1-2$ keV, while $\Gamma(f_0(1500) \to \gamma\gamma) \sim 0.1-0.5$ keV. In general it was found that for three states, such as the $f_0(1370, 1500, 1710)$, that are mixtures of $n\bar{n}$, $s\bar{s}$ and G, the pattern of relative $\gamma\gamma$

widths for the three states will be $f_0(1370) : f_0(1500) : f_0(1710)$ = large: small: medium, typically 10:≤1:3 in order of magnitude.

This needs to be tested for these scalar states. This relative ordering of $\gamma\gamma$ widths is a common feature of mixings for all initial configurations for which the bare G does not lie nearly degenerate with the $n\bar{n}$ state. As such it is a robust test of the general idea of $n\bar{n}$ and $s\bar{s}$ mixing with a nearby G. If, say, the $\gamma\gamma$ width of the $f_0(1710)$ were to be smaller than that of the $f_0(1500)$ or comparable or greater than that of the $f_0(1370)$, the general hypothesis of three-state mixing would be disproved. The corollary is that qualitative agreement may be used to begin isolating in detail the mixing pattern.

Combined with the radiative analyses described elsewhere in this chapter, these could be critical tests for identifying the presence and mixing dynamics of glueballs.

4.3.6 $\gamma\gamma$ amplitudes and relation to $V\gamma$

The $\gamma\gamma$ decay widths of $0^{++}/2^{++}$ being in the ratio of 15/4 arises non-relativistically and it is instructive to see how this is related to the essential $q\bar{q}$ structure and to the electromagnetic transition. The $2J + 1$ factor accounts for a factor 5; the essential spin dynamics yields 3/4 and it is this latter that we need to understand.

The approach of a quark-model practitioner is to consider [98,84,99]

$$\langle q\bar{q}, S = 1, L = 1; J = 0, 2|H_{em}|q\bar{q}, S = 1, L = 0; J = 1\rangle \tag{4.97}$$

for transitions between 0^{++} or 2^{++} and a vector meson ($^3S_1$), the latter then turning into the second photon by vector-meson dominance. In this

$$H_{em} \sim AL_+ + BS_+ + CS_zL_+ \tag{4.98}$$

is the most general transformation structure for a positive helicity γ that can flip the $q\bar{q}$ system's L_z, S_z by one unit (with arbitrary strength A, B respectively) or flip L_z weighted by S_z (strength C). The magnitudes of A, B, C may be calculated in a specific model but in general the relative matrix elements for $J = 0, 2$ may be related to Clebsch–Gordan coefficients driven by the L_+, S_+, S_zL_+ folded into the $L \otimes S = J$ of the $q\bar{q}$ states. The result is

$$T_2 = A + C, \quad T_0 = \frac{1}{\sqrt{6}}(A + 2B - C), \quad S_0 = \frac{1}{\sqrt{3}}(A - B - C), \tag{4.99}$$

where T, S refer to tensor or scalar meson and the subscript denotes J_z. See also page 370 in [96].

The static limit of the electric-dipole approximation corresponds to retaining only the A terms. However, both electric and magnetic transitions can occur in the non-relativistic limit of the positronium calculation, corresponding to [129]

$$A = -2B, \quad C \equiv 0. \tag{4.100}$$

This immediately gives

$$T_0 \equiv 0, \quad S_0 \equiv \frac{\sqrt{3}}{2}T_2. \tag{4.101}$$

Hence we see that the helicity-zero selection rule and the $\frac{15}{4}$ ratio of widths ($\frac{15}{4} \equiv (2J + 1)(\sqrt{\frac{3}{2}})^2$) emerges.

That the 'magic' relation at (4.101) follows may be seen by disentangling the positronium calculation. The key is to study its γ-matrix structure and to interpret it in the time-ordered way [129]. The essential structure of the decay in the non-relativistic limit is

$$(p \cdot \epsilon_2 + i\sigma \cdot k \times \epsilon_2 k \cdot p)\sigma \cdot \epsilon_1 \tag{4.102}$$

and hence the structure at (4.100).

But we can now go further and consider the effect of relativistic corrections, specifically p^2/m^2 effects [129]. Note that even for $c\bar{c}$ systems $v^2/c^2 \sim 1/4$ and so these effects cannot simply be ignored. The structure now is more general than (4.102).

Full details are given in [129] but we may approximate the effects here by replacing (4.100) by $A \sim -2B; C \neq 0$. Equation (4.99) then becomes

$$T_2 = A + C, \quad T_0 = \frac{1}{\sqrt{6}}(0 - C), \quad S_0 = \frac{1}{\sqrt{3}}\left(\frac{3A}{2} - C\right), \tag{4.103}$$

which shows that in the rates $|T_0|^2 \sim 0$ in leading order in the relativistic correction. If A and C are positive (as in specific models), then the scalar rate is reduced and the tensor relatively enhanced, that is

$$\frac{\sigma(S \to \gamma\gamma)}{\sigma(T \to \gamma\gamma)} < \frac{15}{4}, \tag{4.104}$$

though still

$$T_0 \sim 0. \tag{4.105}$$

This prediction of a vanishing amplitude for T_0 is thus robust and has even been used as a constraint in some amplitude analyses [142]. In summary, the relative $\gamma\gamma$ amplitudes for $C = +$ mesons complement their radiative transitions to and from vector-mesons, such as $M \to \gamma(\rho, \omega, \phi)$ and provide a sharp measure of the flavour content of mesons. These promise to come into practical focus with the advent of high-statistics data at ψ factories.

References

[1] Y S Tsai, Physical Review D4 (1971) 2821; *ibid* D13 (1976) 1771(E)
[2] W J Marciano and A Sirlin, Physical Review Letters 61 (1988) 1815
E Braaten and C S Li, Physical Review D42 (1990) 3888
E Braaten, S Narison and A Pich, Nuclear Physics B373 (1992) 581
R Decker and M Finkemeier, Nuclear Physics B438 (1995) 17
J Erler, hep-ph/0211345
[3] A B Arzubov *et al*, Journal of High Energy Physics 9812 (1998) 009
[4] S Binner, J H Keuhn and K Melnikov, Physics Letters B459 (1999)279
[5] M Benayoun *et al*, Modern Physics Letters A14 (1999) 2605
[6] J P Alexander *et al*, CLEO Collaboration, Physical Review D64 (2001) 092001
[7] Z Ligeti, M Luke and M B Wise, Physics Letters B507 (2001) 142
[8] S Eidelman *et al*, Particle Data Group, Physics Letters B592 (2004) 1
[9] R R Akhmetshin *et al*, CMD-2 Collaboration, Physics Letters B527 (2002) 161
[10] M N Achasov *et al*, SND Collaboration, Physical Review D68 (2003) 052006
[11] S Okubo, Physics Letters 5 (1966) 1173
G Zweig, CERN report TH-412, Geneva (1964)
G Zweig in *Developments in the Quark Theory of Hadrons* volume 1, D B
Lichtenberg and S P Rosen eds, Hadronic Press, Nonantum (1980)
J Iizuka, Progress in Theoretical Physics Supplement 37/38 (1966) 21
[12] N N Achasov and A A Kozhevnikov, Physical Review D52 (1995) 3119; Physics of
Atoms and Nuclei 59 (1996) 144
[13] C Erkal and M G Olsson, Zeitschrift für Physik C31 (1986) 615
[14] A Donnachie and H Mirzaie, Zeitschrift für Physik C33 (1987) 407
[15] A Donnachie and A B Clegg, Zeitschrift für Physik C51 (1987) 689
[16] A B Clegg and A Donnachie, Zeitschrift für Physik C62 (1994) 455
A Donnachie and A B Clegg, Physical Review D51 (1995) 4979
[17] N N Achasov and A A Kozhevnikov, Physical Review D55 (1997) 2663
[18] B Aubert *et al*, BABAR Collaboration, Physical Review D71 (2005) 052001
[19] D Bisello *et al*, DM2 Collaboration, Physics Letters B220 (1989) 321
[20] S Anderson *et al*, CLEO Collaboration, Physical Review D61 (2000) 112002
[21] B Aubert *et al*, BABAR Collaboration, Physical Review D70 (2004) 072004
[22] D Bisello *et al*, DM2 Collaboration, Physics Letters 107B (1981) 145
[23] R Baldini-Ferroli, Proceedings of "Fenice" Workshop, Frascati, INFN, Frascati
(1988)
[24] A B Clegg and A Donnachie, Zeitschrift für Physik C45 (1990) 677
[25] P L Frabetti *et al*, E637 Collaboration, Physics Letters B514 (2001) 240
[26] P L Frabetti *et al*, E637 Collaboration, Physics Letters B578 (2004) 290
[27] A Antonelli *et al*, FENICE Collaboration, Physics Letters B365 (1996) 427
[28] T Barnes, F E Close, P R Page and E S Swanson, Physical Review D55 (1997)
4157
[29] J M Link *et al*, FOCUS Collaboration, Physics Letters B545 (2002) 50
[30] D Bisello *et al*, DM2 Collaboration, Nuclear Physics B21 (Proc.Supp.) (1991) 111
[31] M N Achasov *et al*, SND Collaboration, Physics Letters B486 (2000) 29
[32] K W Edwards *et al*, CLEO Collaboration, Physical Review D61 (2000) 072003
[33] A Abele *et al*, Crystal Barrel Collaboration, Physics Letters B391 (1997) 191
[34] A Abele *et al*, Crystal Barrel Collaboration, European Physical Journal C21 (2001)
261
[35] D V Bugg, A V Sarantsev and B S Zou, Nuclear Physics B471 (1996) 59
[36] B Hyams *et al*, Nuclear Physics B64 (1973) 134

[37] P Eugenio *et al*, Physics Letters 97B (2001) 190
[38] S Godfrey and N Isgur, Physical Review D32 (1985) 189
[39] A LeYaouanc, L Oliver, O Pene and J Raynal, Physical Review D8 (1973) 2223
    G Busetto and L Oliver, Zeitschrift für Physik C20 (1983) 247
    R Kokoski and N Isgur, Physical Review D35 (1987) 907
    P Geiger and E S Swanson, Physical Review D50 (1994) 6855
    H G Blundell and S Godfrey, Physical Review D53 (1996) 3700
    E S Ackleh, T Barnes and E S Swanson, Physical Review D54 (1996) 6811
[40] A Donnachie and Yu S Kalashnikova, Physical Review D60 (1999) 114011
[41] R R Akhmetshin *et al*, CMD-2 Collaboration, Physics Letters B466 (1999) 392
[42] A Donnachie and Yu S Kalashnikova, Zeitschrift für Physik C59 (1993) 621
[43] F E Close and P R Page, Physical Review D56 (1997) 1584
[44] Yu S Kalashnikova, Zeitschrift für Physik C62 (1994) 323
[45] A Antonelli *et al*, DM2 Collaboration, Zeitschrift für Physik C56 (1992) 15
[46] R R Akhmetshin *et al*, CMD-2 Collaboration, Physics Letters B489 (1999) 125
[47] D V Bugg, Physics Reports 397 (2004) 257
[48] G Shaw, Physics Letters B39 (1972) 255
[49] A Donnachie and G Shaw, in *Electromagnetic Interactions of Hadrons* Vol.2, A Donnachie and G Shaw eds, Plenum Press, New York (1978)
[50] G Veneziano, Il Nuovo Cimento 57A (1968) 190
[51] D G S Leith, in *Electromagnetic Interactions of Hadrons* Vol.1, A Donnachie and G Shaw eds, Plenum Press, New York (1978)
[52] M Atkinson *et al*, Omega Photon Collaboration, Nuclear Physics B243 (1984) 1
[53] J E Brau *et al*, Physical Review D37 (1988) 2379
[54] B Pick, in *Proceedings Ninth International Conference on Hadron Spectroscopy*, D Amelin and A M Zaitsev eds, American Institute of Physics, New York (2002)
[55] W Dünnweber and F Meyer-Wildhagen, in *Proceedings Tenth International Conference on Hadron Spectroscopy*, E Klempt, H Koch and H Orth eds, American Institute of Physics, New York (2004)
[56] S Bartalucci *et al*, Il Nuovo Cimento 49A (1979) 207
[57] D Aston *et al*, LASS Collaboration, SLAC-PUB-5657 (1991)
[58] J P Alexander *et al*, CLEO Collaboration, Physical Review D64 (2001) 092001
[59] G Penso and T N Truong, Physics Letters 95B (1980) 143
[60] V W Hughes and T Kinoshita, Reviews of Modern Physics 71 (1999) S133
[61] A Czarnecki and W J Marciano, Nuclear Physics (Proc.Sup.) B76 (1999) 245
[62] A Czarnecki, B Krause and W J Marciano, Physical Review Letters 76 (1995) 3267; Physical Review D52 (1995) 2619
    A Czarnecki, W J Marciano and A Vainshtein, Physical Review D67 (2003) 073006, erratum ibid D73 (2006), 119901
    S Peris, M Perrottet and E de Rafael, Physics Letters B355 (1995) 523
    M Knecht *et al*, Journal of High Energy Physics 0211 (2002) 003
[63] S Narison, Physics Letters B513 (2001) 53; Erratum *ibid* B526 (2002) 414
[64] J F Trocóniz and F Yndurain, Physical Review D65 (2002) 093001
[65] M Davier, S Eidelman, A Höcker and Z Zhang, European Physical Journal C27 (2003) 497
[66] B Krause, Physics Letters B390 (1995) 392
[67] M Knecht *et al*, Physical Review D65 (2002) 073034
[68] J Bijnes, E Pallante and J Prades, Nuclear Physics B626 (2002) 410
[69] A D Martin, J Outhwaite and M G Ryskin, Physics Letters B492 (2000) 69
[70] S J Brodsky and E de Rafael, Physical Review 168 (1968) 1620

[71] G W Bennett *et al*, Muon g-2 Collaboration, Physical Review Letters 89 (2002) 101804; Erratum *ibid* 89 (2002) 129903

[72] S Ghozzi and F Jegerlehner, Physics Letters B583 (2004) 222

[73] M Beneke, G Buchalla, M Neubert and C T Sachrajda, Physical Review Letters 83 (1999) 1914; Nuclear Physics B591 (2000) 313

[74] G 't Hooft, Nuclear Physics B72 (1974) 461; *ibid* B75 (1974) 461

[75] M Neubert and B Stech, hep-ph/9705292

[76] G S Bali, Physics Reports 343 (2001) 1

[77] R MacClary and N Byers, Physical Review D28 (1983) 1692

[78] S J Brodsky and J R Primack, Annals of Physics (New York) 52 (1969) 315

[79] F E Close and H Osborn, Physics Letters 34B (1971) 400

[80] A Le Yaouanc, L Oliver, O Pene and J C Raynal, Zeitschrift für Physik C40 (1988) 77

[81] J Linde and H Snellman, Nuclear Physics A619 (1997) 346

[82] P Maris and P C Tandy, Physical Review C65 (2002) 045211

[83] P Bicudo, J E F T Ribeiro and R Fernandes, Physical Review C59 (1999) 1107

[84] J Babcock and J Rosner, Physical Review D14 (1976) 1286

[85] G Karl, S Meshkov, and J L Rosner, Physical Review Letters 45 (1980) 215

[86] F Coester and D O Riska, Annals of Physics (New York) 234 (1994) 141

[87] T A Lahde, Nuclear Physics A714 (2003) 183

[88] L A Copley, G Karl, and E Obryk, Nuclear Physics B13 (1969) 303

[89] R P Feynman, M Kislinger and F Ravndal, Physical Review D3 (1971) 2706

[90] S Capstick, Physical Review D46 (1992) 2864

[91] E Eichten, K Gottfried, T Kinoshita, K D Lane and T-M Yan, Physical Review D17 (1978) 3090

[92] E Eichten, K Gottfried, T Kinoshita, K D Lane and T-M Yan, Physical Review D21 (1980) 203

[93] E J Eichten, K Lane and C Quigg, Physical Review Letters 89 (2002) 162002

[94] E J Eichten, K Lane and C Quigg, hep-ph/0401210

[95] F E Close and G J Gounaris, Physics Letters B343 (1995) 392

[96] F E Close, *Introduction to Quarks and Partons*, Academic Press, London (1979)

[97] F E Close, A Donnachie, and Yu S Kalashnikova, Physical Review D67 (2003) 074031

[98] F J Gilman and I Karliner, Physical Review D10 (1974) 2194

[99] F E Close, A Donnachie and Yu S Kalashnikova, Physical Review D65 (2002) 092003

[100] C McNeile and C Michael, Physical Review D63 (2001) 114503

[101] D Coffman *et al*, MarkIII Collaboration, Physical Review D41 (1990) 1410

[102] D Barberis *et al*, WA102 Collaboration, Physics Letters B440 (1998) 225

[103] F E Close and A Kirk, European Journal of Physics C21 (2001) 531

[104] W Lee and D Weingarten, Physical Review D61 (2000) 014015

[105] N Isgur and J Paton, Physical Review D31 (1985) 2910

[106] F E Close and J J Dudek, Physical Review Letters 91 (2003) 142001

[107] F E Close and J J Dudek, Physical Review D69 (2004) 034010

[108] N Isgur, Physical Review D60 (1999) 114016

[109] T Barnes, F E Close and E S Swanson, Physical Review D52 (1995) 5242

[110] F E Close and J J Dudek, Physical Review D70 (2004) 094015

[111] F E Low, Physical Review 110 (1958) 974

[112] A Aloisio *et al*, KLOE Collaboration, Physics Letters B536 (2002) 209; *ibid* B537 (2002) 21

[113] N N Achasov and V N Ivanchenko, Nuclear Physics B315 (1989) 465

[114] F Close, N Isgur and S Kumano, Nuclear Physics B389 (1993) 513

[115] N N Achasov, Nuclear Physics A728 (2003) 425

[116] D Morgan and M R Pennington, Zeitschrift für Physik C39 (1988) 590

[117] F E Low, Physical Review 96 (1954) 1428

[118] M Gell-Mann and M L Goldberger, Physical Review 96 (1954) 1433

[119] M R Pennington *Second DAΦNE Physics Handbook*, L Maiani, G Pancheri and N Paver eds, INFN, Frascati (1995) pp 531–558

[120] D Morgan and M R Pennington, Zeitschrift für Physik C48 (1990) 620

[121] M Boglione and M R Pennington, European Physical Journal C9 (1999) 11

[122] R Barbieri, R Gatto and R Kogerler, Physics Letters 60B (1976) 183

[123] R Barbieri, R Gatto and E Remiddi, Physics Letters 61B (1976) 465

[124] J D Anderson, M H Austern and R N Cahn, Physical Review D43 (1991) 2094

[125] E S Ackleh, T Barnes and F E Close, Physical Review D46 (1992) 2257

[126] E S Ackleh and T Barnes, Physical Review D45 (1992) 232

[127] F E Close and Z P Li, Physical Review D42 (1990) 2194

[128] T Barnes, *Proc of IX International Workshop on γγ Collisions*, D O Caldwell and H P Paar eds, World Scientific, Singapore (1992) p 263

[129] T Barnes, F E Close and Z P Li, Physical Review D43 (1991) 2161

[130] J Weinstein and N Isgur, Physical Review Letters 48 (1982) 659; Physical Review D27 (1983) 588

[131] N N Achasov, S A Devyanin and G N Shestakov, Physics Letters 88B (1979) 367

[132] F E Close and N Tornqvist, Journal of Physics G28 (2002) R249

[133] T Barnes, Physics Letters 165B (1985) 43

[134] N N Achasov, S A Devyanin and G N Shestakov, Physics Letters 108B (1982) 134

[135] F E Close, Y Dokshitzer, V Gribov, V Khoze and M Ryskin, Physics Letters B319 (1993) 291

[136] C Amsler and F E Close, Physics Letters B353 (1995) 385

[137] C Amsler and F E Close, Physical Review D53 (1996) 295

[138] F E Close and A Kirk, Physics Letters B483 (2000) 345

[139] F E Close, G Farrar and Z P Li, Physical Review D55 (1997) 5749

[140] ALEPH Collaboration, Physics Letters B472 (2000) 189

[141] M Chanowitz, in *Proceedings VI International Workshop on γγ Collisions* (1984); LBL Report LBL-18701, LBL, Berkeley

[142] M Boglione and M R Pennington, Physical Review Letters 79 (1997) 1998

5

Intermediate-energy photoproduction

F E Close and A Donnachie

The principal objective of intermediate-energy photoproduction is to determine the spectrum of higher-mass mesons, particularly the little-known $s\bar{s}$ states and exotic $q\bar{q}g$ or $q\bar{q}q\bar{q}$ mesons discussed in chapter 1. The ability to provide linearly-polarized photon beams is an important feature as this can be used as a filter for exotics, in contrast to pion or kaon beams. Understanding the production mechanism has been shown to be an essential feature in extracting spectroscopic information and this requires an understanding of the concept of complex angular momentum and the Regge-pole formalism.

This formalism is first introduced briefly as a description of high-energy scattering and photoproduction processes in general. Details can be found in [1] and [2]. An appreciation of this formalism and its limitations is essential to an understanding of medium-energy and high-energy two-body and quasi-two-body reactions at small $|t|$. The formalism is applied to the photoproduction of pseudoscalar and vector mesons to illustrate these points explicitly. The prospects for exploring the spectrum of exotic mesons are then discussed.

At large $|t|$ it is more appropriate to use a partonic approach to describe the data. At fixed centre-of-mass angles and large $|t|$, the constituent-quark counting rule makes specific predictions for the energy dependence of exclusive processes. The rule was first derived from simple dimensional counting arguments [3] and was later confirmed in a short-distance perturbative QCD (pQCD) approach [4]. Pseudoscalar and vector meson production at $\theta_{c.m.} = 90°$ appear to be compatible with the predictions of the model from the surprisingly low energy of $E_\gamma \approx 3.3$ GeV.

In the extreme backward direction, the cross sections are dominated by Reggeized u-channel baryon exchange. The main interest here focusses on the ϕ as a guide to a possible intrinsic $s\bar{s}$ component in the wave function of the proton.

Electromagnetic Interactions and Hadronic Structure, eds Frank Close, Sandy Donnachie and Graham Shaw.
Published by Cambridge University Press. © Cambridge University Press 2007

5.1 Regge poles

Consider the two-body process $1 + 2 \to 3 + 4$ at large s and small t. The final state need not be the same as the initial state. Most such processes exhibit a strong forward peak and there is a correlation between the existence of a forward peak (small $t < 0$) in s-channel processes and the exchange of particles or resonances in the t-channel. This is connected directly to the quantum numbers of the exchange and reactions which do not have resonances allowed in the t-channel have appreciably smaller cross sections than those of similar ones which do. For example [2] $\sigma(K^- p \to \pi^- \Sigma^+) \gg \sigma(K^- p \to \pi^+ \Sigma^-)$ and $\sigma(\bar{p} p \to \bar{\Sigma}^- \Sigma^+) \gg \sigma(\bar{p} p \to \bar{\Sigma}^+ \Sigma^-)$.

Particle exchange and resonance exchange in the crossed channel thus appear as an important part of high-energy scattering. However, it is straightforward to show [2] that the scattering amplitude cannot be dominated by the exchange of just a few t-channel resonances. For simplicity consider the elastic scattering of equal-mass spinless particles. The t-channel partial-wave series is

$$A(s, t) = 16\pi \sum_{l=0}^{\infty} (2l + 1) A_l(t) P_l(z_t), \tag{5.1}$$

where

$$z_t = \cos \theta_t = 1 + \frac{2s}{t - 4m^2}. \tag{5.2}$$

Suppose that only one resonance is exchanged with a given spin, σ say, so that only one partial wave in (5.1) will contribute. Dropping the other terms in the sum, then for large s

$$A(s, t) = 16\pi (2\sigma + 1) A_\sigma(t) P_\sigma \left(1 + \frac{2s}{t - 4m^2} \right)$$
$$\sim f(t) s^\sigma. \tag{5.3}$$

The optical theorem relates the total cross section σ^{tot} to the imaginary part of the forward amplitude. At large $\sqrt{s}$ the theorem is

$$\sigma^{tot}(s) \sim s^{-1} \operatorname{Im} A(s, t = 0). \tag{5.4}$$

Applying the optical theorem to (5.3) then gives $\sigma^{tot} \sim s^{\sigma-1}$ at large s. Thus exchange of a spin-0 particle, for example the pion, would give a total cross section that decreases as s^{-1}. The exchange of a spin-1 meson, such as the ρ, would give a constant total cross section and the exchange of a spin-2 meson like the f_2 would require the total cross section to increase linearly with s. None of this is observed in total cross sections. Above the s-channel resonance region, say for centre-of-mass energies $\sqrt{s} \geq 2.5$ GeV, total cross sections initially decrease with increasing energy, approximately as $s^{-0.5}$, and then ultimately increase but only slowly, very much more slowly than would be implied by the exchange of a single spin-2 particle in the t channel.

If we are to retain the picture of particle exchange, then all the resonance contributions in the t channel must act collectively and combine in some way to give the observed energy dependence. Thus any meson, the f_2 for example, has to be considered as a member of a whole family of resonances of increasing spin and mass, and we must consider the exchange of all this family simultaneously and their contributions must be correlated with each other. The mathematical framework for adding the resonances together is based on a formalism initially developed by Regge [5–7] for non-relativistic potential scattering. His formalism involved making the orbital angular momentum l, which is initially defined only for non-negative integer values, into a continuous complex variable. He showed that the radial Schrödinger equation with a spherically-symmetric potential can be solved for complex l. That is, the partial-wave amplitudes $A_l(t)$ can be considered as functions $A(l, t)$ of complex l, such that

$$A(l, t) = A_l(t) \qquad l = 0, 1, 2, \ldots. \tag{5.5}$$

Regge found that if the potential $V(r)$ is a superposition of Yukawa potentials, then the singularities of $A(l, t)$ in the complex l-plane are poles whose locations vary with t:

$$l = \alpha(t). \tag{5.6}$$

These poles are known as Regge poles, or reggeons, and as t is varied they trace out paths defined by (5.6) in the complex l-plane. The functions $\alpha(t)$ are called Regge trajectories. In relativistic scattering theory they are associated with the exchanges of families of particles. Values of t such that $\alpha(t)$ is a non-negative integer correspond to the squared mass of a bound state or resonance having that spin. The theory allows us to sum the whole family of exchanges corresponding to the particles associated with the Regge trajectory $\alpha(t)$.

In relativistic theory it is necessary [2] to introduce two amplitudes $A^{\pm}(l, t)$, such that

$$A^{+}(l, t) = A_l(t), \quad l \text{ even}, \quad A^{-}(l, t) = A_l(t), \quad l \text{ odd} \tag{5.7}$$

and write the t-channel partial-wave expansion (5.1) as

$$A(s, t) = A^{+}(s, t) + A^{-}(s, t) \tag{5.8}$$

with

$$A^{\pm}(s, t) = 8\pi \sum_{l=0}^{\infty} (2l + 1) A_l(t)(P_l(z_t) \pm P_l(-z_t)). \tag{5.9}$$

Because $P_l(-z) = (-1)^l P_l(z)$, $A^+(s, t)$ receives contributions only from even l and $A^-(s, t)$ only from odd l. Therefore in $A^+(s, t)$ we can replace $A_l(t)$ by $A_l^+(t)$ and in $A^-(s, t)$ we can replace $A_l(t)$ by $A_l^-(t)$. The amplitudes $A^\pm$ are known as even- and odd-signatured amplitudes.

The final Regge representation for $A^\pm(s, t)$ is [2]

$$A^\pm(s, t) = -\pi \sum_i \frac{\beta_i^\pm(t)}{\Gamma(\alpha_i^\pm(t) + 1)\sin(\pi\alpha_i^\pm(t))} \left(1 \pm e^{-i\pi\alpha_i^\pm(t)}\right) \left(\frac{s}{s_0}\right)^{\alpha_i^\pm(t)}$$

$$= \sum_i \beta_i^\pm(t)\Gamma(-\alpha_i^\pm(t)) \left(1 \pm e^{-i\pi\alpha_i^\pm(t)}\right) \left(\frac{s}{s_0}\right)^{\alpha_i^\pm(t)}, \qquad (5.10)$$

where we have used the property $\Gamma(\alpha + 1)\Gamma(-\alpha) = -\pi/\sin(\pi\alpha)$. The sum in (5.10) is over allowed exchanges, the $\alpha_i^\pm(t)$ are the corresponding Regge trajectories, the $\beta_i^\pm(t)$ the residues at the poles and s_0 an arbitrary fixed scale. The form (5.10) gives the behaviour of $A^\pm(s, t)$ for large s and small t, with $|t| \ll s$. This latter requirement is a very important constraint in applications. Frequently the variable $\nu = \frac{1}{4}(s - u)$ is used instead of s as consideration of s-channel $\leftrightarrow$ u-channel crossing is important. Crossing simply takes $\nu \to -\nu$. It is also common for the Γ-functions in (5.10) to be absorbed into the (unknown) residues $\beta_i(t)$.

The factors

$$\xi_{\alpha_i}^\pm(t) = 1 \pm e^{-i\pi\alpha_i(t)} \qquad (5.11)$$

are called signature factors. It can be shown [2] that for values of t in the s-channel physical region, the amplitudes $A^\pm(l, t)$ are real for real l and for these physical values of t both the positions $\alpha^\pm(t)$ of the poles of the amplitudes and their residues $\beta_i^\pm(t)$ are real, so the phase of the high-energy behaviour of a Regge-pole contribution to $A^\pm(s, t)$ is given by the signature factor. Similar considerations apply to the u-channel [8].

The Regge formalism tells us nothing about the trajectories $\alpha_i^\pm(t)$, which can only be obtained from experiment.

The $\Gamma(-\alpha_i^\pm(t))$ in (5.10) have poles for values of t when one of the $\alpha_i^\pm(t)$ has a non-negative integer value. The signature factors $\xi_{\alpha_i(t)}^+$ vanish when α is an odd integer, so that the even-signatured amplitude has poles at those values of t for which an $\alpha_i^+(t)$ is a non-negative even integer σ^+. Similarly, the odd-signatured amplitude has poles at the values of t for which an $\alpha_i^-(t)$ is a positive odd integer σ^-. These poles can be identified with the exchanges of particles of spin $\sigma^\pm$, whose squared mass is the corresponding value of t.

The real part of the trajectory for the meson states $\rho(770)$, $\rho_3(1690)$ and $\rho_5(2350)$ is shown in figure 5.1, together with its extension to negative t, extracted from

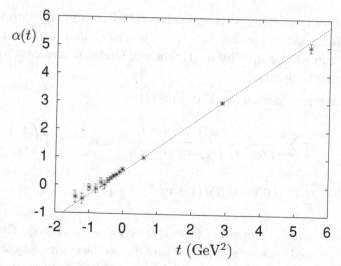

Figure 5.1. Data for $\alpha_\rho(t)$ obtained at $t < 0$ from $\pi^- p$ charge-exchange scattering data and the physical states at $t > 0$, from [2].

the $\pi^- p \to \pi^0 n$ differential cross section. A plot such as this is known as a Chew–Frautschi plot [9]. The near linearity of the trajectory allows a simple extrapolation from the t-channel physical region $t > 4m_\pi^2$ to the s-channel scattering region $t < 0$. That is, the trajectory has the form

$$\alpha(t) = \alpha_0 + \alpha' t. \tag{5.12}$$

The four trajectories $\{f_2(1270), \quad f_4(2050), \ldots\}$, $\{a_2(1320), \quad a_4(2040), \ldots\}$, $\{\omega(780), \quad \omega_3(1670), \ldots\}$ and $\{\rho(770), \quad \rho_3(1690), \ldots\}$ are nearly degenerate [2]. The first pair are even signature, the second pair odd. Each pair contains an isospin-0 and an isospin-1 trajectory, so effectively the 'leading trajectory' contains four near-degenerate trajectories all of which are nearly linear.

Although the real parts of the Regge trajectories $\alpha(t)$ cannot be exactly linear, as above the physical threshold they have a non-zero imaginary part, it turns out that the linear approximation is remarkably good and is sufficient for many purposes, as is the assumption of degeneracy. However, neither is strictly true and the consequences of deviations from both have been extensively explored.

Applying the optical theorem (5.4) to (5.12), a Regge trajectory $\alpha(t)$ gives a term with energy dependence

$$\sigma^{ab}(s) \sim \left(\frac{s}{s_0}\right)^{\alpha_0 - 1}. \tag{5.13}$$

From figure 5.1 we see that the intercept of the trajectories at $t = 0$ is close to 0.5 and this accounts for the observed decrease in hadronic total cross sections at low and intermediate energies. However, all hadronic total cross sections, pp, $\bar{p}p$, pn, $\bar{p}n$ $\pi^{\pm}p$ and $K^{\pm}p$, ultimately increase with energy [10]. The meson exchanges cannot account for this rise, which requires the introduction of a new Regge exchange, the pomeron, which has isospin 0 and $C = +1$. It seems consistent with experiment to assume that the pomeron trajectory $\alpha_{I\!P}(t)$ is linear in t:

$$\alpha_{I\!P}(t) = 1 + \epsilon_{I\!P} + \alpha'_{I\!P}. \tag{5.14}$$

The value of $\epsilon_{I\!P}$ can be obtained from fits [11–14] to the total cross section and lies in the range 0.081–0.112. The value of $\alpha'_{I\!P}$ is determined by fitting the small-t pp cross section and is found [15] to be 0.25 with a high degree of accuracy. As the pomeron is natural-parity exchange, it follows from the signature factor (5.11) that this small value of ϵ means that the pomeron contribution is almost purely imaginary at $t = 0$.

It is generally believed that the pomeron is gluonic in origin and many of its features can be understood qualitatively in terms of the exchange of two non-perturbative gluons [16]. However, it is reasonable to ask whether the pomeron should be interpreted as particle exchange, but in terms of glueballs not $q\bar{q}$ mesons. This suggestion is strongly supported by lattice gauge theory. In [17], the lightest $J = 0, 2, 4, 6$ glueball masses have been calculated in $D = 3 + 1$ SU(3) gauge theory in the quenched approximation and extrapolated to the continuum limit. Assuming that the masses lie on a linear trajectory, the leading glueball trajectory is found to be $\alpha(t) = (0.93 \pm 0.024) + (0.28 \pm 0.02)\alpha'_{I\!R}t$, where $\alpha'_{I\!R} \approx 0.93$ is the slope of the mesonic Regge trajectories. Thus this glueball trajectory has an intercept and slope very similar to that of the phenomenological pomeron trajectory. The results of the calculation and the linear fit are shown in figure 5.2.

Parametrizing high-energy total cross section data by a fixed power $\epsilon_{I\!P} > 0$ of s will ultimately violate the Froissart–Lukaszuk–Martin bound [18,19]:

$$\sigma^{tot}(s) \leq \frac{\pi}{m_{\pi}^2} \log^2(s/s_0) \tag{5.15}$$

for some fixed, but unknown, value of s_0. At currently accessible energies this is not a significant constraint, but it does imply that the power $\epsilon_{I\!P}$ cannot be exactly constant. It must decrease as s increases, but probably rather slowly. A mechanism for this is provided by the exchange of two or more pomerons which moderates the leading power behaviour and, in the case of the eikonal model for example, asymptotically gives a $\log^2 s$ behaviour [2].

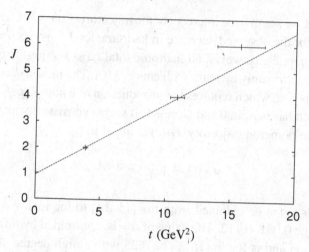

Figure 5.2. Pomeron trajectory from lattice gauge theory [17].

Linearity in the Chew–Frautschi plot is also apparent in the baryon trajectories. The slope is comparable to that for the meson trajectories. An excellent review of the Regge phenomenology of baryon trajectories can be found in [8].

It is familiar that the residue $G(t_0)$ of a particle pole at $t = t_0$ in a partial-wave amplitude $A_\sigma(t)$ factorizes. That is, when it occurs in an s-channel scattering amplitude $1 + 2 \rightarrow 3 + 4$,

$$G^{1+2 \rightarrow 3+4}(t_0) = g^{12}g^{34}.$$ (5.16)

The unitarity relations in the t channel can be used to show that $G(t)$ factorizes similarly, even when t is not close to t_0. Consequently, the $\beta_i^\pm(t)$ in (5.10) have a similar factorization.

5.1.1 Regge cuts

Of course it is possible to exchange more than one reggeon. These multiple exchanges give rise to cuts. Unfortunately the theory of Regge cuts is still not well developed [2]. We do know [20] that the exchange of two reggeons, each having a linear trajectory, yields a cut with a linear $\alpha_c(t)$:

$$\alpha_c(t) = \alpha_c(0) + \alpha_c' t$$ (5.17)

with

$$\alpha_c(0) = \alpha_1(0) + \alpha_2(0) - 1,$$
$$\alpha_c' = \frac{\alpha_1' \alpha_2'}{\alpha_1' + \alpha_2'}.$$ (5.18)

Thus a reggeon–pomeron cut will have an intercept slightly larger and a slope appreciably smaller than those of the reggeon trajectory and so the cut can easily dominate over the pole contribution. Even for a reggeon–reggeon cut, which has a lower intercept than either of the poles, for sufficiently large $|t|$ the cut contribution to the large-s behaviour of the amplitude will dominate over those from the poles, because the slope α'_c is smaller than either of α'_1 and α'_2. So cuts can become important at large t even if they are relatively small at small t.

The cut contributions do not factorize and the leading behaviour for large s is of the form

$$A^c(s, t) \sim s^{\alpha_c(t)}(\log(s))^{-\gamma(t)}. \tag{5.19}$$

The logarithmic factor in (5.19) is highly model-dependent. Although there is no well-formulated theory that allows cut contributions to be calculated, there are models, the most popular one being the eikonal model. This reproduces the general expressions for $\alpha_c(t)$ and allows the cut contributions to be calculated explicitly in terms of the single-reggeon-exchange amplitude [2].

We need to know the quantum numbers associated with a two-reggeon system. Internal quantum numbers, such as isospin and G-parity, combine exactly as if the reggeons were elementary particles. For example, the exchange of the f_2 Regge pole ($I = 0, G = +1$) and the a_2 Regge pole ($I = 1, G = -1$) will give a cut with $I = 1, G = -1$. However, a Regge cut will appear in amplitudes of both parities because of the angular momentum associated with the two-reggeon system, and there is also the important question of signature. It has been shown [21], for external particles both with spin and without spin, that the signature of the cut is

$$\tau = \tau_1\tau_2\eta, \tag{5.20}$$

where τ_1 and τ_2 are the signatures of the two exchanged reggeons and $\eta = -1$ if both reggeons are fermions and $\eta = +1$ otherwise.

A Regge trajectory is said to have natural parity, or naturality $+1$, if the spin and parity of the mesons on it are given by $J^P = J^{(-1)^J}$, for example $0^+, 1^-, 2^+, 3^-, \ldots$, and to have unnatural parity, or naturality -1, if the spin and parity of the mesons on it are given by $J^P = J^{(-1)^{J+1}}$, for example $0^-, 1^+, 2^-, 3^+, \ldots$. The parity and naturality of Regge cuts are discussed in [22], where it was shown that if the exchanged reggeons have naturalities n_1 and n_2, then amplitudes of naturality $-n_1n_2$ are suppressed relative to amplitudes of naturality n_1n_2 and that this suppression grows with increasing energy. As a consequence, for cuts where the two reggeons are any of ρ, ω, a_2 or f_2, all of which are natural parity, the natural-parity cut will dominate over the unnatural parity one. The only exception to this is in reactions for which there is a constraint relation between amplitudes at $t = 0$ which makes the natural and unnatural parity amplitudes equal there. Then the suppression of one is accompanied, at small t, by the suppression of the other, although away from $t = 0$ the $+n_1 n_2$ contribution can recover from this suppression.

5.1.2 Daughter trajectories

For internal consistency, Regge theory requires an infinite set of daughter trajectories [1,2], related to the leading trajectory $\alpha_0(t)$ by $\alpha_1(t) = \alpha_0(t) - 1$, $\alpha_2(t) = \alpha_0(t) - 2, \ldots$ The trajectory $\alpha_1(t)$ is known as the first daughter of $\alpha_0(t)$; $\alpha_2(t)$ is the second daughter and so on. The concept of daughter trajectories was developed at a time when little was known about meson spectroscopy. There was no experimental evidence for them and no theoretical proof of their existence and alternative viewpoints were proposed. Meson spectroscopy is now much more developed, and the non-strange meson resonances do appear to lie [23] on nearly linear, parallel, Regge trajectories. The evidence for daughter trajectories is very convincing.

Because they are low-lying, the contributions from daughter trajectories to cross sections decrease much more rapidly than do the contributions from the leading trajectories and so normally they are unimportant except at very low energies.

5.1.3 Spin

It is straightforward to incorporate spin, the most convenient way being to use the helicity formalism [24]. Each particle is labelled by its helicity λ and we consider helicity amplitudes, that is matrix elements taken between helicity states, in the s-channel centre-of-mass frame. For the process $1 + 2 \rightarrow 3 + 4$ this gives

$$\langle P_3, \lambda_3; P_4, \lambda_4 | T | P_1, \lambda_1; P_2, \lambda_2 \rangle \equiv T_{\lambda_3 \lambda_4; \lambda_1 \lambda_2}(s, t) \tag{5.21}$$

with

$$\frac{d\sigma_{\lambda_3 \lambda_4; \lambda_1 \lambda_2}}{d\Omega} = \frac{|\mathbf{p}_3|}{64\pi^2 |\mathbf{p}_1| s} |T_{\lambda_3 \lambda_4; \lambda_1 \lambda_2}|^2. \tag{5.22}$$

If the spins of the particles are s_1, s_2, s_3, s_4, then the number of amplitudes appears to be $(2s_1 + 1)(2s_2 + 1)(2s_3 + 1)(2s_4 + 1)$. However, because of parity and time-reversal invariance the number of independent amplitudes will be less than this. For example for πp scattering there are two, for $\gamma N \rightarrow \pi N$ there are four, and for pp or $p\bar{p}$ elastic scattering there are five.

Apart from the simplicity of this formula, with no cross terms, the usefulness of the helicity formalism stems from the fact that the partial-wave series is a straightforward generalization of the series for the scattering of two spin-0 particles.

It is

$$T_{\lambda_3\lambda_4;\lambda_1\lambda_2}(s,t) = 16\pi \sum_{J \geq \mu}^{\infty} (2J+1) T^J_{\lambda_3\lambda_4;\lambda_1\lambda_2}(s) \, d^J_{\lambda\lambda'}(\theta_s), \tag{5.23}$$

where J is the total angular momentum in the t channel,

$$\lambda = \lambda_1 - \lambda_2 \quad \lambda' = \lambda_3 - \lambda_4 \quad \mu = \max\left(|\lambda|, |\lambda'|\right) \tag{5.24}$$

and $d^J_{\lambda\lambda'}(\theta_s)$ is an element of the rotation matrix [25]. Using $d^J_{00}(\theta_s) = P_J(\cos\theta_s)$ we immediately recover the partial-wave series for spinless particles, with $J = l$.

In the forward direction $\theta_s = 0$, λ and λ' are projections of the total angular momentum in the same direction and the conservation of angular momentum demands that the amplitude vanishes unless $\lambda = \lambda'$. Thus for $\theta_s \approx 0$, the behaviour of $d^J_{\lambda\lambda'}(\theta_s)$ makes $T_{\lambda_3\lambda_4;\lambda_1\lambda_2}(s,t)$ vanish at least as fast as

$$T_{\lambda_c\lambda_d;\lambda_a\lambda_b} \sim \left(\sin \tfrac{1}{2}\theta_s\right)^n, \tag{5.25}$$

where $n = |\lambda - \lambda'|$ is known as the net helicity flip.

The Reggeization procedure can be applied to the helicity amplitudes and, as in the spinless case, we need to use the t-channel helicity amplitudes $T_{\tilde{\lambda}_2\tilde{\lambda}_4;\tilde{\lambda}_1\tilde{\lambda}_3}(s,t)$. These are related to the s-channel helicity amplitudes by a complicated crossing matrix. Since Regge poles have definite parity, it is appropriate to use t-channel amplitudes of definite parity, so that only poles of that parity contribute to them. This is done analogously to the $A^{\pm}$ of (5.10) by taking sums and differences of partial-wave helicity amplitudes:

$$T^{J\pm}_{\tilde{\lambda}_2\tilde{\lambda}_4;\tilde{\lambda}_1\tilde{\lambda}_3}(t) = T^J_{\tilde{\lambda}_2\tilde{\lambda}_4;\tilde{\lambda}_1\tilde{\lambda}_3}(t) \pm \eta_1\eta_3(-1)^{-s_1-s_3} T^J_{\tilde{\lambda}_2\tilde{\lambda}_4;-\tilde{\lambda}_1,-\tilde{\lambda}_3}(t), \tag{5.26}$$

where the superscript $(\pm)$ specifies the parity and η_1 and η_3 are intrinsic parities of the particles concerned. Note that this is in addition to constructing amplitudes of definite signature. To take account of kinematic constraints for s-channel scattering, for example (5.25), it is easiest to go back to the s-channel helicity amplitudes. It is found that a Regge pole makes s-channel helicity amplitudes behave for large s as $s^{\alpha(t)}$, as in the case of spinless particles, though it need not contribute to all of the amplitudes.

We have seen in (5.25) that the conservation of angular momentum makes s-channel helicity amplitudes vanish at $\theta_s = 0$ at least as fast as a certain power of $\sin \tfrac{1}{2}\theta_s$. However, parity conservation may give a more stringent requirement on the behaviour of the contribution from a given Regge pole. A contribution from any t-channel reggeon has definite parity in the t channel. Therefore it satisfies

$$T_{\lambda_3\lambda_4;\lambda_1\lambda_2} = \pm T_{-\lambda_3\lambda_4;-\lambda_1\lambda_2}. \tag{5.27}$$

The behaviour (5.25) is therefore not achievable and must be modified to

$$T_{\lambda_3\lambda_4;\lambda_1\lambda_2} \sim \left(\sin \tfrac{1}{2}\theta_s\right)^{n+n'}, \tag{5.28}$$

where

$$n + n' = |\lambda_1 - \lambda_3| + |\lambda_2 - \lambda_4|$$
$$= \max\left[|\lambda_1 - \lambda_3 - \lambda_2 + \lambda_4|, \, |\lambda_1 + \lambda_2 - \lambda_3 - \lambda_4|\right]. \tag{5.29}$$

5.1.4 Duality

It is convenient to start by recalling the origins of duality in pion–nucleon scattering through finite-energy sum rules (FESRs) [26–29]. These relate an integral over the resonance region at fixed t to a sum over the Regge-pole terms appropriate to higher energies. An amplitude $A^A(v, t)$ which is odd under crossing, $A^A(v, t) = -A^A(-v, t)$, is represented at high energy by a sum over poles of odd signature:

$$A^A(v, t) = -\pi \sum_i \frac{\beta_i(t)}{\Gamma(\alpha_i^{\pm}(t) + 1)\sin(\pi\alpha_i^{\pm}(t))}\left(1 \pm e^{-i\pi\alpha_i^{\pm}(t)}\right)(2v)^{\alpha_i^{\pm}(t)}. \tag{5.30}$$

The corresponding FESR is

$$\int_{v_0}^{v_c} dv \, \mathrm{Im}A^A(v, t) = -\tfrac{1}{2}\pi \sum_i \frac{\beta_i(t)}{\Gamma(\alpha_i(t) + 2)}(2v_c)^{\alpha_i(t)+1}. \tag{5.31}$$

For an amplitude that is even under crossing, $A^S(v, t) = A^S(-v, t)$, the FESR is written for $vA^S(v, t)$:

$$\int_{v_0}^{v_c} dv \, v\mathrm{Im}A^S(v, t) = \tfrac{1}{4}\pi \sum_i \frac{\beta_i(t)}{\Gamma(\alpha_i(t) + 3)}(2v_c)^{\alpha_i(t)+2}. \tag{5.32}$$

Provided that the convergence criteria are satisfied, FESRs can be written for $v^{2n}A^A(v, t)$ and $v^{2n+1}A^S(v, t)$, where n is an integer, giving sum rules of different moments. Not all amplitudes are crossing even or crossing odd, so for a general amplitude both odd- and even-moment sum rules must be considered [30].

As the FESRs are written for amplitudes, knowledge of the low-energy amplitudes, from partial-wave analysis, provides important information about the Regge-pole amplitudes. The upper limit of integration, v_c, must be sufficiently high for the Regge-pole expression to be valid. However, in practice v_c is determined by the upper limit of phase-shift analysis, corresponding typically to values of $\sqrt{s}$ close to 2 GeV.

The principal applications of FESRs were to pion–nucleon scattering [31,32] and to pion photoproduction [33–36]. It is necessary to assume that the Regge-pole

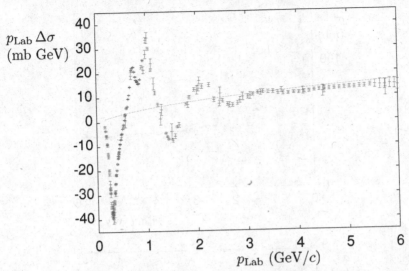

Figure 5.3. Comparison of $p_{Lab}(\sigma^{tot}(\pi^- p) - \sigma^{tot}(\pi^+ p))$ with the extrapolation of Regge fits to higher-energy data [2].

amplitude describes the real physical amplitude at low energy on the average and that this averaging takes place over intervals appreciably smaller than the range of integration. This does happen in practice, as illustrated in figure 5.3 which compares

$$p_{Lab}(\sigma^{tot}(\pi^- p) - \sigma^{tot}(\pi^+ p)) \tag{5.33}$$

with the Regge fit to high-energy data. In this case the Regge exchange is the ρ trajectory, extrapolated to low energies.

The $\pi^- p$ and $\pi^+ p$ high-energy elastic scattering amplitudes receive equal contributions from pomeron exchange, which cancels in the difference (5.33) between the two total cross sections. This implies that the non-pomeron Regge-pole t-channel exchanges are dual to the s-channel resonances. Figure 5.4 shows that the extrapolations to low energy of the Regge fits to the high-energy $\pi^+ p$ and $\pi^- p$ total cross sections give good descriptions of the average low-energy cross sections in each case. The resonances sit on a non-resonance background, so assuming that the non-pomeron t-channel exchanges are dual to the low-energy s-channel resonances leads to the assumption that pomeron exchange is dual to the low-energy s-channel non-resonance background.

This assumption of two-component duality is explicitly realized in the individual partial-wave amplitudes in πN scattering [32]. The linear combinations of partial-wave amplitudes $f_{l\pm}^I$, where the superscript I labels the s-channel isospin and the

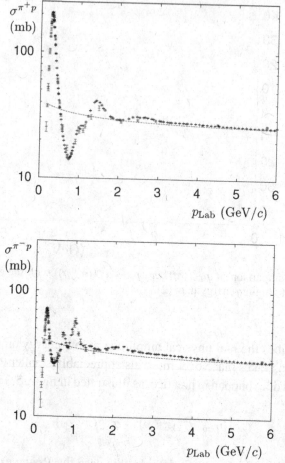

Figure 5.4. Comparison of the low-energy $\pi^+ p$ and $\pi^- p$ total cross sections with extrapolated Regge fits to higher-energy data [2].

subscript $l_{\pm}$ refers to total angular momentum $J = l \pm \frac{1}{2}$,

$$f_{l\pm}^0 = \frac{1}{3}\left(f_{l\pm}^{\frac{1}{2}} + 2f_{l\pm}^{\frac{3}{2}}\right) \quad f_{l\pm}^1 = \frac{1}{3}\left(f_{l\pm}^{\frac{1}{2}} - f_{l\pm}^{\frac{3}{2}}\right) \tag{5.34}$$

correspond to isospin-0 and isospin-1 exchange in the t channel. As the pomeron does not contribute to the t-channel $I = 1$ exchange amplitude, two-component duality predicts that the $f_{l\pm}^1$ should be given entirely by s-channel resonances. On the other hand, the $f_{l\pm}^0$ should not be given by s-channel resonances alone, but should have a predominantly-imaginary smooth background on which the s-channel resonances are superimposed. With the exception of the S-waves, this is what πN partial-wave analysis shows. The t-channel $I = 1$ amplitudes are represented by clear resonance circles in the complex phase-shift plane, with very little background. In contrast the resonance circles in the t-channel $I = 0$ amplitudes are superimposed

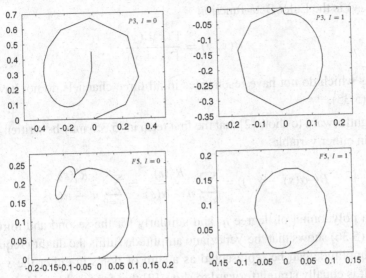

Figure 5.5. $I = 0$ and $I = 1$ exchange amplitudes for $P3$ and $F5$ pion–nucleon partial waves.

on a large and predominantly-imaginary background. The P_3 and F_5 partial waves, from the partial-wave analysis of [37], are given as examples in figure 5.5.

Despite these successes, duality is not a precise concept. A counter example is provided by pp scattering which does have a non-zero Regge contribution, although this is appreciably less than that in $\bar{p}p$ scattering.

5.1.5 *The Veneziano model*

A theoretical representation of the scattering amplitude that is explicitly crossing symmetric and analytic, has Regge behaviour, satisfies the FESRs and exhibits duality, although it violates unitarity, is provided by the Veneziano model [38]. The model contains strictly-linear trajectories, zero-width resonances and an infinite set of daughter trajectories. For a reaction that is identical in all three channels, for example $\pi^0\pi^0$ scattering, the amplitude has the form

$$A(s, t) = \bar{\beta}\,[B(-\alpha(s), -\alpha(t)) + B(-\alpha(s), -\alpha(u)) + B(-\alpha(t), -\alpha(u))],$$
(5.35)

which is explicitly $s \leftrightarrow t$, $s \leftrightarrow u$, $t \leftrightarrow u$ crossing-symmetric. In (5.35), $\bar{\beta}$ is a constant, α is a real linear trajectory

$$\alpha(s) = \alpha(0) + \alpha's,$$
(5.36)

and $B(x, y)$ is the Euler Beta-function

$$B(x, y) = \frac{\Gamma(x)\Gamma(y)}{\Gamma(x + y)}.$$ (5.37)

Reactions which do not have resonances in all three channels do not have all the terms in (5.35).

It is straightforward to show [2] that the first term in (5.35) may be written as a sum of poles in either variable:

$$B(-\alpha(s), -\alpha(t)) = \sum_{n=0}^{\infty} \frac{R_n(t)}{n - \alpha(s)} = \sum_{n'=0}^{\infty} \frac{R'_n(s)}{n' - \alpha(t)},$$ (5.38)

with R_n a polynomial of degree n, and similarly for the second and third terms. Equation (5.36) shows that the Veneziano amplitude fulfils the duality requirement that the amplitude can be represented as a sum of poles in either the s or the t channel. It is equally straightforward to show [2] that for fixed t and large s

$$B(-\alpha(s), -\alpha(t)) = \Gamma(-\alpha(t))(-\alpha' s)^{\alpha(t)} (1 + O(s^{-1}))$$ (5.39)

and in this limit

$$A(s, t) \sim \bar{\beta}\, \Gamma(-\alpha(t)) (1 + e^{-i\pi\alpha(t)}) (\alpha' s)^{\alpha(t)},$$ (5.40)

which is the Regge behaviour for an even-signatured trajectory.

The Veneziano amplitude is important because it demonstrates that simple functions exist which satisfy the theoretical requirements of analyticity, crossing and duality. However, it also demonstrates that there is no unique function as the $B(-\alpha(s), -\alpha(t))$ of (5.36) can be replaced by

$$B(m - \alpha(s), n - \alpha(t))$$ (5.41)

for any integers m, $n \geq 1$, and similarly for the s, u and t, u terms.

5.2 Photoproduction of pseudoscalar mesons

In principle it is possible to perform a complete set of measurements in pseudoscalar-meson photoproduction, that is to determine the amplitudes at fixed t up to an overall phase [39–41]. The reaction is given in terms of four complex amplitudes, which suggests that determining four magnitudes and three phases requires only seven experiments. However, seven experiments are not sufficient to resolve discrete ambiguities. The discussion is most conveniently carried out in terms of transversity amplitudes, which use the normal to the scattering plane as the quantization axis [42,43]. The cross section plus the three single-spin observables (beam, target and recoil baryon polarization) determine the magnitudes of the transversity amplitudes. There are twelve independent double-polarization measurements, four each of beam-target, target-recoil and beam-recoil. With careful selection of four of these it is possible to extract all the requisite phases without discrete ambiguities [41]. In practice, however, such a complete set of measurements has not yet been made and the use of FESRs to provide supplementary information has been very important and informative.

5.2.1 Pion photoproduction

Pion photoproduction is usually discussed either in terms of the four independent s-channel helicity amplitudes $T_{\lambda_N \lambda'_N}^{\lambda_\gamma}$, where λ_γ is the photon helicity and λ_N, λ'_N are the helicities of the initial and final nucleons, or in terms of four independent t-channel amplitudes F_i. These amplitudes, the relations between them and the definition of the cross section and polarization parameters are given in appendix D of [2]. Here we summarize the salient features.

The s-channel helicity amplitudes satisfy the conditions

$$T_{+-}^1 = T_{-+}^{-1}, \qquad T_{--}^1 = T_{++}^{-1},$$
$$T_{++}^1 = T_{--}^{-1}, \qquad T_{-+}^1 = -T_{+-}^{-1}, \tag{5.42}$$

where the usual convention of denoting the nucleon helicities only by their sign has been used as there is no ambiguity. Because of the relations (5.42) it is also conventional to use the $\lambda_\gamma = 1$ helicity amplitudes and omit the photon-helicity label. As the pion has zero helicity there is automatic helicity flip at the photon–pion–reggeon vertex, so the net helicity flip is defined by the nucleon helicities. Then T_{-+} is a non-flip amplitude (N), T_{++} and T_{--} are both single-flip amplitudes ($S1$ and $S2$ respectively) and T_{+-} is double-flip (D).

The amplitudes F_i have definite parity in the t channel. At large s, F_3 and F_4 are respectively natural- and unnatural-parity t-channel amplitudes while F_1 and F_2 are respectively natural- and unnatural-parity t-channel amplitudes at all energies. Further, the amplitudes F_2 and F_3 are constrained at $t = 0$ by

$$F_3 = 2m_N F_2, \tag{5.43}$$

where m_N is the nucleon mass.

5.2.2 π^0 photoproduction

We start with $\gamma p \to \pi^0 p$, for which the principal reggeon exchange is expected to be the ω. The contribution from ρ exchange is small as the γ–π–ω coupling is about three times the γ–π–ρ coupling and the coupling of the ω to the nucleon is

Parity	Amplitude	Trajectory	J^{PC}
Natural	F_1	ω	$(2l+1)^{--}$
Unnatural	F_2	b_1	$(2l+1)^{+-}$
Natural	F_3	ω	$(2l+1)^{--}$
Unnatural	F_4	Unknown	$(2l)^{--}$

Table 5.1. Leading Regge trajectories contributing to π^0 photoproduction.

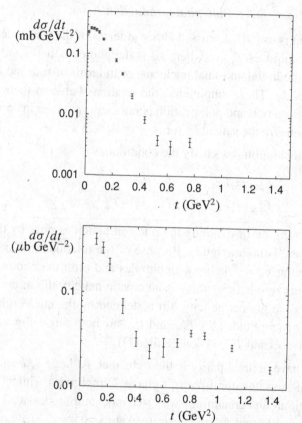

Figure 5.6. Differential cross sections for $\pi^- p \to \pi^0 n$ at $p_{Lab} = 9.8$ GeV [44] (top) and $\gamma p \to \pi^0 p$ at $E_\gamma = 12$ GeV [45] (bottom).

about four times the coupling of the ρ [2]. As the ω trajectory has natural parity it contributes to the F_1 and F_3 amplitudes. A contribution from the lower-lying $b_1(1235)$ trajectory is allowed, as the dominant decay of the $b_1(1235)$ is $\omega\pi$, but its coupling to the nucleon is unknown and has to be determined from experiment. The $b_1(1235)$ contributes only to the F_2 amplitude. The lowest-mass states on the allowed Regge-pole trajectories are given in table 5.1. As cuts do not have a specific parity, the ω-pomeron cut can contribute to F_1, F_2 and F_3.

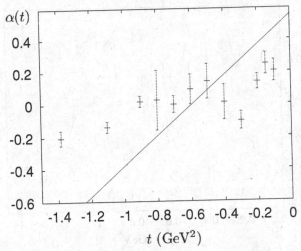

Figure 5.7: Effective ω trajectory obtained from the π^0-photoproduction differential cross section. The line is the same as in figure 5.1.

The differential cross section for $\gamma p \to \pi^0 p$ looks very like that for $\pi^- p \to \pi^0 n$, as shown in figure 5.6. Just as the former reaction is dominated by the ρ-exchange single-flip amplitude, π^0 photoproduction is dominated by the ω-exchange single-flip amplitudes. The origin of the minimum in the cross section at $t \approx -0.55$ GeV^2 is the zero in ω exchange due to the vanishing of the signature factor there, just as for ρ exchange in $\pi^- p$ charge exchange. However, π^0 photoproduction is more complicated [34] than $\pi^- p$ charge exchange, as can be seen from the effective trajectory in figure 5.7. The line is the extrapolated fit to the physical states on the ω trajectory. This figure should be compared with the corresponding one for ρ exchange, figure 5.1. It is clear that for π^0 photoproduction additional contributions are required.

However, without going into detail one can get a general picture of the process directly from the data. As the dominant ω exchange is natural parity, for plane-polarized photons one would expect that the polarized beam asymmetry

$$\Sigma = \frac{\sigma_\perp - \sigma_\parallel}{\sigma_\perp + \sigma_\parallel} \tag{5.44}$$

should be predominantly equal to 1. This is very nearly true, as can be seen in figure 5.8. The slight dip at $t \approx -0.6$ necessarily requires unnatural parity exchange. From table 5.1 it can be seen that there is no known contribution to F_4, implying that this amplitude should be zero or, at most, very small. This in turn requires that the recoil polarization asymmetry R and the polarized target asymmetry T should

F E Close and A Donnachie

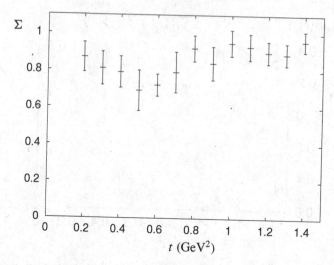

Figure 5.8. Polarized-beam asymmetry for $\gamma p \rightarrow \pi^0 p$ at $E_\gamma = 6.0$ GeV [45].

be equal as, to leading order in s [2],

$$R\frac{d\sigma}{dt} = \frac{1}{16\pi}\frac{\sqrt{-t}}{4m_p^2 - t}\text{Im}\left(F_1 F_3^* - \left(4m_p^2 - t\right)F_4 F_2^*\right),$$

$$T\frac{d\sigma}{dt} = \frac{1}{16\pi}\frac{\sqrt{-t}}{4m_p^2 - t}\text{Im}\left(F_1 F_3^* + \left(4m_p^2 - t\right)F_4 F_2^*\right) \qquad (5.45)$$

and so

$$(R - T)\frac{d\sigma}{dt} = \frac{\sqrt{-t}}{8\pi}\text{Im}(F_2 F_4^*). \qquad (5.46)$$

Comparison of R and T is rather difficult as the data are at different energies and an interpolation in t is also necessary. Additionally F_2 is small, so in practice a comparison of R and T is not a sensitive test of whether $F_4 = 0$. Further information, in the form of FESRs, is necessary.

A model [46] consisting of the ω, ρ and b_1 reggeon exchanges, with non-degenerate ω and ρ trajectories, provides a qualitative description of the data. Although it is quantitatively rather poor except at very small t and does not reproduce accurately the dip structure seen in the differential cross section, it is strongly constrained. The magnitudes of the ω and ρ contributions are fixed by the widths of the radiative decays $\omega \rightarrow \pi\gamma, \rho^0 \rightarrow \pi^0\gamma$ and the known ωNN, ρNN couplings. The only 'real' parameter is the magnitude of the b_1 contribution as its coupling to the nucleon is unknown. The comparative success of this approach is a good indicator of how well simple models can describe data.

An effective trajectory lying above the ω trajectory, clearly required from figure 5.7, cannot be produced by b_1 exchange so cuts are an essential and important part of the reaction. As the high-energy data alone do not determine the amplitudes it is essential to use FESRs. There are five principal conclusions from a combined fit [33,34] to the data and the sum rules. The F_1 amplitude dominates the cross section, as expected, but the zero in the ω contribution is masked by a strong cut. However, the cut does not move the minimum in $d\sigma/dt$ because it is predominantly real, whereas the pole is mainly imaginary there. The FESRs show that F_2 and F_3 are finite at $t = 0$ so it is necessary for the cuts, which are large, to 'conspire' to satisfy the constraint (5.43). The zero of the ω-pole contribution in F_3 is completely masked by the cut. The F_4 amplitude is non-zero and has strong energy dependence, but whether it is due to a low-lying trajectory or cuts or a combination of both cannot be determined. No simple cut model can explain the pattern of cuts observed. The residues found for the ω pole agree reasonably well with what is known about the $\omega\pi\gamma$ and ωNN couplings, so it is reasonable to believe that the correct cut amplitudes have been obtained.

These results confirm our qualitative expectations for ω exchange, but the cut contributions do not follow the conventional picture and are not understood. Thus π^0 photoproduction provides a considerable contrast to $\pi^- p$ charge exchange. Not only are cut effects important in determining the structure of the flip amplitudes, but also it is clear that no simple cut model can account for all the structure observed. In general π^0 photoproduction sets rather difficult problems for high-energy models, even though the basic structure reflects the expected behaviour.

5.2.3 $\pi^{\pm}$ *photoproduction*

The isospin decomposition for charged pion photoproduction is

$$F_i(\gamma p \to \pi^+ n) = \sqrt{2}\left(F_i^{(0)} + F_i^{(-)}\right),$$
$$F_i(\gamma n \to \pi^- p) = \sqrt{2}\left(F_i^{(0)} - F^{(-)}\right) \tag{5.47}$$

and the crossing properties of $F_i^{(0)}$ and $F_i^{(-)}$ under s–u interchange are

$$F_1^{(-)}, \; F_2^{(-)}, \; F_3^{(-)} \text{ and } F_4^{(0)} \text{ are odd,}$$
$$F_1^{(0)}, \; F_2^{(0)}, \; F_3^{(0)} \text{ and } F_4^{(-)} \text{ are even.} \tag{5.48}$$

The lowest-mass states on the allowed Regge-pole trajectories are given in table 5.2. Since the two reactions $\gamma p \to \pi^+ n$ and $\gamma n \to \pi^- p$ are related by line reversal, exchanges with different G-parity enter with different relative sign in the two cases.

Parity	Amplitude	$G = +1$ trajectory	$G = -1$ trajectory
Natural	F_1	ρ	a_2
Unnatural	F_2	b_1	π
Natural	F_3	ρ	a_2
Unnatural	F_4	Unknown	a_1

Table 5.2. Leading Regge trajectories contributing to π^+ photoproduction.

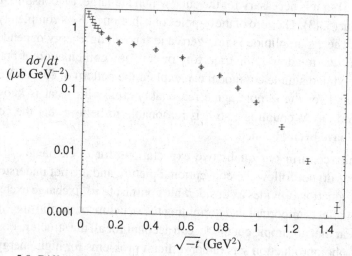

Figure 5.9. Differential cross section for $\gamma p \to \pi^+ n$ at $E_\gamma = 8.0$ GeV [47].

An obvious feature of the differential cross section [47] for $\gamma p \to \pi^+ n$ is the narrow forward peak. The cross section at $E_\gamma = 8$ GeV is shown in figure 5.9, where $d\sigma/dt$ is plotted as a function of $\sqrt{-t}$ to highlight the forward peak. The differential cross sections at lower energies are very similar in shape [47] and there is strong energy dependence: the cross section decreases approximately as s^{-2}. This is evident in the effective trajectory in figure 5.10, which shows that $\alpha_{eff} \approx 0$ over the whole t range. Another obvious feature of these data is that they are structureless, apart from the peak, and there is no dip at $t \approx -0.55$ GeV^2.

The forward peak, in which the differential cross section rises by more than a factor of two between $t = -m_\pi^2$ and $t = t_{min}$, is naturally associated with pion exchange. Pion exchange is unique among reggeon exchanges. The $\Gamma(-\alpha(t))$ in (5.10) has a pole near the physical region, so for small t one can approximate it by the pion-pole term. The pion-exchange term vanishes in the forward direction, even though it occurs in the non-flip amplitude T_{+-} as well as in the double-flip amplitude T_{-+}, and so by itself it cannot produce a forward peak.

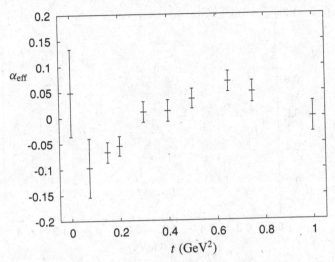

Figure 5.10. Effective trajectory for $\gamma p \to \pi^+ n$.

If we consider pion-exchange as a Feynman diagram, it is well known that the amplitude is not gauge-invariant by itself. When using pseudoscalar (PS) πN coupling, this has to be complemented by the s-channel nucleon diagram for $\gamma p \to \pi^+ n$ and by the u-channel nucleon-exchange diagram for $\gamma n \to \pi^- p$. These are the minimum terms necessary for gauge invariance and correspond to the electric coupling. The anomalous magnetic part is gauge-invariant by itself (there is no corresponding pion term) and need not be included. If it is retained, its influence is negligible. For pseudovector (PV) πN coupling it is necessary to add the Kroll–Ruderman contact term. The two coupling schemes lead to exactly the same gauge-invariant amplitudes.

The problem then arises about Reggeizing these nucleon terms. For the pion pole itself there is no problem [46]: one simply replaces the Feynman propagator $1/(t - m_\pi^2)$ by the appropriate Regge term, P_{Regge} say, of (5.10). In other words the pole-like propagator is multiplied by $(t - m_\pi^2)P_{Regge}$. It is suggested in [46] that one does precisely the same for the nucleon terms, so symbolically

$$M_\pi(\gamma p \to \pi^+ n) \to \left(t - m_\pi^2\right) P_{Regge}^\pi \left(M_{t\text{-exch}}^\pi + M_{s\text{-exch}}^N\right),$$
$$M_\pi(\gamma n \to \pi^- p) \to \left(t - m_\pi^2\right) P_{Regge}^\pi \left(M_{t\text{-exch}}^\pi + M_{u\text{-exch}}^N\right). \quad (5.49)$$

The effect of these additional terms is to create the sharp rise observed in the forward differential cross section over the range $\Delta t \approx m_\pi^2$.

The experimental ratio of the $\gamma n \to \pi^- p$ and $\gamma p \to \pi^+ n$ differential cross sections is shown in figure 5.11. The pion-pole contribution is essentially the same in both reactions which accounts for the ratio being close to unity at small t. Away from

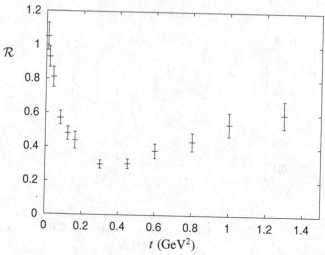

Figure 5.11. Ratio $\mathcal{R} = \frac{d\sigma}{dt}(\gamma n \to \pi^- p)/\frac{d\sigma}{dt}(\gamma p \to \pi^+ n)$ at $E_\gamma = 8.0$ GeV [47].

the very forward direction, the other contributions come into play. In an obvious notation,

$$\frac{d\sigma}{dt}(\gamma p \to \pi^+ n) \propto |(\pi + b_1) + (\rho + a_2)|^2,$$
$$\frac{d\sigma}{dt}(\gamma n \to \pi^- p) \propto |-(-\pi + b_1) + (\rho - a_2)|^2, \qquad (5.50)$$

where π stands for the complete gauge-invariant pion-exchange term and the changes in sign between the two reactions are a consequence of the negative G-parity of the π and a_2 exchanges. As the (ρ, a_2) and (π, b_1) trajectories are nearly degenerate, and as the positive and negative G-parity exchanges have different signature factors, this change in sign leads to rotating phases $\exp(-i\pi\alpha(t))$ for one reaction and to constant phases for the other. This difference is crucial in explaining the data in figure 5.11, and the success in doing so is a remarkable achievement for Regge theory. Fixing the magnitude of the ρ contribution from the width of the radiative decay $\rho^\pm \to \pi^\pm \gamma$ and the known ρNN coupling, parameter-free predictions [46] can be made for the $\gamma p \to \pi^+ n$ and $\gamma n \to \pi^- p$. Although still qualitative, it is better than the corresponding description of $\gamma p \to \pi^0 p$ and in particular gives a precise description of the cross section ratio $\mathcal{R} = \frac{d\sigma}{dt}(\gamma n \to \pi^- p)/\frac{d\sigma}{dt}(\gamma p \to \pi^+ n)$.

To improve on this description, it is necessary to parametrize the amplitudes and to impose FESRs, fitting them and the data rather than attempting a prediction a priori [35,48,36]. In this approach the s-channel and u-channel nucleon poles are not included and are replaced by Regge cuts. The simplest prescription is to put a pion pole directly into the s-channel helicity amplitudes that have no kinematic factors that are not required by s-channel angular-momentum conservation [49,50]. This means replacing the factorizing form $t/(t - m_\pi^2)$ by $m_\pi^2/(t - m_\pi^2)$ in the non-flip amplitude. As we can rewrite $m_\pi^2/(t - m_\pi^2)$ as $t/(t - m_\pi^2) - 1$ we see that we are simply adding a specific smooth background which interferes destructively with the pole and which should be associated with a cut. The pion-exchange contribution to the double-flip amplitude is unaltered as s-channel angular momentum conservation requires that the t dependence is unchanged. This simple prescription, known as poor man's absorption (PMA), describes the near-forward charged-pion-photoproduction data rather well, but fails at larger values of $|t|$ [48]. An interesting feature of the FESRs is that the two amplitudes to which pion exchange contributes are determined, to a good approximation, by the Born terms as the higher-mass baryon resonances cancel in the forward direction. This explains the success of the Born-term model and it is interesting to note that this cancellation agrees with quark-model predictions for the coupling of the baryon resonances to the different helicity states in $\gamma N \to \pi N$ [51].

Pion exchange contributes only to the t-channel amplitude F_2, which we have seen must satisfy the constraint $F_3 = 2m_p F_2$ at $t = 0$. This constraint is satisfied trivially by pion exchange as it vanishes at $t = 0$. However, this is not the case for the cut. The constraint equation automatically requires a cut in F_2 to be accompanied by a cut in F_3, that is the cut occurs in both the unnatural- and natural-parity amplitudes. This immediately predicts the behaviour of the polarized beam asymmetry, Σ, at small t as there [2]

$$\Sigma \sim \left(|F_3|^2 - 4m_p^2|F_2|^2\right) \big/ \left(|F_3|^2 + 4m_p^2|F_2|^2\right), \qquad (5.51)$$

so $\Sigma = 0$ at $t = 0$ from the constraint equation (5.43). As $|t|$ increases F_2 decreases rapidly but F_3 varies slowly. Hence Σ will increase rapidly to a value close to 1, in conformity with the data.

The fact that $\alpha_{eff} \approx 0$ over the whole t range is evidence for strong cuts. It is not surprising that the reggeon-exchange contributions, other than the pion, are not obvious in the differential-cross-section data. Both ρ and a_2 exchange are allowed, but we know that their couplings to the nucleon are weak. We also know from π^0 photoproduction that ρ exchange is not important and that b_1 exchange does not make a significant contribution. However, FESRs [35] show that ρ and a_2 exchange, and their associated cuts, cannot be completely excluded. Their presence is not seen directly in the data but through interference with the strong pion cut in the natural-parity amplitude. The absence of a dip in the differential cross section could be taken to imply that the ρ pole does not have a zero there. However, the presence of the strong cuts, evident from the energy dependence of the cross section, invalidates that argument. The evidence from sum rules is that the zero is required [35].

The principal conclusion to be drawn from pion photoproduction at small t is that predictions of the reactions are at best qualitative, except at very small t. A precise quantitative description, particularly of polarization data, requires the incorporation of Regge cuts. As there is no simple prescription for calculating the cuts, these data provide a salutary reminder that the application of Regge theory is not straightforward.

At large $|t|$ it is more appropriate to use a partonic approach to describing the data. At fixed centre-of-mass angles and large t, the constituent-quark counting rule makes specific predictions for the energy dependence of exclusive processes. The rule was first derived from simple dimensional counting arguments [3] and was later confirmed in a short-distance pQCD approach [4]. The rule is qualitatively consistent with many measurements, although there are a few anomalies in pp scattering and the prediction of conservation of hadron helicity is not confirmed. For pion photoproduction the rule predicts that

$$\frac{d\sigma}{dt} = f(\theta_{c.m.})s^{-7} \tag{5.52}$$

at fixed centre-of-mass angle $\theta_{c.m.}$. Data [52] for $\gamma p \to \pi^+ n$ and $\gamma n \to \pi^- p$ at $\theta_{c.m.} = 90°$ are shown in figure 5.12. The differential cross section has been multiplied by s^7 and the data appear to indicate the onset of scaling from $s \approx 7 \text{ GeV}^2$. Scaling is also apparent from about the same energy at $\theta_{c.m.} = 70°$ in both channels, but not at $\theta_{c.m.} = 50°$. This is not really surprising as these latter data lie in the range $0.696 < |t| < 1.248 \text{ GeV}^2$ for $\sqrt{s} \geq 2.35 \text{ GeV}$, which is still very much in the Regge region.

5.2.4 η^0 photoproduction

The data for $\gamma p \to \eta^0 p$, shown in figure 5.13, pose a severe problem for Regge models. The reaction has the same exchanges as $\gamma p \to \pi^0 p$, but exchanges of different isospin enter in different combinations. If we write

$$F_i(\pi^0 p) = F_i^0 + F_i^1, \tag{5.53}$$

where the superscripts denote the t-channel isospin, then for η photoproduction we can use vector-meson dominance (VMD) arguments [35] to write

$$F_i(\eta p) = \tfrac{1}{3} x \sqrt{3} \left(\tfrac{1}{r} F_i^0 + r F_i^1 \right). \tag{5.54}$$

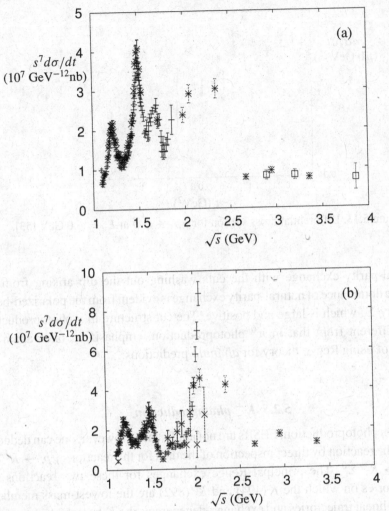

Figure 5.12. Differential cross section [52] for: (a) $\gamma p \rightarrow \pi^+ n$ and (b) $\gamma n \rightarrow \pi^- p$ at $\theta_{c.m.} = 90°$.

Here $r = \gamma_\rho / \gamma_\omega \approx 2.8$ and $x \approx 1.23$ depends on the $\eta - \eta'$ mixing angle. The effect of (5.54) is to suppress ω exchange and enhance ρ exchange. However, whatever the combination of ω and ρ exchange, one would expect a minimum in the cross section around $|t| \sim 0.55$ GeV^2, just as in $\pi^- p$ charge exchange and in π^0 photoproduction.

In this case it is not possible to use FESRs as, although they can be written down formally, there is an unphysical cut from the πp threshold to the ηp threshold whose contribution cannot be evaluated. The 'solution' once again is to invoke cuts and the most plausible description [34] is to have the cross section dominated

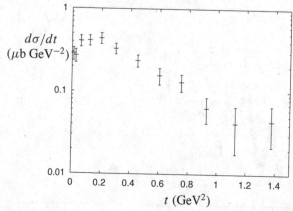

Figure 5.13. Differential cross section for $\gamma p \rightarrow \eta^0 p$ at $E_\gamma = 6.0$ GeV [53].

by natural-parity exchange with the cuts washing out the dip arising from the poles. The dominance of natural-parity exchange is evident from the polarized-beam asymmetry Σ, which is large and positive. The cut structure in η photoproduction is very different from that in π^0 photoproduction, emphasizing once again the difficulty of using Regge theory for *ab initio* predictions.

5.2.5 K^+ photoproduction

Just as for η photoproduction, FESRs are not applicable. However, one can deduce a lot about the reaction by direct inspection of the data for the reactions $\gamma p \rightarrow K^+ \Lambda^0$ and $\gamma p \rightarrow K^+ \Sigma^0$. The principal Regge exchanges for these two reactions are the trajectories on which the $K(494)$ and $K^*(892)$ are the lowest-mass members. Assuming linear trajectories and exchange-degeneracy, the K trajectory is specified in principle by the $K(494)$, $K_1(1270)$, $K_2(1770)$, $K_3(2320)$ and two higher-mass states still in need of confirmation [10], $K_3(2320)$ and $K_4(2500)$. A further complication is that the $K_1(1270)$ and $K_1(1400)$ are mixtures of the $K_{1B}(^1P_1)$ (the natural candidate for the trajectory) and the $K_{1A}(^3P_1)$, so the physical mass is presumably not the correct one to use. Ignoring this problem and fitting to the three established states gives the trajectory

$$\alpha_K \approx 0.7\left(t - m_K^2\right). \tag{5.55}$$

This choice allows satisfactory agreement with the small-t data, although the data are not able to establish definitively the precise K trajectory. There are five states available to establish the K^* trajectory, $K^*(892)$, $K_2^*(1430)$, $K_3^*(1780)$, $K_4^*(2045)$ and $K_5^*(2380)$ although the latter is another state still in need of confirmation [10].

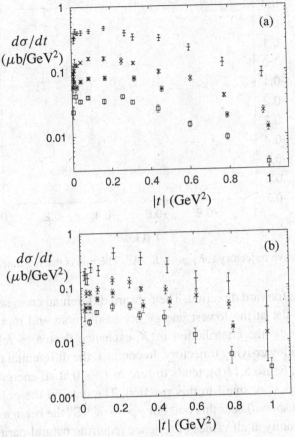

Figure 5.14. Cross sections for (a) $K\Lambda$ and (b) $K\Sigma$ photoproduction [54]. The data, from top to bottom, are at photon laboratory energies of 5, 8, 11 and 16 GeV.

Fitting the four established states gives the trajectory

$$\alpha_{K^*} \approx 0.25 + 0.89t. \tag{5.56}$$

The differential cross sections for $\gamma p \to K^+ \Lambda^0$ and $\gamma p \to K^+ \Sigma^0$ are given in figure 5.14 and the effective trajectory for $\gamma p \to K^+ \Lambda^0$ in figure 5.15. The line in the latter figure is that of (5.56).

In the middle of the t range shown it is clear that the effective trajectory is consistent with exchange of the $K^*(892)$ trajectory. At very small t the effective trajectory is consistent with the dominance of K exchange, from (5.55) the intercept of the trajectory at $t = 0$ is approximately -0.17. Finally, at the larger values of $-t$ there is an indication of a deviation from the K^* trajectory, presumably a K^*–pomeron cut which will have an intercept slightly larger and a slope appreciably smaller than that of the K^* trajectory: recall (5.17) and (5.18). This interpretation is reflected in the data. Just as for ω and ρ exchange in pion photoproduction, K^* exchange

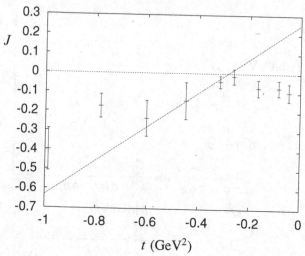

Figure 5.15. Effective trajectory for $\gamma p \rightarrow K^+\Lambda^0$. The line is the K^* trajectory (5.56).

goes to zero in the forward direction. The forward differential cross section, figure 5.14(a), is rather flat at the lowest energy but tends more and more to zero as the energy increases; the contribution of K exchange decreases relative to K^* exchange as it is a lower-lying trajectory. In contrast, the differential cross section for $\gamma p \rightarrow K^+\Sigma^0$, figure 5.14(b), tends to zero as $t \rightarrow 0$ at all energies, implying that K exchange is very small in this reaction. The data on the polarized beam asymmetry Σ (not shown) bear this out. For $\gamma p \rightarrow K^+\Sigma^0$ the beam asymmetry Σ is consistent with unity at all values of t, hence requiring natural-parity exchange. For $\gamma p \rightarrow K^+\Lambda^0$ the beam asymmetry Σ is consistent with unity only for $|t| \geq 0.2$ GeV^2 and requires some unnatural-parity exchange at smaller values of $|t|$.

It is rather straightforward to model this. The most obvious approach is the analogue of the model for $\pi^\pm$ photoproduction discussed above [46]. For K^* exchange, the coupling constant at the γ–K–K^* vertex can be estimated from the electromagnetic width of the K^* and for K exchange the coupling is simply the electric charge. The strong coupling constants involved are much less well known experimentally than the coupling constants in pion photoproduction but can be estimated through SU(3). In practice only the contribution of the K is relatively well determined and it is necessary to treat the hadronic couplings of the K^* as parameters, apart from their sign which is known. This simple model gives a rather good description of the $\gamma p \rightarrow K^+\Lambda^0$ and $\gamma p \rightarrow K^+\Sigma^0$ differential cross sections and shows that K^* exchange completely dominates $\gamma p \rightarrow K^+\Sigma^0$ and, except for $|t| \leq 0.2$ GeV^2, dominates $\gamma p \rightarrow K^+\Lambda^0$. However, the model is not so successful in reproducing the recoil baryon asymmetry in $\gamma p \rightarrow K^+\Lambda^0$, particularly at larger values of $|t|$. This can be ascribed to a K^*–pomeron cut, as already inferred from the effective trajectory.

5.3 Vector-meson photoproduction

Unlike pseudoscalar-meson photoproduction it is not straightforward, even in principle, to perform a complete set of measurements to determine the amplitudes of vector-meson photoproduction at fixed t up to an overall phase. For example, it can be shown [55] that measuring the two-meson decay of a photoproduced ρ or ϕ does not determine the meson's vector polarization, only its tensor polarization. The decay of the vector meson into lepton pairs is also insensitive to the polarization of the vector meson unless the spin of one of the leptons is measured. Nonetheless, despite this and in contrast to the complexity of pseudoscalar-meson photoproduction, the description of light-quark vector-meson photoproduction is surprisingly simple.

5.3.1 ρ and ω photoproduction

A particular feature of ρ photoproduction is the distortion of the line shape of the ρ due to interference between the amplitude for direct ρ production and the amplitude for the Drell–Söding mechanism [56,57]. In the latter, the photon dissociates into two oppositely-charged pions, each of which scatters off the target. This is a particular example of the more general Deck effect [58]. The interference term changes sign from positive to negative in passing through the ρ mass and results in a skewing of the $\pi^+\pi^-$ mass spectrum and an apparent shift in the mass and width of the ρ. At low energies the $\pi^+\pi^-$ mass spectrum is further complicated by contamination from the $\pi\Delta(1232)$ channel, hence it is necessary to use high-energy data to see the Drell–Söding mechanism cleanly. In figure 5.16 the two-pion invariant mass in photoproduction [59] at $\sqrt{s} = 70$ GeV is compared with the naive expectation of a P-wave Breit–Wigner. The skewing is t- and Q^2-dependent and the model gives a good description of both this and the angular correlation of the photoproduced pion pairs.

A direct connection between πp scattering and ρ and ω photoproduction is provided via the assumption of VMD [60]. In its simplest form, VMD says that the cross section for $\gamma p \rightarrow \rho p$ is given by

$$\frac{d\sigma}{dt}(\gamma p \rightarrow \rho p) = \alpha \frac{4\pi}{\gamma_\rho^2} \frac{d\sigma}{dt}(\rho^0 p \rightarrow \rho^0 p), \tag{5.57}$$

where $4\pi/\gamma_\rho^2$ is the ρ–photon coupling, which can be found from the e^+e^- width of the ρ:

$$\Gamma_{\rho \rightarrow e^+e^-} = \frac{\alpha^2}{3} \frac{4\pi}{\gamma_\rho^2} m_\rho. \tag{5.58}$$

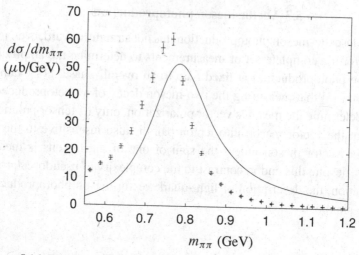

Figure 5.16. The line shape of the ρ in photoproduction. The data are from [59] and the curve is a P-wave Breit–Wigner.

In the additive-quark model, the amplitude for $\rho^0 p \to \rho^0 p$ is simply given by the average of the amplitudes for $\pi^- p$ and $\pi^+ p$ elastic scattering. In this combination of πp scattering amplitudes the $C = -1$ exchanges cancel (as they should) leaving the pomeron and f_2 exchanges. There are two omissions in this procedure. One is a contribution from a_2 exchange, which is forbidden in πp elastic scattering by G-parity but is allowed for ρ^0 photoproduction. However, this is expected to be extremely small. Firstly, the photon is necessarily isoscalar (ω-like) for a_2 exchange, while for f_2 exchange the photon is isovector (ρ-like). Thus $g_{f_2\rho\gamma} \approx 3g_{a_2\rho\gamma}$. Secondly, the a_2–nucleon coupling is much weaker than the f_2–nucleon coupling [2]: phenomenologically $g_{f_2NN} \approx 8g_{a_2NN}$. The net effect is that the amplitude for a_2 exchange is a small percentage of that for f_2 exchange. The near-degeneracy of the f_2 and a_2 trajectories means that any contribution from a_2 exchange will have minimal impact on the energy and t dependence of the cross section. The other omission is pion exchange, but this is again small as the photon is once more necessarily ω-like. Also, its contribution decreases rapidly with increasing energy, so it is relevant only close to threshold. The trajectories of the pomeron, f_2 and a_2 are well known from hadronic scattering, as is the mass scale by which we must divide s before raising it to the Regge power, namely the inverse of the trajectory slope [2]. It is well established that the trajectories couple to the proton through the Dirac electric form factor $F_1(t)$, which can be represented by the dipole form

$$F_1(t) = \frac{4m_p^2 - 2.79t}{4m_p^2 - t} \frac{1}{(1 - t/0.71)^2}. \tag{5.59}$$

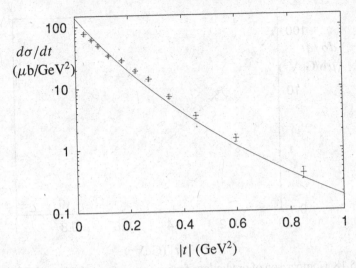

Figure 5.17. Comparison of predictions from (5.61) for $\gamma p \to \rho p$ with data [61] at $\sqrt{s} = 4.3$ GeV.

Wherever it can be experimentally checked, the differential cross section for ρ^0 photoproduction is found to have the same slope at small t as the $\pi^{\pm} p$ elastic differential cross sections, so it is natural to assume that the form factor of the ρ, $F_\rho(t)$, is the same as that of the pion. As discussed in chapter 2, the pion form factor is well described by

$$F_\pi(t) = \frac{1}{1 - t/0.5}. \tag{5.60}$$

Using this gives an excellent description of high-energy πp elastic scattering. Assuming that the spatial structure of the ρ^0 is the same as that of the $\rho^{\pm}$, the amplitude for $\gamma p \to \rho^0 p$ is then

$$T(s, t) = i F_1(t) F_\rho(t) \big(A_P (\alpha'_P s)^{\alpha_P(t)-1} e^{-\frac{1}{2} i \pi (\alpha_P(t)-1)}$$
$$+ A_R (\alpha'_R s)^{\alpha_R(t)-1} e^{-\frac{1}{2} i \pi (\alpha_R(t)-1)} \big). \tag{5.61}$$

Normalizing the amplitude such that $d\sigma/dt = |T(s, t)|^2/(16\pi)$ in μb GeV^{-2}, then $A_P = 48.8$ and $A_R = 129.6$ using (5.57), the PDG value for the $\rho \to e^+ e^-$ width and a standard fit to the $\pi^{\pm} p$ total cross sections [2]. The resulting predictions for the $\gamma p \to \rho^0 p$ differential cross section are compared with data at $\sqrt{s} = 4.3$ GeV [61] in figure 5.17 and at $\sqrt{s} = 71.7$ GeV [62] in figure 5.18. The predictions agree equally well [2] with data at $\sqrt{s} = 6.9$ and 10.8 GeV [63] (not shown). This model has implications for polarization effects in ρ photoproduction. It is well known in ρ photoproduction that the helicity of the ρ is the same as that of the photon, the

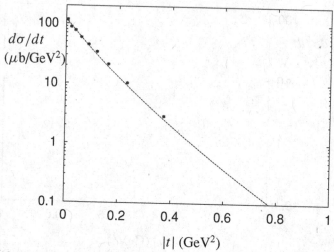

Figure 5.18. Comparison of predictions for $\gamma p \to \rho p$ from (5.61) with ZEUS data [62] at $\sqrt{s} = 71.7$ GeV.

phenomenon of s-channel helicity conservation (SCHC). What is less well known is that pomeron and f_2 exchange also conserve helicity equally well at the nucleon vertex. This is readily seen in pion–nucleon elastic scattering. Applying FESRs shows [31,32] that much the largest $C = +1$ exchange amplitude is the helicity non-flip amplitude A_{++}, that is s-channel helicity is conserved in $C = +1$ exchange. The helicity-flip amplitude A_{+-} is primarily $C = -1$ ρ exchange. These results are reflected in the $\pi^- p$ and $\pi^+ p$ polarization data which are given by the interference of A_{++} and A_{+-}. These are almost mirror-symmetric as a function of t, $P(\pi^- p) \approx -P(\pi^+ p)$, as the ρ-exchange contribution has the opposite sign in $\pi^- p$ and $\pi^+ p$ scattering. The polarization should decrease with energy approximately as $s^{\alpha_\rho - \alpha_P}$ and the data are compatible with this. There is a small deviation from complete mirror symmetry. The t-channel isoscalar–isovector interference terms cancel in the sum of $P d\sigma/dt$ for $\pi^+ p$ and $\pi^- p$ scattering, which allows the ratio of the t-channel isoscalar helicity-flip amplitude to the helicity non-flip amplitude to be determined [64]. It is small, but non-zero. As the amplitude necessarily involves a combination of f_2 and pomeron exchange these cannot be distinguished experimentally.

Target-polarization effects in ρ photoproduction will arise primarily from the interference of the dominant pomeron and f_2 exchange with the a_2 and other small unknown exchanges. However, this does not necessarily mean that polarized-target asymmetries will be small: they are rather large at low and intermediate energies in $\pi^\pm p$ elastic scattering. It does mean that they are not predictable and that they essentially measure the small amplitudes.

Is (5.57) extendable to charged-ρ photoproduction, relating $\gamma p \to \rho^+ n$ to $\pi^- p \to \pi^0 n$ via $\rho^0 p \to \rho^+ n$? Although a_2 exchange is very small compared to pomeron

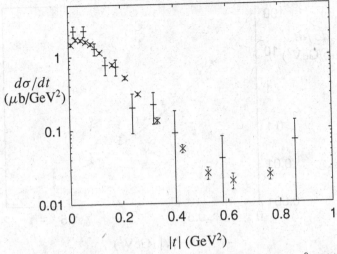

Figure 5.19. Comparison of $\gamma p \to \rho^+ n$ [65] and $\pi^- p \to \pi^0 n$ [44].

and f_2 exchange, it is not so small compared to ρ exchange as their couplings to the nucleon are comparable [2]. One would expect the t dependence and energy dependence of $\gamma p \to \rho^+ n$ and $\pi^- p \to \pi^0 p$ to be the same because of the near-degeneracy of the ρ and a_2 trajectories, but the absolute normalization to be unspecified. The two reactions are compared in figure 5.19 at $\sqrt{s} = 4.3$ GeV. The shapes are certainly in reasonable agreement, but the $\pi^- p$ charge-exchange data have been reduced by a factor of 160 rather than the nominal 278 from the factor $4\pi\alpha/\gamma_\rho^2$ in (5.57). Note the very different scale between $\gamma p \to \rho^0 p$ and $\gamma p \to \rho^+ n$, showing that ρ and a_2 exchange are really small compared to pomeron and f_2 exchange.

Figure 5.18 shows good agreement with the small-t ZEUS data at $\sqrt{s} = 71.7$ GeV. However at $\sqrt{s} = 94$ GeV, where the data are at larger values of $|t|$, the predicted cross section is too low, the discrepancy increasing with increasing $|t|$. A clue to explaining this is provided by the proton structure function $F_2(x, Q^2)$ at small x. Within the framework of conventional Regge theory it is necessary to introduce [66] a second pomeron, the hard pomeron, with intercept a little greater than 1.4. This concept is also compatible [67] with the ZEUS data [68] for the charm component $F_2^c(x, Q^2)$ of $F_2(x, Q^2)$ which seem to confirm its existence. The slope of the trajectory can be deduced [2,67] from the H1 data [69] for the differential cross section for the process $\gamma p \to J/\psi p$:

$$\alpha_{P_h}(t) = 1.44 + \alpha'_{P_h} t \qquad \alpha'_{P_h} = 0.1 \text{ GeV}^{-2}. \tag{5.62}$$

The hard-pomeron contribution to the amplitude for $\gamma p \to J/\psi p$ is then

$$i F_1(t)\left(A_{P_h}\left(\alpha'_{P_h}s\right)^{\alpha_{P_h}(t)-1} e^{-\frac{1}{2}i\pi(\alpha_{P_h}(t)-1)}\right). \tag{5.63}$$

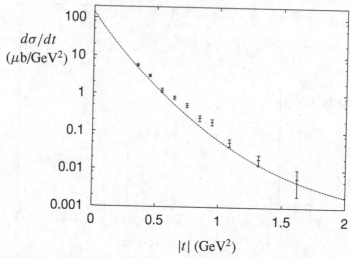

Figure 5.20. Comparison of the two-pomeron model [67] with ZEUS data [70] on $\gamma p \rightarrow \rho^0 p$ at $\sqrt{s} = 94$ GeV.

The contribution from the soft pomeron must be added to this and the coefficients A_{P_h} and A_P, the corresponding coefficient for the soft-pomeron contribution to $J/\psi p$ photoproduction is obtained by fitting the H1 [69] and ZEUS [62] differential cross sections. The fits to the data for F_2 at small x and for F_2^c suggest that the coupling of the hard pomeron to quarks is flavour-blind. Thus the hard-pomeron contribution to $\gamma p \rightarrow \rho p$ can be obtained from that in $\gamma p \rightarrow J/\psi p$ by including the effect of the vector-meson wave functions. Assuming that the point-like component of the photon, rather than the hadron-like component, is responsible, then the strength of the hard-pomeron coupling depends only on the magnitude of the vector-meson wave function at the origin and the relevant quark charges. This means that it is proportional to $\sqrt{\Gamma_{V \rightarrow e^+ e^-}/m_V}$. Figure 5.20 shows that adding this hard-pomeron contribution to the amplitude for $\gamma p \rightarrow \rho p$ provides a good description of the large-$|t|$ data.

At large-$|t|$ at low energy the model falls well below the data, as shown in figure 5.21. As for pion photoproduction a partonic description is more appropriate than one in terms of Regge poles, but unlike pion photoproduction, two-gluon exchange is allowed in addition to quark exchange. We shall see in the case of ϕ photoproduction that pomeron exchange does saturate the large-$|t|$ cross section at the same energy, so there is no need for an additional term. However, it may be that this agreement is accidental and that an interpretation in terms of two-gluon exchange at large $|t|$ matched to the pomeron at small $|t|$ is more appropriate [72,73]. Neither of these is sufficient to explain the data and it is clear that quark-exchange mechanisms are important. The $\theta_{c.m.} = 90°$ data are shown in figure 5.22. The curve is s^{-7}, arbitrarily normalized, implying that the onset of scaling at this angle apparently

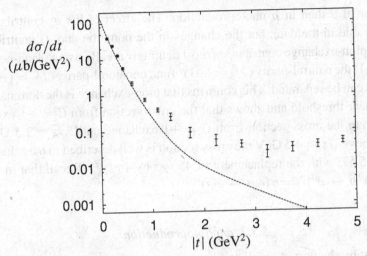

Figure 5.21. Comparison of the two-pomeron model [67] with CLAS data [71] on $\gamma p \to \rho^0 p$ at $E_\gamma = 3.8$ GeV.

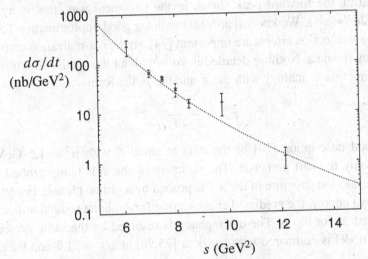

Figure 5.22. The differential cross section for $\gamma p \to \rho^0 p$ at 90° [71]. The curve is s^{-7}, arbitrarily normalized.

occurs already by $s \approx 6$ GeV^2. In the extreme backward direction, Reggeized u-channel baryon exchange becomes the dominant mechanism.

The discussion on ρ^0 photoproduction can be applied directly to ω photoproduction with three modifications. Firstly, the cross section is approximately a factor of 9 smaller due to the difference between $4\pi/\gamma_\rho^2$ and $4\pi/\gamma_\omega^2$. Secondly, the a_2 contribution is larger, both in absolute as well as relative terms, as the photon is now ρ-like for a_2 exchange. Thirdly, the cross section from pion exchange is larger by a factor of 9 than in ρ photoproduction. The effect of the a_2 contribution is not observable in the data, but the changes in the pomeron and f_2 contributions and in the pion-exchange contribution most definitely are. By using plane-polarized photons [61] the natural-parity $(J^P = (-1)^J)$ and unnatural-parity $(J^P = (-1)^{J+1})$ exchanges can be separated. This confirms that pion-exchange is the dominant contribution near threshold and shows that the cross section from $C = -1$ exchange is larger than the cross section from $C = +1$ exchange until $\sqrt{s} \approx 2.5$ GeV. At energies above $\sqrt{s} \approx 4.5$ GeV the cross section is well described in magnitude and shape by (5.57) with the replacement of $4\pi/\gamma_\rho^2$ by $4\pi/\gamma_\omega^2$. (Recall that in simple VMD, $\sigma(\rho^0 p \to \rho^0 p) = \sigma(\omega p \to \omega p)$.)

5.3.2 ϕ photoproduction

For ϕ photoproduction at small $|t|$, because of the pomeron dominance arising from Zweig's rule, the cross section should behave as $s^{2\epsilon}/b$, where b is the near-forward t slope. The data are compatible with this, but are not sensitive to constant b or to letting the forward peak shrink in the canonical way, that is by taking $b = b_0 + 2\alpha' ln(\alpha' s)$. We know that VMD is not a good approximation for the ϕ and that wave-function effects are important [74], so the normalization can only be specified by the data. Nothing detailed is known about the spatial structure of the ϕ. Modelling this in analogy with the π and the ρ, the form

$$F_\phi = \frac{1}{1 - t/\mu^2} \tag{5.64}$$

gives a good description [74] of the data at small t, with $\mu^2 = 1.5$ GeV^2 and including only the soft pomeron. The structure of the ϕ is being probed by the pomeron just as the structure of the $\pi^\pm$ is probed by a virtual photon. For $|t| \geq 0.5$ GeV^2 at high energy, the predicted cross section for ϕ photoproduction lies below the data, just as for the ρ. The discrepancy is resolved by the same prescription [67]. The model is compared with the data [75,76] at $\sqrt{s} = 71.7$ and 94 GeV in figure 5.23.

There are data [77] at $\langle E_\gamma \rangle = 3.5$ GeV to much larger values of $|t|$, indeed almost to the kinematical limit $|t| = 4.6$ GeV^2. As for the ρ, the hard-pomeron contribution is negligible at this energy. The cross section due to the soft pomeron is compared with the low-energy data in figure 5.24, with the surprising result that it agrees with the data out to $|t| = 2.5$ GeV^2. Of course this result is crucially dependent on the choice (5.64) of form factor for the ϕ. An alternative explanation for these data has been proposed in terms of two-gluon exchange [72,73], matched to the pomeron-exchange contribution around $|t| \sim 1$ GeV^2. This model works extremely well and agrees with the data out to $|t| \sim 3$ GeV^2. At high energy and for $|t| \geq 1$ GeV^2

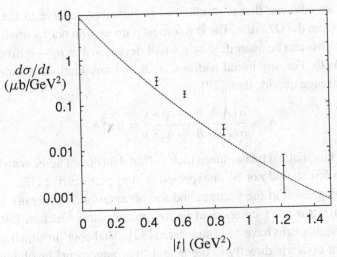

Figure 5.23. Comparison of two-pomeron model [67] with data on $\gamma p \to \phi p$ at $\sqrt{s} = 71.7$ and 94 GeV [75,76].

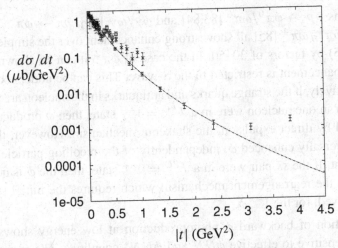

Figure 5.24. Comparison of two-pomeron model [67] with data on $\gamma p \to \phi p$ at $E_\gamma = 3.5$ GeV.

the predictions of the two models differ dramatically, and by $|t| = 3\,\text{GeV}^2$ at $\sqrt{s} = 94$ GeV the two-gluon exchange model predicts a cross section more than a factor of 50 larger than the pomeron-exchange model.

The highest $|t|$ bin in the $E_\gamma = 3.5$ GeV data is certainly due to u-channel N and N^* exchange. Its presence raises the question of an intrinsic $s\bar{s}$ component in the wave function of the proton. This is not a new question and has been studied for some time in $\bar{p}p$ interactions. The first comparison [78] of the reactions $\bar{p}p \to \phi\pi^+\pi^-$ and

$\bar{p}p \to \omega\pi^+\pi^-$ showed that ϕ production is enhanced relative to the ω beyond expectation from the OZI rule. The ϕ is almost pure $s\bar{s}$ with only a small admixture of $u\bar{u} + d\bar{d}$. This can be quantified by a small deviation $\delta = \theta_v - \theta_i$ from the ideal mixing angle θ_i. For any initial hadrons A, B and any final state hadrons X not containing strange quarks, then [79]

$$R = \frac{\sigma(A + B \to \phi X)}{\sigma(A + B \to \omega X)} = \tan^2 \delta \cdot f, \qquad (5.65)$$

where f is a kinematical phase-space factor. That data do not agree with the simple prediction (5.65) should not be unexpected as non-perturbative effects can cause $\bar{s}s$ pairs to be present in the vacuum and the observation is generally interpreted as due to an intrinsic $\bar{s}s$ component in the proton wave function [80,81]. Two production mechanisms have been suggested [82], 'shakeout' in which an intrinsic $\bar{s}s$ component converts directly to the ϕ and 'rearrangement' involving an s and $\bar{s}$ from different $\bar{s}s$ pairs combining to form the ϕ with their $\bar{s}$ and s partners annihilating. These provide a reaction-dependent, non-universal modification of the naive OZI prediction.

The reactions $\bar{p}p \to \phi\pi^0/\omega\pi^0$ [83,84] and $\phi\gamma/\omega\gamma$ [84], $\bar{p}n \to \phi\pi^-/\omega\pi^-$ [85] and $\bar{n}p \to \phi\pi^+/\omega\pi^+$ [85] all show strong enhancement over the simple OZI prediction (5.65) by factors of 30–50. In the case of $\phi\pi^+$ it can be shown [86] that the large enhancement is restricted to the S-wave. This can be understood [82], at least qualitatively, if the strange quarks and antiquarks in the nucleon are polarized. If the $s\bar{s}$ pair in one nucleon were in a $J^{PC} = 1^{--}$ state, then ϕ production could be explained by direct expulsion, the shakeout mechanism. However, this would imply a universally enhanced ϕ, independently of the recoiling particle, contrary to experiment. If the $s\bar{s}$ pair were in a $J^{PC} = 0^{++}$ state, then the ϕ is necessarily produced by the rearrangement mechanism which requires the initial state to be $^3S_1$, as observed, for the $1^{--}\phi$.

The observation of backward ϕ photoproduction at low energy shows that the reaction is sensitive to effective ϕNN and ϕNN^* couplings. The rearrangement mechanism is allowed in backward ϕ photoproduction, but other mechanisms are not excluded so the reaction lacks discrimination.

5.3.3 Higher-mass vector mesons

For a final vector state V, the cross section for $\gamma p \to Vp$ is related by VMD to that for $e^+e^- \to V$ by [87]

$$\frac{d^2\sigma_{\gamma p \to Vp}(s, m^2)}{dt\, dm^2} = \frac{\sigma_{e^+e^- \to V}(m^2)}{4\pi^2\alpha} \frac{d\sigma_{Vp \to Vp}(s, m^2)}{dt}. \qquad (5.66)$$

The amplitude at $t = 0$ is related to the total cross section for Vp scattering by the optical theorem. Using this and integrating over t gives

$$\frac{d\sigma_{\gamma p \to Vp}(s, m^2)}{dm} = \frac{m\sigma_{e^+e^- \to V}(m^2)}{32\pi^3\alpha b} \left(\sigma_{Vp \to Vp}^{tot}(s)\right)^2, \qquad (5.67)$$

where $b \approx 7 \text{ GeV}^{-2}$ is the slope of the near-forward differential cross section.

The shapes of the $\sigma(e^+e^- \to \pi^+\pi^-\pi^+\pi^-)$ and of the differential cross section $d\sigma(\gamma p \to \pi^+\pi^-\pi^+\pi^-)/dm$, in the photon energy range 20–70 GeV, with $\langle\sqrt{s}\rangle = 8.6 \text{ GeV}$, are reasonably well matched via (5.67) [88]. In contrast, the shapes of $d\sigma(\gamma p \to \omega\pi^0)/dm$, over the same photon energy range as the $\pi^+\pi^-\pi^+\pi^-$ data, and $\sigma(e^+e^- \to \omega\pi^0)$ do not agree, the photoproduction data having a pronounced enhancement centred at about 1.25 GeV. A spin-parity analysis [89] of the $\omega\pi^0$ photoproduction data concluded that the enhancement is consistent with the production of $b_1(1235)$, that is $J^P = 1^+$, with $\sim 20\%$ $J^P = 1^-$ background if only these two spin-parity states are included in the analysis. This conclusion was subsequently confirmed [90] at $\sqrt{s} = 6.2$ GeV. The $J^P = 1^-$ background is consistent in shape with the e^+e^- $\omega\pi$ data. If a $J^P = 0^-$ amplitude is added, then $J^P = 1^-$ becomes the largest amplitude [89], with the $\omega\pi$ mass distribution peaking at around 1.3 GeV, albeit with large errors. The enhancement is still present at $\sqrt{s} = 200$ GeV [91], but a spin-parity analysis of these data cannot be performed because of limited acceptance. It has been argued [92] that, if this $\omega\pi$ enhancement is due to the photoproduction of the $b_1(1235)$, then these data provide evidence for quark spin-flip in pomeron exchange as the $\gamma \to b_1(1235)$ transition is from a quark spin-triplet state to a quark spin-singlet one. Alternatively it could be evidence for a $\rho(1300)$, discussed in chapter 4. In this case there is an obvious breakdown of the simple VMD argument that works rather well for the $\pi^+\pi^-\pi^+\pi^-$ channel.

The data on $\gamma p \to \pi^+\pi^- p$ provide a clue to this apparent breakdown of VMD. The deep dip due to destructive interference seen in $e^+e^- \to \pi^+\pi^-$, figure 4.2, becomes a peak in photoproduction, shown in figure 5.25. The interference is now constructive so there has been a relative sign change between the ρ and the higher-mass states responsible for the structure. It has been suggested [93] that this is due to mixing between a 2^3S_1 $q\bar{q}$ state and a hybrid vector meson which cannot be excited diffractively in lowest-order QCD. This model is based on the dipole picture, see chapter 8, and the effect arises partly from the wave function of the 2^3S_1 state and partly from the mixing that has to be quite strong. An alternative explanation is that the effect is due primarily to the Drell–Söding mechanism [56,57] discussed in section 5.3.1. Above the ρ peak the interference term is negative, as is apparent from figure 5.16, and could be responsible for the phase change. The two mechanisms have different implications for the $\omega\pi$ channel in photoproduction. If

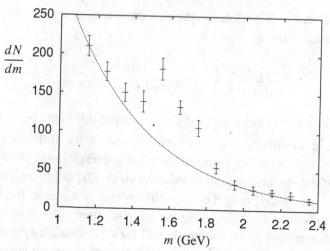

Figure 5.25. The $\pi^+\pi^-$ mass distribution in the reaction $\gamma p \to \pi^+\pi^- p$ [89]. The curve is an estimate of the background under the peak.

the effect is due to the behaviour of the wave functions of the excited states, then it is channel independent and the same phase change will be present in the $\omega\pi$ channel. The Drell–Söding mechanism is specific to the $\pi^+\pi^-$ channel so has no direct relevance for $\omega\pi$, although the equivalent Deck mechanism [58] in the $\omega\pi$ channel could cause a similar phase change. However, to evaluate this requires a full understanding of $e^+e^- \to \pi\omega$ that we do not yet have: see chapter 4.

A further example of a possible phase change is seen in the $\eta\rho$ channel. The cross section for $e^+e^- \to \eta\rho$ rises rapidly from threshold and peaks at about 1.5 GeV. In contrast the invariant mass distribution for the $J^P = 1^-$ state in photoproduction decreases rapidly and monotonically from threshold. If a change of phase between e^+e^- annihilation and photoproduction is assumed, then it is possible to describe both data sets simultaneously with constructive interference between the ρ and $\rho(1450)$ in e^+e^- annihilation and destructive interference in photoproduction [94].

Data on the photoproduction of $J^P = 1^-$ isoscalar states exhibit a number of anomalies [95,96] when compared with the corresponding channels in e^+e^- annihilation or with the equivalent $J^P = 1^-$ isovector channels in photoproduction. The most notable anomaly occurs in the data on photoproduction of the $J^P = 1^-$ $\rho\pi$ final state when compared with that of the $J^P = 1^-$ $\omega\pi^0$ final state, the latter being an order of magnitude larger than expected. The $\omega\pi^+\pi^-$ invariant mass distribution has a double-peaked structure, the dip coinciding with the single peak of the cross section for $e^+e^- \to \omega\pi^+\pi^-$ that is dominated by the $\omega(1650)$. The photoproduced $\eta\omega$ mass spectrum has a clear peak, consistent with dominance by the $\omega(1650)$, in complete contrast to the corresponding photoproduced $\eta\rho$ mass spectrum that decreases monotonically, as discussed above.

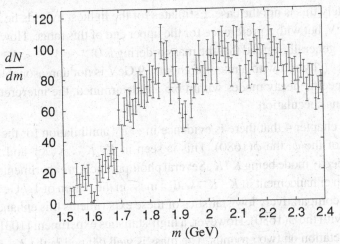

Figure 5.26. The $3\pi^+3\pi^-$ mass distribution in the reaction $\gamma p \to 3\pi^+3\pi^- p$ [97,98].

These anomalies, of course, are based on the implicit assumption that the dominant exchanges in photoproduction are the pomeron and the $f_2(1250)$ Regge trajectory. There is good reason for this. The only other leading Regge trajectory that can contribute is that of the $a_2(1320)$ as the ω and ρ trajectories are excluded by C-parity. The coupling of the a_2 trajectory to the nucleon is known to be an order of magnitude smaller than that of the f_2 [2], so it appears unlikely that a_2 exchange could give rise to such dramatic effects.

The narrow structure seen in $e^+e^- \to 6\pi$ near 1.9 GeV, discussed in chapter 4, is also apparent in $\gamma p \to 3\pi^+3\pi^- p$ [97,98], see figure 5.26. A broad structure is seen in the $\omega\rho\pi$ mass spectrum in photoproduction [99]. Fitting this peak with a Breit–Wigner, assuming a fixed width, gives a mass of 2.28 ± 0.05 GeV and a width of 0.44 ± 0.11 GeV. However, as there is evidence that the principal mode is ωa_1, threshold effects will distort the resonance and reduce the mass to a value compatible with that found in the analysis of $e^+e^- \to 6\pi$, namely 2.11 GeV [100]. The structure in the $3\pi^+3\pi^-$ photoproduction data has been interpreted as due to the interference of a narrow resonance either with a continuum or a continuum plus a second resonance. This second resonance has a mass of 1.730 ± 0.034 GeV and a width of 315 ± 100 MeV so can be identified with the $\rho(1700)$. The narrow resonance has a mass of 1.91 ± 0.01 GeV, a width of 37 ± 13 MeV and a small e^+e^- width: it is only visible because of the interference.

A possible interpretation of this narrow state, if that is the correct interpretation of the data, is in terms of a hybrid $q\bar{q}g$ meson. Three candidates exist [101] for mesons with the exotic quantum numbers $J^{PC} = 1^{-+}$ at 1.4, 1.6 and 2.0 GeV, although their interpretation in terms of resonances remains ambiguous. Ideally we would have mass predictions for light-quark hybrids comparable to those for $q\bar{q}$ states, but unfortunately this is not the case. Estimates for the lightest hybrids lie between 1.3 and 1.9 GeV, but with a preference for the upper end of this range. However, it does seem to be generally agreed that the mass ordering is $0^{-+} < 1^{-+} < 1^{--} < 2^{-+}$, so a $J^{PC} = 1^{--}$ hybrid with a mass of about 1.9 GeV is not unreasonable. However, until the specific decay modes within 6π are determined, the interpretation of the data remains speculation.

We saw in chapter 4 that there is evidence in e^+e^- annihilation for the 2^3S_1 radial excitation of the ϕ, the $\phi(1680)$. This is seen in K^+K^-, $K^0_S K^0_L$ and $K^0_S K\pi$, the dominant decay mode being K^*K. Several photoproduction experiments [102] have observed an enhancement in K^+K^- with a mass in the region of 1.7 GeV or higher. Due to the comparatively low statistics of these experiments this enhancement was identified with the $\phi(1680)$. However, a high-statistics experiment [103] challenges this interpretation on two grounds. The mass is well defined in the K^+K^- channel as $1753.5 \pm 1.5 \pm 2.3$ MeV, clearly inconsistent with 1680 MeV, and no evidence is found for an enhancement in K^*K. In both respects these data contradict not only the low-statistics photoproduction data but also the e^+e^- data, thus adding more confusion to an already confused picture.

5.4 Exotic mesons

Although estimates of the electromagnetic transition amplitudes between conventional and hybrid mesons now exist [104,105], precise rates for the photoproduction of exotic (hybrid) mesons are presently uncalculable as there is no clear understanding of the Regge exchange contributions to these cross sections. Indeed, what is required are detailed data on hybrid photoproduction so that they can give new insights into Regge dynamics and duality. Ideas on duality were originally developed based on the production and classification of 'normal' hadrons; what is the pattern of duality in a world with glueballs and hybrids – which states are dual to what? A strategic study of exotic hadron production, its energy and t dependences for example, may give insight to such questions. However, in the absence of such information it is hard to quantify the production of hybrids.

Nonetheless, there are qualitative, and semiquantitative, reasons to suppose that their production rates (at least for light flavours) are not dramatically less than those of conventional hadrons.

The initial estimates were based on VMD. Afanasev and Page [106] computed π_1 photoproduction rates under the assumption that the process was driven by π exchange. The photon converted to ρ in a vector dominance model and the observed rate for $\pi_1 \to \pi\rho$ was used to model the production vertex. There are two problems here. One is that a large coupling of $\pi_1 \to \pi\rho_H$ may compensate for the small $\gamma \rightarrow \rho_H$ whereby $\gamma \rightarrow \rho_H$ followed by $\rho_H \pi \rightarrow \pi_1$ can compete and interfere with the ρ contribution. The second, potentially serious concern, is that strong criticisms have been made about the claimed observation of π_1 in the $\rho \pi$ mode [107].

Photoproduction of the exotic hybrid 2^{+-} is expected to be significant on rather general grounds as $\gamma N \rightarrow 2^{+-} N$ can occur by diffractive (pomeron) and by f_2 exchange.

Diffractive scattering is known to give a significant cross section for $\pi N \rightarrow a_1 N$ and also apparently $\gamma N \rightarrow b_1 N$ [92], thus $1^{--} N \rightarrow 2^{+-} N$ may also be expected. Furthermore, whereas $\pi N \rightarrow a_1 N$ and $\gamma N \rightarrow b_1 N$ involve spin flip at quark level, which empirically leads to some suppression [92], the $1^{--} \rho$ and the 2^{+-} hybrid are both spin triplets, and so no such penalty is expected in $\gamma N \rightarrow 2^{+-} N$. Hence 2^{+-} production is expected to be copious at high energy and probably also at lower energies where f_2 exchange can contribute [108].

Isgur had conjectured [109] that light hybrids have photoproduction cross sections $\sim 50\%$ those of normal hadrons. Close and Dudek [104,105] have shown how to compute rates for both conventional and hybrid mesons in the lattice-based flux-tube model. These computations endorse Isgur's conjecture. Hence there is no reason to suppose that photoproduction rates of hybrids are insignificant: the challenge will be to identify the exotic $J^{PC} 2^{+-}$ or 1^{-+} in the many-particle final states. JLab at 12 GeV would provide an ideal environment where the meson states at small t may be separated from decays at the baryon vertex.

References

[1] P D B Collins, *An Introduction to Regge Theory*, Cambridge University Press, Cambridge (1977)

[2] A Donnachie, H G Dosch, P V Landshoff and O Nachtmann, *Pomeron Physics and QCD*, Cambridge University Press, Cambridge (2002)

[3] S J Brodsky and G R Farrar, Physical Review Letters 31 (1973) 1153
S J Brodsky and G R Farrar, Physical Review D11 (1975) 1309
V A Matveev, R M Muradyan and A N Tavkhelidze, Nuovo Cimento Letters 7 (1973) 719

[4] G P Lepage and S J Brodsky, Physical Review D22 (1980) 215

[5] T Regge, Il Nuovo Cimento 14 (1959) 951

[6] T Regge, Il Nuovo Cimento 18 (1960) 947

[7] A Bottino, A M Longhoni and T Regge, Il Nuovo Cimento 18 (1962) 954

[8] J K Storrow, Physics Reports 103 (1984) 317

[9] G F Chew and S C Frautschi, Physical Review Letters 8 (1962) 41

[10] S Eidelman *et al*, Particle Data Group, Physics Letters B592 (2004) 1

[11] A Donnachie and P V Landshoff, Physics Letters B296 (1992) 227

[12] N A Amos *et al*, E710 Collaboration, Physical Review Letters 63 (1989) 2784

[13] F Abe *et al*, CDF Collaboration, Physical Review D50 (1994) 5550

[14] J R Cudell *et al*, Physical Review D61 (1998) 034019
[15] G A Jaroskiewicz and P V Landshoff, Physical Review D10 (1974) 170
[16] P V Landshoff and O Nachtmann, Zeitschrift für Physik C35 (1987) 405
[17] H B Meyer and M J Teper, Physics Letters B605 (2005) 344
[18] M Froissart, Physical Review 123 (1961) 1053
[19] L Lukaszuk and A Martin, Il Nuovo Cimento 52 (1967) 122
[20] P V Landshoff and J C Polkinghorne, Physics Reports 5C (1972) 1
[21] D Branson, Physical Review 179 (1969) 1608
[22] L M Jones and P V Landshoff, Nuclear Physics B94 (1975) 145
[23] D V Bugg in *HADRON '01*, D Amelin and A M Zaitsev eds, American Institute of Physics, New York (2002)
[24] M Jacob and G C Wick, Annals of Physics 7 (1959) 404
[25] A R Edmonds, *Angular Momentum in Quantum Mechanics* Princeton University Press, Princeton (1966)
[26] K Igi, Physical Review Letters 9 (1962) 76
[27] R Dolen, D Horn and C Schmid, Physical Review Letters 19 (1967) 402
[28] K Igi and S Matsuda, Physical Review Letters 18 (1967) 625
[29] A A Logunov, L D Soloviev and A N Tavkhelidze, Physics Letters 24B (1967) 181
[30] C Schmid, Proceedings of the Royal Society 318A (1970) 257
[31] V Barger and E J N Phillips, Physical Review 187 (1969) 2210
[32] H Harari and Y Zarmi, Physical Review 187 (1969) 2230
[33] I S Barker, A Donnachie and J K Storrow, Nuclear Physics B79 (1974) 431
[34] I S Barker and J K Storrow, Nuclear Physics B137 (1978) 413
[35] R Worden, Nuclear Physics B37 (1972) 253
[36] M Rahnama and J K Storrow, Journal of Physics G17 (1991) 243
[37] The SAID database, http://gwdac.phys.gwu.edu/
[38] G Veneziano, Il Nuovo Cimento 57A (1968) 190
[39] I S Barker, A Donnachie and J K Storrow, Nuclear Physics B95 (1975) 347
[40] G Keaton and R Workman, Physical Review C53 (1996) 1434
[41] W-T Chiang and F Tabakin, Physical Review C55 (1997) 2054
[42] A Kotanski, Acta Physica Polonica 29 (1966) 699
[43] J K Storrow in *Electromagnetic Interactions of Hadrons* Vol 1, A Donnachie and G Shaw eds, Plenum Press, New York (1978)
[44] P Sonderegger *et al*, Physics Letters 20 (1966) 75
[45] R L Anderson *et al*, Physical Review D4 (1971) 1937
[46] M Guidal, J-M Laget and M Vanderhaeghen, Nuclear Physics A627 (1997) 645
[47] A M Boyarski *et al*, Physical Review Letters 20 (1968) 300
[48] M Rahnama and J K Storrow, Journal of Physics G8 (1982) 455
[49] P K Williams, Physical Review D1 (1970) 1312
[50] G C Fox and C Quigg, Annual Review of Nuclear Science 23 (1973) 219
[51] I M Barbour, W Malone and R G Moorhouse, Physical Review D4 (1971) 1521
[52] L Y Zhu *et al*, Physical Review C71 (2005) 044603
[53] W Braunschweig *et al*, Physics Letters 61B (1970) 479
[54] A M Boyarski *et al*, Physical Review Letters 22 (1969) 1131
[55] W M Kloet, W-T Chiang and F Tabakin, Physical Review C58 (1998) 1086
[56] S D Drell, Physical Review Letters 5 (1960) 278
[57] P Söding, Physics Letters 19 (1965) 702
[58] R T Deck, Physical Review Letters 13 (1964) 169

[59] J Breitweg *et al*, ZEUS Collaboration, European Physical Journal C2 (1998) 247

[60] A Donnachie and G Shaw in *Electromagnetic Interactions of Hadrons* Vol 2, A Donnachie and G Shaw eds, Plenum Press, New York (1978)

[61] J Ballam *et al*, Physical Review D7 (1973) 3150

[62] J Breitweg *et al*, ZEUS Collaboration, European Physical Journal C1 (1998) 81

[63] D Aston *et al*, Omega-Photon Collaboration, Nuclear Physics B209 (1982) 56

[64] N H Buttimore *et al*, Physical Review D59 (1999) 114010

[65] J Abramson *et al*, Physical Review Letters 36 (1976) 1432

[66] A Donnachie and P V Landshoff, Physics Letters B437 (1998) 408

[67] A Donnachie and P V Landshoff, Physics Letters B478 (2000) 146

[68] J Breitweg *et al*, ZEUS Collaboration, European Physical Journal C12 (2000) 35

[69] C Adloff *et al*, H1 Collaboration, Physics Letters B483 (2000) 23

[70] J Breitweg *et al*, ZEUS Collaboration, European Physical Journal C14 (2000) 213

[71] M Battaglieri *et al*, CLAS Collaboration, Physical Review Letters 87 (2001) 172002

[72] J M Laget, Nuclear Physics A699 (2002) 184

[73] F Cano and J M Laget, Physical Review D65 (2002) 074022

[74] A Donnachie and P V Landshoff, Physics Letters B348 (1995) 213

[75] M Derrick *et al*, ZEUS Collaboration, Physics Letters B377 (1996) 259

[76] J Breitweg *et al*, ZEUS Collaboration, European Physical Journal C14 (2000) 21

[77] E Anciant *et al*, CLAS Collaboration, Physical Review Letters 85 (2000) 213

[78] A M Cooper *et al*, Nuclear Physics B146 (1978) 1

[79] H J Lipkin, Physics Letters 60B (1976) 371

[80] J Ellis, E Gabathuler and M Karliner, Physics Letters B217 (1989) 173

[81] E M Henley, G Krein and A G Williams, Physics Letters B281 (1992) 178

[82] J Ellis, M Karliner, D E Kharzeev and M G Sapozhnikov, Physics Letters B353 (1995) 319

[83] J Reifenrother *et al*, ASTERIX Collaboration, Physics Letters B267 (1991) 299

[84] C Amsler *et al*, Crystal Barrel Collaboration, Physics Letters B346 (1995) 343

[85] V G Ableev *et al*, OBELIX Collaboration, Nuclear Physics A585 (1995) 577

[86] A Filippi *et al*, OBELIX Collaboration, Nuclear Physics A655 (1999) 453

[87] J J Sakurai and D Schildknecht, Physics Letters 40B (1972) 121

[88] D Aston *et al*, Omega Photon Collaboration, Nuclear Physics B189 (1981) 15

[89] M Atkinson *et al*, Omega Photon Collaboration, Nuclear Physics B243 (1984) 1

[90] J E Brau *et al*, Physical Review D37 (1988) 2379

[91] T Golling and K Meier, H1 Collaboration, H1-preliminary-010117; T Berndt, H1 Collaboration, in *Proceedings International Conference on High Energy Physics*, Elsevier Science, Amsterdam (2002); T Berndt, H1 Collaboration, Acta Physica Polonica B33 (2002) 3499

[92] A Donnachie, Physics Letters B611 (2005) 255

[93] H G Dosch, T Gousset, G Kulzinger and H J Pirner, Physical Review D55 (1997) 2602

[94] A Donnachie and A B Clegg, Zeitschrift für Physik C34 (1987) 257

[95] A Donnachie and A B Clegg, Zeitschrift für Physik C42 (1989) 663

[96] A Donnachie and A B Clegg, Zeitschrift für Physik C48 (1990) 111

[97] P L Frabetti *et al*, E637 Collaboration, Physics Letters B514 (2001) 240

[98] P L Frabetti *et al*, E637 Collaboration, Physics Letters B578 (2001) 290

[99] M Atkinson *et al*, Omega Photon Collaboration, Zeitschrift für Physik C29 (1985) 333

[100] A B Clegg and A Donnachie, Zeitschrift für Physik 45 (1990) 677

[101] J Kuhn in *HADRON '03*, E Klempt, H Koch and H Orth eds, American Institute of Physics, New York (2003)

[102] D Aston *et al*, Omega Photon Collaboration, Physics Letters B104 (1981) 231
M Atkinson *et al*, Omega Photon Collaboration, Zeitschrift für Physik C27 (1985) 233
J Busenitz *et al*, E401 Collaboration, Physical Review D40 (1989) 1

[103] J M Link *et al*, FOCUS Collaboration, Physics Letters B545 (2002) 50

[104] F E Close and J J Dudek, Physical Review Letters 91 (2003) 142001-1

[105] F E Close and J J Dudek, Physical Review D69 (2004) 034010

[106] A Afanasev and P R Page, Physical Review D57 (1998) 6771

[107] A R Dzierba, C A Meyer and A P Szczepaniak, Journal of Physics Conference Series 9 (2004) 192

[108] F E Close and J J Dudek, Physical Review D70 (2004) 094015

[109] N Isgur, Physical Review D60 (1999) 114016

6

Chiral perturbation theory

M Birse and J McGovern

In the Standard Model the up and down quarks – the constituents of normal matter – couple very weakly to the electroweak Higgs field. The resulting 'current' masses for these quarks are of the order of 10 MeV. This is much smaller than the typical energies of hadronic states (several hundred MeV or more). In the context of low-energy QCD, this means that these quarks are nearly massless and so, to a good approximation, their chiralities are preserved as they interact with vector fields (gluons, photons, Ws and Zs). This conservation of chirality has important consequences for hadronic physics at low energies.

If chirality is conserved, then right- and left-handed quarks are independent and we have two copies of the isospin symmetry which relates up and down quarks. The QCD Lagrangian therefore is symmetric under the chiral symmetry group, $SU(2)_R \times SU(2)_L$. However, the hadron spectrum shows no sign of this larger version of isospin symmetry; in particular, particles do not come in doublets with opposite parities. Moreover, constituent quarks, as the building blocks of hadrons, appear to have masses that are much larger than their current masses.

This conundrum can be resolved if the chiral symmetry of the theory is hidden (or spontaneously broken) in the QCD vacuum. A non-vanishing condensate of quark–antiquark pairs is present in the vacuum, and this is not invariant under $SU(2)_R \times SU(2)_L$ transformations. The energy spectrum of quarks propagating in this vacuum has a gap (like that for electrons in a superconductor), implying that masses have been dynamically generated.

In the limit of exact chiral symmetry (vanishing current masses for the up and down quarks) pions would be massless 'Goldstone bosons'. Their properties reflect the fact that the dynamics is chirally symmetric even though the vacuum is not. For example, there is a non-vanishing pion-to-vacuum matrix element of the axial

Electromagnetic Interactions and Hadronic Structure, eds Frank Close, Sandy Donnachie and Graham Shaw.
Published by Cambridge University Press. © Cambridge University Press 2007

current: f_π, the decay constant measured in weak decays of charged pions. In addition, all strong interactions of pions would vanish at threshold in the chiral limit.

In the real world, the small current masses of the up and down quarks explicitly break this symmetry. As a result pions do have small masses and their interactions do not vanish exactly. Nonetheless the forms of the interactions of pions with each other and with other hadrons remain strongly constrained by the approximate chiral symmetry of QCD. The weakness of pionic interactions at low energies makes a new kind of perturbative expansion possible: one where we expand in powers of momenta and pion masses. The success of such an expansion relies on the fact that the current quark masses make small contributions to hadron properties and interactions. (Contrary to the claims of those who say that the elusive Higgs boson is the origin of mass in the universe, if quarks were massless the proton and neutron would still have at least 90% of the masses we observe in our world.) This expansion can be obtained from an effective field theory known as chiral perturbation theory (χPT). A comprehensive review of it and its applications to mesons and baryons can found in [1]. Also very useful are two older reviews: [2] for the mesonic sector, and [3] for the single-nucleon sector. Extensions of these ideas to systems of two or more nucleons (which we do not discuss here) can be found in [4,5].

6.1 Chiral symmetry

The QCD Lagrangian with two flavours of light quark (up and down) has the form

$$\mathcal{L}(x) = \bar{q}(x)\gamma^\mu \left(i\partial_\mu - g\frac{\lambda^a}{2} G_\mu^a(x) \right) q(x) - \bar{q}(x)m_q q(x) - \tfrac{1}{4} G^{a\mu\nu} G_{\mu\nu}^a, \quad (6.1)$$

where $G_\mu^a(x)$ are the gluon fields and

$$m_q \equiv \begin{pmatrix} m_u & 0 \\ 0 & m_d \end{pmatrix} \qquad\qquad (6.2)$$

is the matrix of current masses for the quarks.

For equal current masses ($m_u = m_d$) the theory is symmetric under isospin rotations,

$$q(x) \to \exp(i\boldsymbol{\alpha} \cdot \boldsymbol{\tau})q(x). \qquad\qquad (6.3)$$

In the limit of vanishing current masses ($m_u = m_d = 0$), the right- and left-handed quarks decouple and the theory is symmetric under two copies of isospin symmetry [1],

$$q(x) \to \exp\left(i\frac{1+\gamma_5}{2} \boldsymbol{\alpha}_R \cdot \boldsymbol{\tau} \right) q(x), \qquad q(x) \to \exp\left(i\frac{1-\gamma_5}{2} \boldsymbol{\alpha}_L \cdot \boldsymbol{\tau} \right) q(x).$$

$$(6.4)$$

These chiral transformations form a representation of the global symmetry group $SU(2)_R \times SU(2)_L$. The smallness of the masses compared to typical hadronic energy scales means that, to a good approximation, QCD respects this chiral symmetry. The extension of this symmetry to three flavours is also useful, although the much larger strange-quark mass ($m_s \sim 100$ MeV) means that the $SU(3)_R \times SU(3)_L$ is more strongly broken.

The Noether currents corresponding to the chiral transformations (6.4) are the isospin currents

$$J_\mu^a(x) = \overline{q}(x)\gamma_\mu \frac{\tau^a}{2}q(x) \qquad (6.5)$$

and the axial currents

$$J_{5\mu}^a(x) = \overline{q}(x)\gamma_\mu\gamma_5 \frac{\tau^a}{2}q(x). \qquad (6.6)$$

The currents and their conserved charges obey the algebra [1,6]

$$[Q^a, J_\mu^b(x)] = i\epsilon^{abc} J_\mu^c(x),$$

$$[Q^a, J_{5\mu}^b(x)] = i\epsilon^{abc} J_{5\mu}^c(x),$$

$$[Q_5^a, J_{5\mu}^b(x)] = i\epsilon^{abc} J_\mu^c(x). \qquad (6.7)$$

The strong interactions between quarks and gluons in QCD lead to a complicated vacuum state, with non-zero expectation values for colourless, scalar combinations of quark and gluon fields – 'condensates'. Even in the absence of current masses, this vacuum is symmetric under only the ordinary (vector) isospin symmetry, not the full chiral symmetry. The axial charges do not annihilate the vacuum, but instead create unnormalizable states. This indicates that the chiral symmetry of the Lagrangian is hidden or 'spontaneously broken'. Goldstone's theorem implies that there must be massless excitations with the same quantum numbers as the axial charges, spin-0, odd parity and isospin-1: the quantum numbers of the pions. Another signal of this special nature of the pion is the non-zero matrix element of the axial current between a pion and the vacuum,

$$\langle 0|J_{5\mu}^a(x)|\pi^b(q)\rangle = if_\pi q_\mu e^{-iq\cdot x}\delta_{ab}, \qquad (6.8)$$

where $f_\pi \simeq 92.5$ MeV is the pion decay constant, determined empirically from the weak decay of the charged pions.

In the real world, chiral symmetry is explicitly broken by the current masses of the quarks and so the pions are not exactly massless. Nonetheless, the approximate chiral symmetry of QCD explains why the pions are so much lighter than all other hadrons. Taking the divergence of (6.8), we get

$$\partial^\mu \langle 0|J_{5\mu}^a(x)|\pi^b(q)\rangle = f_\pi m_\pi^2 e^{-iq\cdot x}\delta_{ab}, \tag{6.9}$$

and so we can use the divergences of the axial currents as interpolating fields which create and destroy pions. At low energies, the small pion masses mean that matrix elements of these currents will be dominated by their pionic parts. Consequently, many properties and interactions of low-momentum pions can be determined from chiral symmetry alone, and so satisfy chiral low-energy theorems (LETs).

6.1.1 Partial conservation of the axial current (PCAC)

The principle of PCAC was used to derive many of these results in the 1960s [6]. It takes $\partial^\mu J_{5\mu}^a(x)$ as a pion field to extrapolate pion scattering amplitudes off-shell to the point where the pion four-momentum vanishes, known as the soft-pion limit.

As a first example, consider the pion propagator (the expectation value of the time-ordered product of two of these fields):

$$\int d^4x\, e^{iq\cdot x}\langle 0|\mathrm{T}(\partial^\mu J_{5\mu}^a(x), \partial^\nu J_{5\nu}^b(0))|0\rangle = i\frac{f_\pi^2 m_\pi^4}{q^2 - m_\pi^2} F(q^2)\delta_{ab}. \tag{6.10}$$

This has a pole at the pion mass, with a residue given by the square of the strength with which the interpolating field couples to the pion (6.9). Other states with the same quantum numbers as the pion also contribute, but at low energies these multi-pion states are suppressed by phase space. The first resonance in this channel lies above 1 GeV. Close to the pion pole, therefore, these states provide a background that varies slowly with q^2. This background has been absorbed into the function $F(q^2)$, which is normalized to $F(m_\pi^2) = 1$ at the pion pole.

In the soft-pion limit (defined to avoid ambiguity by setting $\mathbf{q} = 0$ and then taking $q^0 \to 0$) we can integrate the left-hand side of (6.10) by parts to get

$$-\langle 0|[Q_5^a, \partial^\nu J_{5\nu}^b(0)]|0\rangle = i\langle 0|[Q_5^a, [Q_5^b, \mathcal{H}(0)]]|0\rangle. \tag{6.11}$$

Here we have also used Noether's theorem: $\partial^\nu J_{5\nu}^b(x) = -i\left[Q_5^b, \mathcal{H}(x)\right]$, where $\mathcal{H}(x)$ is the Hamiltonian density. The crucial assumption underlying PCAC is that the scales Λ_χ controlling the energy and momentum dependences of amplitudes are all much larger than m_π, and so differences between amplitudes at $q^2 = m_\pi^2$ and $q^2 = 0$ are suppressed by powers of m_π/Λ_χ. The same assumption of a wide separation of scales underlies χPT, as we shall see. Under this assumption, we have $F(0) \simeq F(m_\pi^2)$, and hence the soft-pion limit of (6.10) becomes

$$\langle 0|[Q_5^a, [Q_5^b, \mathcal{H}(0)]]|0\rangle \simeq -f_\pi^2 m_\pi^2 \delta_{ab}, \tag{6.12}$$

up to terms of order m_π^4. This is the Gell-Mann–Oakes–Renner (GOR) relation [7], which shows that to leading order the square of the pion mass is proportional to the strength of the explicit symmetry breaking.

In QCD, where the symmetry breaking term has the form $\mathcal{H}_{SB}(x) = \overline{q}(x)m_q q(x)$, the GOR relation is

$$\overline{m}_q \langle 0|\overline{q}(x)q(x)|0\rangle \simeq -f_\pi^2 m_\pi^2, \tag{6.13}$$

where $\overline{m}_q$ is the average of the up- and down-quark masses and the vacuum expectation value of the scalar density of quarks is known as the 'quark condensate'. Taking a typical estimate for the light-quark masses of $\overline{m}_q \sim 7$ MeV, we find a condensate of about 3 fm^{-1} (about six times the ordinary vector density of quarks at the centre of a nucleus).

These ideas can also be applied to the interactions between low-momentum pions. For example, to order m_π^2, the amplitude for elastic scattering of two pions, $\pi^a(p_1)\pi^b(p_2) \to \pi^c(p_3)\pi^d(p_4)$, is [8]

$$T_{\pi\pi}^{cd,ab} \simeq \delta_{ab}\delta_{cd}\frac{s - m_\pi^2}{f_\pi^2} + \delta_{ac}\delta_{bd}\frac{t - m_\pi^2}{f_\pi^2} + \delta_{ad}\delta_{bc}\frac{u - m_\pi^2}{f_\pi^2}, \tag{6.14}$$

where the Mandelstam variables are

$$s = (p_1 + p_2)^2, \qquad t = (p_1 - p_3)^2, \qquad u = (p_1 - p_4)^2. \tag{6.15}$$

Note that in the chiral limit, $m_\pi = 0$, this amplitude vanishes at threshold. This illustrates an important property of Goldstone bosons: their interactions vanish in the limit of exact chiral symmetry and zero four-momentum.

Chiral symmetry also controls the forms of the interactions between pions and nucleons. In particular the pion–nucleon coupling is related to the nucleon's axial coupling constant, $g_A(0)$. This is defined by

$$\langle N(p_2)|J_{5(\mu}^a x)|N(p_1)\rangle = \overline{u}(p_2)\left[g_A(q^2)\gamma_\mu\gamma_5 + g_P(q^2)q_\mu\gamma_5\right]\tfrac{1}{2}\tau^a u(p_1)e^{iq\cdot x}, \tag{6.16}$$

where $g_A(q^2)$ is the axial form factor of the nucleon, $g_P(q^2)$ is its 'induced' pseudoscalar form factor, and $q = p_2 - p_1$. If we take the divergence of (6.16), we find that the left-hand side contains the pion pole

$$\frac{if_\pi m_\pi^2 i g_{\pi NN}}{q^2 - m_\pi^2}\,\overline{u}(p_2)q_\mu\gamma_5\tau^a u(p_1), \tag{6.17}$$

where $g_{\pi NN}$ is the physical pion–nucleon coupling constant, defined at the pion pole. Under the PCAC assumption that all other terms vary smoothly between $q^2 = m_\pi^2$ and $q^2 = 0$, the $q \to 0$ limit of the divergence gives us the Goldberger–Treiman relation,

$$g_{\pi NN} f_\pi \simeq M_N g_A(0), \tag{6.18}$$

up to corrections of order m_π^2. In the chiral limit, this would provide an exact connection between $g_{\pi NN}$ and $g_A(0)$. For realistic quark masses, corrections to it are of the order of 5% or less.

There are also important chiral constraints on pion–nucleon scattering at low energies. The starting point for deriving these is the amplitude for forward scattering, $N(p)\pi^b(k) \rightarrow N(p)\pi^a(k)$, written in terms of a matrix element of two PCAC pion fields. By manipulating this along similar lines to the derivation of the GOR relation, the terms in this amplitude up to first order in the pion energy can be related to the chiral properties of the nucleon.

The PCAC scattering amplitude $T_{\pi N}^{ba}$ is defined for off-shell pions by

$$i \int d^4x \, e^{ik \cdot x} \langle N(p)| \mathrm{T} \left(\partial^\mu J_{5\mu}^a(x), \partial^\nu J_{5\nu}^b(0) \right) |N(p)\rangle$$

$$= \left(\frac{i f_\pi m_\pi^2}{k^2 - m_\pi^2} \right)^2 T_{\pi N}^{ba}. \tag{6.19}$$

This can be integrated by parts and for soft pions, with $\mathbf{k} = 0$ and small $\omega \equiv k^0$, it becomes

$$-\left(\frac{f_\pi m_\pi^2}{m_\pi^2 - \omega^2} \right)^2 T_{\pi N}^{ba} = -i \langle N(p)| \left[Q_5^a, \partial^\nu J_{5\nu}^b(0) \right] |N(p)\rangle$$

$$+ \omega \int d^4x \, e^{i\omega x^0} \langle N(p)| \mathrm{T} \left(J_{50}^a(x), \partial^\nu J_{5\nu}^b(0) \right) |N(p)\rangle. \tag{6.20}$$

If translational invariance is used to shift the origin to x, the matrix element in the second term becomes

$$\langle N(p)| \mathrm{T} \left(J_{50}^a(0), \partial^\nu J_{5\nu}^b(-x) \right) |N(p)\rangle. \tag{6.21}$$

This term can then be integrated by parts again, giving

$$-\left(\frac{f_\pi m_\pi^2}{m_\pi^2 - \omega^2} \right)^2 T_{\pi N}^{ba} = -i \langle N(p)| \left[Q_5^a, \partial^\nu J_{5\nu}^b(0) \right] |N(p)\rangle$$

$$- \omega \langle N(p)| \left[Q_5^a, J_{50}^b(0) \right] |N(p)\rangle$$

$$+ i\omega^2 \langle N(p)| \mathrm{T} \left(J_{50}^a(x), J_{50}^b(0) \right) |N(p)\rangle. \tag{6.22}$$

Some care is needed with the final term of (6.22). This could contain contributions from intermediate ground-state nucleons, which would give rise to rapid energy dependence in the scattering amplitude. These do not appear for forward scattering of physical pions or in the soft-pion limit defined here but, more generally, such 'Born terms' need to be isolated before the PCAC approach can be applied.

The second term of (6.22) is linear in ω as $\omega \to 0$. It is isospin-odd and the current algebra of (6.7) shows that it is just proportional to the (vector) isospin of the nucleon. At the physical point, $\omega = m_\pi$, this corresponds to a term of order m_π. Other contributions are at least of order m_π^2 and so this term is the leading-order piece of the scattering amplitude at low energies. Hence, to order m_π, the scattering amplitude has the form found by Weinberg and Tomozawa [8,9],

$$T_{\pi N}^{ba} \simeq i \frac{m_\pi}{f_\pi^2} \epsilon^{abc} \bar{u}(p) \gamma_0 t^c u(p). \tag{6.23}$$

This isospin-odd amplitude may have a similar form to ρ-exchange, but its presence follows purely from chiral symmetry.

The isospin-even amplitude at threshold is much smaller than the isospin-odd one. It consists of various contributions of order ω^2 or m_π^2. One of these is of particular interest since it contains information about the scalar density of quarks in the nucleon. This piece is analogous to the GOR relation for the vacuum and it is obtained in a similar way from the soft-pion limit of the pion–nucleon scattering amplitude. In the limit $\omega \to 0$, only the first term of (6.22) survives, leaving a result proportional to the pion–nucleon sigma commutator [10],

$$\sigma_{\pi N} \equiv \frac{1}{3} \sum_a \langle N(p)| \left[Q_5^a, \left[Q_5^b, \mathcal{H}(0) \right] \right] |N(p)\rangle = \bar{m}_q \langle N(p)|\bar{q}(0)q(0)|N(p)\rangle. \tag{6.24}$$

Estimates of $\sigma_{\pi N}$ from pion–nucleon scattering are typically in the range 45 MeV [11] to 64 MeV [12].

Similar results can also be obtained for electromagnetic processes involving pions. For example, the amplitude for pion photoproduction $\gamma(q)N(p) \to \pi(k)N(p')$ is given by

$$\int d^4x \, e^{ik \cdot x} \langle N(p')|T\left(\partial^\mu J_{5\mu}^a(x), J_\nu^{em}(0)\right)|N(p)\rangle = \frac{f_\pi m_\pi^2}{k^2 - m_\pi^2} T_\nu^a. \tag{6.25}$$

In the soft-pion limit, current algebra can be used to relate this amplitude to a matrix element of the axial current,

$$T_\nu^a = \frac{1}{f_\pi} \langle N(p')| \left[Q_5^a, J_\nu^{em}(0) \right] |N(p)\rangle = \frac{i}{f_\pi} \epsilon^{a3c} \langle N(p')|J_{5\nu}^c(0)|N(p)\rangle. \tag{6.26}$$

This term, first derived by Kroll and Ruderman [13], is the leading piece of the amplitude for photoproduction of charged pions. Its strength is proportional to $g_A(0)/f_\pi$ or, making use of the Goldberger–Treiman relation, $g_{\pi NN}/M_N$.

For neutral pions, $a = 3$ in (6.26) and the Kroll–Ruderman term vanishes. The amplitude for low-energy π^0 photoproduction obtained using PCAC contains two pieces, both arising from the Born terms [14]: one of order m_π, which can be thought of as the coupling of the photon to the electric dipole moment of the $\pi^0 N$ system, and one of order m_π^2 that is proportional to the magnetic moment of the nucleon.

However, this result ignores contributions from other states with small energy denominators, in particular πN states whose denominators are of order m_π. As a result it fails to describe data for π^0 photoproduction at threshold [15].

Indeed a general problem with the PCAC approach is that, while it successfully embodies the symmetry constraints on tree-level amplitudes, it is poorly designed to handle intermediate states where virtual particles are created and destroyed. These are much more conveniently handled within a field-theoretic framework, as proposed by Weinberg [16]. The corresponding effective field theory of low-energy pions, χPT, incorporates all of the older LETs obtained using PCAC and provides the tools for systematically extending them to include the effects of virtual pions.

6.2 Effective field theory of Goldstone bosons

The field theory describing the low-energy interactions of pions is clearly not a fundamental one. Instead it is an 'effective' theory whose Lagrangian contains all possible local terms consistent with the symmetries of the underlying theory, QCD. What makes this theory tractable is the same separation of scales that was responsible for the successes of PCAC: the fact that the pion mass is much smaller than all other energy scales in hadron physics.

Focussing on pions as the Goldstone bosons of the strong interaction and restricting our sights to low momenta, say of the order of m_π, this means that we can expand the theory in powers of ratios of low-energy scales Q (pion masses or momenta) to underlying scales Λ_χ. The latter arise from the degrees of freedom that have been integrated out, and are of the order of 700 MeV to 1 GeV. They include the masses of heavier hadrons, such as the ρ-meson or the nucleon, and also the scale associated with the hidden chiral symmetry, $4\pi f_\pi$. (The latter is a more appropriate scale than f_π itself because of the factors of 4π that typically appear in the denominators of loop diagrams.)

This separation of scales makes it possible to classify the infinite number of terms in the effective theory by counting the powers of the ratio Q/Λ they contain. In general this power counting can be obtained with the help of the renormalization group. In the case of pions, however, the counting is simplified by the fact that all the interactions of Goldstone bosons are weak at low energies, suppressed by powers of their masses or momenta.

For the pions, therefore, it is enough just to count powers of small scales, as pointed out by Weinberg [16]. The terms in the Lagrangian are classified by their order $d = 2, 4, \ldots$: the number of derivatives and powers of the pion mass they contain. Each internal pion line in a Feynman diagram represents a propagator, $1/(q^2 - m_\pi^2)$, which is of order Q^{-2}. Each loop integration over an unconstrained four-momentum is formally of order Q^4. Hence a diagram with N_d vertices of order d, N_I internal lines and N_L loops contributes to an amplitude at order

$$D = \sum_d N_d \, d - 2N_I + 4N_L. \qquad (6.27)$$

Since each vertex provides a δ-function constraining the internal momenta (apart from the δ-function for conservation of the external momenta), the number of loops is

$$N_L = N_I - \sum_d N_d + 1. \qquad (6.28)$$

Using this in (6.27), we can express the order in the form

$$D = \sum_d N_d(d - 2) + 2N_L + 2. \qquad (6.29)$$

Such a diagram may contain a logarithmic divergence, but this can be cancelled against a counter term: the coefficient of a term in the Lagrangian of order D. Since D is of higher order than all of the vertices appearing in the divergent integral, it is possible to renormalize the theory order-by-order in powers of the small scales.

One might worry that diagrams containing vertices involving powers of momenta could also give rise to power-law divergences. In principle they do, but these are always suppressed by additional powers of the large-scale Λ_χ in their denominators. Hence their contributions are irrelevant, in the language of the renormalization group, provided the running regulator or cut-off scale is kept well below Λ_χ. This makes it very convenient to evaluate the loop integrals using dimensional regularization, where the logarithmic divergences can be cleanly isolated and subtracted.

This power counting can be extended to include baryons. As discussed below, at low energies these particles can be treated non-relativistically, in which case their propagators are of order Q^{-1}. Hence the order of a diagram is given by

$$D = \sum_d \left(N_d^{(M)} + N_d^{(MB)} \right) d - 2N_I^{(M)} - N_I^{(B)} + 4N_L, \qquad (6.30)$$

where it has $N_I^{(M)}$ internal meson lines, $N_I^{(B)}$ internal baryon lines, $N_d^{(M)}$ mesonic vertices and $N_d^{(MB)}$ meson–baryon vertices of order d. The number of loops is now given by

$$N_L = N_I^{(M)} + N_I^{(B)} - \sum_d \left(N_d^{(M)} + N_d^{(MB)}\right) + 1. \tag{6.31}$$

In addition, if we restrict our attention to processes involving a single baryon, we have the constraint

$$\sum_d N_d^{(MB)} - N_I^{(B)} = 1. \tag{6.32}$$

These allow us to express the order in the form

$$D = \sum_d N_d^{(M)}(d - 2) + \sum_d N_d^{(MB)}(d - 1) + 2N_L + 1. \tag{6.33}$$

6.3 Mesonic chiral perturbation theory

The effective Lagrangian describing the pions can be constructed in terms of an SU(2) matrix field $U(x)$, which transforms linearly under chiral symmetry:

$$U(x) \rightarrow RU(x)L^\dagger, \tag{6.34}$$

where $(R, L) \in \mathrm{SU}(2)_R \times \mathrm{SU}(2)_L$. This can be represented in various ways as a non-linear function of three canonical pion fields. For example, one commonly used form is

$$U(x) = \exp\left(i\frac{\boldsymbol{\tau} \cdot \boldsymbol{\phi}(x)}{f}\right). \tag{6.35}$$

An alternative, used by Bernard *et al* [3], is the square-root form,

$$U(x) = \sqrt{1 - \frac{\phi(x)^2}{f^2}} + i\frac{\boldsymbol{\tau} \cdot \boldsymbol{\phi}(x)}{f}. \tag{6.36}$$

The form in (6.35) can easily be extended to the three-flavour chiral symmetry, $\mathrm{SU}(3)_R \times \mathrm{SU}(3)_L$, by writing

$$U(x) = \exp\left(i\frac{\lambda^a \phi^a(x)}{f}\right), \tag{6.37}$$

where the λ^a are the eight Gell-Mann matrices and the fields $\phi^a(x)$ now represent the particles that would be Goldstone bosons in the limit of three massless flavours. Apart from the pions, these are the kaons and η-mesons.

The larger masses of the pseudoscalar mesons containing strange quarks (~ 500 MeV) mean that the chiral expansion will converge more slowly than that with pions alone. Nonetheless, the three-flavour version of χPT, at least in the mesonic sector, does seem to converge rapidly enough to be useful.

6.3.1 Leading order

At lowest order, the Lagrangian contains kinetic and mass terms for the pions,

$$\mathcal{L}_\pi^{(2)} = \tfrac{1}{4} f^2 \, \mathrm{tr}\left[(\nabla_\mu U)^\dagger \nabla^\mu U\right] + \tfrac{1}{4} f^2 \, \mathrm{tr}\left[\chi U^\dagger + \chi^\dagger U\right]. \tag{6.38}$$

Here f is the pion decay constant at this order, and χ is proportional to the quark mass matrix, which explicitly breaks the chiral symmetry. The covariant derivative,

$$\nabla_\mu U = \partial_\mu U - i(v_\mu + a_\mu)U + iU(v_\mu - a_\mu), \tag{6.39}$$

contains couplings to external vector and axial fields, $v_\mu(x)$ and $a_\mu(x)$. The combinations $v_\mu \pm a_\mu$ correspond to right- and left-handed gauge transformations. These gauge fields include, in particular, electromagnetism,

$$v_\mu(x) = eQ A_\mu(x), \tag{6.40}$$

where

$$Q = \tfrac{1}{6}\lambda_8 + \tfrac{1}{2}\lambda_3 \tag{6.41}$$

is the quark charge matrix and we have followed the convention [3] that $e = -|e|$ is the charge of the electron, $-\sqrt{4\pi\alpha}$. There are also the W and Z fields of the weak interaction, but in practice in determining the weak decay constant of the pion, it is more convenient to use an external axial vector source,

$$a_\mu(x) = \tfrac{1}{2} \tau \cdot \mathbf{a}_\mu(x). \tag{6.42}$$

If we expand $\mathcal{L}_\pi^{(2)}$ up to second order in the pion fields of (6.35), we find the expected kinetic and mass terms as well as the electromagnetic couplings of the charged pions,

$$\mathcal{L}_\pi^{(2)} \simeq \tfrac{1}{2} \partial_\mu \phi \cdot \partial^\mu \phi + e A_\mu \, \epsilon_{3ij} \, \phi_i \partial^\mu \phi_j + \tfrac{1}{2} e^2 A_\mu A^\mu \left(\phi_1^2 + \phi_2^2\right)$$
$$- \tfrac{1}{4} \mathrm{tr}\left[\chi + \chi^\dagger\right] \phi^2 + \cdots . \tag{6.43}$$

This shows that the mass of the pions at lowest order, denoted by m, is given by

$$m^2 = \tfrac{1}{2} \mathrm{tr}\left[\chi + \chi^\dagger\right]. \tag{6.44}$$

The non-linear nature of $U(x)$ means that $\mathcal{L}^{(2)}$ also contains interactions between the pions. Expanding it to fourth order in the pion fields we get

$$\mathcal{L}_\pi^{(2)} \simeq \cdots + \frac{1}{6f^2} \left[(\phi \cdot \partial_\mu \phi)(\phi \cdot \partial^\mu \phi) - \phi^2 \partial_\mu \phi \cdot \partial^\mu \phi \right]$$

$$+ \frac{m^2}{24f^2} (\phi^2)^2 + \cdots . \tag{6.45}$$

Note that for terms with three or more pion fields, the choice of representation for $U(x)$ does make a difference to the vertices and Feynman rules. However, all on-shell scattering amplitudes are the same. For example, the amplitude obtained from (6.45) for the scattering process $\pi^a(p_a)\pi^b(p_b) \to \pi^c(p_c)\pi^d(p_d)$ is

$$T_{\pi\pi}^{cd,ab} \simeq \delta_{ab}\delta_{cd} \frac{s - m^2}{f^2} + \delta_{ac}\delta_{bd} \frac{t - m^2}{f^2} + \delta_{ad}\delta_{bc} \frac{u - m^2}{f^2}$$

$$- \frac{1}{3f^2} \left(\delta_{ab}\delta_{cd} + \delta_{ac}\delta_{bd} + \delta_{ad}\delta_{bc} \right) \left(p_a^2 + p_b^2 + p_c^2 + p_d^2 - 4m^2 \right) . \tag{6.46}$$

The final term in this expression is absent for the square-root representation of $U(x)$, but this term vanishes when all particles are on-shell, leaving a physical amplitude that agrees with Weinberg's form, (6.14).

6.3.2 Next-to-leading order

At next order the Lagrangian consists of all possible terms involving four derivatives, two derivatives and one power of χ, or two powers of χ [17]. Since the symmetry-breaking strength χ is proportional to the square of the pion mass, all of these are of order Q^4. The resulting Lagrangian can be written in the form [18] (see [1], appendix D.1),

$$\mathcal{L}_\pi^{(4)} = \frac{l_1}{4} \left(\text{tr} \left[(\nabla_\mu U)^\dagger \nabla^\mu U \right] \right)^2 + \frac{l_2}{4} \text{tr} \left[(\nabla_\mu U)^\dagger \nabla_\nu U \right] \text{tr} \left[(\nabla^\mu U)^\dagger \nabla^\nu U \right]$$

$$+ \frac{l_3 + l_4}{16} \left(\text{tr} \left[\chi U^\dagger + \chi^\dagger U \right] \right)^2 + \frac{l_4}{8} \text{tr} \left[(\nabla_\mu U)^\dagger \nabla^\mu U \right] \text{tr} \left[\chi U^\dagger + \chi^\dagger U \right]$$

$$+ l_5 \text{tr} \left[U^\dagger F_R^{\mu\nu} U F_{L\mu\nu} \right] + i \frac{l_6}{2} \text{tr} \left[F_R^{\mu\nu} \nabla_\mu U (\nabla_\nu U)^\dagger + F_L^{\mu\nu} (\nabla_\mu U)^\dagger \nabla_\nu U \right]$$

$$- \frac{l_7}{16} \left(\text{tr} \left[\chi U^\dagger - \chi^\dagger U \right] \right)^2 + \frac{h_1 + h_3 - l_4}{4} \text{tr} \left[\chi^\dagger \chi \right]$$

$$+ \frac{h_1 - h_3 - l_4}{16} \left\{ \left(\text{tr} \left[\chi U^\dagger + \chi^\dagger U \right] \right)^2 + \left(\text{tr} \left[\chi U^\dagger - \chi^\dagger U \right] \right)^2 \right.$$

$$\left. - 2 \text{tr} \left[\chi U^\dagger \chi U^\dagger + \chi^\dagger U \chi^\dagger U \right] \right\}$$

$$- \frac{4h_2 + l_5}{2} \text{tr} \left[F_{R\mu\nu} F_R^{\mu\nu} + F_{L\mu\nu} F_L^{\mu\nu} \right] . \tag{6.47}$$

Note that although the coupling constants are defined as in [17], the form of the Lagrangian is different from the one given there, and corresponds instead to that of [18]. This reflects the fact that there are many different but equivalent ways to write down an effective Lagrangian. These are related by transformations of the field operators, as discussed in [19,1]. This Lagrangian contains a number of coefficients, l_i, known as low-energy constants (LECs). At present, these are either fitted to data on mesonic scattering or decay processes, or they can be estimated by assuming resonance saturation [20]. Ultimately one would like to be able to determine them directly from lattice QCD. A summary of empirical values for them can be found in appendix D of [1]. There are also three constants, h_i, that multiply combinations of external fields only.

At this order, the Lagrangian contains pieces that are quadratic in the field tensors for the external vector and axial-vector fields:

$$F_R^{\mu\nu} = \partial^\mu(v^\nu + a^\nu) - \partial^\nu(v^\mu + a^\mu) - i[v^\mu + a^\mu, v^\nu + a^\nu],$$
$$F_L^{\mu\nu} = \partial^\mu(v^\nu - a^\nu) - \partial^\nu(v^\mu - a^\mu) - i[v^\mu - a^\mu, v^\nu - a^\nu]. \tag{6.48}$$

In the case of electromagnetism, the field tensors are

$$F_{\mu\nu}^R = F_{\mu\nu}^L = -eQF_{\mu\nu}, \tag{6.49}$$

and one of these pieces provides a wave-function renormalization for the photon.

The analogous Lagrangian for the three-flavour case is [21]

$$\begin{aligned}
\mathcal{L}_{\pi K \eta}^{(4)} &= L_1 \left(\text{tr}\left[(\nabla_\mu U)^\dagger \nabla^\mu U\right]\right)^2 + L_2 \text{tr}\left[(\nabla_\mu U)^\dagger \nabla_\nu U\right] \text{tr}\left[(\nabla^\mu U)^\dagger \nabla^\nu U\right] \\
&\quad + L_3 \text{tr}\left[(\nabla_\mu U)^\dagger \nabla^\mu U (\nabla_\nu U)^\dagger \nabla^\nu U\right] \\
&\quad + L_4 \text{tr}\left[(\nabla_\mu U)^\dagger \nabla^\mu U\right] \text{tr}\left[\chi U^\dagger + \chi^\dagger U\right] \\
&\quad + L_5 \text{tr}\left[(\nabla_\mu U)^\dagger \nabla^\mu U(\chi U^\dagger + \chi^\dagger U)\right] + L_6 \left(\text{tr}\left[\chi U^\dagger + \chi^\dagger U\right]\right)^2 \\
&\quad + L_7 \left(\text{tr}\left[\chi U^\dagger - \chi^\dagger U\right]\right)^2 + L_8 \text{tr}\left[\chi U^\dagger \chi U^\dagger + \chi^\dagger U \chi^\dagger U\right] \\
&\quad - iL_9 \text{tr}\left[F_R^{\mu\nu} \nabla_\mu U(\nabla_\nu U)^\dagger + F_L^{\mu\nu}(\nabla_\mu U)^\dagger \nabla_\nu U\right] \\
&\quad + L_{10} \text{tr}\left[U^\dagger F_R^{\mu\nu} U F_{L\mu\nu}\right] + H_1 \text{tr}\left[F_{R\mu\nu} F_R^{\mu\nu} + F_{L\mu\nu} F_L^{\mu\nu}\right] \\
&\quad + H_2 \text{tr}\left[\chi^\dagger \chi\right].
\end{aligned} \tag{6.50}$$

This contains more independent terms than the two-flavour version, because the group $SU(3)_R \times SU(3)_L$ has more invariants. However, since some of the data used to determine the LECs comes from kaon decays [22], the three-flavour version is useful even in the context of purely pionic physics.

As already mentioned, these LECs in $\mathcal{L}^{(4)}$ play the vital role of absorbing all the one-loop divergences. In dimensional regularization, logarithmic divergences show up

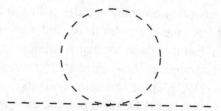

Figure 6.1. The basic one-loop diagram contributing to the pion self-energy at order Q^4.

in loop integrals as poles at the physical number of space-time dimensions, $d = 4$, and the LECs can be renormalized to cancel these. It is convenient to use a modified minimal subtraction, which also cancels some finite numerical terms that naturally appear in dimensional regularization. The resulting renormalized couplings, l_i^r, are defined by

$$l_i = l_i^r(\lambda) + \gamma_i L(\lambda), \tag{6.51}$$

where

$$L(\lambda) = \frac{\lambda^{d-4}}{16\pi^2}\left(\frac{1}{d-4} + \tfrac{1}{2}(\gamma_E - 1 - \log 4\pi)\right). \tag{6.52}$$

Here d is the number of space-time dimensions, γ_E is Euler's constant and λ is the arbitrary scale introduced in this regularization.

6.3.3 Pion propagator

At order Q^4, the pion propagator receives contributions from loop diagrams involving interactions from $\mathcal{L}^{(2)}$ and from insertions of terms from $\mathcal{L}^{(4)}$. The pion-loop contribution arises from the diagram shown in figure 6.1, where the interaction vertex is just the lowest-order $\pi\pi$ scattering amplitude in (6.46). For a pion with momentum p, this gives

$$-i\Sigma_{\pi\,\text{loop}}^{(4)}(p^2) = \int \frac{d^d l}{(2\pi)^d}\, i\left[\frac{(p+l)^2 - m^2}{f^2} + 3\frac{(-m^2)}{f^2} + \frac{(p-l)^2 - m^2}{f^2}\right.$$
$$\left. - \frac{5}{3f^2}(2p^2 + 2l^2 - 4m^2)\right]\frac{i}{l^2 - m^2 + i\epsilon}. \tag{6.53}$$

Using the fact that in dimensional regularization

$$i\int \frac{d^d l}{(2\pi)^d} = 0, \tag{6.54}$$

all of the integrals can be expressed in terms of one basic 'tadpole' integral,

$$\Delta_\pi \equiv \int \frac{d^d l}{(2\pi)^d} \frac{i}{l^2 - m^2 + i\epsilon}$$

$$= 2m^2 \left(L(\lambda) + \frac{1}{16\pi^2} \log \frac{m}{\lambda} \right) + \mathcal{O}(d-4), \tag{6.55}$$

and the pion-loop contributions to the self-energy can be written

$$-i\Sigma^{(4)}_{\pi\,\text{loop}}(p^2) = \frac{i}{6f^2}(-4p^2 + m^2)\Delta_\pi. \tag{6.56}$$

This self-energy is of order Q^4, in accordance with Weinberg's power counting. It also contains a logarithmic divergence, signalled by the pole at dimension $d = 4$. This has the same structure as the self-energy insertions arising from the $\mathcal{O}(Q^4)$ Lagrangian, and can be renormalized using the constants L_i. These 'counter terms' contribute a self-energy

$$-i\Sigma^{(4)}_{C.T.}(p^2) = i\frac{2m^2}{f^2}\left[l_4 p^2 - (l_3 + l_4)m^2\right], \tag{6.57}$$

where we have used (6.44) to express χ in terms of the leading-order pion mass m. The full self-energy is thus

$$-i\Sigma^{(4)}(p^2) = \frac{i}{f^2}\left[\left(-\tfrac{2}{3}\Delta_\pi + 2l_4 m^2\right)p^2 + \left(\tfrac{1}{6}\Delta_\pi - 2(l_3 + l_4)m^2\right)m^2\right], \tag{6.58}$$

and the pion propagator is

$$\frac{i}{p^2 - m^2 - \Sigma^{(4)}(p^2) + i\epsilon}. \tag{6.59}$$

The physical pion mass is defined by the position of the pole which, to this order, is given by

$$m_\pi^2 = m^2 + \Sigma^{(4)}(m^2)$$

$$= m^2 + \frac{m^2}{f^2}\left(\tfrac{1}{2}\Delta_\pi + 2l_3 m^2\right). \tag{6.60}$$

The renormalized LEC $l_3^r(\lambda)$ is given by (6.51) with $\gamma_3 = -\tfrac{1}{2}$ [17]. Using this we can write the pion mass as

$$m_\pi^2 = m^2 + \frac{m^4}{f^2}\left(\frac{1}{16\pi^2}\log\frac{m}{\lambda} + 2l_3^r(\lambda)\right). \tag{6.61}$$

The pion mass is finite, as it should be. It is also independent of the arbitrary renormalization scale λ, since the $\log \lambda$ term in $l_3^r(\lambda)$ exactly cancels the corresponding term from the pion loop. A common choice of renormalization scale is $\lambda = m$. This is convenient in phenomenological applications since the chiral logarithm $\log(m/\lambda)$ then vanishes. However, it should be used with caution in comparisons with lattice QCD results, since m^2 is an adjustable parameter there and it is important to keep all dependence on it explicit.

This result for m_π^2 also illustrates some general features of χPT. It contains a piece $m^4 \log m$ which is non-analytic in the symmetry-breaking strength m^2 and so can only arise from low-momentum virtual pions (or the long-distance part of the pion cloud). The other piece, m^4, is analytic in m^2, but involves an LEC. This arises from short-distance physics which has been integrated out of the effective theory, including high-momentum virtual pions and heavy mesons, not to mention the quark substructure of the pion.

The pion wave function renormalization at this order is given by

$$\frac{1}{Z_\pi} = 1 - \left.\frac{\partial \Sigma}{\partial p^2}\right|_{p^2 = m^2}$$

$$= 1 - \frac{1}{f^2}\left(\tfrac{2}{3}\Delta_\pi - 2l_4 m^2\right). \tag{6.62}$$

Since $\gamma_4 = 2$ [17], this is neither finite nor independent of λ. However, this is not a problem, since Z_π is not an observable. (In fact it also depends on the choice of Lagrangian, and the representation used for $U(x)$ [1].)

6.3.4 Pion form factor

At lowest order, the pion is simply a point particle. Its coupling to a photon with polarization ϵ is represented by the Feynman rule

$$e\epsilon^{a3b}\epsilon \cdot (p + p'), \tag{6.63}$$

where p, a are the momentum and isospin of the initial pion, and p', b are those of the final one. However, at higher orders, pion loops generate an electromagnetic form factor for the charged pion [17]. This contains a non-trivial dependence on the momentum transfer $q = p' - p$. Those parts of the form factor where the momentum scale is set by m_π can be predicted from χPT, while the rest must be parametrized by LECs. At order Q^4, an external photon can be coupled either to the internal pion line or to the vertex in the basic one-loop diagram of figure 6.1. There is also a contribution from the bare vertex multiplied by a factor of $\sqrt{Z_\pi}$ for each external pion line. The insertion on the internal line leads to the following pieces: a divergent term proportional to $m^2\epsilon \cdot (p + p')$, a divergent term proportional to $q^2\epsilon \cdot (p + p') - \epsilon \cdot q(p'^2 - p^2)$ and a term with this structure multiplied by a finite function of q^2/m^2. With the exponential representation of $U(x)$, the insertion at the vertex gives only a divergent term proportional to $m^2\epsilon \cdot (p+p')$. (For the square-root form, this contribution vanishes.)

When all terms of order Q^2 are put together, the divergences proportional to $m^2\epsilon \cdot (p+p')$ are found to cancel against the divergent piece of the wave function renormalization mentioned above. Also the structure $q^2\epsilon \cdot (p+p') - \epsilon \cdot q(p'^2 - p^2)$ is precisely that of the $\pi\pi\gamma$ vertex from the l_6 term in $\mathcal{L}_\pi^{(4)}$ and so it can be absorbed by renormalizing that coefficient. The resulting form factor is then [17,1]

$$F_\pi(q^2) = 1 - \frac{q^2}{f^2}\left\{l_6^r(\lambda) + \frac{1}{96\pi^2}\left[2\log\frac{m}{\lambda} + \frac{1}{3} + \left(1 - 4\frac{m^2}{q^2}\right)J^{(0)}\left(\frac{q^2}{m^2}\right)\right]\right\},$$
(6.64)

where, at least for $x < 0$ (corresponding to space-like q: $q^2 = -Q^2 < 0$),

$$J^{(0)}(x) = -2 + \sqrt{\frac{x-4}{x}}\log\frac{\sqrt{4-x} + \sqrt{-x}}{\sqrt{4-x} - \sqrt{-x}}.$$
(6.65)

The corresponding charge radius of the pion is given by

$$\langle r^2\rangle_\pi = 6\left.\frac{\partial F_\pi}{\partial q^2}\right|_{q^2=0}$$

$$= -\frac{6}{f^2}\left[l_6^r(\lambda) + \frac{1}{96\pi^2}\left(1 + 2\log\frac{m}{\lambda}\right)\right].$$
(6.66)

Note that, as often happens in χPT, the quantity of most physical interest cannot be predicted since it contains an undetermined LEC. Instead, the pion charge radius must be used to fix the LEC $l_6^r(\lambda)$. The full form factor $F_\pi(q^2)$ is then predicted to this order. In order to do this, we need to replace the leading-order quantities f and m by their physical values. Although this introduces an error, this is beyond the order to which we are working. In the three-flavour theory, the relevant LEC is $L_9^r(\lambda)$ [23]. Once this has been determined from the pion charge radius, the radius of the charged kaon can be predicted. The result is consistent with experimental determinations, although the uncertainties are rather large.

6.3.5 Anomalous processes

The terms in the Lagrangian discussed so far are symmetric under the interchange $U \leftrightarrow U^\dagger$, $a^\mu \leftrightarrow -a^\mu$. As a result they all involve only even numbers of pions. This symmetry is a natural one for a purely bosonic theory, but it is not one that is respected by theories like QCD that contain fermions. Fermion loops can violate the Ward identities corresponding to local chiral invariance [24,25], giving rise to processes like $\pi^0 \to \gamma\gamma$ and $K^+ K^- \to \pi^+ \pi^0 \pi^-$. To describe these in our effective field theory, we need to include terms that involve odd numbers of meson fields, and hence have odd intrinsic parity.

The necessary term in the effective action was first constructed by Wess and Zumino [26]. Later, Witten [27] showed that it is a topological invariant which has a non-local structure in four dimensions but which can be expressed in a local form in five dimensions. Its coefficient contains an integer which can be identified with the number of colours of quark, $N_c = 3$.

The full Wess–Zumino–Witten (WZW) action can be written as a local functional of $U(x)$ only in five dimensions but, when expanded in powers of the pion field, each term can be integrated to give a local four-dimensional term. The local terms all contain an odd number of pseudoscalar meson fields, and four derivative operators or vector fields contracted using a totally antisymmetric Levi–Civita tensor. For more details, see [1]. Perhaps the most important of these terms in the context of electromagnetism is the one,

$$-i \frac{N_c e^2}{96\pi^2 f} \epsilon^{\mu\nu\rho\sigma} F_{\mu\nu} F_{\rho\sigma} \phi_3(x), \qquad (6.67)$$

which gives the leading contribution to the decay of the neutral pion, $\pi^0 \to \gamma\gamma$.

Since it is a topological invariant whose LEC contains the integer N_c, one would not expect the WZW action to be renormalized [28,29]. There are, however, other odd-intrinsic-parity terms of higher order which are chirally invariant. These are needed to renormalize loop diagrams containing vertices from the WZW action [29]. For example, the order-Q^6 Lagrangian, which is needed for the one-loop diagrams, contains 23 independent terms with odd intrinsic parity [30]. These higher-order terms can contribute corrections of order m^2 to processes like $\pi^0 \to \gamma\gamma$ but they do not affect the anomalous Ward identities. More discussion of them can be found in [2].

6.3.6 Next-to-next-to-leading order

The state-of-the-art in mesonic χPT is centred on calculations at order Q^6, which involve the evaluation of two-loop diagrams with vertices from $\mathcal{L}_\pi^{(2)}$. One particularly important application is to $\pi\pi$ scattering, but there are also a number of interesting results for electromagnetic processes. These calculations are formidably complicated, in part because of the number of diagrams that need to be evaluated, but also because of the number of LECs: more than a hundred in the $\mathcal{O}(Q^6)$ Lagrangian. Reviews of the renormalization of two-loop diagrams in χPT can be found in [31–33].

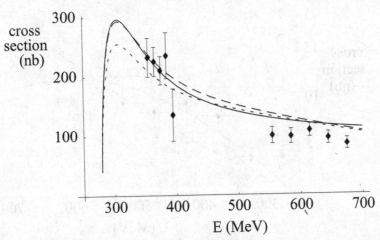

Figure 6.2. Cross section integrated over $|\cos\theta| < 0.6$ for the process $\gamma\gamma \to \pi^+\pi^-$, plotted against centre-of-mass energy E. The curves show the results of χPT at order Q^2 (short-dashed line), Q^4 (long-dashed) and Q^6 (solid). They are taken from Bürgi [36], using the same parameters and LECs. The contributions of irreducible two-loop integrals have been neglected since they are very small for $\gamma\gamma \to \pi\pi$ in this energy range, as discussed in [40,36]. The data points are from the Mark II Collaboration [35].

One of the first electromagnetic processes to be studied at this order was the production of charged pions in two-photon collisions, $\gamma\gamma \to \pi^+\pi^-$. Calculations of the cross section at order Q^4 [34] had showed an enhancement over the Born (order-Q^2) result, bringing it into agreement with the low-energy data from the Mark II Collaboration [35], as can be seen in figure 6.2. Bürgi extended these calculations to order Q^6 [36] and found that the chiral expansion was converging well for this process. Unfortunately, there are no data in the regions where the two-loop contributions have most effect, although the result is at least in good agreement with dispersion theory [37]. One should note that the $\mathcal{O}(Q^6)$ result does depend on three poorly-determined LECs from $\mathcal{L}_\pi^{(6)}$, which Bürgi estimated using resonance saturation. The resulting values suggest that these LECs make rather small contributions to the total cross section.

In the case of $\gamma\gamma \to \pi^0\pi^0$, there are no tree-level contributions at either Q^2 or Q^4. At leading order, therefore, the amplitude can be calculated just from one-loop diagrams [34,38]. The result is a chiral prediction, only involving pion and kaon masses and f_π. It falls significantly below the data from the Crystal Ball Collaboration [39] for energies between threshold and about 400 MeV. The inclusion of two-loop contributions [40] improves the picture considerably, giving a cross section in good agreement with the available data even up to energies of about 650 MeV, see figure 6.3. In addition, the behaviour near threshold matches well

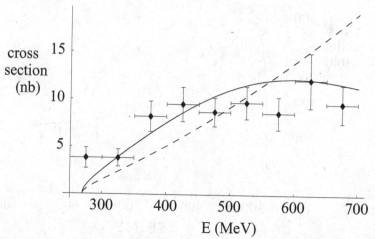

Figure 6.3. Cross section integrated over $|\cos\theta| < 0.8$ for the process $\gamma\gamma \rightarrow \pi^0\pi^0$, plotted against centre-of-mass energy E. The curves show the results of χPT at order Q^4 (dashed line) and Q^6 (solid). They are taken from Bellucci *et al* [40], but use the updated parameters and LECs listed in [36]. The contributions of irreducible two-loop integrals have been neglected, as in figure 6.2. The data points are from the Crystal Ball Collaboration [39].

with Pennington's dispersion-theoretic results [41]. As in the charged-pion case, the result does involve three $\mathcal{O}(Q^6)$ LECs but, again, estimates using resonance saturation suggest that their contributions are small.

Also important is Compton scattering from pions, either charged or neutral, since this can give information on the electric and magnetic polarizabilities of the pions. To second order in the photon energies (ω_1, ω_2) the amplitude for $\gamma(q_1)\pi^\pm(p_1) \rightarrow \gamma(q_2)\pi^\pm(p_2)$ can be written in the form

$$T = \left(-\frac{e^2}{m_\pi} + 4\pi\alpha_\pi\omega_1\omega_2\right)\epsilon_1 \cdot \epsilon_2 + 4\pi\beta_\pi\omega_1\omega_2(\epsilon_1 \times \hat{q}_1) \cdot (\epsilon_2 \times \hat{q}_2), \quad (6.68)$$

where α_π and β_π are the electric and magnetic polarizabilities, respectively. For scattering from a neutral pion, the energy-independent term is absent. The results (theoretical and experimental) for these polarizabilities were reviewed in [42], and an updated survey can be found in [43]. For the charged pions, the experimental situation is somewhat confused. The extraction of α_π from the Mark II data on $\gamma\gamma \rightarrow \pi^+\pi^-$ leads to a value of the order 2×10^{-4} fm^3, but with a large uncertainty of nearly 100%. In contrast determinations from radiative pion scattering on nuclei $(\pi N \rightarrow \gamma\pi N)$ or radiative pion photoproduction from the proton $(\gamma p \rightarrow \gamma\pi^+ n)$ lead to much larger values, in the region of $(7-20) \times 10^{-4}$ fm^3. The result, from a $\gamma p \rightarrow \gamma\pi^+ n$ experiment at Mainz [43], is

$$\alpha_{\pi^+} - \beta_{\pi^+} = (11.6 \pm 1.5 \pm 3, 0 \pm 0.5) \times 10^{-4} \text{ fm}^3, \quad (6.69)$$

where the three errors are: statistical, systematic, and an estimate of the model dependence of the nucleonic processes that contribute to the observed cross section.

Since Compton scattering and pair production are related by crossing symmetry, their amplitudes can be calculated from the same set of diagrams. In χPT at one-loop level, the electric and magnetic polarizabilities are related by

$$\alpha_\pi + \beta_\pi = 0 \tag{6.70}$$

for both charged and neutral pions. At this level, the polarizability of the charged pion comes entirely from contact interactions and can be expressed in terms of two $\mathcal{O}(Q^4)$ LECs [34,37]. In terms of the couplings in the three-flavour Lagrangian (6.50), it has the form

$$\alpha_{\pi^+} = \frac{e^2}{\pi f_\pi^2 m_\pi} \left(L_9^r + L_{10}^r\right). \tag{6.71}$$

The combination of LECs appearing here is one which can be determined from radiative pion decay. In contrast, the polarizability of the neutral pion comes entirely from loops at this level, and so it is predicted in terms of f_π and m_π to be

$$\alpha_{\pi^0} = -\frac{e^2}{384\pi^3 f_\pi^2 m_\pi}. \tag{6.72}$$

These calculations were extended to order Q^6, by Bürgi for the charged pion [36] and by Bellucci *et al* [40] for the neutral pion. The polarizabilities are much more sensitive to the $\mathcal{O}(Q^6)$ LECs than the total cross sections for pair production, and so they have significant uncertainties coming from the use of resonance saturation. The results are

$$
\begin{aligned}
\alpha_{\pi^+} + \beta_{\pi^+} &= (0.3 \pm 0.1) \times 10^{-4}\ \text{fm}^3, \\
\alpha_{\pi^+} - \beta_{\pi^+} &= (4.4 \pm 1.0) \times 10^{-4}\ \text{fm}^3, \\
\alpha_{\pi^0} + \beta_{\pi^0} &= (1.15 \pm 0.30) \times 10^{-4}\ \text{fm}^3, \\
\alpha_{\pi^0} - \beta_{\pi^0} &= (-1.90 \pm 0.20) \times 10^{-4}\ \text{fm}^3.
\end{aligned}
\tag{6.73}
$$

The effects of newer determinations of the $\mathcal{O}(Q^4)$ LECs on the charged pion polarizabilities are discussed in [43]. The new values all lie within the old error bars: for example, the central value for $\alpha_{\pi^+} - \beta_{\pi^+}$ changes from 4.4 to 4.9 in the usual units. It is interesting to note that there is a very significant deviation between the two-loop χPT prediction for $\alpha_{\pi^+} - \beta_{\pi^+}$ and the experimental result from Mainz (6.69). One obvious comment is that the experiment uses a proton target as a source of pions, and so the scattering involves a virtual pion. There are also nucleonic processes that contribute to the cross section, which the Mainz group had to estimate using two models. Kao *et al* have examined the process

$\gamma p \to \gamma \pi^+ n$ using heavy-baryon χPT and found that two combinations of πN LECs contribute significantly [44]. They suggest that results covering a wider range of scattering angles may be needed to disentangle the pion polarizabilities from πN physics.

6.4 Including nucleons

6.4.1 Introduction

Interesting though purely mesonic processes are, we clearly also want to be able to consider cases with one or more baryons present. It is obvious from the start that such a theory poses more questions about the appropriate low-energy degrees of freedom than does the purely mesonic case. In the latter case there is a clear mass gap between the pions and heavier mesons (ρ...) and so χPT, in which these heavier mesons are integrated out, should be valid for a significant range of energies, up to several times the pion mass. However, in the baryon sector there are many resonances, with the mass gap between the nucleon and its lowest lying resonance, the $\Delta(1232)$, being only around twice the pion mass. The problem is even more acute in the case of SU(3), with the mass for instance of the $\overline{K}\Sigma$ system lying at or above the mass of a dozen nucleon or Δ resonances. In fact, in sharp contrast to the mesonic sector, rather little work has been done including strange baryons, and we will here address only the non-strange sector. For low-enough energies it should be permissible to work only with pions and nucleons as explicit degrees of freedom, but even here we may find unexpectedly large LECs generated by integrating out the $\Delta(1232)$. For processes where energies significantly exceed the pion mass, explicit Δs will be mandatory.

The simplest possible Lagrangian which describes the coupling of a pseudoscalar pion to a nucleon would be

$$\overline{\Psi}(i\not{D} - M + g\gamma_5\phi \cdot \tau)\Psi, \qquad (6.74)$$

but this is clearly not chirally invariant and generates, for instance, a large πN scattering length in contradiction to experiment. One way of restoring the symmetry is to include the scalar partner of the pion, the σ; the σ-exchange tree graph then cancels against the direct pion scattering graphs and gives, in the chiral limit, the vanishing scattering length demanded by symmetry. However, in an effective field theory a heavy degree of freedom like the σ should not be present. Integrating it out will generate a pion–nucleon seagull, but also terms in which four or more pions couple directly to the nucleon. Of course the sigma model is just that, a model, but it does help to explain the fact that chiral symmetry demands that the form of the pion–nucleon coupling is more complicated than above.

In the most widely used representation, the nucleon field itself transforms under chiral rotations in a way that depends on the pion field: $\Psi \to K(L, R, U)\Psi$. Recalling that the pion field described by U transforms as $U \to RUL^\dagger$, we can introduce u which satisfies $u^2 = U$ and $u \to RuK^\dagger \equiv KuL^\dagger$. Skipping the details of the constraints this imposes on K, we can see that a Lagrangian built of structures which transform homogeneously under chiral transformations, $X \to KXK^\dagger$, will be chirally invariant. Two such structures are the covariant derivative,

$$
\begin{aligned}
D_\mu &= \partial_\mu + \tfrac{1}{2}[u^\dagger, \partial_\mu u] - \frac{i}{2}u^\dagger(v_\mu + a_\mu)u - \frac{i}{2}u(v_\mu - a_\mu)u^\dagger \\
&= \partial_\mu - ieQA_\mu \\
&\quad - \frac{i}{4f^2}\epsilon^{abc}(\partial_\mu\phi_a)\phi_b\tau_c - \frac{ie}{8f^2}A_\mu(\delta^{a3}\tau^b + \delta^{b3}\tau^a - 2\delta^{ab}\tau^3)\phi_a\phi_b + \cdots,
\end{aligned}
\tag{6.75}
$$

where v_μ and a_μ are external vector and axial-vector fields (in terms of the photon field, $v_\mu = eQA_\mu$) and

$$
\begin{aligned}
u_\mu &= i(u^\dagger\nabla_\mu u - u\nabla_\mu u^\dagger) \\
&= -\frac{1}{f}(\boldsymbol{\tau} \cdot \partial_\mu\phi + e\epsilon^{a3b}\tau_a\phi_b A_\mu + \cdots).
\end{aligned}
\tag{6.76}
$$

In each case the expansion to order one photon and two pions is given. The leading order pion–nucleon Lagrangian is then

$$
\overline{\Psi}\left(i\slashed{D} - M + \frac{g}{2}\gamma^\mu\gamma_5 u_\mu\right)\Psi.
\tag{6.77}
$$

The derivative coupling of the pion field ensures that, as required by chiral symmetry, soft pions do not interact with nucleons. At lowest order, the pseudovector coupling of the pion to the nucleon is equal to the axial coupling constant g, as required by the Goldberger–Treiman relation. Other chiral LETs are also automatically incorporated in this Lagrangian: the term with two pion fields gives the Weinberg–Tomozawa piece [8,9] of πN scattering, and the one with one pion and one photon gives the Kroll–Ruderman term in charged-pion photoproduction.

Beyond leading order four further structures enter that transform homogeneously and are second order in derivatives or the pion mass; these allow coupling to vector and axial-vector, and to scalar and pseudoscalar external fields respectively:

$$
\begin{aligned}
f^\pm_{\mu\nu} &= u^\dagger F^R_{\mu\nu}u \pm uF^L_{\mu\nu}u^\dagger, \\
\chi_\pm &= u^\dagger\chi u^\dagger \pm u\chi^\dagger u.
\end{aligned}
\tag{6.78}
$$

Expanded to second order in the pion fields, the vector and scalar structures become

$$f_{\mu\nu}^+ = e(\partial_\mu A_\nu - \partial_\nu A_\mu)\left(2Q + \frac{1}{4f^2}(\delta^{a3}\tau^b + \delta^{b3}\tau^a - 2\delta^{ab}\tau^3)\phi_a\phi_b\right) + \cdots,$$

$$\chi_+ = 2m^2\left(1 - \frac{1}{4f^2}\phi^2 + \cdots\right) + \mathcal{O}((m_u - m_d)\tau_3), \tag{6.79}$$

where $F_{\mu\nu}^{R,L}$ are field tensors defined in terms of $v_\mu + a_\mu$ and $v_\mu - a_\mu$ respectively, so that for photons $F_{\mu\nu}^R = F_{\mu\nu}^L = eQF_{\mu\nu}$. Symmetry-breaking effects due to the quark masses enter the Lagrangian through the second of these structures, $\chi_\pm$. It is clear that, in contrast to the mesonic sector where only terms with even numbers of derivatives or powers of meson masses enter, in the nucleon sector there are terms with both odd and even powers. This is a further reason why high accuracy is a harder goal to achieve in the baryonic sector.

Initial attempts to go beyond tree level with this Lagrangian immediately hit a snag. If the integrals are regulated with dimensional regularization, one-loop integrals, which one would hope to be of third order, actually generate corrections at lower orders. For instance in the chiral limit the one-loop correction to the nucleon mass is

$$\delta M_N = -\frac{3g^2 M^3}{f^2}\left(L(\lambda) + \frac{1}{16\pi^2}\log\frac{M}{\lambda}\right) + \mathcal{O}(d-4), \tag{6.80}$$

where λ is the scale introduced by dimensional regularization and $L(\lambda)$ is defined in (6.52). To retain M as the chiral-limit nucleon mass, a counter term is required at leading order. Similarly, away from the chiral limit there are corrections proportional to Mm^2/f^2 which need to be absorbed in second-order counter terms. This is repeated at every order, with lower-order counter terms continually requiring adjustment to absorb contributions from higher-order graphs, and so the power counting is not manifest. The problem is clear: the nucleon mass does not disappear in the chiral limit, so allowing the nucleon to propagate provides another scale to disrupt the power-counting.

Historically the solution adopted was the heavy-baryon approach (HBχPT), which starts from the observation that since M_N is of the order of the underlying heavy scale of the effective field theory ($\Lambda_\chi \approx m_\rho \approx 4\pi f_\pi$), a consistent treatment will expand amplitudes in powers of p/M_N as well as p/Λ_χ. This can be achieved by decomposing the Dirac nucleon field into 'large' and 'small' components (in the rest frame, upper and lower components) and integrating out the latter so that antinucleons no longer propagate. In leading order the remaining field acts as an infinitely heavy source for the pion field; corrections of successively higher powers of $1/M_N$ are present at each higher order. With M_N no longer present as a dynamical scale, the power-counting is restored. The price to be paid is an extra set of terms in the Lagrangian at each order, and hence more diagrams to calculate. Lorentz invariance is realized only to the given order in $1/M_N$, and the analytic structure of amplitudes as a function of M_N is destroyed. For example, the photoproduction threshold occurs at $\omega = m_\pi$ to any finite order (though in this case the resummation of higher-order terms to obtain the correct threshold is easy to carry out).

More recently it has been realized that other regularization schemes exist which allow power-counting to be maintained within a relativistic approach. Essentially these regularizations work by automatically absorbing all terms with positive powers of M_N into LECs, but they differ as to how they treat the remainder. Chiral symmetry ensures that the results of such calculations, if expanded in powers of m_π/M_N, must give the same results for the non-analytic terms, but they may differ in the analytic terms (and hence have different definitions of the LECs.) Since fourth-order calculations have already been done in HBχPT for most quantities of interest, the main impact of these alternative schemes will be whether they will allow fifth-order calculations (including two-loop calculations) to be done routinely. In this chapter we will deal mainly with heavy-baryon pion–nucleon χPT, with some further comments on other versions at the end.

6.4.2 Heavy baryon chiral perturbation theory

The details of the heavy-baryon reduction are well covered in [3], and will not be repeated here. The essential point is that the nucleon four-momentum is written as $P_\mu = M v_\mu + p_\mu$, where v_μ (not to be confused with an external vector field!) is the velocity vector, $v^2 = 1$, and p_μ is a small residual momentum which, unlike P_μ, is of order Q. After eliminating the 'small' or lower components, the nucleon propagator depends only on p_μ. The first two orders of the Lagrangian read

$$\mathcal{L}_{\pi N}^{(1)} = \psi^\dagger (iv \cdot D + gu \cdot S)\psi,$$

$$\mathcal{L}_{\pi N}^{(2)} = \psi^\dagger \left(\frac{1}{2M}((v \cdot D)^2 - D^2 - ig\{S \cdot D, v \cdot u\}) + c_1 \langle \chi_+ \rangle \right.$$

$$+ \left(c_2 + \frac{g^2}{8M} \right) (v \cdot u)^2 + c_3 u \cdot u + \left(c_4 + \frac{1}{4M} \right) [S \cdot u, S \cdot u] + c_5 \tilde{\chi}_+$$

$$\left. - \frac{i}{4M}[S^\mu, S^\nu] \left(\tfrac{1}{2}(1 + \mathring{k}_s)\langle f_{\mu\nu}^+ \rangle + (1 + \mathring{k}_v)\tilde{f}_{\mu\nu}^+ \right) \right) \psi, \tag{6.81}$$

where operators have been split into diagonal and traceless parts via $X = \frac{1}{2}\langle X \rangle + \tilde{X}$. The spin operator S^μ obeys $v \cdot S = 0, S \cdot S = (1 - d)/4$ and $[S^\mu, S^\nu] = i\epsilon^{\mu\nu\alpha\beta}v_\alpha S_\beta$ with $\epsilon^{0123} = -1$. The LECs c_i have dimensions of inverse mass; the isoscalar and isovector anomalous magnetic moments are related to the parameters in the relativistic Lagrangian of [45,18] by $\mathring{k}_v = Mc_6$ and $\mathring{k}_s = M(c_6 + 2c_7)$. The magnetic moment terms are a good example of the workings of HBχPT: the Dirac magnetic moments come in as a $1/M$ correction, while the anomalous magnetic moments, arising from resonance contributions or quark substructure according to one's point of view, come in proportional to a priori unknown LECs. Note that at this order all LECs are finite as loops do not contribute until third order.

There are three terms in $\mathcal{L}_{\pi N}^{(2)}$ which have terms proportional to $1/M$ only. Some of these could be eliminated using the equation of motion from $\mathcal{L}_{\pi N}^{(1)}$, at the expense of generating higher-order corrections, and this was done by Ecker and Mojžiš [46]. Although the resulting Lagrangian is shorter, terms with a physically obvious meaning are disguised, and keeping track of higher-order terms is laborious. Although versions with and without equation-of-motion simplifications exist at third order, only the unsimplified version exists at fourth order. It is important to stick to one version or the other in a single calculation.

At third order there are 23 independent terms which can be constructed in the relativistic Lagrangian using the structures already introduced, and hence 23 LECs. These are denoted b_i in [46] and d_i in [47]. The non-relativistic reduction contains further terms with coefficients proportional to $1/M_N^2$ or c_i/M_N. At this order pion loops enter also. In general these loops, evaluated in dimensional regularization, are divergent, and amplitudes are rendered finite by requiring the LECs to have infinite parts: $d_i = d_i^r(\lambda) + (\kappa_i/16\pi^2 f^2)L$, where the finite part $d_i^r(\lambda)$ is dependent on the renormalization scale. LECs evaluated at different scales are related by $d_i^r(\lambda') = d_i^r(\lambda) + (\kappa_i/16\pi^2 f^2)\log(\lambda/\lambda')$.

Some comment is in order about the different versions of the third-order Lagrangian that exist in the literature. The full list of relativistic terms was first written down by Krause [45]. Ecker found a set of 22 of these together with the corresponding β-functions which was sufficient for renormalization [48], and this set was also listed in [3], with the corresponding LECs labelled B_i and their β-functions β_i. This basis is still widely used. Subsequently Ecker and Mojžiš constructed the minimal full non-relativistic Lagrangian, using the equations of motion to reduce the number of terms [46]; they termed the LECs b_i but retained the notation β_i though there is no direct correspondence with the β-functions of the earlier paper. Some quantities which though not observable are still useful, such as the wave function renormalization, are not finite in the second approach.

Finally Fettes *et al* repeated this work, discovering that the fourth operator on Ecker and Mojžiš's list was redundant but otherwise keeping the same ordering [47]. However, they chose not to use the equations of motion to simplify the terms which come from the non-relativistic reduction, nor those required for renormalization. As a result there are another eight operators, labelled d_i, where $i = 24$–31, whose LECs can be chosen to vanish at some renormalization point (they used $\lambda = m_\pi$), and the β-functions (labelled κ_i) are not the same as the β_i.

This basis is convenient for calculations with $\lambda = m_\pi$ but, as already noted, it requires some care if the non-analytic structures of amplitudes are of interest. In that case care must be taken not to cancel, say, the LEC $d^r_{31}(\lambda)$, whose numerical value happens to be $(g^2/16\pi^2 f^2)\log(\lambda/m_\pi^{phys.})$, against a corresponding term $(g^2/16\pi^2 f^2)\log(m_\pi/\lambda)$, which comes from a loop calculation after cancellation of the divergence.

At fourth order there are 118 independent terms in the relativistic Lagrangian, with LECs denoted by e_i. These, together with the additional terms generated in the non-relativistic reduction, are listed in [49]. The corresponding β-functions are listed in a differently ordered, non-minimal basis in [50].

6.4.3 Loop integrals: the nucleon mass

The nucleon propagator obtained from $\mathcal{L}^{(1)}_{\pi N}$ is $i/v \cdot p$, where p is the residual nucleon momentum after subtracting Mv. If a chiral expansion is made of $v \cdot p$ the leading term is usually a loop momentum or an external energy of order m_π. However, at higher order $v \cdot p$ also contains the mass shift $(M_N - M)v$ and the external nucleon kinetic energy. These are cancelled by contributions to the nucleon propagator from terms in $\mathcal{L}^{(2,3,\dots)}_{\pi N}$ and from loops, to leave only a dependence on the off-shell energy. This rather cumbersome procedure is an example of the practical complications introduced by the heavy-baryon reduction.

The nucleon mass is determined by the pole of the nucleon propagator; if the self-energy at residual energy $v \cdot p = \omega$ is denoted by $\Sigma(\omega)$, this implies (for a nucleon at rest)

$$\delta M - \Sigma(\delta M) = 0, \tag{6.82}$$

where $M_N = M + \delta M$. Up to third order this gives $\delta M = \Sigma(0)$. Furthermore, the wave function renormalization is given by

$$Z_N = 1 + \Sigma'(0). \tag{6.83}$$

A further point to note is that the overall wave function normalization in the heavy baryon approach corresponds to a normalized upper component for a Dirac spinor.

At second order, the first correction to the chiral-limit nucleon mass M appears: the term proportional to c_1 in $\mathcal{L}^{(2)}_{\pi N}$ contributes $-4m_\pi^2 c_1$ to the nucleon mass. Here m_π^2 comes from an explicit occurrence of the quark mass in the Lagrangian: this is not due to long-range pion loops, but an effect from the quark structure of the nucleon. At third order, however, there is a contribution from the pion cloud round the nucleon, coming from the diagram shown in figure 6.4. (Tadpoles vanish because the lowest-order $\pi\pi NN$ seagull vertex is isovector.) The resulting self-energy as a

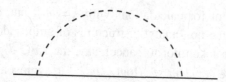

Figure 6.4. The pion-loop contribution to the nucleon self-energy.

function of $\omega = v \cdot p$ is

$$\Sigma_{\text{loop}}^{(3)}(\omega) = i \frac{g^2}{f^2} \int \frac{d^d l}{(2\pi)^d} S \cdot (-l) \tau_b \frac{i}{v \cdot (p - l) + i\eta} S \cdot l \tau_a \frac{i \delta^{ab}}{l^2 - M^2 + i\epsilon}$$

$$= -\frac{3g^2}{2f^2} \{S^\mu, S^\nu\} \frac{1}{i} \int \frac{d^d l}{(2\pi)^d} \frac{l_\mu l_\nu}{(v \cdot l - \omega - i\eta)(M^2 - l^2 - i\epsilon)}.$$
(6.84)

The integral in this expression can be expressed in terms of a simpler one with no factors of l_μ in the numerator [3]. That is termed $J_0(\omega, m)$, and can be evaluated by combining the denominators using the Feynman parametrization

$$\frac{1}{AB} = 2 \int_0^\infty dy \frac{1}{(A + 2yB)^2},$$
(6.85)

giving

$$J_0(\omega, M) = -4L\omega - \frac{\omega}{8\pi^2} \left(2 \log \frac{M}{\lambda} - 1 \right) - \frac{1}{4\pi^2} \sqrt{m^2 - \omega^2} \arccos \left(-\frac{\omega}{m} \right)$$
$$+ O(d - 4).$$
(6.86)

By means of a variety of standard manipulations, all integrals encountered can be expressed in terms of J_0. For instance, the integral in (6.84) gives $g_{\mu\nu} J_2(\omega, M) + v_\mu v_\nu J_3(\omega, m)$, where $J_2 = ((m^2 - \omega^2) J_0 - \omega \Delta_\pi)/(d - 1)$ etc. Integration-by-parts identities also allow derivatives of the J_i with respect to ω or m^2 to be simplified, for instance $\partial J_2/\partial m^2 = \frac{1}{2} J_0$.

Thus the pion–nucleon loop nucleon self-energy is

$$\Sigma_{loop}^{(3)}(\omega) = -\frac{3g^2 m^3}{32\pi f^2} - \frac{3(d - 1)g^2}{4f^2} \omega \Delta_\pi + O(\omega^2 m),$$
(6.87)

the first term giving the mass shift due to the pion cloud, and the second the loop contribution to the wave function renormalization. The mass shift is non-analytic in m_π^2, and (of necessity) finite. There can be no contribution from the third-order Lagrangian of this form, so this is the full mass shift at this order.

Whether there are other third-order contributions affecting the wave function renormalization and terms of order ω^3 depends on the Lagrangian. With Ecker's original version there are contributions from two terms in $\mathcal{L}_{\pi N}^{(3)}$, both with undetermined LECs (B_{15} and B_{20}). In Ecker and Mojžiš' version there are no further contributions proportional to the b_i. Finally in the form used by Fettes *et al* the terms which contribute have no finite parts (at $\lambda = m_\pi$). In the first and last approaches the LECs serve to make $\Sigma^{(3)}(\omega)$ finite; in the latter, for instance, the counterterm contribution is

$$\Sigma_{C.T.}^{(3)} = d_{24}\omega^3 - 8d_{28}m^2\omega, \tag{6.88}$$

where $\kappa_{28} = -9g^2/16$. This gives

$$Z_N = 1 - \frac{3(d-1)g^2}{4f^2}\Delta_\pi - 8d_{28}m^2$$

$$= 1 - \frac{3g^2m^2}{32\pi^2 f^2}\left(1 + 3\log\frac{m}{\lambda}\right) - 8m^2 d_{28}^r(\lambda), \tag{6.89}$$

which is finite and determined. (Numerically, the logarithm and the d_{28}^r terms cancel). For completeness we also give the first two terms of the expansion of g_A:

$$g_A = g + m^2\left(4d_{16}^r(\lambda) - 8gd_{28}^r(\lambda) - \frac{g(2g^2+1)}{8\pi^2 f^2}\log\frac{m}{\lambda} - \frac{g^3}{16\pi^2 f^2}\right). \tag{6.90}$$

Finally, the leading correction to the Goldberger–Trieman relation in Fettes' scheme is given by

$$\frac{g_{\pi NN}}{M_N} = \frac{g_A M_N}{f_\pi} - \frac{2m_\pi^2 d_{18}}{f_\pi}, \tag{6.91}$$

where the LEC d_{18} is finite and independent of λ.

For comparison, Z_N, g_A and $g_{\pi NN}$ in Ecker's scheme are obtained by replacing $d_{16}^r(\lambda)$ with $B_9^r(\lambda)/(4\pi f)^2$, and similarly for d_{18} and B_{23}, and d_{28} and B_{20}.

6.4.4 Electromagnetic interactions: Compton scattering

Calculations of the chiral expansion of the nucleon mass or axial coupling constant are of limited interest since these terms already enter at lowest order: they are primarily determined by non-pionic physics and the higher terms are only of interest as ingredients in other calculations or for carrying out chiral extrapolations of lattice QCD data. The magnetic moment, too, has a contribution from short-range (quark substructure or resonance) physics at lowest order (expressed in the LECs c_6 and c_7). There are other properties of the nucleon, however, which in χPT are dominated by pion effects, and hence these quantities are *predicted* in χPT to a given order.

Perhaps the most famous of these are the electromagnetic polarizabilities of the nucleon.

The scattering of real, unpolarized, long-wavelength photons from, for instance, a proton is required by gauge and Lorentz invariance to depend only on the mass and charge of the target: this is the Thomson limit of Compton scattering ($\nu \to 0$). As the photon energy is increased, deviations from this cross section arise because the proton is not perfectly rigid, but can deform in response to electric and magnetic fields. This is characterized by the electric and magnetic polarizabilities, α and β, which enter the scattering amplitude through

$$T = \epsilon' \cdot \epsilon \left(-\frac{Z^2 e^2}{M} + 4\pi \alpha \omega \omega' \right) + 4\pi \beta \, \epsilon' \times k' \cdot \epsilon \times k + \cdots , \qquad (6.92)$$

where ϵ and ϵ' are the polarization vectors of the initial- and final-state photons, whose momenta and energy are k, ω and k', ω' respectively. Spin-dependent terms have been omitted. The PDG [51] give the following values for the proton

$$\alpha_p = 12.0 \pm 0.6 \times 10^{-4} \, \text{fm}^3,$$
$$\beta_p = 1.9 \pm 0.5 \times 10^{-4} \, \text{fm}^3. \qquad (6.93)$$

The neutron values are less accurately known but the data are consistent with vanishing isovector polarizabilities. These values are surprisingly small in view of the roughly 1 fm^3 volume of the proton.

The sum of α and β can also be obtained from a dispersion relation relating the forward Compton scattering amplitude to the unpolarized photon absorption cross section. This is the Baldin sum rule and it gives $\alpha_p + \beta_p = 14.2 \pm 0.5 \times 10^{-4} \, \text{fm}^3$.

In HBχPT the first diagram to contribute is the seagull term from the first two terms of $\mathcal{L}^{(2)}$; these arise from the non-relativistic reduction, contain no LECs and exactly reproduce the Thomson amplitude. At third order the first loop diagrams enter, but crucially there are no contributions from LECs. Thus to third order the full Compton scattering amplitude is obtained as a function of ω/m_π, and the polarizabilites can be obtained from a Taylor series in powers of ω. The results are the same for the proton and neutron [52]:

$$\alpha = \frac{5e^2 g_A^2}{384\pi^2 f_\pi^2 m_\pi} = 12.2 \times 10^{-4} \, \text{fm}^3,$$

$$\beta = \frac{e^2 g_A^2}{768\pi^2 f_\pi^2 m_\pi} = 1.2 \times 10^{-4} \, \text{fm}^3. \qquad (6.94)$$

Note that in the chiral limit the polarizabilities, being proportional to $1/m_\pi$, would diverge; nonetheless the physical values are in extremely good agreement with the (surprisingly small) experimental values.

The agreement of the lowest-order prediction for α is actually embarrassingly good: one would expect higher-order corrections to be at least 10% rather than 1%. The next-order calculation can also be done, but at this order the pion loops are divergent and there is a fourth-order seagull with four unknown LECs contributing to the isoscalar and isovector α and β: the effective seagull, which has contributions from six chirally-invariant terms, is given by

$$\mathcal{L}^{(4)} = 2\pi N^\dagger \Big\{ \tfrac{1}{2}[\delta\beta_s + \delta\beta_v \tau_3] g_{\mu\nu}$$
$$- [(\delta\alpha_s + \delta\beta_s) + (\delta\alpha_v + \delta\beta_v)\tau_3] v_\mu v_\nu \Big\} F^{\mu\rho} F^\nu{}_\rho N + \cdots .$$
(6.95)

Attempts to estimate the LECs by assuming that they are generated from pole graphs on integrating out the Δ give contributions to α and β of around 5 fm^3 and spoil the agreement found at lowest order [53]: this is still an unsolved problem.

Of course, a calculation in HBχPT to a given order does not just give the polarizabilities; it gives a full prediction for the cross section as a function of photon energy and scattering angle. To third order there are no free parameters; to fourth order some of the second-order LECs c_i enter, along with the contributions to α and β mentioned above. If the c_i are taken from fits to pion–nucleon scattering, the remaining LECs can be fit to proton Compton scattering data [54]. Results are obtained that are consistent with the PDG values [51] (in fact they were used as input in the 2004 PDG values) and the fit is good provided the energy and the magnitude of the momentum transfer are restricted to be below 200 MeV. This is consistent with the expectation that the $\Delta(1232)$ will enter as an active degree of freedom at around that scale, and this will be mentioned again when we discuss including the $\Delta(1232)$ explicitly. Figure 6.5 shows the agreement between theory and experiment [54].

In spite of the generally satisfactory situation for unpolarized scattering, there are other aspects of Compton scattering that are not so satisfactory. In writing the scattering amplitude of (6.92) we omitted four structures dependent on the proton spin. Four spin-dependent polarizabilities are associated with the low-energy behaviours of these structures. For forward scattering the relevant combination is called γ_0; for backward scattering, γ_π. One would expect these to be important for scattering of polarized photons and targets, but such experiments have not yet been done. However, dispersion relations link the spin-dependent forward Compton scattering amplitude to polarized photon absorption cross sections: the most famous is the Gerasimov–Drell–Hearn (GDH) sum rule [55]

$$\frac{\pi e^2 \kappa^2}{2M_N^2} = \int_{\nu_0}^{\infty} d\nu' \frac{\sigma_{3/2}(\nu') - \sigma_{1/2}(\nu')}{\nu'},$$
(6.96)

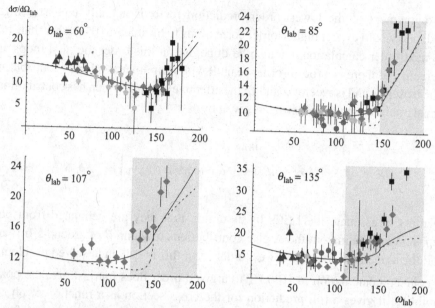

Figure 6.5. HBχPT results for the Compton scattering differential cross section (in nb/sr) at four different laboratory angles, compared to data from various experiments. The regions shaded in grey correspond to $\sqrt{-t} > 200$ MeV, and so were excluded from the fit when extracting α and β. The solid lines are the fourth-order result; the dashed lines are third order. (The figure is taken from [54]; a list of experiments may be found therein.)

where 3/2 and 1/2 refer to parallel and antiparallel photon and proton spins; an analogous integral weighted by $1/\nu'^3$ gives γ_0. These cross sections have been measured at Mainz and Bonn, and the proton forward spin polarizability has been accurately determined to be $\gamma_0 = -0.94 \pm 0.15 \times 10^{-4}$ fm^4. This quantity is predicted in HBχPT at both third and fourth order. Unfortunately the results are in sharp contrast to the success of the spin-independent case, with the third- and fourth-order results being 4.5×10^{-4} and -3.9×10^{-4} fm^4 respectively [56,57]. Clearly the lack of convergence makes the actual value irrelevant. The reason for this failure is not clear; discussion can be found in [58,59].

Cross sections for inelastic scattering of electrons on protons (that is, proton structure functions) can similarly be used to construct the amplitudes for forward 'doubly virtual Compton scattering', including Q^2-dependent versions of the GDH sum rule and of $\alpha + \beta$ and γ_0. These quantities are typically of interest because the high-Q^2 region – well-established from deep-inelastic scattering – is very different from the $Q^2 = 0$ point. HBχPT can also be used at non-zero Q^2, and so can make interesting predictions about the low-Q^2 evolution. To date, though, rather few calculations have been done. Most work has focussed on the generalized GDH sum rule $I_1(Q^2)$,

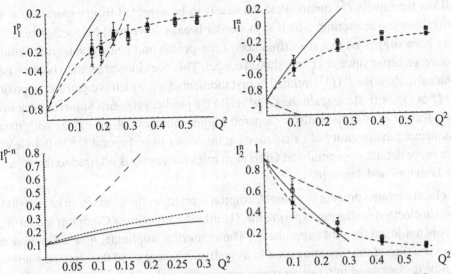

Figure 6.6. I_1 is a generalized GDH sum rule and is related to the integral of the spin-dependent structure function g_1; I_2 is similarly related to g_2. Q^2 is measured in GeV^2. The proton data are from [61], while the neutron data (from ^{3}He) are from [62]; empty symbols represent the experimental data and filled symbols are corrected for the high-energy part of the integral which is not measured in these experiments. The third panel shows the difference between I_1 for the proton and neutron (the integral which at high Q^2 gives the Bjorken sum rule). The solid curve is $HB\chi PT$ [60]. The short-dashed curve is a simple polynomial fit to all the data (not just the points shown) constrained to have the right value at $Q^2 = 0$. The long-dashed curve is the relativistic χPT result (discussed later) [63].

which satisfies $I_1(0) = -\kappa^2/4$ and which can be expressed in terms of the first moment of the spin-dependent nucleon form factor g_1:

$$I_1(Q^2) = \frac{2M^2}{Q^2} \int_0^{x_0} dx\, g_1(x, Q^2). \qquad (6.97)$$

This has been calculated in $HB\chi PT$ by Ji, Kao and collaborators [60]. It vanishes at $\mathcal{O}(p^3)$ so the $\mathcal{O}(p^4)$ result is the leading one, and no comment can be made about the convergence of the chiral expansion. Results are shown in figure 6.6.

Data for $I_1^{p,n}(Q^2)$ have been obtained in electron scattering experiments [61,62]; results for the neutron have been extracted from data on the deuteron and on ^{3}He. To the experimental data must be added a high-energy contribution estimated from deep-inelastic scattering. Although the GDH sum rule gives $I_1^p(0) = -0.80$, the data are small and positive down to rather small Q^2, and only those from JLab, which go down to 0.15 GeV^2 ($Q = 390$ GeV), dip below the axis. The slope at $Q^2 = 0$ cannot therefore be clearly extracted, but it is very compatible with that given by $HB\chi PT$, as is that for the neutron. However, the $HB\chi PT$ curve badly overshoots all but the lowest Q^2 points. Perhaps that is to be expected from experience with real Compton scattering, which at this order breaks down well below that scale. It has been suggested that the difference of the proton and neutron integrals should converge better since Δ effects should cancel. This does indeed seem to be the case. Data also exist for $I_2^n(Q^2)$, related to first moment of g_2, which is equal and opposite to I_1^n at $Q^2 = 0$. Here again the $\mathcal{O}(p^4)$ HBχPT prediction seems valid for very low Q^2. Kao *et al* [64] have also compared higher moments of g_1 and g_2 with data, confirming the inability of calculations at this order to go beyond $Q^2 \approx 0.1$ GeV^2. For more details of generalized GDH sum rules the reader is referred to the review by Drechsel and Tiator [65].

A closely related process is virtual Compton scattering, the production of real photons in electron scattering experiments. (Unlike doubly-virtual Compton scattering, this is not just a thought experiment.) The scattering amplitude, now a function of the virtuality of the incoming photon as well as its energy and the scattering angle, can be decomposed into twelve scalar amplitudes, only six of which survive in the limit $Q^2 \to 0$. Generalized polarizabilities, which are functions of Q^2, are defined by taking the limit in which the real photon energy goes to zero. In the standard notation two are generalizations of α and β, and in fact there are only four others which are independent though eight can be defined. Only two of these four are non-vanishing in the limit $Q^2 \to 0$, and these correspond to γ_3 and $\gamma_2 + \gamma_4$.

The sole relevant experiment has been done at MAMI, at fixed $Q^2 = 0.33$ GeV^2 and photon beam energy of 0.6 GeV but with a range of outgoing photon energies from 30 to 110 MeV [66]. The experimental cross section is dominated by the Bethe–Heitler process in which the real photon is radiated by the initial- or final-state electron, so direct cross section comparisons are not particularly illuminating. From analysis of the full data set, two terms in the low-energy expansion of the residual (non-Born or Bethe–Heitler) spin-independent amplitude are extracted, termed $P_{LL} - P_{TT}/\epsilon$ and P_{LT}. At $Q^2 = 0$ these are related to α and β respectively, though they are not a theorist's natural extension of these polarizabilities to finite Q^2. These can also be extracted from an HBχPT calculation of the virtual Compton scattering amplitude; so far this has only been done at $\mathcal{O}(p^3)$ [67], but surprisingly, given the relatively large Q^2, the agreement is perfect.

More theoretical effort has been applied to the generalized spin-dependent polarizabilities, which have all now been calculated to two chiral orders [67,68]. However, there are no relevant data. As might have been expected from the situation with the polarizabilities defined in real scattering, there is no convergence, and about all one can say about the agreement with estimates based on dispersion relations and the MAID pion electroproduction database is that the order of magnitude is right (which is at least more than can be said for some quark models [69]).

For more information the reader is referred to the review of real and virtual Compton scattering, with particular reference to dispersion relations but also covering χPT, by Drechsel *et al* [69].

6.4.5 *Electromagnetic interactions: pion photoproduction*

Threshold neutral pion photoproduction from the proton is one of the most celebrated applications of χPT. In 1970 a 'low-energy theorem' was derived [14] which gave the threshold value of the S-wave amplitude E_{0+} as an expansion in powers of m_π/M_N. As discussed above in section 6.1.1 the leading term of order m_π/M_N simply reflects the magnitude of the electric dipole moment of the $p\pi^0$ system. However, when the first precision measurements were done at Saclay and Mainz [70,71], they disagreed with the 'LET' by a factor of about two. With the advent of χPT, however, Bernard *et al* [72] recognized that the so-called LET was in fact just the Born term, and that pion loops contributed at order $(m_\pi/M_N)^2$. Furthermore, the contribution was numerically large, calling into question the convergence, so the χPT prediction at this order, while disproving the old LET, does not really constitute a new one. At one order higher unknown LECs enter [73]. For some considerable time this finding was controversial, but by now there is general agreement with the position that in general the LETs of the 1960s and 1970s are simply lowest-order predictions of χPT, and that χPT is indeed the correct framework in which to go beyond lowest order. Further experiments, including some with polarized photons at Mainz [74], have confirmed that the threshold value of E_{0+} does not agree with the old LET. However, when the two LECs which enter the full amplitude are fitted to the data, very satisfactory agreement with total cross section, differential cross section and photon asymmetry is obtained; in particular the cusp in E_{0+} at the charged pion production threshold is reproduced. In addition, two of the P-wave amplitudes, P_1 and P_2, are predicted at leading order independently of the LECs, and hence constitute new LETs which are both convergent and in good agreement with the data [73,76]. Figures 6.7 and 6.8 show the full fourth-order HBχPT results and the TAPS data to which the five LECs are fit. (The curves shown are from the authors' fit following the procedures of Bernard *et al* [76].)

6.4.6 *Chiral perturbation theory and lattice QCD*

There are many quantities of interest in hadronic physics that cannot be predicted by χPT. Most fundamental perhaps is the nucleon mass, which simply enters as a parameter in the lowest-order Lagrangian. (More accurately, what enters is the value that the mass would have in the chiral limit, that is if quarks were massless.) In general, any property of the nucleon which is not primarily due to the pion cloud

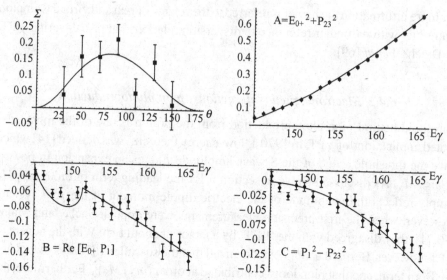

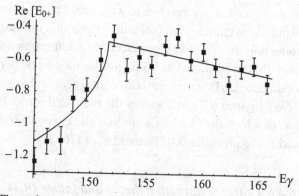

Figure 6.7. TAPS data on neutral pion photoproduction [74,75]. The first figure shows the photon asymmetry Σ. The other three summarize the differential cross section: at each energy the three parameters A, B and C (in units of 10^{-4} fm^2) are given by the best fit curve $d\sigma/d\Omega = (q/k)(A + B\cos\theta + C\cos^2\theta)$, where q and k are the pion and photon three-momenta respectively. Such a quadratic form will hold if only the four s- and p-wave amplitudes E_{0+} and P_1, P_2 and P_3 are important; $P_{23}^2 \equiv \frac{1}{2}(P_2^2 + P_3^2)$. For more details see [73]. The curves are the fourth-order HBχPT results, with the five free parameters fit to the full data set.

Figure 6.8. The real part of the s-wave amplitude E_{0+}, in units of $10^{-3}/m_{\pi^+}$: 'data' points, from [74], are extracted from a fit to the raw data with certain assumptions about the energy dependence of the p-waves; the curve is the fourth-order HBχPT result based on the fit shown in figure 6.7.

will also not be predicted; examples include the anomalous magnetic moment and the axial coupling constant, as well as many others which are less well known. In a few cases one can argue that the dominant contribution is from hadronic resonances, and greater predictive power may be obtained in a theory which contains the Δ as an explicit degree of freedom, as will be discussed below. But in general these properties are due intrinsically to quark-level physics. The only way we currently have to predict these is to use the computer simulations of the underlying theory of quarks and gluons: lattice QCD.

For all the immense progress that has been made in lattice QCD, computer power still places severe limitations on how closely the simulations can approach the real-world limit. (For a review of the state-of-the-art in 2003 and further references, see [77].) By definition all calculations are done in finite volumes with finite lattice spacing, but in addition so far all have used quark masses heavier than the real ones – often much heavier, so that the pion mass might be around 500 MeV [78–80]. A further commonly-used shortcut is to exclude the contributions from quark loops – the 'quenched approximation' [81].

Usually simulations (quenched or unquenched) are done at several values of these parameters and some form of extrapolation to infinite volume, zero lattice spacing and physical pion mass is performed. Polynomial extrapolation is well motivated for the first two, but it was common to see linear extrapolation in the quark mass as well. However, χPT tells us how nucleon properties must vary as the quark mass goes to zero, and that behaviour is not in general linear.

It is now realized that lattice QCD and χPT give us largely complementary information, the former dealing well with short-distance properties about which χPT is silent, and the latter describing the pion cloud which, due to large quark masses and finite volumes, still eludes lattice calculation. The possibility therefore exists of using extrapolation formulae given by the χPT expansion of the quantity being calculated, fitting the LECs which enter to the lattice data. It is even possible to provide alternative χPT extrapolations which allow for the effects of quenching [82–85]. The effects of unquenched pion loops may then be added in by hand. These chiral extrapolations have a double benefit, enabling more accurate lattice QCD predictions and providing first-principles extractions of the LECs. At the current time, however, their full potential has yet to be realized, since most lattice data involve pion masses beyond the likely radius of convergence of χPT. (How large that is is also disputed.) Two approaches have had some success so far. One is naively fitting the χPT formula to data with pion masses as high as 600 MeV [86] or more [80] (although in some cases an extended version with explicit Δs is used [87]). The other is to use some theoretically-motivated function of pion mass which has the correct low-mass expansion. An example is the inclusion of form factors in the one-loop diagrams [88]. It is unlikely that either approach will win full acceptance until lattice data exist for masses sufficiently low that the chiral expansion is trustworthy and the data show convincingly that a linear extrapolation cannot be correct.

6.4.7 *Extensions and variants of chiral perturbation theory*

In this section, we will briefly cover two topics: calculations that avoid the heavy-baryon reduction, and the inclusion of Δ baryons in the theory.

The first attempts to include baryons in χPT, naturally enough, used relativistic nucleons, but it was quickly recognized that the integrals which arise, regulated with dimensional regularization in the standard way, did not respect any power counting and progress was only possible with the application of the heavy-baryon reduction. Amplitudes so calculated respect all symmetries including Lorentz invariance order by order, that is any violations will be higher order in powers of $1/M_N$ than the order to which one is working. However, various properties of the full amplitude are no longer manifest – crossing symmetry for instance – and sometimes by-hand corrections are required to ensure that thresholds occur at the correct place. These corrections amount to resumming successive terms in the $1/M_N$ expansion. Further complications arise when intermediate nucleons can go on-shell. In addition the heavy-baryon Lagrangian with all its fixed terms is much lengthier than the relativistic one, and calculations are correspondingly more involved with more diagrams entering. For these reasons relativistic treatments would seem more attractive [89–91].

The best-known relativistic framework is called 'infrared regularization' [89,90]. Without going into details, the integrals that arise from pion–nucleon loops are split in two; one part is UV finite and contains all the non-analyticities in m_π^2, while the other is IR regular (though possibly UV divergent), generating an expression which can be expanded as a regular polynomial in m_π^2. This polynomial is rendered finite by the LECs in the usual fashion. The resulting amplitude respects power counting and the non-analytic part agrees with the heavy baryon result.

Discussion of the advantages of the IR scheme tend to focus on its superior treatment of kinematic effects. Surprisingly, however, in several cases the predictions of the IR and heavy baryon theories differ quite significantly, even though they agree to, say, $\mathcal{O}(m_\pi/M_N)^2$. In figure 6.6 the long-dashed curve is the relativistic χPT result [63], while the solid curve is the HBχPT result [60]. Clearly they are very different, and HBχPT is much closer to the experimental situation. At first sight this is very worrying and suggests that the HBχPT results are only coincidentally accurate; higher-order terms, it would seem, must spoil the agreement. However, this is not necessarily correct. From the point of view of HBχPT the higher-order terms in the IR result are only a subsection of the total. All the higher-order LEC contributions are missing. For instance, the Dirac and anomalous magnetic moments are treated differently, since one is a $1/M_N$ effect and the other is due to an LEC. Furthermore, there are reasons to believe that the relativistic and LEC contributions might tend to cancel. The IR results contain not only kinematic corrections, but also contributions from nucleon-level 'z-graphs'. These, however, are clearly strongly suppressed by the large size of the nucleon. This suppression might be implemented by the inclusion of LECs order-by-order in the HBχPT scheme, but the connection is lost in the IR scheme. This is a topic where much work remains to be done.

A more fundamental modification of baryon χPT is to include the $\Delta(1232)$ baryon as an explicit field in the Lagrangian [92–94]. The motivation is clear. Since the nucleon–$\Delta(1232)$ mass difference is only twice the pion mass, integrating out the $\Delta(1232)$ (as is done implicitly when writing down a Lagrangian of pions and nucleons only) will put an absolute upper limit of $2m_\pi$ on the radius of convergence for any process in which the $\Delta(1232)$ can be excited – probably significantly lower in practice.

It is straightforward to include a $\Delta(1232)$ field in a heavy-baryon Lagrangian. However, a new scale, $M_\Delta - M_N$, is introduced, which does not vanish in the chiral limit. (While the pion cloud contributes some of the extra energy of the $\Delta(1232)$, quark-level effects such as the hyperfine splitting are also expected to be important.) A decision has to be taken as to the power-counting to be used. The most common one is the 'small-scale expansion' [94] which treats $M_\Delta - M_N$ as being of order m_π, which may be numerically correct in this world but clearly is wrong in the chiral limit. In this scheme $\pi\Delta$ loops contribute at the same order as πN loops, whereas if the $\Delta(1232)$ is integrated out its effects are first manifest (through LECs) at one order higher. To simplify a somewhat complicated picture, this tends to spoil the good predictions of HBχPT for static properties such as the electric polarizability [95]. However, if corrected by hand (by 'promoting' a single higher-order term) the energy dependence is then improved, and in Compton scattering the problem encountered at moderate energies and backward angles can be solved [96]. On balance the introduction of the $\Delta(1232)$ does improve the predictions of lowest-order χPT, although the large number of extra LECs which enter at higher orders has so far prevented calculations being pushed beyond lowest order.

Finally, we should mention that, while the state-of-the-art for HBχPT currently consists of one-loop calculations at order Q^4, some first two-loop (order-Q^5) results have been obtained [97–99]. These will be important for testing the convergence of the expansions of quantities such as the spin-polarizabilities, which first appear at order Q^4 [100].

References

[1] S Scherer, Advances in Nuclear Physics 27 (2003) 27

[2] G Ecker, Progress in Particle and Nuclear Physics 36 (1996) 71

[3] V Bernard, N Kaiser and U-G Meissner, International Journal of Modern Physics E4 (1995) 193

[4] S R Beane, P F Bedaque, W C Haxton, D R Phillips and M J Savage, *At the Frontier of Particle Physics: Handbook of QCD*, M Shifman ed, Vol. 1, p. 133, World Scientific, Singapore (2001)

[5] P F Bedaque and U van Kolck, Annual Review of Nuclear and Particle Science 52 (2002) 339

[6] V de Alfaro, S Fubini, G Furlan and C Rossetti, *Currents in Hadron Physics*, North Holland, Amsterdam (1973)

[7] M Gell-Mann, R J Oakes and B Renner, Physical Review 175 (1968) 2195

[8] S Weinberg, Physical Review Letters 17 (1966) 616

[9] Y Tomozawa, Nuovo Cimento 46A (1966) 707

[10] E Reya, Reviews of Modern Physics 46 (1974) 545

[11] J Gasser, H Leutwyler and M E Sainio, Physics Letters B253 (1991) 252

[12] M M Pavan, I I Strakovsky, R L Workman and R A Arndt, πN Newsletter 16 (2002) 110

[13] N M Kroll and M A Ruderman, Physical Review 93 (1954) 233

[14] P de Baenst, Nuclear Physics B24 (1970) 633

[15] A M Bernstein and B R Holstein, Comments on Nuclear and Particle Physics 20 (1991) 197

[16] S Weinberg, Physica 96A (1979) 327

[17] J Gasser and H Leutwyler, Annals of Physics 158 (1984) 142

[18] J Gasser, M E Sainio and A Švarc, Nuclear Physics B307 (1988) 779

[19] S Scherer and H W Fearing, Physical Review D52 (1995) 6445

[20] G Ecker, J Gasser, A Pich and E de Rafael, Nuclear Physics B321 (1989) 311

[21] J Gasser and H Leutwyler, Nuclear Physics B250 (1985) 465

[22] J Bijnens, G Ecker and J Gasser, in *The Second DAΦNE Physics Handbook*, L Maiani, G Pancheri and N Paver eds, INFN, Frascati (1995)

[23] J Gasser and H Leutwyler, Nuclear Physics B250 (1985) 517

[24] S L Adler, Physical Review 177 (1969) 2426

[25] J S Bell and R Jackiw, Nuovo Cimento A60 (1969) 47

[26] J Wess and B Zumino, Physics Letters 37B (1971) 95

[27] E Witten, Nuclear Physics B223 (1983) 422

[28] J F Donoghue and D Wyler, Nuclear Physics B316 (1989) 289

[29] D Issler, SLAC-PUB-4943 (1990)

[30] J Bijnens, L Ghirlanda and P Talavera, European Physical Journal C23 (2002) 539

[31] H W Fearing and S Scherer, Physical Review D53 (1996) 315

[32] J Gasser and M E Sainio, European Physical Journal C6 (1999) 297

[33] J Bijnens, G Colangelo and G Ecker, Annals of Physics 280 (2000) 100

[34] J Bijnens and F Cornet, Nuclear Physics B296 (1988) 557

[35] J Boyer *et al*, Mark II Collaboration, Physical Review D42 (1990) 1350

[36] U Bürgi, Physics Letters B377 (1996) 147; Nuclear Physics B479 (1996) 392

[37] J F Donoghue and B R Holstein, Physical Review D48 (1993) 137

[38] J F Donoghue, B R Holstein and Y C R Lin, Physical Review D37 (1988) 2423

[39] H Marsiske *et al*, Crystal Barrel Collaboration, Physical Review D41 (1990) 3324

[40] S Bellucci, J Gasser and M E Sainio, Nuclear Physics B423 (1994) 80; erratum: *ibid* B431 (1994) 413

[41] M R Pennington, in *The DAΦNE Physics Handbook*, p.379, L Maiani, G Pancheri and N Paver eds, INFN, Frascati (1992);
      M Boglione and M R Pennington, European Physical Journal C9 (1999) 11

[42] J Portolés and M R Pennington, in *The Second DAΦNE Physics Handbook*, p.579, L Maiani, G Pancheri and N Paver eds, INFN, Frascati (1995)

[43] J Ahrens *et al*, European Physical Journal A23 (2005) 113

[44] C-W Kao, B E Norum and K Wang, hep-ph/0409081

[45] A Krause, Helvetica Physica Acta 63 (1990) 3

[46] G Ecker and M Mojžiš, Physics Letters B365 (1996) 312

[47] N Fettes, U-G Meissner and S Steininger, Nuclear Physics A640 (1998) 199

[48] G Ecker, Physics Letters B336 (1994) 508

[49] N Fettes, U-G Meissner, M Mojžišand and S Steininger, Annals of Physics 283 (2000) 273; erratum: *ibid* 288 (2001) 249

[50] U-G Meissner, G Müller and S Steininger, Annals of Physics 279 (2000) 1

[51] Particle Data Group, Physics Letters B592 (2004) 1

[52] V Bernard, N Kaiser, J Kambor and U-G Meissner, Nuclear Physics B388 (1992) 315

[53] V Bernard, N Kaiser, A Schmidt and U-G Meissner, Physics Letters B319 (1993) 269

[54] S R Beane, M Malheiro, J A McGovern, D R Phillips and U van Kolck, Physics Letters B567 (2003) 200; Nuclear Physics A747 (2005) 311

[55] S B Gerasimov, Soviet Journal of Nuclear Physics 2 (1966) 430;
      S D Drell and A C Hearn, Physical Review Letters 16 (1966) 908

[56] X Ji, C-W Kao and J Osborne, Physical Review D61 (2000) 074003

[57] K B V Kumar, J A McGovern and M C Birse, Physics Letters B479 (2000) 167

[58] G C Gellas, T R Hemmert, U-G Meissner, Physical Review Letters 85 (2000) 14; *ibid* 86 (2001) 3205

[59] M C Birse, X Ji and J A McGovern, Physical Review Letters 86 (2001) 3204

[60] X Ji, C-W Kao and J Osborne, Physics Letters B472 (2000) 1;
      C-W Kao, T Spitzenberg and M Vanderhaeghen, Physical Review D67 (2003) 016001

[61] R Fatemi *et al*, CLAS Collaboration, Physical Review Letters 91 (2003) 222002;
      R Fatemi, private communication

[62] M Amarian *et al*, JLab E94-010 Collaboration, Physical Review Letters 92 (2004) 022301;
      Z-E Meziani, private communication

[63] V Bernard, T R Hemmert and U-G Meissner, Physics Letters B545 (2002) 105;
      Physical Review D67 (2003) 076008

[64] C-W Kao, D Drechsel, S Kamalov and M Vanderhaeghen, Physical Review D69 (2004) 056004

[65] D Drechsel and L Tiator, Annual Review of Nuclear and Particle Science 54 (2004) 69

[66] J Roche *et al*, Physical Review Letters 85 (2000) 708

[67] T R Hemmert, B R Holstein, G Knöchlein and S Scherer, Physical Review D55 (1997) 2630; Physical Review Letters 79 (1997) 22;
      T R Hemmert, B R Holstein, G Knöchlein and D Drechsel, Physical Review D62 (2000) 014013

[68] C-W Kao and M Vanderhaeghen, Physical Review Letters 89 (2002) 272002;
      C-W Kao, B Pasquini and M Vanderhaeghen, Physical Review D70 (2004) 114004

[69] D Drechsel, B Pasquini and M Vanderhaegen, Physics Reports 378 (2003) 99

[70] E Mazzucato *et al*, Physical Review Letters 57 (1986) 3144

[71] R Beck *et al*, Physical Review Letters 65 (1990) 1841

[72] V Bernard, J Gasser, N Kaiser and U-G Meissner, Physics Letters B268 (1991) 219;
V Bernard, N Kaiser and U-G Meissner, Nuclear Physics B383 (1992) 442

[73] V Bernard, N Kaiser and U-G Meissner, Zeitschrift für Physik C70 (1996) 483

[74] A Schmidt *et al*, Physical Review Letters 87 (2001) 232501

[75] A Schmidt, PhD thesis, Universität Mainz (2001);
R Beck, private communication

[76] V Bernard, N Kaiser and U-G Meissner, European Physical Journal A11 (2001) 209

[77] T DeGrand, International Journal of Modern Physics A19 (2004) 1337

[78] A Ali Khan *et al*, CP-PACS Collaboration, Physical Review D65 (2002) 054505;
erratum: *ibid* D67 (2003) 059901

[79] S Aoki *et al*, JLQCD Collaboration, Physical Review D68 (2003) 054502

[80] A Ali Khan *et al*, QCDSF-UKQCD Collaboration, Nuclear Physics B689 (2004) 175

[81] S Aoki *et al*, CP-PACS Collaboration, Physical Review D67 (2003) 034503

[82] C Bernard and M Golterman, Physical Review D46 (1992) 853;
S R Sharpe, Physical Review D46 (1992) 3146

[83] J N Labrenz and S R Sharpe, Physical Review D54 (1996) 4595

[84] S R Sharpe and N Shoresh, Physical Review D62 (2000) 094503; *ibid* D64 (2001) 114510

[85] J-W Chen and M J Savage, Physical Review D65 (2002) 094001

[86] M Procura, T R Hemmert and W Weise, Physical Review D69 (2004) 034505

[87] T R Hemmert, M Procura and W Weise, Physical Review D68 (2003) 075009

[88] D B Leinweber, A W Thomas, K Tsushima and S V Wright, Physical Review D61 (2000) 074502; D B Leinweber, A W Thomas and R D Young, Physical Review Letters 92 (2004) 242002

[89] P J Ellis and H-B Tang, Physical Review C57 (1998) 3356

[90] T Becher and H Leutwyler, European Physical Journal C9 (1999) 643

[91] T Fuchs, J Gegelia, G Japaridze and S Scherer, Physical Review D68 (2003) 056005

[92] E Jenkins and A V Manohar, Physics Letters B259 (1991) 353; in *Effective Field Theories of the Standard Model*, U-G Meissner ed, World Scientific, Singapore (1992)

[93] M N Butler, M J Savage and R P Springer, Nuclear Physics B399 (1993) 69

[94] T R Hemmert, B R Holstein and J Kambor, Physics Letters B395 (1997) 89; Journal of Physics G24 (1998) 1831

[95] T R Hemmert, B R Holstein and J Kambor, Physical Review D55 (1997) 2630

[96] R P Hildebrandt, H W Griesshammer, T R Hemmert and B Pasquini, European Physical Journal A20 (2004) 293

[97] M C Birse and J A McGovern, Physics Letters B446 (1999) 300

[98] N Kaiser, Nuclear Physics A720 (2003) 157

[99] N Kaiser, Physical Review C67 (2003) 027002

[100] C-W Kao and J A McGovern, in preparation

7

Spin structure functions

P Mulders

This chapter deals with properties of hadrons in high-energy scattering processes with special emphasis on spin dependence. We consider electroweak interactions, specifically lepton–hadron scattering that allows a separation of the scattering amplitude for the reaction into a leptonic part and a hadronic part, where the leptonic part involving elementary particles is known. The structure of the hadronic part is constrained by its Lorentz structure and fundamental symmetries and can be parametrized in terms of a number of structure functions. The resulting expression for the scattering amplitude can be used to calculate the cross sections in terms of these structure functions and, in turn, a theoretical study of the structure functions can be made. Part of this can be done rigorously with the only input, or assumption if one prefers, the known interactions of the hadronic constituents, quarks and gluons, within the standard model. For this both the electroweak couplings of the quarks needed to describe the interactions with the leptonic part via the exchange of photon, Z^0 or $W^{\pm}$ bosons, as well as the strong interactions of the quarks among themselves via the exchange of gluons are important. For a general reference see the books of Roberts [1] or Leader [2].

7.1 Leptoproduction

In this section we discuss the basic kinematics of a particular set of hard electroweak processes, namely the scattering of a high-energy lepton, for example an electron, muon or neutrino, from a hadronic target, $\ell(k)H(P) \to \ell'(k')X$. In this process at least one hadron is involved. If one does not care about the final state, counting every event irrespective of what is happening in the scattering process, one talks about an inclusive measurement. If one detects specific hadrons in coincidence

Electromagnetic Interactions and Hadronic Structure, eds Frank Close, Sandy Donnachie and Graham Shaw.
Published by Cambridge University Press. © Cambridge University Press 2007

with the scattered lepton one talks about semi-inclusive measurements or more specifically one-particle inclusive, two-particle inclusive etc., depending on the number of hadrons that are detected.

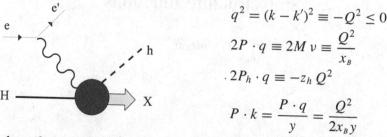

$$q^2 = (k - k')^2 \equiv -Q^2 \leq 0$$

$$2P \cdot q \equiv 2M \nu \equiv \frac{Q^2}{x_B}$$

$$2P_h \cdot q \equiv -z_h\, Q^2$$

$$P \cdot k = \frac{P \cdot q}{y} = \frac{Q^2}{2x_B y}$$

In the above figure, k, k' are the four-momenta of the initial and scattered leptons, P that of the initial hadron, and so forth. In this scattering process a hadron is probed with a space-like (virtual) photon, for which one can consider the Breit frame in which the photon four-momentum has a spatial component only. This shows that the spatial resolving power of the probing photon is of the order $\lambda \approx 1/Q$. Roughly speaking, one probes a nucleus (1–10 fm) with $Q \approx 10$–100 MeV, baryon or meson structure (with sizes in the order of 1 fm) with $Q \approx 0.1$–1 GeV and deep into the nucleon (<0.1 fm) with $Q > 2$ GeV. Electroweak interactions with the constituents of the hadrons, namely the quarks, are known. This opens the way to study how quarks are embedded in hadrons (for example in leptoproduction or in the Drell–Yan process, $A(P_A)B(P_B) \to \ell(k)\bar{\ell}(k')X$) or to study how quarks fragment into hadrons (in leptoproduction and e^+e^- annihilation into hadrons).

For inclusive unpolarized electron scattering the cross section, assuming one-photon exchange, is given by

$$E' \frac{d\sigma}{d^3k'} = \frac{1}{s - M^2} \frac{\alpha^2}{Q^4} L^{(S)}_{\mu\nu} 2MW^{\mu\nu}, \tag{7.1}$$

where $s = (k + P)^2$, $L^{(S)}_{\mu\nu}$ is the symmetric lepton tensor,

$$L^{(S)}_{\mu\nu}(k, k') = \mathrm{Tr}[\gamma_\mu(\slashed{k}' + m)\gamma_\nu(\slashed{k} + m)] = 2k_\mu k'_\nu + 2k_\nu k'_\mu - Q^2 g_{\mu\nu}, \tag{7.2}$$

and $W_{\mu\nu}$ is the hadron tensor, which contains the information on the hadronic part of the scattering process:

$$2M\, W_{\mu\nu}(P, q) = \frac{1}{2\pi} \sum_n \int \frac{d^3 P_n}{(2\pi)^3\, 2E_n}\, \langle P|J_\mu^\dagger(0)|P_n\rangle$$

$$\times \langle P_n|J_\nu(0)|P\rangle\, (2\pi)^4\, \delta^4(P + q - P_n), \tag{7.3}$$

where $|P\rangle$ represents a target with four-momentum P, and the sum is over intermediate hadronic states.

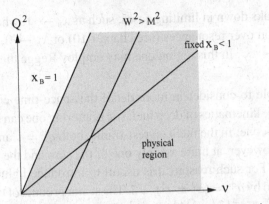

Figure 7.1. The physical region in deep inelastic scattering.

Using $(2\pi)^4\delta^4(P + q - P_n) = \int d^4x \, \exp(iP \cdot x + iq \cdot x - iP_n \cdot x)$, shifting the argument of the current, $J_\mu(x) = \exp(iP_{op} \cdot x)J_\mu(0)\exp(-iP_{op} \cdot x)$ and using completeness for the intermediate states, the hadron tensor can be written as the expectation value of the product of currents: $J_\mu(x)J_\nu(0)$. If a second term $\propto J_\nu(0)J_\mu^\dagger(0)\delta^4(P - q - P_n)$ is added, which is zero in the physical region ($\nu > 0$) because of the spectral conditions of the intermediate states n ($P_n^0 > M$), then after a similar procedure one can combine the terms to give

$$2M \, W_{\mu\nu} = \frac{1}{2\pi} \int d^4x \, e^{iq \cdot x} \langle P|[J_\mu^\dagger(x), J_\nu(0)]|P\rangle. \qquad (7.4)$$

This expression remains valid when electron polarization is included provided that summing and averaging over spins is understood.

What is probed in leptoproduction? In the reaction $e + H \to e' + X$ the unobserved final state X can be the target (elastic scattering) or an excitation. In a plot of the two independent variables ν and Q^2, see figure 7.1, elastic scattering corresponds to the line $\nu = Q^2/2M$, where M is the mass of the target. The behaviour of the cross section along this line, where the ratio $x_B = 1$, is proportional to the square of a form factor: see chapter 2. Exciting the nucleon gives rise to inelastic contributions in the cross section at $\nu > Q^2/2M$, starting at the threshold $W = M + m_\pi$. Note that as a function of x_B any resonance contribution will move closer to the elastic limit when Q^2 increases: see chapter 10. When Q^2 and ν are both large enough, the cross section becomes the incoherent sum of elastic scattering off the point-like constituents of the nucleon. This is known as the deep inelastic scattering region, in which one finds (approximate) Bjorken scaling. In this simple picture the structure functions become functions of one kinematic variable, x_B, that is identified with the momentum fraction of the struck quark in the nucleon, enabling measurement of quark distributions. We will make this explicit below.

The picture breaks down in limiting cases, such as $x_B \to 1$, where it becomes dual to the summation over resonances (see chapter 10) or $x_B \to 0$, corresponding, for fixed Q^2, to $\nu \to \infty$. In this region one may employ Regge theory (see chapters 5 and 8).

It is also possible to consider in more detail the space-time correlations that are probed. From the kinematics of deep inelastic scattering one can see that the process probes the light-cone. In the nucleon rest-frame, both $q^0 = \nu$ and $q^3 = \sqrt{Q^2 + \nu^2}$ go to infinity. However, at finite x_B only one of the sum and the difference of them goes to infinity. For such reasons it is useful to introduce light-cone coordinates. These are defined by $a^\pm = (a^0 \pm a^3)/\sqrt{2}$. The scalar product of two vectors is given by $a \cdot b = a^+ b^- + a^- b^+ - a^1 b^1 - a^2 b^2$. Choosing q along the negative z-axis, $q^- = (\nu + |q|)/\sqrt{2} \to \infty$ and $q^+ = (\nu - |q|)/\sqrt{2} \approx -M x_B/\sqrt{2}$. In the hadron tensor, which involves a Fourier transform of the product of currents, this corresponds to $|x^+| \approx 1/q^- \to 0$ and $|x^-| \approx 1/|q^+| \to 1/M x_B$. Thus, although the distances and times probed depend on x_B and are not necessarily small, $x^2 = x^+ x^- - x_\perp^2 \approx -x_\perp^2 \leq 0$. On the other hand, causality requires that $x^2 \geq 0$, so that deep inelastic scattering probes the light-cone, $x^2 \approx 0$.

7.2 Structure functions and cross section

The simplest thing one can do with the hadron tensor is to express it in standard tensors and structure functions that depend on the invariants. Instead of the traditional choice of tensors, $g_{\mu\nu}$, $P_\mu P_\nu$ and $\epsilon_{\mu\nu\rho\sigma} q^\rho P^\sigma$ multiplying structure functions W_1, W_2 and W_3 that depend on ν and Q^2, we go immediately to a dimensionless representation. First define a Cartesian basis of vectors [3], starting with the natural space-like momentum defined by q. Using the target-hadron momentum P^μ, one can construct a four-vector $\tilde{P}^\mu = P^\mu - (P \cdot q/q^2) q^\mu$, orthogonal to P and q, which is time-like with length $\tilde{P}^2 = \kappa P \cdot q$, where

$$\kappa = 1 + \frac{M^2 Q^2}{(P \cdot q)^2} = 1 + \frac{4 M^2 x_B^2}{Q^2}. \tag{7.5}$$

The quantity κ takes into account mass corrections proportional to M^2/Q^2. Thus define

$$Z^\mu \equiv -q^\mu, \tag{7.6}$$

$$T^\mu \equiv -\frac{q^2}{P \cdot q} \tilde{P}^\mu = q^\mu + 2x_B P^\mu. \tag{7.7}$$

For these vectors $Z^2 = -Q^2$ and $T^2 = \kappa Q^2$ and we shall often use the normalized vectors $\hat{z}^\mu = -\hat{q}^\mu = Z^\mu/Q$ and $\hat{t}^\mu = T^\mu/Q\sqrt{\kappa}$. With respect to these vectors one can also define transverse tensors,

$$g_\perp^{\mu\nu} \equiv g^{\mu\nu} + \hat{q}^\mu\hat{q}^\nu - \hat{t}^\mu\hat{t}^\nu, \tag{7.8}$$

$$\epsilon_\perp^{\mu\nu} \equiv \epsilon^{\mu\nu\rho\sigma}\hat{t}_\rho\hat{q}_\sigma. \tag{7.9}$$

To get the parametrization of hadronic tensors, such as in (7.4), including for generality also a spin four-vector S, we use the general symmetry property

$$W_{\mu\nu}(q, P, S) = W_{\nu\mu}(-q, P, S) \tag{7.10}$$

as well as properties following from hermiticity, parity and time-reversal invariance,

$$W_{\mu\nu}^*(q, P, S) = W_{\nu\mu}(q, P, S), \tag{7.11}$$

$$W_{\mu\nu}(q, P, S) = \overline{W}_{\nu\mu}(\bar{q}, \bar{P}, -\bar{S}) \quad \text{(parity)}, \tag{7.12}$$

$$W_{\mu\nu}^*(q, P, S) = \overline{W}_{\mu\nu}(\bar{q}, \bar{P}, \bar{S}) \quad \text{(time reversal)}, \tag{7.13}$$

where $\bar{p} = (p^0, -\boldsymbol{p})$. Finally we use current conservation implying $q^\mu W_{\mu\nu} = W_{\mu\nu}q^\nu = 0$. For inclusive unpolarized leptoproduction the most general form for the symmetric tensor is

$$M\, W^{\mu\nu\,(S)}(P, q) = \left(\frac{q^\mu q^\nu}{q^2} - g^{\mu\nu}\right) F_1(x_B, Q^2) + \frac{\tilde{P}^\mu\tilde{P}^\nu}{P\cdot q} F_2(x_B, Q^2)$$

$$\equiv -g_\perp^{\mu\nu}\, \underbrace{F_1(x_B, Q^2)}_{F_T} + \hat{t}^\mu\hat{t}^\nu \underbrace{\left(-F_1 + \frac{\kappa}{2x_B} F_2\right)}_{F_L}, \tag{7.14}$$

where the structure functions F_1, F_2, or equivalently the transverse and longitudinal structure functions, F_T and F_L, depend only on the variables Q^2 and x_B. This is the structure for the electromagnetic part of the electroweak interaction. For the weak W- or Z-exchange part, both vector and axial-vector currents with different parity behaviour occur. In that case the antisymmetric tensor

$$M\, W^{\mu\nu\,(A)}(q, P) = \frac{i\epsilon^{\mu\nu\rho\sigma} P_\rho q_\sigma}{(P\cdot q)} F_3(x_B, Q^2)$$

$$= i\kappa\,\epsilon_\perp^{\mu\nu} F_3(x_B, Q^2) \tag{7.15}$$

is also allowed. This arises from that part of the tensor in which one of the currents in the product is a vector current and the other is an axial-vector current.

The cross section is obtained from the contraction of the lepton and hadron tensors. It is convenient to expand the lepton momenta k and $k' = k - q$ in $\hat{t}$, $\hat{z}$ and a perpendicular component using the scaling variable $y = P\cdot q/P\cdot k$. (In the target rest-frame this reduces to $y = \nu/E$.) The result, including target-mass

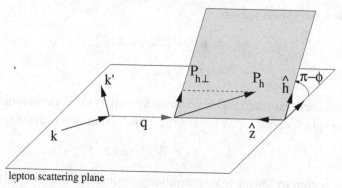

Figure 7.2. Kinematics for lepton–hadron scattering. Transverse directions indicated with a $\perp$ index are orthogonal to P and q, for example the orthogonal component of the momentum of a produced hadron has been indicated. Similarly one can consider the orthogonal component of the spin vector of the target.

corrections, is

$$k^\mu = \frac{2-y}{y}\frac{1}{\kappa}T^\mu - \frac{1}{2}Z^\mu + k_\perp^\mu$$

$$= \frac{Q}{2}\hat{q}^\mu + \frac{(2-y)}{2y}\frac{Q}{\sqrt{\kappa}}\hat{t}^\mu + \frac{\sqrt{1-y+\frac{1}{4}(1-\kappa)y^2}}{y}\frac{Q}{\sqrt{\kappa}}\hat{\ell}^\mu$$

$$\xrightarrow{Q^2\to\infty} \frac{Q}{2}\hat{q}^\mu + \frac{(2-y)Q}{2y}\hat{t}^\mu + \frac{Q\sqrt{1-y}}{y}\hat{\ell}^\mu, \tag{7.16}$$

where $\hat{\ell}^\mu = k_\perp^\mu/|k_\perp|$ is a space-like unit vector in the perpendicular direction lying in the lepton scattering plane. The kinematics in the frame where the virtual photon and the target are collinear, including the target rest-frame, is illustrated in figure 7.2. With this definition of $\hat{\ell}$ and neglecting mass corrections ($\kappa = 1$), for unpolarized leptons we obtain the symmetric lepton tensor

$$L^{\mu\nu(S)} = \frac{Q^2}{y^2}\Big[-2\left(1-y+\tfrac{1}{2}y^2\right)g_\perp^{\mu\nu} + 4(1-y)\hat{t}^\mu\hat{t}^\nu$$

$$+ 4(1-y)\left(\hat{\ell}^\mu\hat{\ell}^\nu + \tfrac{1}{2}g_\perp^{\mu\nu}\right)$$

$$+ 2(2-y)\sqrt{1-y}\,(\hat{t}^\mu\hat{\ell}^\nu + \hat{t}^\nu\hat{\ell}^\mu)\Big]. \tag{7.17}$$

For electromagnetic scattering the explicit contraction of lepton and hadron tensors gives the result

$$\frac{d\sigma^{ep}}{dx_B dy} = \frac{4\pi\,\alpha^2\,x_B\,s}{Q^4}\Big[\left(1-y+\tfrac{1}{2}y^2\right)F_T(x_B,Q^2) + (1-y)F_L(x_B,Q^2)\Big]$$

$$= \frac{2\pi\,\alpha^2\,s}{Q^4}[(1-y)F_2(x_B,Q^2) + x_B\,y^2\,F_1(x_B,Q^2)]. \tag{7.18}$$

We have used the known photon coupling to the lepton and parametrized in the structure functions our ignorance of hadron structure. The fact that we know how the photon interacts with the quark constituents of the hadrons will be used later to relate the structure functions to quark properties. In the same way it is known how the Z^0 and W couple to quarks. To describe weak interactions the antisymmetric part of the lepton tensor is needed also. We shall encounter this when we discuss polarization.

7.3 Virtual photon cross sections

The tensor $W_{\mu\nu}$ also appears in the total cross section for $\gamma^* H \to$ everything, where γ^* indicates a virtual photon. For a given virtuality Q^2 of the photon, this cross section depends on only one variable, $W^2 = (P + q)^2$, or equivalently on the variable $\nu = P \cdot q/M$,

$$\sigma^{\gamma^* H}(\nu) = \frac{4\pi^2 \alpha}{K} \epsilon^{\mu*} W_{\mu\nu} \epsilon^\nu, \tag{7.19}$$

where $4MK$ is the photon flux factor. This flux factor is physical only for real photons ($Q^2 = 0$). For $Q^2 \neq 0$ several conventions have been used: (i) define K by $4MK = 4\sqrt{(p \cdot q)^2 - p^2 q^2}$, that is $K = \sqrt{\nu^2 + Q^2}$; (ii) take the real photon result $4MK = 4P \cdot q$ or $K = \nu$; (iii) use the result $4MK = 2(W^2 - M^2)$ for a massless photon and equate W^2 to the invariant mass in the case of a virtual photon, that is $W^2 = 2P \cdot q + M^2 - Q^2$ or $K = \nu - Q^2/2M$.

Being a total cross section for virtual photoabsorption, the hadronic tensor is related to the forward virtual Compton amplitude through the optical theorem,

$$W_{\mu\nu} = \frac{1}{\pi} \operatorname{Im} T_{\mu\nu}, \tag{7.20}$$

where

$$2M T_{\mu\nu}(P, q) = i \int d^4x \, e^{iq \cdot x} \langle P | T J_\mu(x) J_\nu(0) | P \rangle. \tag{7.21}$$

Using the photon polarization vectors ϵ_α^μ, where α indicates one of the polarization directions orthogonal to q^μ,

$$\epsilon_\pm^\mu = \sqrt{\tfrac{1}{2}} (0, \mp 1, -i, 0) = \mp \sqrt{\tfrac{1}{2}} (\epsilon_x \pm i \, \epsilon_y), \tag{7.22}$$

$$\epsilon_L^\mu = \frac{1}{\sqrt{Q^2}} (q^3, 0, 0, q^0), \tag{7.23}$$

one gets two transverse structure functions and one longitudinal structure function, $F_\alpha = \epsilon_\alpha^{\mu*} M W_{\mu\nu} \epsilon_\alpha^\nu$. Because they are proportional to cross sections for virtual photons, the structure functions F_+, F_- and F_L are non-negative. We have

$$F_T = \frac{1}{2}(F_+ + F_-) = F_1, \tag{7.24}$$

$$F_L = \frac{F_2}{2x_B} - F_1, \tag{7.25}$$

$$F_3 = F_+ - F_-. \tag{7.26}$$

7.4 Symmetry properties of the structure functions

For the virtual Compton scattering process, $\gamma^*(q) + H(P) \to \gamma^*(q') + H(P')$, the amplitude $T_{\mu\nu}(P, q, q')$ can be expanded in terms of amplitudes T_i that depend on the invariants in the scattering process. These invariants are two of the three Mandelstam variables for the $\gamma^* N$ process, $s = (P + q)^2$, $t = (P - q)^2$ and $u = (P - q')^2$. Note that $s + t + u = 2M^2 - 2Q^2$.

In the forward direction only two such amplitudes survive for spin-averaged scattering when $Q^2 \neq 0$. First we review the symmetries and analytic properties of the amplitudes $T_{1,2}$ and the structure functions $W_{1,2}$. Crossing symmetry relates the amplitudes

$$T^{ab \to cd}(p_a, p_b, p_c, p_d) = T^{\bar{c}b \to \bar{a}d}(-p_c, p_b, -p_a, p_d). \tag{7.27}$$

For the virtual Compton amplitude, $T_1^{\gamma^* \bar{H} \to \gamma^* \bar{H}}(u, s) = T_1^{\gamma^* H \to \gamma^* H}(s, u)$. For the elastic $\gamma^* H$ and $\gamma^* \bar{H}$ amplitudes, the crossing properties imply $T_1(s, u) = T_1(u, s)$. From the optical theorem the relation between the total cross section for $\gamma^* H \to X$ and the imaginary part of the forward elastic $\gamma^* H \to \gamma^* H$ amplitude is the discontinuity over the cut in the physical region, namely the imaginary part of the forward amplitude T_1. For the forward amplitude $\nu = (s - u)/4M$ and T_1 is a symmetric function of ν. One has

$$\sigma_T^{\gamma^* H}(\nu) = \frac{4\pi\alpha}{K} \operatorname{Im} T_1(s + i\epsilon, u) = \frac{4\pi\alpha}{K} \operatorname{Im} T_1(\nu + i\epsilon), \tag{7.28}$$

$$\sigma_T^{\gamma^* \bar{H}}(\nu) = \frac{4\pi\alpha}{K} \operatorname{Im} T_1(u + i\epsilon, s) = \frac{4\pi\alpha}{K} \operatorname{Im} T_1(-\nu + i\epsilon)$$

$$= -\frac{4\pi\alpha}{K} \operatorname{Im} T_1(\nu + i\epsilon), \tag{7.29}$$

where the last equality follows from the symmetry result $T_1(\nu) = T_1(-\nu)$ and the fact that T_1 is a real analytic function, $T_1(\nu) = T_1^*(\nu^*)$. Note that as $W_{\mu\nu}$ is defined as the commutator of the currents, the cross section for the crossed part enters with a negative sign. As an analytic function of ν, however, one has

$$W_1(\nu, Q^2) = \frac{1}{\pi} \operatorname{Im} T_1(\nu + i\epsilon, Q^2) = -W_1(-\nu, Q^2), \tag{7.30}$$

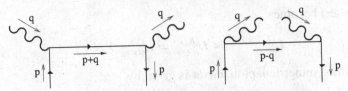

Figure 7.3. The Compton amplitude for a point fermion.

which can also be derived from the translation invariance properties of the commutator defining $W^{\mu\nu}$. The analytic behaviour and symmetry in $x_B = Q^2/2M\nu$ is exactly the same as that in ν, that is $W_1(x_B) = -W_1(-x_B)$.

A simple example in which all of these properties are illustrated is the forward Compton amplitude for scattering off a point fermion with mass m and charge $Q/e = e_f$, illustrated in figure 7.3. The amplitude T_1 is the coefficient of $g_{\mu\nu}$ in the amplitude $T_{\mu\nu}$,

$$2m\, T_{\mu\nu} = e_f^2 \frac{\bar{u}(p)\gamma_\mu(\not{p} + \not{q} + m)\gamma_\nu u(p)}{(p+q)^2 - m^2 + i\epsilon} + [\mu \leftrightarrow \nu,\, q \leftrightarrow -q]. \qquad (7.31)$$

This equals

$$2m\, T_1 = e_f^2 \left(\frac{s - u}{2(s - m^2 + i\epsilon)} + \frac{u - s}{2(u - m^2 + i\epsilon)} \right)$$

$$= e_f^2 \left(\frac{1}{1 - x + i\epsilon} + \frac{1}{1 + x + i\epsilon} \right), \qquad (7.32)$$

the imaginary part of which is the structure function for a point-like fermion,

$$2m\, W_1 = e_f^2 \left[\delta(1 - x_B) - \delta(1 + x_B) \right]. \qquad (7.33)$$

Analogous arguments can be applied to T_2 and W_2 [1].

7.5 Polarized leptoproduction

For polarized leptons in the initial state (7.1) becomes

$$L_{\mu\nu}^{(s)} = \mathrm{Tr}\left[\gamma_\mu(\not{k}' + m)\gamma_\nu(\not{k} + m)\frac{1 \pm \gamma_5 \not{s}}{2} \right]$$

$$= 2 k_\mu k_\nu' + 2 k_\mu' k_\nu - Q^2 g_{\mu\nu} \pm 2i\, m\, \epsilon_{\mu\nu\rho\sigma} q^\rho s^\sigma, \qquad (7.34)$$

where m is the lepton mass. Note that for light particles, or particles at high energies, the helicity states ($\hat{s} = \hat{k}$) become chirality eigenstates. For $L_{\mu\nu}$ the equivalence is easily seen because for $s^\mu = (|k|/m,\, E\hat{k}/m)$ ($s^2 = -1$ and $s \cdot p = 0$) in the limit $E \approx |k|$ one obtains the result $s^\mu \approx k^\mu/m$. Then the leptonic tensor for helicity states ($\lambda_e = \pm$) becomes

$$L_{\mu\nu}^{(\lambda_e = \pm 1/2)} \approx L_{\mu\nu}^{(R/L)} = L_{\mu\nu}^{(S)} + \lambda_e L_{\mu\nu}^{(A)}, \tag{7.35}$$

where the antisymmetric lepton tensor is given by

$$\begin{aligned} L_{\mu\nu}^{(A)}(k, k') &= \text{Tr}[\gamma_\mu \gamma_5 \slashed{k}' \gamma_\nu \slashed{k}] \\ &= 2i\, \epsilon_{\mu\nu\rho\sigma} q^\rho k^\sigma. \end{aligned} \tag{7.36}$$

Expanding in the Cartesian set $\hat{\imath}$, $\hat{2}$ and the vector $\hat{\ell}$ in the same way as for the symmetric part, the antisymmetric part of the lepton tensor has the form

$$L^{\mu\nu(A)} = \frac{Q^2}{y^2}\left[-i\,y(2-y)\,\epsilon_\perp^{\mu\nu} - 2i\,y\sqrt{1-y}\left(\hat{\imath}^\mu \epsilon_\perp^{\nu\rho} - \hat{\imath}^\nu \epsilon_\perp^{\mu\rho}\right)\hat{\ell}_\rho\right]. \tag{7.37}$$

Polarized leptons in deep inelastic scattering $\vec{e}p \to X$ can be used to probe the antisymmetric tensor for unpolarized hadrons, measuring the F_3 structure function. This contribution comes via the interference between the γ and the Z exchange terms.

When the target is polarized there are several more structure functions compared to the case of an unpolarized target. For a spin-$\frac{1}{2}$ particle the initial state is described by a two-dimensional spin-density matrix $\rho = \sum_\alpha |\alpha\rangle p_\alpha \langle\alpha|$ describing the probabilities p_α for a variety of spin possibilities. This density matrix is Hermitian with $\text{Tr}\,\rho = 1$. In the target rest-frame it can be expanded in terms of the unit matrix and the Pauli matrices,

$$\rho_{ss'} = \tfrac{1}{2}(1 + \boldsymbol{S} \cdot \boldsymbol{\sigma}_{ss'}), \tag{7.38}$$

where $\boldsymbol{S}$ is the spin vector. When $|\boldsymbol{S}| = 1$ there is only one state $|\alpha\rangle$ and $\rho^2 = \rho$. When $|\boldsymbol{S}| \leq 1$ there is an ensemble of states. The case $|\boldsymbol{S}| = 0$ is simply an average over spins, corresponding to an unpolarized ensemble. To include spin one, the hadron tensor can be generalized to a matrix in spin space, $\tilde{W}_{s's}^{\mu\nu}(q, P) \propto \langle P, s'|J^\mu|X\rangle\langle X|J^\nu|P, s\rangle$ depending only on the momenta, or one can look at the tensor $\sum_\alpha p_\alpha \tilde{W}_{\alpha\alpha}^{\mu\nu}(q, P)$. The latter is given by

$$W^{\mu\nu}(q, P, S) = \text{Tr}(\rho(P, S)\tilde{W}^{\mu\nu}(q, P)), \tag{7.39}$$

with the space-like spin vector S appearing linearly, satisfying $P \cdot S = 0$ in an arbitrary frame and having invariant length $-1 \leq S^2 \leq 0$. It is convenient to write

$$S^\mu = \frac{S_L}{M}\left(P^\mu - \frac{M^2}{P \cdot q}q^\mu\right) + S_\perp^\mu, \tag{7.40}$$

with

$$S_L \equiv \frac{M\,(S \cdot q)}{(P \cdot q)}. \tag{7.41}$$

For a pure state $S_L^2 + S_\perp^2 = 1$. Parity requires that the polarized part of the tensor, that is the part containing the spin vector, enters in an antisymmetric tensor of the form

$$M W^{\mu\nu(A)}(q, P, S) = S_L \frac{i\epsilon^{\mu\nu\rho\sigma} q_\rho P_\sigma}{(P \cdot q)} g_1 + \frac{M}{P \cdot q} i\epsilon^{\mu\nu\rho\sigma} q_\rho S_{\perp\sigma} (g_1 + g_2)$$

$$= -i S_L \epsilon_\perp^{\mu\nu} g_1 - i \frac{2M}{Q} \left[\hat{\imath}^\mu \epsilon_\perp^{\nu\rho} - \hat{\imath}^\nu \epsilon_\perp^{\mu\rho}\right] S_{\perp\rho} x_B (g_1 + g_2).$$

$$(7.42)$$

Equation (7.42) contains two structure functions, $g_1(x_B, Q^2)$ and $g_2(x_B, Q^2)$. The combination $g_T \equiv g_1 + g_2$ is also used. The resulting cross section is

$$\frac{d\Delta\sigma_{LL}}{dx_B \, dy} = \frac{4\pi \alpha^2}{Q^2} \lambda_e \left[S_L (2 - y) g_1 - |S_\perp| \cos\phi_S^\ell \frac{2M}{Q} \sqrt{1 - y} \, x_B (g_1 + g_2) \right].$$

$$(7.43)$$

In all of the above formulae mass corrections proportional to M^2/Q^2 have been neglected.

A special case of inclusive scattering is elastic scattering. Then the final state four-momentum is $P' = P + q$ and $(P + q)^2 = M^2$, that is $x_B = 1$. The formalism for inclusive leptoproduction can still be used but the hadron tensor becomes

$$2M W_{\mu\nu}(q, P) = \underbrace{\langle P|J_\mu(0)|P'\rangle \langle P'|J_\nu(0)|P\rangle}_{H_{\mu\nu}(P;P')} \frac{1}{Q^2} \delta(1 - x_B). \qquad (7.44)$$

7.6 The parton model

7.6.1 The intuitive approach

In the intuitive derivation of the parton model the γ^*-quark cross section is convoluted with a momentum distribution of quarks in the nucleon. The $\gamma^* q$ cross section is given by

$$\hat{\sigma}(\gamma^* q) = \frac{4\pi^2 \alpha}{2p \cdot q} \epsilon_\mu^* w^{\mu\nu} \epsilon_\nu,$$

$$w_{\mu\nu}(p, q) = \frac{1}{2M} \left[\left(\frac{q_\mu q_\nu}{q^2} - g_{\mu\nu}\right) Q^2 + 4 \tilde{p}_\mu \tilde{p}_\nu \right] \delta(2 p \cdot q + q^2), \quad (7.45)$$

which can be separated into the transverse and longitudinal cross sections:

$$\hat{\sigma}_T(\gamma^* q) = 4\pi^2 \alpha \, e_q^2 \, \delta(2p \cdot q - Q^2), \qquad (7.46)$$

$$\hat{\sigma}_L(\gamma^* q) = 4\pi^2 \alpha \, e_q^2 \, \frac{4m_q^2}{Q^2} \delta(2p \cdot q - Q^2) \ll \hat{\sigma}_T. \qquad (7.47)$$

Note that there are, in principle, ambiguities here because of the flux factor for virtual photons and that $\hat{\sigma}_L(\gamma^*q) \ll \hat{\sigma}_T$ at large Q^2.

This parton cross section is then folded with the probability function for finding partons in the target. For this purpose it is convenient to give the explicit momenta as light-cone components, $p = [p^-, p^+, p_\perp]$, where $p^\pm = (p^0 \pm p^3)/\sqrt{2}$ or $p = p^- n_- + p^+ n_+ + p_T$ in terms of two light-like vectors satisfying $n_-^2 = n_+^2 = 1$ and $n_+ \cdot n_- = 0$. Thus $p^\pm = p \cdot n_\mp$. Note that $n_\pm$ are not unique. The light-cone expansion for the external vectors P and q is

$$\left.\begin{array}{c} q^2 = -Q^2 \\ P^2 = M^2 \\ 2P \cdot q = \dfrac{Q^2}{x_B} \end{array}\right\} \longleftrightarrow \left\{\begin{array}{l} q = \sqrt{\tfrac{1}{2}}Q\, n_- - \sqrt{\tfrac{1}{2}}Q\, n_+ \\[2mm] P = \dfrac{x_B M^2}{Q\sqrt{2}} n_- + \dfrac{Q}{x_B \sqrt{2}} n_+ \end{array}\right.$$

The representation with light-like vectors shows that when Q^2 becomes large, the nucleon momentum is on the scale of Q, that is it is light-like. The hard momentum has both components proportional to Q, but this is not the case for P and $P^- \ll q^-$.

A scaling variable that includes finite-Q effects is often used. This is the Nachtmann variable $x_N = -q^+/P^+$. Recalling that $x_B = Q^2/2P \cdot q$,

$$x_N = \frac{2x_B}{1 + \sqrt{1 + 4M^2 x_B^2/Q^2}}. \tag{7.48}$$

Compared with the hard scale Q the parton momentum is also light-like. It is useful to expand $p = p^- n_- + x P^+ n_+ + p_T$, where the light-cone momentum fraction $x \equiv p^+/P^+$ has been introduced. It is straightforward to show that on-shell ($p^2 = m^2$)

$$p^- = \frac{m^2 + p_T^2}{2p^+} = \frac{m_\perp^2}{2x\,P^+}, \tag{7.49}$$

while in a hadron

$$p^- = \frac{2p \cdot P - x M^2}{2P^+},$$

$$p_T^2 = x(1-x) M^2 - x M_R^2 - (1-x)p^2,$$

where $M_R^2 = (P - p)^2$.

Under the assumption that all invariants $p \cdot P \sim M_R^2 \sim p^2 \sim P^2 = M^2$, then for a quark in a hadron the expansion in terms of $n_\pm$ gives $p^+ \sim P^+ \sim Q$, while $p^- \sim M^2/Q$ and $p_T^2 \sim M^2$. This is sufficient to derive the parton model results.

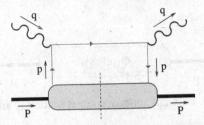

Figure 7.4. The handbag diagram for inclusive deep inelastic scattering off a hadronic target.

The cross section for $\gamma^* q$ elastic scattering can be rewritten as

$$\hat{\sigma}_T = \frac{4\pi^2 \alpha e_q^2}{Q^2} \delta(1 - x_p), \tag{7.50}$$

where $x_p = -q^+/p^+$.

Introducing probabilities $f_i(x)$ for finding partons carrying fraction $x = p^+/P^+ = x_B/x_p$ of the target light-cone momentum, leads to*

$$\sigma_T = \sum_i e_i^2 \int dx \, f_i(x) \frac{4\pi^2 \alpha}{Q^2} \delta\left(1 - \frac{x_B}{x}\right)$$

$$= \frac{4\pi^2 \alpha}{Q^2} \sum_i e_i^2 \, x_B \, f_i(x_B). \tag{7.51}$$

Comparing with

$$\sigma_T = \frac{8\pi^2 \alpha}{Q^2} x_B \, F_1, \tag{7.52}$$

we get

$$F_1(x_B) = \frac{1}{2} \sum_i e_i^2 \, f_i(x_B). \tag{7.53}$$

As $\hat{\sigma}_L \propto 1/Q^2 \to 0$ as $Q^2 \to \infty$, one obtains $F_L = 0$ and the Callan–Gross relation,

$$F_2(x_B) = 2 x_B \, F_1(x_B). \tag{7.54}$$

7.6.2 *The diagrammatic approach*

Another way in which the parton model is obtained is by considering the quark handbag diagram of figure 7.4 and its antiquark equivalent. These diagrams are

---

* Note that the probability actually involves $f_i(x) \, dx/x$, but for the cross section we need also to include counting rates. This requires that we need the cross section multiplied by flux factors. The ratio of the flux factors for quarks and hadrons is $p \cdot q/P \cdot q \approx p^+/P^+ = x$. Hence we need to weight the cross section with $f_i(x) \, dx$.

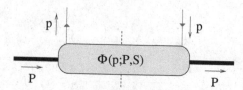

Figure 7.5. Forward antiquark-target scattering amplitude.

the leading ones out of a full set in which the connection to hadrons is left as an unknown quantity [4,5]. The basic expression for the quark part of the handbag diagram is

$$
2M\, W^{\mu\nu}(P,q)
$$
$$
= \sum_q e_q^2 \int dp^-\, dp^+\, d^2p_\perp \,\mathrm{Tr}\,(\Phi(p)\gamma^\mu(\slashed{p}+\slashed{q}+m)\gamma^\nu)\,\delta((p+q)^2 - m^2)
$$
$$
\approx \sum_q e_q^2 \int dp^-\, dp^+\, d^2p_\perp \,\mathrm{Tr}\left(\Phi(p)\gamma^\mu \frac{q^-\gamma^+}{2q^-}\gamma^\nu\right)\delta(p^+ + q^+)
$$
$$
\approx -g_\perp^{\mu\nu}\frac{1}{2}\sum_q e_q^2 \int dp^-\, d^2p_\perp \,\mathrm{Tr}(\gamma^+\,\Phi(p))\Big|_{p^+=x_B P^+} + \cdots, \tag{7.55}
$$

where $\Phi(p)$ is the forward antiquark-target scattering amplitude,

$$
\Phi_{ij}(p,P,S) = \frac{1}{(2\pi)^4}\int d^4\xi\, e^{i\,p\cdot\xi}\,\langle P, S|\overline{\psi}_j(0)\psi_i(\xi)|P,S\rangle, \tag{7.56}
$$

diagrammatically represented by figure 7.5. Including the antiquark part and comparing with the general form of the hadronic tensor, then

$$
2\dot{F}_1(x_B) = 2M\, W_1(x_B, Q^2) = \sum_q e_q^2\,[q(x_B) + \bar{q}(x_B)], \tag{7.57}
$$

with

$$
q(x_B) = \frac{1}{4\pi}\int d\xi^-\, e^{+ix_B P^+\xi^-}\,\langle P, S|\overline{\psi}(0)\gamma^+\psi(\xi)|P,S\rangle\Big|_{\xi^+=\xi_\perp=0}, \tag{7.58}
$$
$$
\bar{q}(x_B) = \frac{1}{4\pi}\int d\xi^-\, e^{-ix_B P^+\xi^-}\,\langle P, S|\overline{\psi}(0)\gamma^+\psi(\xi)|P,S\rangle\Big|_{\xi^+=\xi_\perp=0}, \tag{7.59}
$$

satisfying $\bar{q}(x_B) = -q(-x_B)$. As expected, the result is a light-cone correlation function of quark fields.

7.6.3 *The operator in coordinate-space*

The parton result for the structure functions can also be derived by inserting free currents in the hadron tensor for the current commutator and using the expression for the free field commutator.

By using the anticommutation relations for free quark fields, given by $\{\psi(\xi), \overline{\psi}(0)\} = \not{\partial}\delta(\xi^2)\epsilon(\xi^0)/2\pi$ the $g_{\mu\nu}$ contribution in the current–current commutator for quarks can be derived. It is

$$[J_\mu(\xi), J_\nu(0)] = [: \overline{\psi}(\xi)\gamma_\mu\psi(\xi) :, : \overline{\psi}(0)\gamma_\nu\psi(0) :]$$

$$= \frac{-g_{\mu\nu}}{2\pi}[\partial_\rho \, \delta(\xi^2)\,\epsilon(\xi^0)] : \overline{\psi}(\xi)\gamma^\rho\psi(0) - \overline{\psi}(0)\gamma^\rho\psi(\xi) :. \quad (7.60)$$

An important feature, evident in the free-current commutator, is the light-cone dominance. By sandwiching the commutator between physical states and taking the Fourier transform, it is a straightforward calculation to obtain the same result for the hadron tensor as in the diagrammatic approach above. Details can be found in [6].

7.6.4 *Flavour dependence*

The explicit flavour and spin dependence of the structure functions in electroweak processes depend on the probe being a γ, Z^0 or $W^\pm$ boson. However, the currents are known in terms of the quark fields. Omitting the coupling constants e or $\sqrt{G_F}$, the standard-model currents coupling to fermions are

$$J_\mu^{(\gamma)} = : \overline{\psi}(x)\, Q\, \gamma_\mu\, \psi(x) :, \quad (7.61)$$

$$J_\mu^{(Z)} = : \overline{\psi}(x)\left(I_W^3 - Q \sin^2\theta_W\right)\gamma_{\mu L}\,\psi(x) : - : \overline{\psi}(x)\, Q \sin^2\theta_W\,\gamma_{\mu R}\,\psi(x) :$$

$$= : \overline{\psi}(x)\left(I_W^3 - 2\,Q \sin^2\theta_W\right)\gamma_\mu\,\psi(x) : - : \overline{\psi}(x)\,I_W^3\,\gamma_\mu\gamma_5\,\psi(x) :, \quad (7.62)$$

$$J_\mu^{(W)} = : \overline{\psi}(x)\,I_W^\pm\,\gamma_{\mu L}\,\psi(x) :, \quad (7.63)$$

where $\gamma_{\mu R/L} = \gamma_\mu(1 \pm \gamma_5)$.

In terms of the quark distribution functions $u_p(x)$, $d_p(x)$, etc., the one-photon exchange contribution to ep scattering is

$$\frac{F_2^{ep}(x)}{x} = 2\,F_1^{ep}(x)$$

$$= \tfrac{4}{9}\,(u_p(x) + \overline{u}_p(x)) + \tfrac{1}{9}\,(d_p(x) + \overline{d}_p(x)) + \tfrac{1}{9}\,(s_p(x) + \overline{s}_p(x)) + \cdots$$

$$\equiv \tfrac{4}{9}\,(u(x) + \overline{u}(x)) + \tfrac{1}{9}\,(d(x) + \overline{d}(x)) + \tfrac{1}{9}\,(s(x) + \overline{s}(x)) + \cdots, \quad (7.64)$$

where the last line exhibits the convention of using the proton as the reference hadron for distribution functions. Using flavour symmetry, $u_p = d_n, d_p = u_n, s_p = s_n$, then for en scattering

$$\frac{F_2^{en}(x)}{x} = 2 F_1^{en}(x)$$

$$= \tfrac{1}{9}(u(x) + \bar{u}(x)) + \tfrac{4}{9}(d(x) + \bar{d}(x)) + \tfrac{1}{9}(s(x) + \bar{s}(x)) + \cdots . \quad (7.65)$$

As the difference between quarks and antiquarks contributes to the quantum numbers, it is convenient to divide the quark distribution into a valence part and a sea part

$$q(x) = q_v(x) + q_s(x), \quad (7.66)$$

where

$$q_v(x) \equiv q(x) - \bar{q}(x). \quad (7.67)$$

The quark distributions are positive definite so, for instance,

$$\frac{1}{4} \leq \frac{F_2^{en}}{F_2^{ep}} = \frac{(u + \bar{u}) + 4(d + \bar{d}) + (s + \bar{s}) + \cdots}{4(u + \bar{u}) + (d + \bar{d}) + (s + \bar{s}) + \cdots} \leq 4. \quad (7.68)$$

Near $x_B \approx 0$ the experimental ratio is about 1, indicating dominance of sea quarks. Near $x_B \approx 1$ the valence quarks dominate. Naively one might expect $u = 2d$ and all others zero, that is a ratio of $\tfrac{2}{3}$. The experimentally observed limit for $x_B \to 1$ is consistent with tending towards $\tfrac{1}{4}$, the lower limit, which is reached for $d \ll u$, that is dominance of the u-quark in the proton (and therefore the d-quark in the neutron).

It is clear that in order to determine the quark distributions several processes are needed. At present the various quark and gluon distributions are known over a wide range of x_B and Q^2, for example see figure 7.6. For compilations we refer to [8].

7.7 Properties of quark distributions

7.7.1 Interpretation as densities

To be convinced that the above expressions for $q(x)$ and $\bar{q}(x)$ can be interpreted as momentum densities it is necessary to realize that $\overline{\psi}(\xi)\gamma^+\psi(0) = \sqrt{2}\,\psi_+^\dagger(\xi)\psi_+(0)$, where $\psi_\pm = P_\pm \psi$ are projections obtained with projection operators onto quark

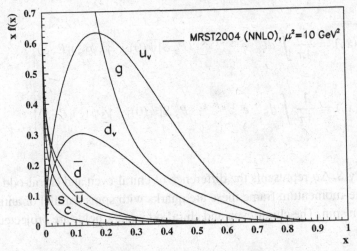

Figure 7.6. Distributions of x times the unpolarized parton distributions $f(x)$ (where $f = u_v, D_v, \bar{u}, \bar{d}, s, c, g$) using the NNLO MRST2004 parametrization [7] at a scale $\mu^2 = 10 \text{ GeV}^2$.

states [9], with $P_\pm = \frac{1}{2}\gamma^\mp\gamma^\pm$. A complete set of states can then be inserted to give

$$q(x) = \int \frac{d\xi^-}{2\pi\sqrt{2}} e^{ip\cdot\xi} \langle P, S|\psi_+^\dagger(0)\psi_+(\xi)|P, S\rangle\Big|_{\xi^+=\xi_T=0}$$

$$= \sqrt{\tfrac{1}{2}} \sum_n |\langle P_n|\psi_+|P\rangle|^2 \, \delta(P_n^+ - (1-x)P^+), \tag{7.69}$$

which represents the probability that a quark is annihilated from $|P\rangle$ giving a state $|n\rangle$ with $P_n^+ = (1-x)P^+$. Since $P_n^+ \geq 0$ one sees that $x \leq 1$. From the antiquark distribution $\bar{q}(x)$ and its relation to $-q(-x)$ one obtains $x \geq -1$, thus showing that the support of the functions is $-1 \leq x \leq 1$ [10,11].

7.7.2 Polarized parton densities

Analogous to the unpolarized structure functions, for the polarized structure functions

$$2\,g_1(x_B) = \sum_q e_q^2 \left[\Delta q(x_B) + \Delta\bar{q}(x_B)\right], \tag{7.70}$$

where

$$S_L \Delta q(x_B) = \frac{1}{4\pi} \int d\xi^- \, e^{+ix_B P^+ \xi^-} \langle P, S | \overline{\psi}(0) \gamma^+ \gamma_5 \psi(\xi) | P, S \rangle \Big|_{\xi^+ = \xi_\perp = 0,}$$

(7.71)

$$S_L \Delta \overline{q}(x_B) = \frac{1}{4\pi} \int d\xi^- \, e^{-ix_B P^+ \xi^-} \langle P, S | \overline{\psi}(0) \gamma^+ \gamma_5 \psi(\xi) | P, S \rangle \Big|_{\xi^+ = \xi_\perp = 0.}$$

(7.72)

The quantity $S_L \Delta q$ represents the difference of chiral-even and chiral-odd quarks; in an infinite-momentum frame these are quarks with spins parallel or antiparallel to the proton spin. The chiral-even and chiral-odd quark fields are projected out by the operators

$$P_{R/L} = \tfrac{1}{2}(1 \pm \gamma_5)$$

(7.73)

that commute with the operators $P_\pm$. Hence $q_R(x_B)$ and $q_L(x_B)$ are obtained, where $q(x_B) = q_R(x_B) + q_L(x_B)$ and $\Delta q(x_B) = q_R(x_B) - q_L(x_B)$.

7.7.3 Sum rules

As probability distributions, the quark distribution functions satisfy a number of obvious sum rules, such as

$$\int_0^1 dx \, [u(x) - \overline{u}(x)] = n_u = 2,$$

(7.74)

$$\int_0^1 dx \, [d(x) - \overline{d}(x)] = n_d = 1,$$

(7.75)

$$\int_0^1 dx \, [s(x) - \overline{s}(x)] = n_s = 0,$$

(7.76)

corresponding to the net number of each of these flavour species in the proton. On the basis of this one finds a number of sum rules for the structure functions, such as the Gottfried sum rule [12] that is based on a flavour-symmetric sea distribution: $\overline{u}(x) = \overline{d}(x)$,

$$S_G = \int_0^1 \frac{dx}{x} \left[F_2^{ep}(x) - F_2^{en}(x) \right]$$

$$= \tfrac{1}{3} \int_0^1 dx [u(x) - d(x) + \overline{u}(x) - \overline{d}(x)] = \tfrac{1}{3}.$$

(7.77)

The experimental result [13] is 0.240 ± 0.016, indicating that $\overline{u}(x) \neq \overline{d}(x)$.

If the explicit quark distributions are found, the quantity obtained by weighting the sum over all quarks with the momentum can be determined:

$$\int_0^1 dx\, x\, \Sigma(x) \equiv \int_0^1 dx\, x\, [u(x) + \bar{u}(x) + d(x) + \bar{d}(x) + s(x) + \bar{s}(x) + \cdots] = \epsilon_q. \tag{7.78}$$

This represents the total momentum fraction of the proton carried by quarks and antiquarks and experimentally is substantially less than 1. This deficit is attributed to the momentum carried by gluons.

Estimates for the polarized structure functions can be obtained using the naive flavour-spin structure of the proton based on SU(6) symmetry:

$$|p \uparrow\rangle = \sqrt{\tfrac{1}{18}}(2\, u_\uparrow u_\uparrow d_\downarrow - u_\uparrow u_\downarrow d_\uparrow - u_\downarrow u_\uparrow d_\uparrow$$
$$+ 2\, d_\uparrow u_\uparrow u_\downarrow - d_\uparrow u_\downarrow u_\uparrow - d_\downarrow u_\uparrow u_\uparrow$$
$$+ 2\, u_\uparrow d_\uparrow u_\downarrow - u_\uparrow d_\downarrow u_\uparrow - u_\downarrow d_\uparrow u_\uparrow). \tag{7.79}$$

From this wave function, the results in terms of a normalized one-quark distribution $q(x)$ are

$$u_\uparrow(x) = \tfrac{5}{3} q(x), \qquad u_\downarrow(x) = \tfrac{1}{3} q(x),$$
$$d_\uparrow(x) = \tfrac{1}{3} q(x), \qquad d_\downarrow(x) = \tfrac{2}{3} q(x), \tag{7.80}$$

or

$$u(x) = 2\, q(x), \qquad \Delta u(x) = \tfrac{4}{3} q(x),$$
$$d(x) = q(x), \qquad \Delta d(x) = -\tfrac{1}{3} q(x), \tag{7.81}$$

and all other distributions (strange quarks or antiquarks) are zero. In this case sum rules are obtained for the polarized distributions:

$$\int_0^1 dx\, [\Delta u(x) + \Delta \bar{u}(x)] = \Delta u = \tfrac{4}{3}, \tag{7.82}$$

$$\int_0^1 dx\, [\Delta d(x) + \Delta \bar{d}(x)] = \Delta d = -\tfrac{1}{3}. \tag{7.83}$$

Note that the sum $\Delta\Sigma = \Delta u + \Delta d + \Delta s + \cdots$ represents (as probability distributions) the total number of quarks parallel to the proton spin. If the proton spin comes from the quark spins, as is the case for the SU(6) wave function (7.80) multiplying a spherically symmetric spatial wave function, one expects this to be 1.

It leads to

$$\Gamma_1^p = \int_0^1 dx \ g_1^p(x) = \tfrac{1}{2}\left[\tfrac{4}{9}\,\Delta u + \tfrac{1}{9}\,\Delta d + \tfrac{1}{9}\,\Delta s\right] = \tfrac{5}{18} \approx 0.28, \qquad (7.84)$$

$$\Gamma_1^n = \int_0^1 dx \ g_1^n(x) = \tfrac{1}{2}\left[\tfrac{4}{9}\,\Delta d + \tfrac{1}{9}\,\Delta u + \tfrac{1}{9}\,\Delta s\right] = 0, \qquad (7.85)$$

which are in disagreement with the experimental results [14], $\Gamma_1^p \approx 0.15$, $\Gamma_1^n \approx -0.04$.

These unexpected results stimulated a vigorous experimental programme, incorporating inclusive and semi-inclusive measurements (see section 7.10) and leading to the picture of the polarizations for the different flavours shown in figure 7.7 (taken from [15]). These indicate a strong deviation from the above naive expectations $\Delta u/u = \tfrac{2}{3}$ and $\Delta d/d = -\tfrac{1}{3}$, especially at small x_B.

The importance of these sum rules becomes clearer when one starts with the expressions for the distribution functions in terms of matrix elements of bilocal operator combinations. One has

$$\int_{-1}^1 dx \ q(x) = \int_0^1 dx \ [q(x) - \bar{q}(x)] = \frac{\langle P|\bar{\psi}(0)\gamma^+\psi(0)|P\rangle}{2P^+} = n_q, \quad (7.86)$$

where n_q is the coefficient in the expectation value $\langle P|\bar{\psi}(x)\gamma^\mu\psi(0)|P\rangle = 2\,n_q\,P^\mu$. The coefficient n_q is precisely the quark number because the vector currents are used to obtain the quantum numbers for flavour (up, down, strange, ...). In general one obtains

$$\int_{-1}^1 dx \ x^{n-1}\,q(x) = \int_0^1 dx \ x^{n-1}\,[q(x) + (-)^n\,\bar{q}(x)]$$

$$= \frac{1}{2P^+}\langle P|\bar{\psi}(0)\,\gamma^+\left(\frac{i\partial^+}{P^+}\right)^{n-1}\psi(0)|P\rangle. \qquad (7.87)$$

The moments of the structure functions are related to expectation values of particular quark operators. In a field theory these matrix elements depend on a renormalization scale μ^2 and thus a similar renormalization scale dependence must be present for the structure functions. In the next sections these QCD corrections will be discussed in more detail. In some cases, such as the rule in (7.86), the result is scale-independent. This is true if the operator combination corresponds to a conserved current. The situation is different for the second moment that appeared in the momentum sum rule (7.78),

$$\int_{-1}^1 dx \ x \ q(x, \log \mu^2) = \int_0^1 dx \ x \ [q(x, \log \mu^2) + \bar{q}(x, \log \mu^2)]$$

$$= \frac{\langle P|\bar{\psi}(0)\,i\gamma^+\partial^+\psi(0)|P\rangle_{(\mu^2)}}{2(P^+)^2} = \epsilon_q(\mu^2), \qquad (7.88)$$

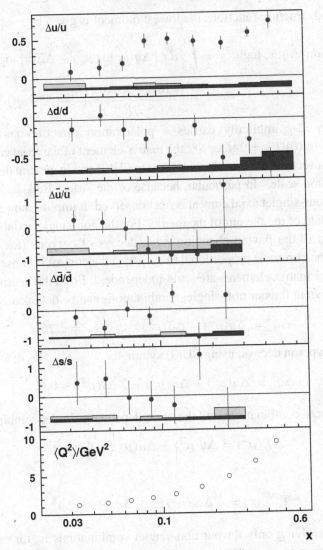

Figure 7.7. The quark polarizations obtained from inclusive and semi-inclusive spin measurements by the HERMES collaboration at DESY [15]. The error bars indicate statistical uncertainties while the bands indicate systematic uncertainties in which the light grey band specifies the part due to errors in the fragmentation process. The lowest plot gives the scales at which the asymmetries in the various bins are measured.

where ϵ_q is defined in $\langle P|\overline{\psi}(0)\,i\gamma_\mu\partial_\nu\psi(0)|P\rangle = 2\epsilon_q\,P_\mu P_\nu + \cdots$ and is the relative contribution of quarks to the energy–momentum tensor of the proton (the dots indicate trace terms $\propto M^2\,g_{\mu\nu}$). Only the first moment of the sum including quark and gluon distributions in the proton is scale-independent as in this case the local operator turns out to be the energy–momentum stress tensor of QCD.

For polarized structure functions the lowest moment is given by

$$\int_{-1}^{1} dx \, \Delta q(x, \log \mu^2) = \int_{0}^{1} dx \, [\Delta q(x, \log \mu^2) + \Delta \bar{q}(x, \log \mu^2)]$$

$$= \frac{\langle P|\bar{\psi}(0)\gamma^+\gamma_5\psi(0)|P\rangle_{(\mu^2)}}{2\,P^+} = \Delta q(\mu^2). \quad (7.89)$$

The quantity Δq implicitly carries a polarization direction and appears in $\langle P|\bar{\psi}(0)\gamma^+\gamma_5\psi(0)|P\rangle = 2M\Delta q\, S^\mu$, the matrix element of the axial current, where S^μ is the spin vector for the nucleon defined in (7.40) and in general depends on the renormalization scale. In particular, because of the Adler–Bell–Jackiw anomaly [16], the flavour-singlet axial current is not conserved. It implies, however, a breaking independent of the flavour of the quarks. For the non-singlet axial currents that are important in the flavour-changing weak decays of baryons (for example for the neutron the current is proportional to $\tau_+ \gamma^\mu\gamma_5$) the current is conserved and the corresponding matrix elements are scale-independent. From the neutron decay the scale-independent flavour non-singlet combinations can be deduced:

$$\Delta q^3 = \Delta u(\mu^2) - \Delta d(\mu^2) = g_A/g_V = 1.26, \quad (7.90)$$

while from hyperon decays, using SU(3) symmetry,

$$\Delta q^8 = \Delta u(\mu^2) + \Delta d(\mu^2) - 2\,\Delta s(\mu^2) \approx 0.6. \quad (7.91)$$

In terms of these combinations and the scale-dependent singlet combination,

$$\Delta\Sigma(\mu^2) = \Delta u(\mu^2) + \Delta d(\mu^2) + \Delta s(\mu^2), \quad (7.92)$$

one has

$$\Gamma_1^{p/n}(x) = \tfrac{1}{9}\,\Delta\Sigma(\mu^2) \pm \tfrac{1}{12}\,\Delta q^3 + \tfrac{1}{36}\,\Delta q^8. \quad (7.93)$$

A sum rule involving only flavour non-singlet combinations is, for example, the polarized Bjorken sum rule [17],

$$\int_{0}^{1} dx \, [g_1^{ep}(x) - g_1^{en}(x)] = \tfrac{1}{6}\Delta q^3 = \tfrac{1}{6}g_A/g_V \approx 0.21, \quad (7.94)$$

in fairly good agreement with experiment (see also (7.95), where $O(\alpha_s)$ effects bring data and theory into excellent agreement). Note that this result is a factor 0.75 smaller than the naive expectation for which $g_A/g_V = \Delta q^3 = \tfrac{5}{3}$ instead of $g_A/g_V \approx 1.26$. A long-known explanation for this reduction is the relativistic nature of quarks in hadrons implying a sizeable P-wave contribution in the lower components of the quark spinor that reduces the spinor densities $\bar{\psi}\gamma_5\gamma_3\psi = \psi^\dagger\sigma_z\psi$.

Using the result for Γ_1^p or Γ_1^n as the third input one can solve for $\Delta\Sigma$, leading to $\Delta\Sigma \approx 0.2$, which is very small compared to the naive expectation that $\Delta\Sigma$ is of the order of 0.75, taking the same reduction factor for relativistic quarks as for g_A/g_V. At present we know that the scale dependence is important and that the deep inelastic measurements imply $\Delta\Sigma(20\,\text{GeV}^2) \approx 0.2$.

7.8 QCD corrections in deep inelastic scattering

The connection of structure functions and quark distribution functions to local operators via sum rules is more formally grounded in the operator product expansion (OPE). See standard texts for details. For example chapter 2 in [18] defines the anomalous dimension, or gamma function, γ_i and coefficient function C_i. We state here the empirical implications.

The Gottfried sum rule gets a very small correction. The polarized Bjorken sum rule gets a correction to the order of α_s^2 as follows [19]:

$$\int_0^1 dx \left[g_1^p(x, Q^2) - g_1^n(x, Q^2) \right] = \tfrac{1}{6} g_A/g_V \left(1 - \frac{\alpha_s}{\pi} + \frac{\alpha_s^2}{\pi^2}(-4.5833 + \tfrac{1}{3}f) \right),$$

$$(7.95)$$

where f is the number of active flavours. This gives an excellent explanation of the experimental result $\Gamma_1^p - \Gamma_1^n \approx 0.19$ being somewhat smaller than $\tfrac{1}{6}g_A/g_V = 0.21$. As we have seen, the g_1^p sum rule by itself involves the singlet combination $\Delta\Sigma$ connected to the singlet axial current. The leading-order result is

$$\Delta\Sigma(Q^2) = \exp\left(\frac{\gamma_1^\Sigma}{8\pi b_0} \left(\alpha_s(Q^2) - \alpha_s\left(Q_0^2\right) \right) \right) \Delta\Sigma(Q_0^2) \qquad (7.96)$$

and the correction for the singlet contribution to the spin sum rules is

$$\Gamma_1^S(Q^2) = \left(1 - \frac{\alpha_s}{\pi} + \frac{\alpha_s^2}{\pi^2}(-4.5833 + 1.16248f) \right) \Delta\Sigma(Q^2). \qquad (7.97)$$

Including this, the deviation of the proton sum rule from the naive result can be understood.

7.9 Evolution equations

The intuitive folding picture that we have used to derive the parton model can be extended to obtain the QCD corrections. This approach also provides a practical way to calculate the coefficient functions $c_i(g)$ and the gamma functions $\gamma_i(g)$.

Extending the result for the transverse structure function written as a delta function contribution,

$$\sigma_T = \sum_q \int_0^1 dz\, q(z) \left[e_q^2 \frac{4\pi^2\alpha}{Q^2} x\, \delta(x - z) + \delta\hat\sigma_q(z, Q^2) + \cdots \right], \qquad (7.98)$$

one obtains contributions from the process $\gamma^* q \to Gq$. As there are two particles in the final state, $\delta\sigma_q(z, Q^2)$ is not simply proportional to $\delta(x - z)$. The amplitude is

$$|\mathcal{M}|^2 = 32\pi^2 \, e_q^2 \, \alpha\alpha_s \, \frac{4}{3} \left(-\frac{\hat{t}}{\hat{s}} - \frac{\hat{s}}{\hat{t}} + \frac{2\hat{u}Q^2}{\hat{s}\hat{t}} \right) \tag{7.99}$$

and contributes

$$\delta\sigma(\hat{s}) = e_q^2 \, \frac{2\pi\alpha\alpha_s}{(\hat{s} + Q^2)^2} \, \frac{4}{3} \left(-\frac{\hat{t}}{\hat{s}} - \frac{\hat{s}}{\hat{t}} - \frac{2\hat{t}Q^2 + 2\hat{s}Q^2 + 2Q^4}{\hat{s}\hat{t}} \right) d\hat{t}. \tag{7.100}$$

Using light-cone components, the momenta for $\gamma^*(q) + \text{quark } (k) \to \text{gluon} \, (p_G) + \text{quark } (p_q)$ are

$$q \equiv \left[\frac{Q^2}{xA\sqrt{2}}, -\frac{xA}{\sqrt{2}}, 0_\perp \right], \tag{7.101}$$

$$k \equiv \left[\frac{m^2}{zA\sqrt{2}}, \frac{zA}{\sqrt{2}}, 0_\perp \right] \approx \left[0, \frac{zA}{\sqrt{2}}, 0_\perp \right], \tag{7.102}$$

$$p_G = \left[\zeta q^-, \frac{p_\perp^2}{2\zeta q^-}, p_\perp \right] \approx \left[\frac{\zeta Q^2}{xA\sqrt{2}}, 0, p_\perp \right], \tag{7.103}$$

$$p_q = \left[(1 - \zeta)q^-, \frac{p_\perp^2}{2(1 - \zeta)q^-}, -p_\perp \right] \approx \left[\frac{(1 - \zeta)Q^2}{xA\sqrt{2}}, 0, -p_\perp \right]. \tag{7.104}$$

Note that z can be written as x/ξ, where $\xi \equiv -q^+/k^+$ is the Bjorken scaling variable for the subprocess. Neglecting all particle masses, in which case $p_q^2 = 0$, gives

$$p_\perp^2 = \frac{\zeta(1 - \zeta)(1 - \xi)}{\xi} \, Q^2. \tag{7.105}$$

The invariants for the subprocess become

$$\hat{s} = (k + q)^2 = \frac{1 - \xi}{\xi} \, Q^2 = \frac{1}{\zeta(1 - \zeta)} \, p_\perp^2, \tag{7.106}$$

$$\hat{t} = (p_G - q)^2 = -\frac{1 - \zeta}{\xi} \, Q^2 = -\frac{1}{\zeta(1 - \xi)} \, p_\perp^2. \tag{7.107}$$

The kinematic range of the process $\hat{s} \geq 0$, $-(\hat{s} + Q^2) \leq t \leq 0$ in principle restricts the ranges of (ξ, ζ) to $0 \leq \xi \leq 1$ and $0 \leq \zeta \leq 1$. The cross section is

$$\delta\hat{\sigma}_q(\xi, Q^2) = e_q^2 \, \frac{2\pi\alpha\alpha_s}{Q^2} \, \frac{4}{3} \, \xi \left[\frac{1 - \zeta - 2\xi}{1 - \xi} + \frac{1 + \xi^2}{(1 - \xi)(1 - \zeta)} \right] d\zeta. \tag{7.108}$$

For inclusive scattering one has to integrate over the final state, hence over ζ. However, there are singular points, specifically for $\zeta = 1$, which correspond to a gluon radiated with $p_\perp = 0$. These divergences are therefore referred to as collinear divergences. There are several ways of regularizing them, for example by giving quarks and gluons masses, by dimensional regularization, or by imposing a $p_\perp$ cutoff. The restriction $p_\perp^2 \geq \mu^2$ modifies the allowed region in the (ξ, ζ) plane. For a given ξ (not too close to unity) the integration is limited to

$$\frac{\xi}{1-\xi}\frac{\mu^2}{Q^2} \leq \zeta \leq 1 - \frac{\xi}{1-\xi}\frac{\mu^2}{Q^2}. \tag{7.109}$$

With this regularization the result becomes

$$\delta\hat{\sigma}_q(\xi, Q^2) = e_q^2 \frac{2\pi\alpha\alpha_s}{Q^2} \xi \left[B(\xi) + P_{qq}(\xi) \log\left(\frac{Q^2}{\mu^2}\right) \right], \tag{7.110}$$

where

$$P_{qq}(\xi) = \frac{4}{3}\frac{1+\xi^2}{(1-\xi)} \tag{7.111}$$

is the splitting function coming from the collinear $1/(1-\zeta)$ singularity. Combining the results of the leading contribution (omitting for now the term $B(\xi)$),

$$\sigma_T = \frac{4\pi^2\alpha}{Q^2} x \sum_q e_q^2 \int_x^1 \frac{dz}{z} q(z) \left[\delta\left(1 - \frac{x}{z}\right) + \frac{\alpha_s}{2\pi} P_{qq}\left(\frac{x}{z}\right) \log\left(\frac{Q^2}{\mu^2}\right) \right]. \tag{7.112}$$

This can be rewritten in a form reminiscent of the parton model as

$$\sigma_T = \frac{4\pi^2\alpha}{Q^2} x \sum_q e_q^2 q(x, \log Q^2), \tag{7.113}$$

where the functions $q(x, \log Q^2)$ satisfy

$$\frac{\partial q(x, \log Q^2)}{\partial \log Q^2} = \frac{\alpha_s(Q^2)}{2\pi} \int_x^1 \frac{dz}{z} q(z) P_{qq}\left(\frac{x}{z}\right). \tag{7.114}$$

As in the case of integrating over ζ, here one again encounters divergences, in this case for $\xi = x/z \to 1$. They are dealt with by considering the $1/(1-\xi)$ appearing in the splitting functions as functionals,

$$\int_0^{1-\delta} dx \frac{f(x)}{1-x} = -\int f(x)\, d\log(1-x)$$

$$= \int_0^{1-\delta} dx \frac{f(x)-f(1)}{1-x} - f(1)\log\delta$$

$$= \int_0^1 dx \frac{f(x)-f(1)}{1-x} - \log\delta \int dx\, f(x)\delta(1-x)$$

$$\equiv \int_0^1 dx \frac{f(x)}{(1-x)_+} - \log\delta \int dx\, f(x)\delta(1-x),$$

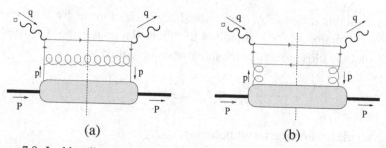

Figure 7.8. Ladder diagrams used to calculate the asymptotic behaviour of the correlation functions. See text.

which defines

$$\frac{1}{1-x} = \frac{1}{(1-x)_+} - \delta(1-x)\log\delta. \qquad (7.115)$$

Including these and other singular contributions (vertex corrections to the process $\gamma^* + q \to q$), then

$$q(x, Q^2) = \int_x^1 \frac{dz}{z} q(z) \left\{ \delta\left(1 - \frac{x}{z}\right) \right.$$

$$\left. + \frac{\alpha_s}{2\pi} \log\left(\frac{Q^2}{\mu^2}\right) \left[a\,\delta\left(1 - \frac{x}{z}\right) + P_{qq}\left(\frac{x}{z}\right) \right] \right\}. \qquad (7.116)$$

The terms between curly brackets can be interpreted as the probability density $\mathcal{P}_{qq}$ of finding a quark inside a quark with fraction $\xi = x/z$ of the parent quark to first order in α_s. Instead of calculating all contributions and checking all cancellations, it is easier to see what the final result must be by using probability conservation:

$$\int d\xi\, \mathcal{P}_{qq}(\xi) = 1. \qquad (7.117)$$

Keeping the form in (7.112), the expression between square brackets in (7.116) can be written as

$$P_{qq}(\xi) = \frac{4}{3}\frac{1+\xi^2}{(1-\xi)_+} + 2\,\delta(1-\xi). \qquad (7.118)$$

While this splitting function describes how QCD corrections arising from $q \to qG$ splitting are incorporated into the parton distributions (figure 7.8(a)), other splitting functions such as P_{qG}, describing how quark and gluon distribution functions mix (figure 7.8(b)) are required. They are calculated from the process $\gamma^* G \to q\bar{q}$. Since gluons are flavour-blind, such corrections do not affect non-singlet and valence distribution functions.

The splitting function for the polarized distribution functions is given by

$$\Delta P_{qq}(x) = P_{qq}(x) = \frac{4}{3} \frac{1 + \xi^2}{(1 - \xi)_+} + 2\,\delta(1 - \xi). \tag{7.119}$$

7.9.1 Solutions of the evolution equations

In solving the evolution equations the moments play an important role, while they also establish the connection with the OPE. Rewriting the evolution equation (7.114) as

$$\frac{dq(x, \tau)}{d\tau} = \frac{\alpha_s(\tau)}{2\pi} \int_0^1 dz \int_0^1 dy\,\delta(x - yz)\, P_{qq}(y)\, q(z, \tau), \tag{7.120}$$

and using the moments

$$M_n(\tau) \equiv \int_0^1 dx\, x^{n-1}\, q(x, \tau), \tag{7.121}$$

$$A_n \equiv \int_0^1 dy\, y^{n-1}\, P_{qq}(y, \tau)$$

$$= \frac{4}{3} \left[-\frac{1}{2} + \frac{1}{n(n+1)} - 2 \sum_{j=2}^{n} \frac{1}{j} \right], \tag{7.122}$$

then

$$\frac{dM_n(\tau)}{d\tau} = \frac{\alpha_s(\tau)}{2\pi} A_n M_n(\tau). \tag{7.123}$$

Using the leading-order QCD result for $\alpha_s(\tau), d\alpha_x/d\tau = -(b_0/4\pi)\,\alpha_s$, this is easily solved, giving

$$\frac{M_n(Q^2)}{M_n(\mu^2)} = \left(\frac{\alpha_s(Q^2)}{\alpha_s(\mu^2)} \right)^{d_n}, \tag{7.124}$$

where $d_n = -2\,A_n/b_0$. The moments of the splitting functions are, up to a factor, equal to the gamma functions γ_{0n}, namely $A_n = -\gamma_{0n}/4$ [18]. As an example, consider the second moment of the quark distributions, for which $A_2 = -\frac{16}{9}$. The result for d_2 is $d_2 = \frac{32}{9} b_0 = \frac{32}{81}$ for three flavours. The fraction of momentum carried by valence quarks thus vanishes for $Q^2 \to \infty$ as

$$M_2(Q^2) = \int dx\, x[q(x, \log Q^2) - \bar{q}(x, \log Q^2)] \propto (\alpha_s(Q^2))^{d_2}. \tag{7.125}$$

Mixing occurs for sea quarks and gluons. As a consequence, the combination corresponding to the total momentum (involving the second moment of the singlet-quark distribution and the second moment of the gluon distribution) is Q^2-independent.

In principle the evolution of the structure functions can be obtained by means of the evolution equations. Using the moments is actually quite convenient, although one needs all of them. A very useful method in practice is to parametrize the distributions at one value of Q^2 with a function for which the moments can be easily calculated, evolve the moments and apply an inverse Mellin transform. One has

$$q(x, Q^2) = -\frac{1}{2\pi i} \int_{c-i\infty}^{c+i\infty} dn \, x^{-n} \, M_n(Q^2), \qquad (7.126)$$

where c must be such that M_n has no singularities in the complex n-plane to the right of the line $\operatorname{Re} n = c$. A similar relation exists for the moments of the splitting functions, $A_n = -\gamma_{0n}/4$,

$$P_{qq}(x) = \frac{1}{8\pi i} \int_{c-i\infty}^{c+i\infty} dn \, x^{-n} \, \gamma_{0n}. \qquad (7.127)$$

While the splitting functions are universal (process-independent), the contributions $B(x)$ of (7.110) are in general process-dependent. An example of a contribution of this type is the longitudinal structure function (for which the dominant parton model result is zero). Calculating both the α_s and $B(x)$ contributions for F_1 and F_2 in electroproduction gives the first non-vanishing contribution in the longitudinal structure function,

$$F_L(x, Q^2) = \frac{\alpha_s(Q^2)}{\pi} \left[\frac{4}{3} \int_x^1 \frac{dz}{z} \left(\frac{x}{z}\right)^2 F_2(z, Q^2) \right.$$

$$\left. + \left(2 \sum_q e_q^2\right) \int_x^1 \frac{dz}{z} \left(\frac{x}{z}\right)^2 \left(1 - \frac{x}{z}\right) z \, G(z, Q^2) \right]. \quad (7.128)$$

7.10 One-particle inclusive leptoproduction

We now consider the case in which one particle is detected in coincidence with the scattered lepton. This is known as one-particle inclusive or 1PI leptoproduction. The kinematics of this process has been shown already in figure 7.2. With a target hadron (momentum P) and a detected hadron h in the final state (momentum P_h) one has a situation in which two hadrons are involved and the operator product expansion cannot be used. Within the framework of QCD and knowing that the photon or Z^0 current couples to the quarks, it is possible to write down a diagrammatic expansion for leptoproduction. The relevant diagrams in the deep inelastic limit ($Q^2 \to \infty$) for 1PI production are those of figure 7.9.

In analogy with the case of inclusive scattering, in 1PI scattering we again parameterize the momenta with the help of two light-like vectors, which are now chosen

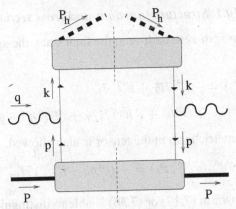

Figure 7.9. The simplest (parton-level) diagrams for semi-inclusive scattering, of which we consider 1PI leptoproduction. Note that also the diagram with opposite fermion flow has to be added.

along the hadron momenta:

$$
\left.\begin{array}{l}
q^2 = -Q^2 \\
P^2 = M^2 \\
P_h^2 = M_h^2 \\
2\,P \cdot q = \dfrac{Q^2}{x_B} \\
2\,P_h \cdot q = -z_h\,Q^2
\end{array}\right\}
\longleftrightarrow
\left\{\begin{array}{l}
P_h = \dfrac{z_h\,Q}{\sqrt{2}}\,n_- + \dfrac{M_h^2}{z_h\,Q\sqrt{2}}\,n_+ \\[2mm]
q = \dfrac{Q}{\sqrt{2}}\,n_- - \dfrac{Q}{\sqrt{2}}\,n_+ + q_T \\[2mm]
P = \dfrac{x_B\,M^2}{Q\sqrt{2}}\,n_- + \dfrac{Q}{x_B\sqrt{2}}\,n_+
\end{array}\right.
$$

An additional invariant z_h comes in. Note that the expansion is appropriate for current fragmentation, when the produced hadron is hard with respect to the target momentum, that is $P \cdot P_h \sim Q^2$. The minus component p^- is irrelevant in the lower soft part of figure 7.9, while the plus component k^+ is irrelevant in the upper soft part. Note that having made the choice of P and P_h, one can no longer omit a transverse component in the momentum transfer q. Up to mass effects, one has the relation

$$
q_T^\mu = q^\mu + x_B\,P^\mu - \frac{P_h^\mu}{z_h} \equiv -Q_T\,\hat{h}^\mu. \tag{7.129}
$$

This relation allows the experimental determination of the 'transverse momentum' effect from the external vectors q, P and P_h, which are in general not collinear. The vector $\hat{h}$ defines the orientation of the hadronic plane in figure 7.2.

An important consequence in the theoretical approach (figure 7.9) is that one can no longer simply integrate over the transverse components of the quark momenta.

7.10.1 Structure functions and cross sections

For an unpolarized or spin-zero hadron in the final state, the symmetric part of the tensor is given by

$$MW_S^{\mu\nu}(q, P, P_h) = -g_\perp^{\mu\nu}\,\mathcal{H}_T + \hat{t}^\mu\hat{t}^\nu\,\mathcal{H}_L$$

$$+ (\hat{t}^\mu\hat{h}^\nu + \hat{t}^\nu\hat{h}^\mu)\,\mathcal{H}_{LT} + (2\,\hat{h}^\mu\hat{h}^\nu + g_\perp^{\mu\nu})\mathcal{H}_{TT}. \quad (7.130)$$

Note that an antisymmetric term in the tensor is also allowed,

$$MW_A^{\mu\nu}(q, P, P_h) = -i(\hat{t}^\mu\hat{h}^\nu - \hat{t}^\nu\hat{h}^\mu)\,\mathcal{H}'_{LT}. \quad (7.131)$$

Clearly the lepton tensor in (7.17) or (7.36) is able to distinguish all the structures in the semi-inclusive hadron tensor.

The symmetric part gives the cross section for an unpolarized lepton and an unpolarized target,

$$\frac{d\sigma_{OO}}{dx_B dy\, dz_h d^2 q_T} = \frac{4\pi\,\alpha^2\, s}{Q^4}\, x_B z_h \left\{ \left(1 - y + \frac{1}{2}\,y^2\right)\mathcal{H}_T + (1 - y)\,\mathcal{H}_L \right.$$

$$\left. - (2 - y)\sqrt{1 - y}\,\cos\phi_h^\ell\,\mathcal{H}_{LT} + (1 - y)\cos 2\phi_h^\ell\,\mathcal{H}_{TT} \right\},$$

$$(7.132)$$

while the antisymmetric part gives the cross section for a longitudinally polarized lepton (note that the target is not polarized)

$$\frac{d\sigma_{LO}}{dx_B dy\, dz_h d^2 q_T} = \lambda_e\,\frac{4\pi\,\alpha^2}{Q^2}\, z_h\,\sqrt{1 - y}\,\sin\phi_h^\ell\,\mathcal{H}'_{LT}. \quad (7.133)$$

Of course, many more structure functions appear for polarized targets or if one considers polarimetry in the final state. In this case the most convenient way to describe the spin vector of the target theoretically is via an expansion of the form

$$S^\mu = -S_L\,\frac{M x_B}{Q\sqrt{2}}\,n_- + S_L\,\frac{Q}{M x_B\sqrt{2}}\,n_+ + S_T. \quad (7.134)$$

Up to $\mathcal{O}(1/Q^2)$ corrections $S_L \approx M\,(S \cdot q)/(P \cdot q)$ and $S_T \approx S_\perp$, where the subscript $\perp$ still refers to the labelled vector being perpendicular to q and P. For a pure state, $S_L^2 + S_T^2 = 1$, but in general this quantity is less than or equal to 1.

7.10.2 The parton model approach

The expression for $\mathcal{W}_{\mu\nu}$ can be rewritten as a non-local product of currents and it is a straightforward exercise to show by inserting the currents $j_\mu(x) = :\overline{\psi}(x)\gamma_\mu\psi(x):$

that for 1PI scattering in the tree approximation

$$
\begin{aligned}
2MW_{\mu\nu}&(q; PS; P_h S_h) \\
&= \frac{1}{(2\pi)^4} \int d^4x \, e^{iq \cdot x} \, \langle PS| : \overline{\psi}_j(x)(\gamma_\mu)_{jk}\psi_k(x) : \sum_X |X; P_h S_h\rangle \\
&\quad \times \langle X; P_h S_h| : \overline{\psi}_l(0)(\gamma_\nu)_{li}\psi_i(0) : |PS\rangle \\
&= \frac{1}{(2\pi)^4} \int d^4x \, e^{iq \cdot x} \, \langle PS|\overline{\psi}_j(x)\psi_i(0)|PS\rangle(\gamma_\mu)_{jk} \\
&\quad \times \langle 0|\psi_k(x) \sum_X |X; P_h S_h\rangle\langle X; P_h S_h|\overline{\psi}_l(0)|0\rangle(\gamma_\nu)_{li} \\
&\quad + \frac{1}{(2\pi)^4} \int d^4x \, e^{iq \cdot x} \, \langle PS|\psi_k(x)\overline{\psi}_l(0)|PS\rangle(\gamma_\nu)_{li} \\
&\quad \times \langle 0|\overline{\psi}_j(x) \sum_X |X; P_h S_h\rangle\langle X; P_h S_h|\psi_i(0)|0\rangle(\gamma_\mu)_{jk} + \cdots \\
&= \int d^4p \, d^4k \, \delta^4(p + q - k) \, \mathrm{Tr}\left(\Phi(p)\gamma_\mu\Delta(k)\gamma_\nu\right) + \left\{ \begin{matrix} q \leftrightarrow -q \\ \mu \leftrightarrow \nu \end{matrix} \right\},
\end{aligned}
$$

$$(7.135)$$

where

$$
\Phi_{ij}(p) = \frac{1}{(2\pi)^4} \int d^4\xi \, e^{ip \cdot \xi} \, \langle PS|\overline{\psi}_j(0)\psi_i(\xi)|PS\rangle,
$$

$$
\Delta_{kl}(k) = \frac{1}{(2\pi)^4} \int d^4\xi \, e^{ik \cdot \xi} \, \langle 0|\psi_k(\xi) \sum_X |X; P_h S_h\rangle\langle X; P_h S_h|\overline{\psi}_l(0)|0\rangle.
$$

Note that in Φ (quark production) a summation over colours is assumed, while in Δ (quark decay) an averaging over colours is assumed. The quantities Φ and Δ correspond to the blobs in figure 7.9 and parametrize the soft physics, leading to the definitions of distribution and fragmentation functions [20,21]. Soft refers to all invariants of momenta being small as compared to the hard scale, that is for $\Phi(p)$ one has $p^2 \sim p \cdot P \sim P^2 = M^2 \ll Q^2$.

In general many more diagrams have to be considered in evaluating the hadron tensors, but in the deep inelastic limit they can be neglected or considered as corrections to the soft blobs. We return to this later.

7.11 Collinear parton distributions

The form of Φ is constrained by hermiticity, parity and time-reversal invariance. The quantity depends not only on the quark momentum p but also on the target momentum P and the spin vector S. It is necessary to satisfy the requirements

$$[\text{hermiticity}] \Rightarrow \Phi^\dagger(p, P, S) = \gamma_0 \, \Phi(p, P, S) \, \gamma_0, \qquad (7.136)$$

$$[\text{parity}] \Rightarrow \Phi(p, P, S) = \gamma_0 \, \Phi(\bar{p}, \bar{P}, -\bar{S}) \, \gamma_0, \qquad (7.137)$$

$$[\text{time reversal}] \Rightarrow \Phi^*(p, P, S)$$
$$= (-i\gamma_5 C) \, \Phi(\bar{p}, \bar{P}, \bar{S}) \, (-i\gamma_5 C), \qquad (7.138)$$

where $C \equiv i\gamma^2\gamma_0$, $-i\gamma_5 C = i\gamma^1\gamma^3$ and $\bar{p} = (p^0, -\mathbf{p})$.

To obtain the leading contribution in inclusive deep inelastic scattering one can integrate over the component p^- and the transverse momenta (see the discussion in section 7.6 where the parton model has been derived). This integration restricts the non-locality in $\Phi(p)$. The relevant soft part then is a particular Dirac trace of the quantity

$$\Phi_{ij}(x) = \int dp^- \, d^2p_T \, \Phi_{ij}(p, P, S)$$

$$= \int \frac{d\xi^-}{2\pi} \, e^{ip\cdot\xi} \, \langle P, S | \bar{\psi}_j(0)\psi_i(\xi) | P, S \rangle \Big|_{\xi^+=\xi_T=0}, \qquad (7.139)$$

depending on the light-cone fraction $x = p^+/P^+$. When one wants to calculate the leading order in $1/Q$ for a hard process, one looks for leading parts in M/P^+ because $P^+ \propto Q$. The leading contribution [22] turns out to be proportional to $(M/P^+)^0$,

$$\Phi(x) = \frac{1}{2} \left\{ f_1(x) \, \slashed{n}_+ + S_L \, g_1(x) \, \gamma_5 \, \slashed{n}_+ + h_1(x) \frac{\gamma_5 [\slashed{S}_\perp, \slashed{n}_+]}{2} \right\}. \qquad (7.140)$$

The precise expression of the functions $f_1(x)$, etc. as integrals over the amplitudes can be easily written down after forming traces with the appropriate Dirac matrix,

$$f_1(x) = \int \frac{d\xi^-}{4\pi} \, e^{ip\cdot\xi} \, \langle P, S | \bar{\psi}(0)\gamma^+\psi(\xi) | P, S \rangle \Big|_{\xi^+=\xi_T=0}, \qquad (7.141)$$

$$S_L \, g_1(x) = \int \frac{d\xi^-}{4\pi} \, e^{ip\cdot\xi} \, \langle P, S | \bar{\psi}(0)\gamma^+\gamma_5\psi(\xi) | P, S \rangle \Big|_{\xi^+=\xi_T=0}, \qquad (7.142)$$

$$S_T^i \, h_1(x) = \int \frac{d\xi^-}{4\pi} \, e^{ip\cdot\xi} \, \langle P, S | \bar{\psi}(0) i\sigma^{i+}\gamma_5 \, \psi(\xi) | P, S \rangle \Big|_{\xi^+=\xi_T=0}. \qquad (7.143)$$

Including flavour indices, the functions $f_1^q(x) = q(x)$ and $g_1^q(x) = \Delta q(x)$ are precisely the functions that we encountered before.

The third function in the above parametrization is known as *transversity* or *transverse spin distribution* [23]. Including flavour indices one also denotes $h_1^q(x) = \delta q(x)$. In the same way as we have seen for $f_1(x)$ and $g_1(x)$, the function h_1

can be interpreted as a density, but instead of the projectors on quark chirality states, $P_{R/L} = \frac{1}{2}(1 \pm \gamma_5)$, one needs those on quark transverse spin states,
$P_{\uparrow/\downarrow} = \frac{1}{2}(1 \pm \gamma^i \gamma_5)$. One has

$$f_1(x) = f_{1R}(x) + f_{1L}(x) = f_{1\uparrow}(x) + f_{1\downarrow}(x), \tag{7.144}$$

$$g_1(x) = f_{1R}(x) - f_{1L}(x), \tag{7.145}$$

$$h_1(x) = f_{1\uparrow}(x) - f_{1\downarrow}(x). \tag{7.146}$$

This results in some trivial bounds such as $f_1(x) \geq 0$ and $|g_1(x)| \leq f_1(x)$. We have
discussed already the support and charge-conjugation properties of $f_1(x)$. The analysis for all these functions shows that the support is in all cases $-1 \leq x \leq 1$,
while the charge conjugation properties of the functions are $\overline{f}(x) = -f(-x)$
(C-even) for f_1 and h_1 and $\overline{f}(x) = +f(-x)$ (C-odd) for g_1.

The Dirac structure for h_1 in terms of chirality states is $\overline{\psi}_R \psi_L$ and $\overline{\psi}_L \psi_R$. Such
functions are called chiral-odd and cannot be measured in inclusive deep inelastic
scattering.

While the evolution equations for $q(x)$ and $\Delta q(x)$ require quark–quark and quark–
gluon splitting functions, the chiral-odd nature of $\delta q(x)$ prevents mixing of this
quantity with gluon distributions. The splitting function is given by

$$\delta P_{qq}(\xi) = \frac{4}{3} \frac{2\xi}{(1-\xi)_+} + 2\,\delta(1-\xi). \tag{7.147}$$

7.12 Bounds on the distribution functions

The trivial bounds on the distribution functions ($|h_1(x)| \leq f_1(x)$ and $|g_1(x)| \leq f_1(x)$) can be sharpened. For instance one can look explicitly at the structure of the
correlation functions Φ_{ij} of (7.139) in Dirac space. Actually, we will look at the cor-
relation functions $(\Phi \gamma_0)_{ij}$, which involves matrix elements $\psi^\dagger_{+j}(0)\psi_{+i}(\xi)$ at leading
order. In the representation where $\gamma^0 = \rho^1, \gamma^i = -i\rho^2\sigma^i, \gamma_5 = i\gamma^0\gamma^1\gamma^2\gamma^3 = \rho^3$,
the matrices may be written as

$$P_+ = \begin{pmatrix} 1 & 0 & 0 & 0 \\ 0 & 0 & 0 & 0 \\ 0 & 0 & 0 & 0 \\ 0 & 0 & 0 & 1 \end{pmatrix},$$

$$P_+\gamma_5 = \begin{pmatrix} 1 & 0 & 0 & 0 \\ 0 & 0 & 0 & 0 \\ 0 & 0 & 0 & 0 \\ 0 & 0 & 0 & -1 \end{pmatrix}, \qquad P_+\gamma^1\gamma_5 = \begin{pmatrix} 0 & 0 & 0 & 1 \\ 0 & 0 & 0 & 0 \\ 0 & 0 & 0 & 0 \\ 1 & 0 & 0 & 0 \end{pmatrix}.$$

The projector $P_+(1 \pm \gamma_5)$ only leaves two independent Dirac spinors, one right-handed (R) and one left-handed (L). In this basis the relevant matrix $(\Phi \not{n}_-)$ for hard scattering processes is given by

$$(\Phi \not{n}_-)_{ij}(x) = \begin{pmatrix} f_1 + S_L\, g_1 & (S_T^1 + i\, S_T^2)\, h_1 \\ (S_T^1 - i\, S_T^2)\, h_1 & f_1 - S_L\, g_1 \end{pmatrix}. \tag{7.148}$$

The S-dependent correlation function Φ can also be turned into a matrix in the nucleon-spin space via the standard spin-$\frac{1}{2}$ density matrix $\rho(P, S)$. The relation is $\Phi(x; P, S) = \text{Tr}\,[\Phi(x; P)\,\rho(P, S)]$. Writing

$$\Phi(x; P, S) = \Phi_O + S_L\, \Phi_L + S_T^1\, \Phi_T^1 + S_T^2\, \Phi_T^2, \tag{7.149}$$

then on the basis of spin-$\frac{1}{2}$ target states with $S_L = +1$ and $S_L = -1$ respectively, one has

$$\Phi_{ss'}(x) = \begin{pmatrix} \Phi_O + \Phi_L & \Phi_T^1 - i\, \Phi_T^2 \\ \Phi_T^1 + i\, \Phi_T^2 & \Phi_O - \Phi_L \end{pmatrix}. \tag{7.150}$$

By generalizing $\Phi(p)$ to matrix elements between states $\langle P, s|$ and $|P, s'\rangle$, the matrix $M = (\Phi \not{n}_-)^T$ (where T means transposed in Dirac space) has the property that $v^\dagger M v \geq 0$ for any direction v in Dirac space.

In the basis $+R$, $-R$, $+L$ and $-L$, the matrix in quark $\otimes$ nucleon spin-space becomes

$$(\Phi(x)\,\not{n}_-)^T = \begin{pmatrix} f_1 + g_1 & 0 & 0 & 2\,h_1 \\ 0 & f_1 - g_1 & 0 & 0 \\ 0 & 0 & f_1 - g_1 & 0 \\ 2\,h_1 & 0 & 0 & f_1 + g_1 \end{pmatrix} \tag{7.151}$$

Any diagonal element of this matrix must always be positive, hence the eigenvalues must be positive, which gives a bound on the distribution functions stronger than the trivial bounds, namely

$$|h_1(x)| \leq \tfrac{1}{2}\,(f_1(x) + g_1(x)). \tag{7.152}$$

This is known as the Soffer bound [24].

Changing to the transverse-quark spin basis gives the quark production matrix

$$(\Phi(x)\,\slashed{n}_-)^T = \begin{bmatrix} f_1 + h_1 & 0 & 0 & g_1 + h_1 \\ 0 & f_1 - h_1 & g_1 - h_1 & 0 \\ 0 & g_1 - h_1 & f_1 - h_1 & 0 \\ g_1 + h_1 & 0 & 0 & f_1 + h_1 \end{bmatrix} \qquad (7.153)$$

7.13 Transverse momentum-dependent correlation functions

Without integration over p_T, the soft part is

$$\Phi(x, p_T) = \int \frac{d\xi^- d^2\xi_T}{(2\pi)^3}\, e^{ip\cdot\xi}\, \langle P, S|\overline{\psi}(0)\psi(\xi)|P, S\rangle\bigg|_{\xi^+=0}. \qquad (7.154)$$

For the leading-order results the parts involving unpolarized targets (O), longitudinally polarized targets (L) and transversely polarized targets (T), up to terms proportional to M/P^+, take the forms [25,26]:

$$\Phi_O(x, p_T) = \tfrac{1}{2}\left\{ f_1(x, p_T)\,\slashed{n}_+ + h_1^\perp(x, p_T)\frac{i[\slashed{p}_T, \slashed{n}_+]}{2M} \right\}, \qquad (7.155)$$

$$\Phi_L(x, p_T) = \tfrac{1}{2}\left\{ S_T\, g_{1T}(x, p_T)\,\gamma_5\,\slashed{n}_+ + S_T\, h_{1L}^\perp(x, p_T)\frac{\gamma_5[\slashed{p}_T, \slashed{n}_+]}{2M} \right\}, \qquad (7.156)$$

$$\begin{aligned}
\Phi_T(x, p_T) = \tfrac{1}{2}\bigg\{ & f_{1T}^\perp(x, p_T)\frac{\epsilon_{\mu\nu\rho\sigma}\gamma^\mu n_+^\nu p_T^\rho S_T^\sigma}{M} \\
& + \frac{p_T \cdot S_T}{M} g_{1T}(x, p_T)\,\gamma_5\,\slashed{n}_+ + h_{1T}(x, p_T)\frac{\gamma_5[\slashed{S}_T, \slashed{n}_+]}{2} \\
& + \frac{p_T \cdot S_T}{M} h_{1T}^\perp(x, p_T)\frac{\gamma_5[\slashed{p}_T, \slashed{n}_+]}{2M} \bigg\}. \qquad (7.157)
\end{aligned}$$

All functions appearing in (7.157) have a natural interpretation as densities, just as for the p_T-integrated functions. Densities such as those of longitudinally-polarized quarks in a transversely-polarized nucleon (g_{1T}) and of transversely-polarized quarks in a longitudinally-polarized nucleon ($h_{1L}^\perp$) are now included.

Not all functions survive after integration over p_T. The p_T integration leaves (7.140) with

$$f_1(x) = \int d^2 p_T \, f_1(x, p_T),$$

$$g_1(x) = \int d^2 p_T \, g_{1L}(x, p_T),$$

$$h_1(x) = \int d^2 p_T \left[h_{1T}(x) + \frac{p_T^2}{2M^2} h_{1T}^\perp(x, p_T) \right].$$

The explicit treatment of transverse momenta also provides a way to include the evolution of quark distribution and fragmentation functions. The assumption that soft parts vanish sufficiently fast as a function of the invariants $p \cdot P$ and p^2, which at constant x implies a sufficiently fast vanishing as a function of p_T^2, turns out to be true. Assuming that the result for $p_T^2 \geq \mu^2$ is given by the diagram shown in figure 7.8, the extra distribution written in terms of p_T becomes

$$f_1\left(x, p_T^2\right) \overset{p_T^2 \geq \mu^2}{\Longrightarrow} \frac{1}{\pi \, p_T^2} \frac{\alpha_s(\mu^2)}{2\pi} \int_x^1 \frac{dy}{y} \, P_{qq}\left(\frac{x}{y}\right) f_1(y; \mu^2), \qquad (7.158)$$

giving

$$f_1(x; \mu^2) \equiv \pi \int_0^{\mu^2} dp_T^2 \, f_1\left(x, p_T^2\right), \qquad (7.159)$$

which results in a logarithmic scale dependence.

Different functions survive after integrating over p_T weighted with p_T^α. For example,

$$\Phi_\partial^\alpha(x) \equiv \int d^2 p_T \, \frac{p_T^\alpha}{M} \, \Phi(x, p_T)$$

$$= \frac{1}{2} \left\{ -g_{1T}^{(1)}(x) \, S_T^\alpha \, \slashed{n}_+ \gamma_5 - S_T \, h_{1L}^{\perp(1)}(x) \frac{[\gamma^\alpha, \slashed{n}_+]\gamma_5}{2} \right.$$

$$\left. - f_{1T}^{\perp(1)} \, \epsilon^\alpha{}_{\mu\nu\rho} \gamma^\mu n_-^\nu S_T^\rho - h_1^{\perp(1)} \frac{i[\gamma^\alpha, \slashed{n}_+]}{2} \right\} \qquad (7.160)$$

involves transverse moments defined as

$$g_{1T}^{(1)}(x) = \int d^2 p_T \, \frac{p_T^2}{2M^2} \, g_{1T}(x, p_T), \qquad (7.161)$$

and similarly for the other functions. The functions $h_1^\perp$ and $f_{1T}^\perp$ are *T-odd*. This is discussed in section 7.16 on colour gauge invariance. Time-reversal invariance cannot be used for the transverse moments and so they are not required to vanish. Fragmentation functions are also not required to vanish. The T-odd functions correspond to unpolarized quarks in a transversely-polarized nucleon ($f_{1T}^\perp$) or transversely-polarized quarks in an unpolarized nucleon ($h_1^\perp$). The easiest way to interpret the functions is by considering their place in the quark production matrix $(\Phi(x, p_T)\,\not{n}_-)^T$, which becomes [27]

$$
\begin{pmatrix}
f_1 + g_{1L} & \frac{|p_T|}{M} e^{i\phi} g_{1T} & \frac{|p_T|}{M} e^{-i\phi} h_{1L}^\perp & 2\,h_1 \\[2mm]
\frac{|p_T|}{M} e^{-i\phi} g_{1T}^* & f_1 - g_{1L} & \frac{|p_T|^2}{M^2} e^{-2i\phi} h_{1T}^\perp & -\frac{|p_T|}{M} e^{-i\phi} h_{1L}^{\perp *} \\[2mm]
\frac{|p_T|}{M} e^{i\phi} h_{1L}^{\perp *} & \frac{|p_T|^2}{M^2} e^{2i\phi} h_{1T}^\perp & f_1 - g_{1L} & -\frac{|p_T|}{M} e^{i\phi} g_{1T}^* \\[2mm]
2\,h_1 & -\frac{|p_T|}{M} e^{i\phi} h_{1L}^\perp & -\frac{|p_T|}{M} e^{-i\phi} g_{1T} & f_1 + g_{1L}
\end{pmatrix}.
$$

In this representation T-odd functions appear as imaginary parts, $f_{1T}^\perp = -\mathrm{Im}\, g_{1T}$ and $h_1^\perp = \mathrm{Im}\, h_{1L}^\perp$.

7.14 Fragmentation functions

Analysis of the soft part describing quark fragmentation can be performed analogously to that for the distribution functions [21]. This needs

$$
\Delta_{ij}(z, k_T) = \sum_X \int \frac{d\xi^+ d^2\xi_T}{(2\pi)^3} \, e^{ik\cdot\xi}\, \mathrm{Tr}\langle 0|\psi_i(\xi)|P_h, X\rangle\langle P_h, X|\overline\psi_j(0)|0\rangle \Bigg|_{\xi^-=0}. \tag{7.162}
$$

For the production of unpolarized (or spin-0) hadrons h in hard processes, to leading order in $1/Q$ one needs the $(M_h/P_h^-)^0$ part of the correlation function,

$$
\Delta_O(z, k_T) = z\, D_1(z, k_T')\, \not{n}_- + z\, H_1^\perp(z, k_T') \frac{i\,[\not{k}_T, \not{n}_-]}{2M_h}. \tag{7.163}
$$

The arguments of the fragmentation functions D_1 and $H_1^\perp$ are $z = P_h^-/k^-$ and $k_T' = -z k_T$. The first is the light-cone momentum fraction of the produced hadron, the second is the transverse momentum of the produced hadron with respect to the quark. The fragmentation function D_1 is the analogue of the distribution function f_1. It can be interpreted as a quark decay function, giving the probability of finding a hadron h in a quark. The quantity $n_h = \int dz\, D_1(z)$ is the number of hadrons and the normalization of the fragmentation functions is given by $\sum_h \int dz\, z\, D_1^{q\to h}(z) = 1$.

The function $H_1^\perp$, interpretable as the difference between the numbers of unpo-larized hadrons produced from a transversely-polarized quark depending on the hadron's transverse momentum, is allowed because of the non-applicability of time-reversal invariance [28]. This is natural for the fragmentation functions [29,30] because of the appearance of out-states $|P_h, X\rangle$ in the definition of Δ, in contrast to the plane-wave states appearing in Φ. The function $H_1^\perp$ is of interest because it is chiral-odd. This means that it can be used to probe the chiral-odd quark distribu-tion function h_1, which can be achieved, for example, by measuring an azimuthal asymmetry of pions produced in the current fragmentation region.

The spin structure of fragmentation functions is also conveniently summarized in an explicit R and L chiral-quark basis. For decay into spin-zero hadrons,

$$(\Delta(z, k_T)\,\not{n}_+)^T = \left[\begin{matrix} D_1 & i\frac{|k_T|e^{-i\phi}}{M_h} H_1^\perp \\[2ex] -i\frac{|k_T|e^{+i\phi}}{M_h} H_1^\perp & D_1 \end{matrix} \right] \begin{matrix} \text{\tiny R} \\[2ex] \text{\tiny L} \end{matrix} . \tag{7.164}$$

$$\qquad\qquad\qquad \text{\tiny R} \qquad\qquad\quad \text{\tiny L}$$

7.14.1 Examples of azimuthal asymmetries

Transverse momentum dependence shows up in the azimuthal dependence in the semi-inclusive deep inelastic scattering (SIDIS) cross section, via $\hat{h}$ or the trans-verse spin vectors, in most cases requiring polarization of beam and/or target or polarimetry [31–34]. Examples of leading azimuthal asymmetries, appearing for polarized leptoproduction, are

$$\left\langle \frac{Q_T}{M} \sin\left(\phi_h^\ell - \phi_S^\ell\right) \right\rangle_{OT} = \frac{2\pi\alpha^2 s}{Q^4} |S_T| \left(1 - y + \tfrac{1}{2} y^2\right)$$
$$\times \sum_{a,\bar{a}} e_a^2 x_B f_{1T}^{\perp(1)a}(x_B) D_1^a(z_h), \tag{7.165}$$

$$\left\langle \frac{Q_T}{M_h} \sin(\phi_h^\ell + \phi_S^\ell) \right\rangle_{OT} = \frac{4\pi\alpha^2 s}{Q^4} |S_T| (1 - y)$$
$$\times \sum_{a,\bar{a}} e_a^2 x_B h_1^a(x_B) H_1^{\perp(1)a}(z_h). \tag{7.166}$$

Here $\langle W \rangle$ denotes the q_T-integrated cross section with weight W. The factor Q_T is included, because it and the direction $\hat{h}$ combine to form q_T. This allows the cross section to be separated into distribution and fragmentation parts, one of them weighted with transverse momentum.

Note that both of these asymmetries involve T-odd functions, which can only appear in single-spin asymmetries. The latter can easily be checked from the

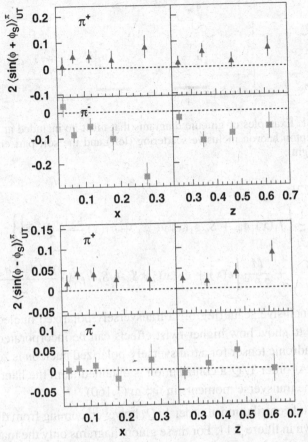

Figure 7.10. Weighted asymmetries for the Collins and Sivers angles [33,34] (see (7.165) and (7.166)) obtained in semi-inclusive single-spin asymmetries measured on a transversely-polarized hydrogen target by the HERMES collaboration at DESY [35]. The error bars represent the statistical uncertainties.

conditions on the hadronic tensor, which are the same as those in (7.11)–(7.13). They require an odd number of spin vectors entering in the symmetric part and an even number of spin vectors entering in the antisymmetric part of the hadron tensor. The results of single-spin asymmetries in SIDIS measurements on a transversely-polarized target from HERMES [35] are shown in figure 7.10. An extended review of transverse momentum-dependent functions and transversity can be found in [36].

7.15 Inclusion of subleading contributions

If one proceeds up to order $1/Q$, one also needs terms in the parametrization of the soft part proportional to M/P^+. Limiting ourselves to the p_T-integrated

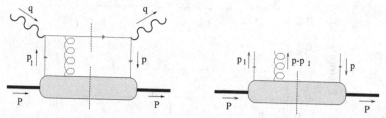

Figure 7.11. Examples of gluonic diagrams that must be included at subleading order in lepton hadron inclusive scattering (left) and the soft part entering this process (right).

correlations [22],

$$\Phi(x) = \frac{1}{2} \left\{ f_1(x) \, \slashed{n}_+ + S_L \, g_1(x) \, \gamma_5 \, \slashed{n}_+ + h_1(x) \frac{\gamma_5 \, [\slashed{S}_T, \slashed{n}_+]}{2} \right\}$$

$$+ \frac{M}{2P^+} \left\{ e(x) + g_T(x) \, \gamma_5 \, \slashed{S}_T + S_L \, h_L(x) \frac{\gamma_5 \, [\slashed{n}_+, \slashed{n}_-]}{2} \right\}. \quad (7.167)$$

We shall use inclusive scattering off a transversely-polarized nucleon ($|S_\perp| = 1$) as an example to show how higher-twist effects can be incorporated in the cross section. The hadronic tensor for a transversely-polarized nucleon is zero to leading order in $1/Q$. At order $1/Q$ a contribution is obtained from the handbag diagram that involves the transverse moments in Φ_∂^α in (7.160).

There is a second contribution at order $1/Q$, however, coming from diagrams such as the one shown in figure 7.11. For these gluon diagrams only the matrix elements of the bilocal combinations $\overline{\psi}(0) \, g A_T^\alpha(\xi) \, \psi(\xi)$ and $\overline{\psi}(0) \, g A_T^\alpha(0) \, \psi(\xi)$ are necessary. The $\Phi_A^\alpha(x)$ and $\Phi_\partial^\alpha(x)$ contributions sum to $\Phi_D^\alpha(x)$ involving matrix elements of bilocal combinations $\overline{\psi}(0) \, i D_T^\alpha \, \psi(\xi)$ for which the QCD equations of motion can be used to relate them to the functions appearing in Φ:

$$\Phi_D^\alpha(x) = \frac{M}{2} \left\{ -\left(x \, g_T - \frac{m}{M} h_1 \right) S_T^\alpha \, \slashed{n}_+ \gamma_5 - S_L \left(x \, h_L - \frac{m}{M} g_1 \right) \frac{[\gamma^\alpha, \slashed{n}_+] \gamma_5}{2} \right\}. \quad (7.168)$$

For example, the distribution function $g_T(x_B, Q^2)$, defined in (7.42), appears in the corresponding structure function of polarized inclusive deep inelastic scattering:

$$2M \, W_A^{\mu\nu}(q, P, S_T) = i \frac{2M x_B}{Q} [\hat{t}^\mu \epsilon_\perp^{\nu\rho} - \hat{t}^\nu \epsilon_\perp^{\mu\rho}] S_{\perp\rho} \, g_T(x_B, Q^2), \quad (7.169)$$

leading to the result

$$g_T(x_B, Q^2) = \frac{1}{2} \sum_q e_q^2 \big(g_T^q(x_B, Q^2) + g_T^{\bar{q}}(x_B, Q^2) \big). \quad (7.170)$$

The process of integrating the correlation functions over p^-, p_T and finally over p^+, consecutively restrains the non-locality to lightfront-separated fields, light-cone-separated fields and local fields. This allows interesting relations to be derived. For example, the correlators $\int dx\, \Phi^{[\gamma^\mu \gamma_5]}(x)$ must yield $g_A\, S_T^\mu$, for any μ, which means that the functions in the non-local correlators $\Phi^{[\gamma^+ \gamma_5]}(x)$ and $\Phi^{[\gamma^\alpha \gamma_5]}(x)$, with α transverse, yield the same result after integration over x, $\int dx\, g_1(x) = \int dx\, g_T(x)$. They also give the Burkhardt–Cottingham sum rule $\int dx\, g_2(x) = 0$ [37]. For quark–quark correlators, similar considerations yield relations between the subleading functions and the transverse momentum-dependent leading functions, referred to as Lorentz invariance relations, such as [38,31]

$$g_T = g_1 + \frac{d}{dx} g_{1T}^{(1)}, \tag{7.171}$$

although these relations may be too naive if one includes gauge links (see section 7.16). An interesting result is obtained by combining the relation (7.171) with an often-used approximation, in which the interaction-dependent part Φ_A^α is set to zero. In that case the difference $\Phi_D^\alpha - \Phi_\partial^\alpha$ vanishes. Using (7.168) and (7.160) this gives

$$x\,\tilde{g}_T = x\,g_T - g_{1T}^{(1)} - \frac{m}{M}\,h_1 = 0. \tag{7.172}$$

As an application of the relations between twist-three functions and transverse momentum-dependent functions in combination with the Lorentz invariance relations, $g_{1T}^{(1)}$ can be eliminated using (7.171) to obtain a relation between g_T, g_1 and $\tilde{g}_T$. The relation for $g_2 = g_T - g_1$ takes the form

$$g_2(x) = -\left[g_1(x) - \int_x^1 dy\, \frac{g_1(y)}{y} \right] + \frac{m}{M}\left[\frac{h_1(x)}{x} - \int_x^1 dy\, \frac{h_1(y)}{y^2} \right]$$
$$+ \left[\tilde{g}_T(x) - \int_x^1 dy\, \frac{\tilde{g}_T(y)}{y} \right]. \tag{7.173}$$

Neglecting the interaction-dependent part, that is setting $\tilde{g}_T(x) = 0$, the Wandzura–Wilczek approximation [39] for g_2 is obtained. This provides a simple and often-used estimate for g_2 when the quark mass term is neglected.

7.16 Colour gauge invariance

So far two issues have been disregarded. The first is that the correlation functions Φ involve two quark fields at different space-time points and hence are not colour gauge invariant. The second issue is that in the gluonic diagrams, such as figure 7.11, there are correlation functions involving matrix elements with longitudinal

(A^+) gluon fields,

$$\overline{\psi}_j(0)\, g A^+(\eta)\, \psi_i(\xi).$$

These do not lead to any suppression. The $+$ index in the gluon field causes the matrix element to be proportional to P^+, p^+ or $M\,S^+$, rather than the proportionality to $M\,S_T^\alpha$ or p_T^α that one gets for a gluonic matrix element with transverse gluons.

A straightforward calculation, however, shows that the gluonic diagrams with one or more longitudinal gluons involve matrix elements (soft parts) of operators $\overline{\psi}\psi$, $\overline{\psi}\,A^+\psi$, $\overline{\psi}\,A^+A^+\psi$, etc. that can be resummed into a correlation function

$$\Phi_{ij}(x) = \int \frac{d\xi^-}{2\pi}\, e^{ip\cdot\xi}\,\langle P,S|\overline{\psi}_j(0)\mathcal{U}(0,\xi)\,\psi_i(\xi)|P,S\rangle\Bigg|_{\xi^+=\xi_T=0}, \qquad (7.174)$$

where $\mathcal{U}$ is a gauge link operator

$$\mathcal{U}(0,\xi) = \mathcal{P}\exp\left(-i\int_0^{\xi^-} d\zeta^-\, A^+(\zeta)\right), \qquad (7.175)$$

a path-ordered exponential with the path along the negative direction. The unsuppressed gluonic diagrams combine into a colour gauge invariant correlation function [40]. We note that at the level of operators one expands

$$\overline{\psi}(0)\psi(\xi) = \sum_n \frac{\xi^{\mu_1}\cdots\xi^{\mu_n}}{n!}\,\overline{\psi}(0)\partial_{\mu_1}\cdots\partial_{\mu_n}\psi(0), \qquad (7.176)$$

in a set of local operators. However, only the expansion of the non-local combination with a gauge link

$$\overline{\psi}(0)\mathcal{U}(0,\xi)\,\psi(\xi) = \sum_n \frac{\xi^{\mu_1}\cdots\xi^{\mu_n}}{n!}\,\overline{\psi}(0)D_{\mu_1}\cdots D_{\mu_n}\psi(0), \qquad (7.177)$$

is an expansion in terms of local gauge-invariant operators. The latter operators are precisely the local (quark) operators that appear in the operator-product expansion applied to inclusive deep inelastic scattering.

For the p_T-dependent functions, one finds that inclusion of A^+ gluonic diagrams leads to a colour gauge invariant matrix element with links running via $\xi^- = \pm\infty$ [41,42]. For instance, in lepton–hadron scattering

$$\Phi(x,p_T) = \int \frac{d\xi^- d^2\xi_T}{(2\pi)^3}\, e^{ip\cdot\xi}\,\langle P,S|\overline{\psi}(0)\mathcal{U}^{[+]}(0,\xi)\,\psi(\xi)|P,S\rangle\Bigg|_{\xi^+=0}, \qquad (7.178)$$

where the link $\mathcal{U}^{[+]}$ is shown in figure 7.12(a).

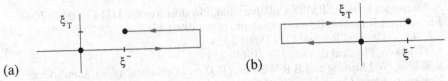

Figure 7.12. The gauge link structure in the quark–quark correlator Φ in SIDIS (a) and Drell–Yan process (b) respectively.

We note that the gauge link involves transverse gluons [43,44], showing that in processes involving more hadrons the effects of transverse gluons are not necessarily suppressed. This has also been demonstrated in explicit model calculations [45].

Moreover, depending on the process the gauge link can also run via minus infinity, involving the link in figure 7.12(b). This is, for instance, the case in Drell–Yan processes. Here again the transverse momentum-dependent distribution functions are no longer constrained by time reversal, as the time-reversal operation inter-changes the $U^{[+]}$ and $U^{[-]}$ links, leading to the appearance of T-odd functions in (7.160). The process dependence of the gauge link, however, points to particular sign changes when single-spin azimuthal asymmetries in semi-inclusive leptoproduction are compared to those in Drell–Yan scattering. For such effects the measurement of transverse momentum dependence is a must, since the specific link structure does not matter in p_T-integrated functions, in which both links in figure 7.12 reduce to the same straight-line link connecting 0 and ξ.

References

[1] R G Roberts, *The Structure of the Proton*, Cambridge University Press, Cambridge (1990)
[2] E Leader, *Spin in Particle Physics*, Cambridge University Press, Cambridge (2001)
[3] R Meng, F I Olness and D E Soper, Nuclear Physics B371 (1992) 79
[4] R K Ellis, W Furmanski and R Petronzio, Nuclear Physics B212 (1983) 29; *ibid* B207 (1982) 1
[5] A V Efremov and O V Teryaev, Soviet Journal of Nuclear Physics 39 (1984) 962
[6] R L Jaffe in *Relativistic Dynamics and Quark Nuclear Physics*, M B Johnson and A Picklesheimer eds, Wiley, New York (1986)
[7] A D Martin, R G Roberts, W J Stirling and R S Thorne, Physics Letters B604 (2004) 61
[8] Durham-RAL databases, http://durpdg.dur.ac.uk/HEPDATA/ Particle Data Group, http://pdg.web.cern.ch/pdg/
[9] J B Kogut and D E Soper, Physical Review D1 (1970) 2901
[10] R L Jaffe, Nuclear Physics B229 (1983) 205
[11] M Diehl and T Gousset, Physics Letters B428 (1998) 359
[12] K Gottfried, Physical Review Letters 18 (1967) 1174
[13] D Allasia *et al*, NMC Collaboration, Physics Letters B249 (1990) 366
[14] J Ashman *et al*, Physics Letters B206 (1988) 364; Nuclear Physics B328 (1989) 1

[15] A Airapetian *et al*, HERMES Collaboration, Physical Review D71 (2005) 012003
[16] J S Bell and R Jackiw, Il Nuovo Cimento 60A (1969) 47;
S L Adler, Physical Revjew 177 (1969) 2426
[17] J D Bjorken, Physical Review 179 (1969) 1547
[18] R K Ellis, W J Stirling and B R Webber, *QCD and Collider Physics*, Cambridge University Press, Cambridge (1996)
[19] J Kodaira *et al*, Physical Review D20 (1979) 627;
S A Larin, F V Tkachev and J A M Vermaseren, Physics Letters B259 (1991) 345;
Physical Review Letters 66 (1991) 862
[20] D E Soper, Physical Review D15 (1977) 1141; Physical Review Letters 43 (1979) 1847
[21] J C Collins and D E Soper, Nuclear Physics B194 (1982) 445
[22] R L Jaffe and X Ji, Nuclear Physics B375 (1992) 527
[23] X Artru and M Mekhfi, Zeitschrift für Physik C45 (1990) 669
J L Cortes, B Pire and J P Ralston, Zeitschrift für Physik C55 (1992) 409
[24] J Soffer, Physical Review Letters 74 (1995) 1292
[25] J P Ralston and D E Soper, Nuclear Physics B152 (1979) 109
[26] R D Tangerman and P J Mulders, Physical Review D51 (1995) 3357
[27] A Bacchetta, M Boglione, A Henneman and P J Mulders, Physical Review Letters 85 (2000) 712
[28] J C Collins, Nuclear Physics B396 (1993) 161
[29] K Hagiwara, K Hikasa and N Kai, Physical Review D27 (1983) 84
[30] R L Jaffe and X Ji, Physical Review Letters 71 (1993) 2547; Physical Review D57 (1998) 3057
[31] P J Mulders and R D Tangerman, Nuclear Physics B461 (1996) 197; *ibid* B484 (1997) 538 (E)
[32] D Boer and J Mulders, Physical Review D57 (1998) 5780
[33] D Sivers, Physical Review D41 (1990) 81; *ibid* D43 (1991) 261
[34] J C Collins, Nuclear Physics B396 (1993) 161
[35] A Airapetian *et al*, HERMES Collaboration, Physical Review Letters 94 (2005) 012002
[36] V Barone, A Drago and P G Ratcliffe, Physics Reports 359 (2002) 1
[37] H Burkhardt and W N Cottingham, Annals of Physics (N.Y.) 56 (1976) 453
[38] A P Bukhvostov, E A Kuraev and L N Lipatov, Soviet Physics JETP 60 (1984) 22
[39] S Wandzura and F Wilczek, Physical Review D16 (1977) 707
[40] A V Efremov and A V Radyushkin, Theoretical and Mathematical Physics 44 (1981) 774
[41] D Boer and P J Mulders, Nuclear Physics B569 (2000) 505
[42] J C Collins, Physics Letters B536 (2002) 43
[43] A V Belitsky, X Ji and F Yuan, Nuclear Physics B656 (2003) 165
[44] D Boer, P J Mulders and F Pijlman, Nuclear Physics B667 (2003) 201
[45] S J Brodsky, D S Hwang and I Schmidt, Physics Letters B530 (2002) 99; Nuclear Physics B642 (2002) 344

8

Diffraction and colour dipoles

J Forshaw and G Shaw

Diffractive photoprocesses are both copious and varied, and their theoretical inter-pretation involves an interplay of perturbative and non-perturbative ideas which presently defies a rigorous treatment in QCD. Nonetheless, considerable progress has been made, and the colour-dipole model has emerged as a unifying frame-work for discussing the wide range of phenomena observed. In this chapter we will discuss the main features of diffractive photoprocesses, their interpretation in the colour-dipole model and the relation between this model and other theoretical approaches.

8.1 Diffractive processes

Diffraction is a high-energy phenomenon, illustrated in figure 8.1 where A and B may be photons or hadrons and Y and Z may be single particles or an inclusive sum over $n \geq 1$ particle states. The dashed line indicates an exchange of energy, momentum and angular momentum, but no non-zero colour, flavour or isospin quantum numbers may be exchanged and charge conjugation is positive. In QCD, this exchange is assumed to be mediated by two or more gluons in a colour-singlet state. High energy means that the square of the centre-of-mass energy $s = W^2$ is much larger than any other energy scale:

$$s \gg -t, m_Y^2, m_Z^2 \ldots .$$

For diffractive processes initiated by virtual photons, the latter include the photon virtuality Q^2, implying

$$x \approx Q^2/s \ll 1.$$

Electromagnetic Interactions and Hadronic Structure, eds Frank Close, Sandy Donnachie and Graham Shaw.
Published by Cambridge University Press. © Cambridge University Press 2007

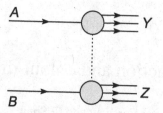

Figure 8.1. The generic mechanism for diffraction, where the dotted line indicates vacuum exchange and the incoming particles A, B can be hadrons or photons. In the final state Y and Z can be single particles or inclusive sums over hadrons.

We note that $-t, m_Y^2, m_Z^2, Q^2$ etc. can themselves become large, provided they remain much smaller than s. This region is of particular theoretical interest, since the presence of a hard scale opens up the subject to perturbative QCD.

Diffractive processes are characterized experimentally by two distinctive features: rising cross sections and rapidity gaps. The two groups of final-state particles in figure 8.1 emerge in roughly the forward and backward directions in the centre-of-mass frame and are well separated in rapidity or pseudo-rapidity:

$$\eta = -\log \tan \left(\tfrac{1}{2}\theta\right),$$

where θ is the polar angle with respect to the beam direction. Such rapidity gaps are characteristic of colour-singlet exchange, which can be contrasted to the typically copious particle production that arises as a result of the breaking of colour strings in colour-exchange processes. Rapidity gaps occur not only in diffractive processes but also, for example, in colour-singlet meson-exchange processes. However, meson exchange gives rise to cross sections which fall with increasing energy: see chapter 5. Nonetheless at finite energies one may need to take account of small contributions from the exchange of flavour-singlet meson-exchange contributions that can interfere in general with the dominant diffractive process.

The precise energy dependence of diffraction depends upon the process and kinematic region observed. For processes involving only hadrons and in the absence of any hard scale (that is an energy scale much larger than the typical hadronic scale $\sim 200\,\mathrm{MeV}$), the energy dependence of all observed diffractive processes can be described by the exchange of a simple Regge pole, the soft pomeron, with the universal trajectory [1]

$$\alpha_P(t) \approx 1.08 + 0.25t, \tag{8.1}$$

discussed in chapter 5. The fact that the intercept $\alpha_P(0)$ is close to unity reflects the fact that these diffractive cross sections rise slowly with energy. The total cross sections for hadron–hadron scattering, for example, all increase as $s^{\alpha_P(0)-1} \approx s^{0.08}$

at high energies. To a good approximation, the same behaviour applies to the total cross section for real photo absorption

$$\gamma + p \to X, \tag{8.2}$$

where X represents an inclusive sum over hadronic states, while the energy dependence of the photoproduction processes

$$\gamma + p \to V + p \qquad V = \rho, \omega, \phi \tag{8.3}$$

is also found to be compatible with the 'soft pomeron' behaviour (8.1). For this and other reasons, early studies of diffractive photoprocesses emphasized the 'hadron-like' behaviour of real photoprocesses, and their interpretation via simple vector dominance ideas [2]. This type of behaviour is now loosely, but conveniently, referred to as 'soft diffraction.'

This picture was changed dramatically with the advent of HERA, which enabled an impressive variety of diffractive photoprocesses to be studied at very high energies. These processes include deep inelastic scattering (DIS)

$$\gamma^* + p \to X \tag{8.4}$$

that is related to elastic virtual Compton scattering

$$\gamma^* + p \to \gamma^* + p \tag{8.5}$$

by the optical theorem, and the diffractive deep inelastic scattering (DDIS) process

$$\gamma^* + p \to X + p, \tag{8.6}$$

where the hadronic state X is separated from the proton by a rapidity gap. The exclusive processes observed include deeply-virtual Compton scattering (DVCS)

$$\gamma^* + p \to \gamma + p, \tag{8.7}$$

where the final-state photon is real, and the vector-meson production processes

$$\gamma^* + p \to V + p, \tag{8.8}$$

where $V = \rho, \omega, \phi$ or J/ψ.

Of even greater significance was the fact that HERA enabled these processes to be studied in the regime where the photon virtuality Q^2 is large compared to the typical hadronic scale, but nonetheless is very small compared to the energy variable $s = W^2$, as required for diffraction. In this region, the cross sections rise much more steeply than would be expected from the soft-pomeron behaviour (8.1), a phenomenon that is loosely but conveniently referred to as 'hard diffraction'.

More generally, if different data sets are each parametrized by a single effective Regge-pole-exchange formula, the intercept is found to vary roughly in the range

$$1.08 \leq \alpha_{eff}(0) \leq 1.4, \tag{8.9}$$

depending on the value of Q^2 and the particular process observed. Of course, this implies that diffractive photoprocesses cannot really be described by a single Regge-pole exchange, since this would require a universal intercept in all reactions. Hard diffraction seems to occur whenever any hard scale is present. For example, even for real photons, J/ψ photoproduction has a significantly steeper energy dependence than ρ or ϕ photoproduction, due to the presence of the relatively hard scale provided by the mass of the J/ψ.

These processes involve an interplay of perturbative and non-perturbative physics that currently defies a rigorous treatment in QCD. Rather there are several models, each of which throws light on different aspects of the problem, with varying degrees of success. Here we focus on the 'colour-dipole model' that provides a unified description of the various processes (8.2)–(8.8) and allows different dynamical scenarios to be explored. Before doing so, we introduce some more general ideas that we hope will help develop physical insight.

8.1.1 Diffraction in the laboratory frame

Throughout this chapter, we shall concern ourselves with the scattering of a photon off a proton. In general the photon may have some space-like virtuality Q^2. Different reference frames are conveniently chosen to emphasize different aspects of the physics. In the infinite-momentum frame, at large Q^2, the parton distribution functions have a simple interpretation: they describe the longitudinal momentum distribution of point-like constituents of the proton target. In this frame the virtual photon is regarded as point-like. However, in the laboratory frame the incoming photon typically develops hadronic fluctuations a long distance from the proton target and the intermediate states into which it converts can reasonably be regarded as constituents of the photon. The typical distance travelled by a vacuum fluctuation of a photon into a hadron of invariant mass m_h is given by the *coherence length* [3], which is just the inverse of the longitudinal momentum required to put it on mass-shell, and is given at high energies $s \gg m_h^2, Q^2$ by

$$l_c = \frac{s}{M\left(Q^2 + m_h^2\right)} \gg M^{-1} \approx 0.2 \text{ fm},$$

where M is the proton mass. In photoproduction $Q^2 = 0$ and the dominant hadronic contributions h are usually assumed to be the light vector mesons, so that typically

$$l_c \approx s/(M m_\rho^2) \qquad Q^2 \approx 0,$$

which, in the diffractive region, is very large compared to the size of the proton or the range of the strong interaction. Hence the contribution from a particular intermediate vector-meson state can to a good approximation be factorized into an amplitude for the photon to convert to the vector meson, times an amplitude describing the interaction of the vector meson with the proton target. This is the conceptual basis of vector-meson-dominance models of diffraction [2] and is well known (see chapter 5).

What is perhaps not so well known is that the same space-time picture and factorization property generalizes to diffractive photoprocesses at large Q^2. If we consider DIS at large Q^2, the typical value of m_h^2 is itself of order Q^2 in models that give scaling. Hence the typical value of the coherence length is

$$\bar{l}_c \sim 1/Mx, \tag{8.10}$$

whether the fluctuations are $q\bar{q}$ pairs (as in the parton model) or hadrons (as in hadron-dominance models). This typical coherence length is again very long in the diffractive region, so that once again the contribution from a particular vacuum fluctuation h can, to a good approximation, be factorized into an amplitude for the photon to convert to the state h times an amplitude describing the interaction of h with the proton target. Different models are then defined by the choice of dominant hadronic states h and the nature of their interactions with the target.

8.2 The dipole framework

In the colour-dipole model [4,5] the dominant states h are assumed to be quark–antiquark pairs, characterized by their transverse size r and by the fraction z of the light-cone momentum of the pair carried by the quark. Such states are called 'colour dipoles' and are assumed to be eigenstates of diffraction; that is they scatter without change of r and z in the diffractive limit. This statement can be justified in perturbative QCD. Beyond that, like the dominance of $q\bar{q}$ pairs, it is a model-dependent assumption that must be tested against experiment. The resulting mechanism for the generic process

$$\gamma^* + p \to A + p \qquad A = \gamma^*, \gamma, \rho, \ldots \tag{8.11}$$

is shown in figure 8.2, and leads to amplitudes of the schematic form

$$\mathcal{A}(\gamma + p \to A + p) = \int d\tau \Psi_\gamma^{in} \, T \, \Psi_A^{out}, \tag{8.12}$$

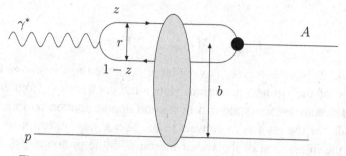

Figure 8.2. The colour-dipole model for $\gamma^* + p \rightarrow A + p$.

where T is the scattering amplitude of a given dipole configuration with the target and the integral runs over all dipole configurations. For the moment we ignore kinematic factors.

The strength of this approach lies in the fact that the dipole scattering amplitude is the same for all processes. However, within this common framework there are many different formulations for the interaction amplitude T that have been applied with varying degrees of success. In the rest of this section we first formulate the dipole model in more detail, obtaining the precise form for relations like (8.12), before discussing different formulations for the form of the dipole scattering amplitude. We start by discussing the dipole fluctuations themselves.

8.2.1 Dipole states

In the light-cone quantization approach [6], the vacuum state of the free Hamiltonian is also an eigenstate of the full Hamiltonian. Hence an eigenstate of the latter can be expanded in terms of the eigenstates of the free light-cone Hamiltonian, in which each free parton i is characterized by its light-cone energy $k_i^+ = (k_i^0 + k_i^3)/2$, together with other appropriate variables. Hence a hadronic state $|\Psi\rangle$ can be expanded in the form

$$|\Psi\rangle = \sum_n \Psi_n |n\rangle, \tag{8.13}$$

where Ψ_n are the light-cone wave functions of the hadronic state and each colour-singlet partonic state $|n\rangle$ is characterized by the transverse momenta $\vec{k}_{\perp i}$ and helicities λ_i of the partons i, together with their longitudinal momenta k_i^+. However, the diffractive eigenstates are believed to be those for which the partons have fixed impact parameters [7]. It will therefore be more convenient for what follows to choose a mixed configuration–momentum space description, in which the transverse momenta $\vec{k}_{\perp i}$ are replaced by their conjugate transverse position vectors $\vec{b}_i$. At the same time, the light-cone energies k_i^+ are replaced by the fractional light-cone energies

$$x_i = \frac{k_i^+}{P^+} \qquad \sum_i x_i = 1, \qquad (8.14)$$

where

$$P^+ = \sum_i k_i^+ \qquad P_\perp = \sum_i \vec{k}_{\perp i}$$

are the light-cone energy and transverse momentum of the hadron state. In other words, the expansion (8.13) takes the form

$$|\Psi\rangle = \sum_i \Psi(x_i, \vec{b}_i, \lambda_i) |x_i, \vec{b}_i, \lambda_i\rangle, \qquad (8.15)$$

where the light-cone wave functions $\Psi(x_i, \vec{b}_i, \lambda_i)$ are invariant under a Lorentz boost in the longitudinal z-direction.

In the colour-dipole model [4,5] we assume that a similar decomposition

$$|\gamma^*\rangle_H = \sum \Psi_{q\bar{q}} |q\bar{q}\rangle + \sum |\Psi_{q\bar{q}g} |q\bar{q}g\rangle + \text{higher Fock states} \qquad (8.16)$$

can be applied to the hadronic fluctuations of the photon, and that, in the diffractive processes (8.11), the $q\bar{q}$ states dominate. We shall discuss this approximation in more detail subsequently and will go beyond it in our discussion of the DDIS process. Denoting their transverse position vectors by $\vec{b}_1$ and $\vec{b}_2$ respectively, we define the impact vector of the pair by

$$\vec{b} = \tfrac{1}{2}(\vec{b}_1 + \vec{b}_2) \qquad b = |\vec{b}| \qquad (8.17)$$

and the relative transverse position vector by

$$\vec{r} = \vec{b}_1 - \vec{b}_2 \qquad r = |\vec{r}|, \qquad (8.18)$$

where r is referred to as the transverse size, or just the size, of the dipole. Finally the expansion (8.16) becomes

$$|\gamma^*, \lambda\rangle_H = \sum_{h\bar{h}} \int dz \, d^2\vec{r} \; \Psi_{h,\bar{h}}^{\gamma,\lambda}(\vec{r}, z) |z, \vec{b}_i, h, \bar{h}\rangle + \text{higher Fock states}, \quad (8.19)$$

where λ is the photon helicity, $h, \bar{h}$ are the helicities of the quark and antiquark respectively, and we have incorporated the fact that the probability of finding a particular $q\bar{q}$ configuration is independent of its impact vector $\vec{b}$.

8.2.2 The dipole cross section

We now consider the diffractive processes (8.11) shown in figure 8.2. Implicit in this diagram is the central assumption that, in the diffractive limit, the dipole states are eigenstates of the scattering operator, with spin-independent eigenvalues:*

$$\hat{T}|z, \vec{b}_i, h, \bar{h}\rangle = \tau(\vec{b}, s; z, \vec{r})|z, \vec{b}_i, h, \bar{h}\rangle. \tag{8.20}$$

The corresponding amplitude for elastically scattering a dipole from the proton is given by

$$A(s, \vec{r}, z, \vec{P}_\perp) = \int d^2\vec{b}\; e^{i\vec{P}_\perp \cdot \vec{b}} \langle z, \vec{b}_i, h, \bar{h}|\hat{T}|z, \vec{b}_i, h, \bar{h}\rangle,$$

where $\vec{P}_\perp$ is the transverse momentum of the outgoing proton.

Appealing to the optical theorem (setting $-t \approx \vec{P}_\perp^2 = 0$), we therefore refer to

$$\sigma(s, r, z) = \int d^2\vec{b}\; \frac{\mathrm{Im}\,\tau(\vec{b}\,s; z, \vec{r})}{s} \tag{8.21}$$

as the 'dipole cross section' and interpret it as the total cross section for scattering a dipole of size r from a proton. Note that our notation makes explicit the assumption that the dipole cross section is independent of the dipole orientation and the helicities of the quark and antiquark.

8.2.3 Reaction cross sections

The dipole cross section is important because, given the light-cone wave functions, it is all that is required to calculate diffractive processes in the forward direction. The differential cross section for the exclusive process $\gamma^* + p \to A + p$ is given by

$$\left.\frac{d\sigma}{dt}\right|_{t=0} = \frac{1}{16\pi s^2}|\langle A|\hat{T}|\gamma^*, \lambda\rangle(s, t = 0)|^2, \tag{8.22}$$

where $\lambda = L, T$ for longitudinal and transverse photons respectively. The forward amplitude can be expressed in terms of the scattering amplitude at fixed impact parameter $\vec{b}$ by

$$\langle A|\hat{T}|\gamma^*, \lambda\rangle(s, t = 0) = \int d^2\vec{b}\; \langle A|\hat{T}|\gamma^*, \lambda\rangle(s, \vec{b}), \tag{8.23}$$

---

* A factor i is often inserted into the right-hand side of this equation, so that $\tau(\vec{b}, s; z, \vec{r})$ is real if the scattering amplitude is purely imaginary. We do not follow that convention here.

which in turn is given by

$$\langle A|\hat{T}|\gamma^*, \lambda\rangle(s, \vec{b})$$
$$= \sum_{h,\bar{h}} \int dz \, d^2\vec{r} \, \langle A|z, \vec{b}_i, h, \bar{h}\rangle\langle z, \vec{b}_i, h, \bar{h}|\hat{T}|z, \vec{b}_i, h, \bar{h}\rangle\langle z, \vec{b}_i, h, \bar{h}|\gamma^*, \lambda\rangle, \quad (8.24)$$

where we have summed over all dipole configurations and neglected all higher Fock states. Using (8.19) and (8.21) together with (8.23), one obtains

$$\text{Im}\langle A|\hat{T}|\gamma^*, \lambda\rangle(s, t = 0)$$
$$= s \sum_{h,\bar{h}} \int dz \, d^2\vec{r} \, \Psi_{h,\bar{h}}^A(\vec{r}, z)^* \, \Psi_{h,\bar{h}}^{\gamma,\lambda}(\vec{r}, z) \, \sigma(s, r, z). \quad (8.25)$$

The forward differential cross section (8.22) can then be calculated by substituting (8.25) into (8.22) and either neglecting the correction from the real part of the amplitude or, preferably, estimating it using dispersion-relation techniques. The total reaction cross section is then usually estimated by assuming an exponential ansatz for the t dependence, giving

$$\sigma_{L,T}(\gamma^* p \to Ap) = \frac{1}{B} \frac{d\sigma^{T,L}}{dt}\Big|_{t=0}, \quad (8.26)$$

and the value of the slope parameter B is taken from experiment. Some of the most interesting results are those obtained for the inclusive processes (8.4) (DIS) and (8.6) (DDIS), for which there exists a wealth of data as functions of x and Q^2 and, in the case of DDIS, of the mass of the diffractively produced state m_X^2. We start by considering the total virtual photoabsorption cross sections $\sigma_{L,T}(\gamma^* p \to X)$ which define the proton structure functions

$$2x F_1(x, Q^2) = \frac{Q^2}{4\pi^2 \alpha_{em}} \sigma_T, \qquad F_2(x, Q^2) = \frac{Q^2}{4\pi^2 \alpha_{em}}(\sigma_L + \sigma_T). \quad (8.27)$$

From the optical theorem

$$\sigma_\lambda = s^{-1}\text{Im}\langle\gamma^*, \lambda|\hat{T}|\gamma^*, \lambda\rangle(s, t = 0), \qquad \lambda = L, T,$$

which, together with (8.25), gives the result

$$\sigma_{L,T}(x, Q^2) = \int dz \, d^2\vec{r} \, |\Psi_\gamma^{L,T}(r, z)|^2 \sigma(s, r, z), \quad (8.28)$$

where

$$|\Psi_\gamma^{L,T}(r, z)|^2 \equiv \sum_{h,\bar{h}} |\Psi_{h\bar{h}}^{\gamma, L,T}(\vec{r}, z)|^2. \quad (8.29)$$

Turning to the diffractive DIS reaction (8.6), we can express the final state as an incoherent sum over the diffractive eigenstates (that is dipole states):

$$\frac{d\sigma^D_{T,L}}{dt}\bigg|_{t=0} = \frac{1}{16\pi s^2} \sum_{all\ dipoles} |\langle \gamma^{T,L}|\hat{T}|z,r\rangle|^2 \tag{8.30}$$

and hence

$$\frac{d\sigma^D_{T,L}}{dt}\bigg|_{t=0} = \frac{1}{16\pi} \int dz d^2\vec{r}\ |\Psi^{T,L}_\gamma(r,z)|^2 \sigma^2(s,r,z). \tag{8.31}$$

Note that this result includes a sum over all forward diffraction products X, and hence over all corresponding masses m_X, whereas experiments also measure the dependence on the mass m_X. To make comparison with this requires more elaborate expressions, which we will introduce later. In addition, only that subset of the diffractive-dissociation final state that is composed exclusively of a quark–antiquark pair has been included in (8.31). As we shall see later, when we consider the case of large values of m_X, we will need to go beyond this approximation.

8.3 Light-cone wave functions

We see from the above results that the crucial ingredients of the model are the colour-dipole cross section (8.21), which will be discussed in the following section, and the light-cone wave functions. Among the wave functions, those of the photon are particularly important, since they are the only wave functions required to calculate several of the most important processes, like DIS (8.4), DDIS (8.6) and virtual Compton scattering (8.7). Furthermore, while the photon light-cone wave functions can be calculated within QED, at least for small dipole sizes, the vector-meson wave functions are not reliably known and must be obtained from models.

The light-cone wave functions $\Psi_{h,\bar{h}}(\vec{r},z)$ in the mixed representation used in the dipole model are obtained from a two-dimensional Fourier transform

$$\Psi_{h,\bar{h}}(\vec{r},z) = \int \frac{d^2\vec{k}}{(2\pi)^2} e^{i\vec{k}\cdot\vec{r}} \Psi_{h,\bar{h}}(\vec{k},z) \tag{8.32}$$

of the momentum space light-cone wave functions $\Psi_{h,\bar{h}}(\vec{k},z)$. For small transverse separations it is reasonable to neglect gluonic interactions between the quark and antiquark in calculating the photon light-cone wave functions. The probability amplitude for the photon to fluctuate into a $q\bar{q}$ pair in momentum space can then be calculated in lowest-order QED:

$$\Psi^{\gamma,\lambda}_{h,\bar{h}}(\vec{k},z) = \sqrt{\frac{N_c}{4\pi}} \frac{\bar{u}_h(\vec{k})}{\sqrt{z}} (ee_f \gamma.\varepsilon^\lambda_\gamma) \frac{v_{\bar{h}}(-\vec{k})}{\sqrt{1-z}} \Phi^\gamma(k,z), \tag{8.33}$$

per fermion of charge ee_f. Here the ε^λ are the polarization vectors of the photons, and the 'scalar' part of the photon light-cone wave function, Φ^γ, is given by

$$\Phi^\gamma(k, z) = \frac{z(1-z)}{z(1-z)Q^2 + k^2 + m_f^2}. \tag{8.34}$$

This would be the photon light-cone wave function in a toy model of scalar quarks and photons.

For the vector mesons, the simplest approach is to assume the same vector current as in the photon case, with an additional (unknown) vertex factor $\Gamma_\lambda(k, z)$:

$$\Psi_{h,\bar{h}}^{V,\lambda}(\vec{k}, z) = \sqrt{\frac{N_c}{4\pi}} \frac{\bar{u}_h(\vec{k})}{\sqrt{z}} (\gamma.\varepsilon_V^\lambda) \frac{v_{\bar{h}}(-\vec{k})}{\sqrt{1-z}} \Phi_\lambda^V(k, z), \tag{8.35}$$

where the scalar part of the meson light-cone wave function is given by

$$\Phi_\lambda^V(k, z) = \frac{z(1-z)\Gamma_\lambda(k, z)}{-z(1-z)M_V^2 + k^2 + m_f^2}. \tag{8.36}$$

Different models are defined by specifying these scalar wave functions. In practice, it is common to choose the same functional form for Φ_T^V and Φ_L^V; perhaps allowing the numerical parameters to differ.

Before considering the different cases in more detail, it is instructive to consider the longitudinal wave functions more explicitly. Using the polarization vectors

$$\varepsilon_\gamma^L = \left(\frac{q^+}{Q}, \frac{Q}{q^+}, \vec{0}\right); \qquad \varepsilon_V^L = \left(\frac{v^+}{M_V}, -\frac{M_V}{v^+}, \vec{0}\right), \tag{8.37}$$

it follows that the longitudinal-photon light-cone wave function is

$$\Psi_{h,\bar{h}}^{\gamma,L}(\vec{k}, z) = \sqrt{\frac{N_c}{4\pi}} \delta_{h,-\bar{h}} ee_f \left(\frac{2z(1-z)Q}{k^2 + m_f^2 + z(1-z)Q^2} - \frac{1}{Q}\right) \tag{8.38}$$

and that of the vector meson is

$$\Psi_{h,\bar{h}}^{V,L}(\vec{k}, z) = \sqrt{\frac{N_c}{4\pi}} \delta_{h,-\bar{h}} \left(\frac{z(1-z)2M_V\Gamma(k, z)}{k^2 + m_f^2 - z(1-z)M_V^2} + \frac{\Gamma(k, z)}{M_V}\right). \tag{8.39}$$

On substituting (8.38) in (8.32) the second term of (8.38) leads to a dipole of vanishing size, which does not contribute to the cross section. This is in accord with gauge invariance. The same argument cannot be used to justify the omission of the second term in the meson wave function (8.39), since the latter has a k dependence. In practice, this term is omitted in some models, but retained in others, as we will see. A discussion of the gauge-invariance issues surrounding this point can be found in [8].

8.3.1 The photon wave functions

The normalized photon light-cone wave functions resulting from (8.32) and (8.33) are

$$\Psi_{h,\bar{h}}^{\gamma,L}(\vec{r}, z) = \sqrt{\frac{N_c}{4\pi}}\,\delta_{h,-\bar{h}}\,ee_f\,2z(1-z)Q\frac{K_0(\epsilon r)}{2\pi} \qquad (8.40)$$

and

$$\Psi_{h,\bar{h}}^{\gamma,T(\gamma=\pm)}(\vec{r}, z) = \pm\sqrt{\frac{N_c}{2\pi}}\,ee_f$$

$$\times [ie^{\pm i\theta_r}(z\delta_{h\pm,\bar{h}\mp} - (1-z)\delta_{h\mp,\bar{h}\pm})\partial_r + m_f\delta_{h\pm,\bar{h}\pm}]\frac{K_0(\epsilon r)}{2\pi}, \qquad (8.41)$$

where

$$\epsilon^2 = z(1-z)Q^2 + m_f^2 \qquad (8.42)$$

and $K_0(x)$ and $K_1(x) = -\partial_x K_0(x)$ are modified Bessel functions [9] with the asymptotic behaviours

$$K_0(x) \approx \left(\frac{\pi}{2x}\right)^{1/2} e^{-x} \approx K_1(x) \qquad x \to \infty, \qquad (8.43)$$

$$K_0(x) \approx -\log x \quad K_1(x) \approx x^{-1} \qquad x \to 0. \qquad (8.44)$$

The dipole size $\vec{r}$ has been written in polar form (r, θ_r) and the notation for the quark and antiquark helicities is such that if the photon helicity is ± 1, then the term proportional to z forces the quark to have helicity $\pm\frac{1}{2}$ and the antiquark to have helicity $\mp\frac{1}{2}$. The term proportional to $1 - z$ simply swaps the quark and antiquark and the term proportional to m_f is the helicity-flip term for which the quark and antiquark have equal helicities.

The corresponding results for the spin-summed wave functions (8.29), which are required to evaluate the structure functions (8.27), are

$$\left|\Psi_\gamma^L(r, z)\right|^2 = \frac{6}{\pi^2}\alpha_{\rm em}\sum_{f=1}^{n_f} e_f^2 Q^2 z^2(1-z)^2 K_0^2(\epsilon r), \qquad (8.45)$$

$$\left|\Psi_\gamma^T(r, z)\right|^2 = \frac{3}{2\pi^2}\alpha_{\rm em}\sum_{f=1}^{n_f} e_f^2 \left\{[z^2 + (1-z)^2]\epsilon^2 K_1^2(\epsilon r) + m_f^2 K_0^2(\epsilon r)\right\}, \quad (8.46)$$

where the sum is over all n_f quark flavours f.

It follows from (8.43) that the wave functions decrease exponentially with r at large Q^2 and fixed z. Specifically

$$\Psi(r, z) \sim \exp[-rQz(1-z)],$$

provided $Q^2 z(1-z) \gg m_f^2$, so that only small dipoles can contribute unless z is close to its end-point values 0 or 1. As can be seen from (8.40) and (8.41), these end-points are suppressed for the longitudinal but not for the transverse case. This is the origin of the statement that longitudinal photon processes are more inherently perturbative than transverse photon processes, other things being equal.

For low Q^2 the situation is different. In particular, for $Q^2 = 0, \epsilon = m_f$, and the wave functions fall off as $\exp(-m_f r)$ for large r. Hence, for light quarks in particular, large dipoles with $r \approx m_f^{-1}$ can contribute.

At this point we need to remind ourselves that the above perturbative calculation of the wave function makes no sense for large dipoles, since for $r > 1$ fm strong forces between the quarks obviously play an important role, leading to confinement. In this region a $q\bar{q}$ picture only makes sense if we consider constituent quarks, and vector-dominance ideas seem a more appropriate guide than perturbative QCD. In particular, it is well known that generalized vector dominance (GVD) models [10] can describe the nucleon structure function data in the transition region $0 \leq Q^2 \leq 10 \text{ GeV}^2$, from photoproduction to scaling. However, as shown in [11], this success only extends to nuclei if the off-diagonal version [12] is used. Frankfurt *et al* [13] have studied the pattern of scattering eigenstates in this model, calculating the probability $P(\sigma, Q^2)$ for the photon to interact with the target with a scattering cross section σ. As we shall see in the next section, for small dipoles where perturbative ideas are applicable, the dipole cross section is approximately proportional to r^2, while the behaviour of the wave functions is controlled by (8.44). Using these properties and changing the integration variable from r to σ, it is easily shown from (8.28) that for transverse photons, $P(\sigma, Q^2) \propto \sigma^{-1}$, independent of Q^2. Frankfurt *et al* [13] found approximately the same behaviour in the GVD model for small σ. However, as σ increases to the size of typical hadronic cross sections, there are peaks in $P(\sigma, Q^2)$ at small Q^2 that die away as Q^2 increases and which it is reasonable to associate qualitatively with vector meson states. For a discussion of the relation between GVD models and the dipole approach, see [14].

In order to reproduce this behaviour qualitatively, Forshaw, Kerley and Shaw (FKS) [15] chose a light-quark mass corresponding roughly to the constituent mass and modified the QED wave function by multiplying it by an adjustable Gaussian enhancement factor:

$$\left|\Psi_\gamma^{T,L}(r, z)\right|^2 \to \left|\Psi_\gamma^{T,L}(r, z)\right|^2 f(r), \tag{8.47}$$

where

$$f(r) = \frac{1 + B \exp(-c^2(r - R)^2)}{1 + B \exp(-c^2 R^2)}. \tag{8.48}$$

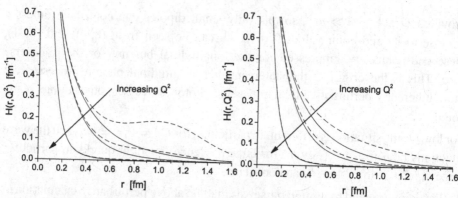

Figure 8.3. The weight function $H(r, Q^2)$ at $Q^2 = 0$, 1 GeV^2 and 15 GeV^2. Left: corresponding to the photon wave function obtained by FKS [15] with $m_q^2 = 0.08$ GeV^2 (dashed lines). Right: corresponding to the perturbative QED photon wave function with $m_q^2 = 0.02$ GeV^2 (dashed lines). In both plots the solid lines correspond to the perturbative QED wave function with $m_q^2 = 0.08$ GeV^2.

This purely phenomenological form enables the width and height of the enhancement to be controlled independently while keeping the enhancement localized at typical hadronic sizes $r \sim 1$ fm. The effect of this is conveniently summarized by integrating out the angular and z dependence in (8.25) and (8.28) to give

$$\sigma_{total}^{\gamma^* p} = \int dz \, d^2r \, (|\Psi_T(z, r)|^2 + |\Psi_L(z, r)|^2)\sigma(s, r, z)$$

$$= \frac{12}{\pi}\alpha_{em} \int dr \, H(r, Q^2)\sigma(s, r), \qquad (8.49)$$

where $\sigma_{total}^{\gamma^* p}$ is the sum of the transverse and longitudinal photon cross sections. In the last line we have made the common assumption that the dipole cross section is independent of z and this is reflected in the slight change in notation. The resulting behaviour of $H(r, Q^2)$ for the final parameter values (which were obtained by fitting to total cross section data) is shown in figure 8.3. As can be seen, the enhancement is important for very low Q^2, but decreases rapidly as Q^2 increases.

Other authors do not in general include an explicit enhancement factor, but achieve a similar but broader enhancement by varying the quark mass. Choosing a smaller quark mass increases the wave function at all large r, while leaving it almost unchanged at small r, as illustrated in figure 8.3. Golec-Biernat and Wüsthoff [16,17], for example, used $m_f^2 = 0.02$ GeV^2, so that m_f is comparable to the pion mass. In practice [18] the difference between these two approaches only becomes important when analysing real photoabsorption data, mainly that from fixed-target experiments [19].

In summary, the photon wave functions are well determined at small r, while their magnitude at large r is model-dependent. It is partly because of this that DDIS data are so important, since they and DIS together involve essentially different combinations of wave function and cross section, as can be seen by comparing (8.28) and (8.31).

8.3.2 Vector-meson wave functions

We next turn to the vector-meson wave functions, which from (8.32) and (8.35) can be written in the forms

$$\Psi_{h,\bar{h}}^{V,T(\gamma=\pm)}(\vec{r},z) = \pm\sqrt{\frac{N_c}{4\pi}}\frac{\sqrt{2}}{z(1-z)}$$
$$\times\left[ie^{\pm i\theta_r}(z\delta_{h\pm,\bar{h}\mp} - (1-z)\delta_{h\mp,\bar{h}\pm})\partial_r + m_f\delta_{h\pm,\bar{h}\pm}\right]\phi_T(r,z) \tag{8.50}$$

and

$$\Psi_{h,\bar{h}}^{V,L}(\vec{r},z) = \sqrt{\frac{N_c}{4\pi}}\delta_{h,-\bar{h}}\frac{1}{M_V z(1-z)}[z(1-z)M_V^2 + \delta\times(m_f^2 - \nabla_r^2)]\phi_L(r,z), \tag{8.51}$$

where $\nabla_r^2 \equiv \frac{1}{r}\partial_r + \partial_r^2$. Note that the second term in square brackets, which occurs in the longitudinal meson case, is a direct consequence of keeping the second term in (8.39) and is omitted or kept depending on whether the parameter $\delta = 0$ or 1.

The scalar parts of the wave functions $\phi_{L,T}(r,z)$ are model-dependent. However, they are subject to two constraints. The first is the normalization condition [20,21]

$$1 = \sum_{h,\bar{h}}\int\frac{d^2\vec{k}}{(2\pi)^2}dz\,|\Psi_{h,\bar{h}}^{V,\lambda}(\vec{k},z)|^2 = \sum_{h,\bar{h}}\int d^2\vec{r}\,dz\,|\Psi_{h,\bar{h}}^{V,\lambda}(\vec{r},z)|^2, \tag{8.52}$$

which embodies the assumption that the meson is composed solely of $q\bar{q}$ pairs. Note that this normalization is consistent with (8.25) and differs by a factor 4π relative to the conventional light-cone normalization.

The second constraint comes from the electronic decay width of the vector meson [21,22], and can be expressed directly in terms of the scalar parts of the wave functions:

$$f_V M_V = \frac{N_c}{\pi}\hat{e}_f\int_0^1\frac{dz}{z(1-z)}[z(1-z)M_V^2 + \delta\times(m_f^2 - \nabla_r^2)]\phi_L(r,z)\big|_{r=0} \tag{8.53}$$

and

$$f_V M_V = -\frac{N_c}{2\pi}\hat{e}_f \int_0^1 \frac{dz}{[z(1-z)]^2} \left[(z^2 + (1-z)^2)\nabla_r^2 - m_f^2\right]\phi_T(r,z)\Big|_{r=0},$$

(8.54)

assuming that $r\phi_{L,T}(r,z) \to 0$ at $r = 0$ and $r = \infty$. The couplings f_V of the mesons to the electromagnetic current are determined from the experimentally measured leptonic decay widths $\Gamma_{V \to e^+ e^-}$ using $3M_V \Gamma_{V \to e^+ e^-} = 4\pi \alpha_{em}^2 f_V^2$. In (8.53) and (8.54) $\hat{e}_f$ is the effective electronic charge arising from the sum over quark flavours in the meson: $\hat{e}_f = \sqrt{\frac{1}{2}}, \frac{1}{3}$ and $\frac{2}{3}$ for the ρ, ϕ and J/ψ respectively.

In what follows we shall confine ourselves to two particular models for the scalar parts of the vector-meson wave functions: the Dosch, Gousset, Kulzinger, Pirner (DGKP) model and the boosted-wave-function model. Our treatment will be brief, and much more detailed formulae can be found in [22], [23] and [24].

DGKP meson wave function In the DGKP approach [22], the r and z dependence of the wave function are assumed to factorize. We note that the theoretical analysis of Halperin and Zhitnitsky [25] shows that such a factorizing ansatz must break down at the end-points of z. However, since the latter are suppressed in the DGKP wave function this has no practical consequence. Specifically, the scalar wave function is given by

$$\phi_\lambda(r,z) = G(r)f_\lambda(z)z(1-z).$$

(8.55)

(Note that DGKP do not actually include the factor $z(1-z)$ in the scalar wave function. This is because they define the scalar wave function to be the right-hand side of (8.36) divided by $z(1-z)$.) A Gaussian dependence on r is assumed,

$$G(r) = \frac{\pi f_V}{N_c \hat{e}_f M_V}e^{-\frac{1}{2}\omega_\lambda^2 r^2},$$

(8.56)

and $f_\lambda(z)$ is given by the Bauer–Stech–Wirbel model [26]:

$$f_\lambda(z) = \mathcal{N}_\lambda \sqrt{z(1-z)}e^{-M_V^2(z-\frac{1}{2})^2/2\omega_\lambda^2}.$$

(8.57)

Dosch *et al* set $\delta = 0$ in (8.51), which is equivalent to neglecting the second term in (8.39).

For a particular quark mass, the free parameters of the DGKP wave function can be determined from the constraints (8.52)–(8.54). The resulting behaviour [24] of the ρ wave functions is shown in figure 8.4. The wave functions peak at $z = \frac{1}{2}$ and $r = 0$, and go rapidly to zero as $z \to 0, 1$ and $r \to \infty$, so that large dipoles are suppressed. From the figures, we see that the transverse wave function has a

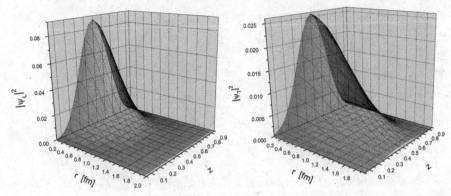

Figure 8.4. The ρ wave functions $|\Psi^L|^2$ (left) and $|\Psi^T|^2$ (right) in the DGKP model with the light quark mass $m_{u,d}^2 = 0.08$ GeV^2. Note the different scales for the ordinate.

broader distribution than the longitudinal wave function. The ϕ wave functions are qualitatively similar to, but slightly more sharply peaked than, the ρ wave functions.

Boosted wave functions In this approach, the scalar part of the wave function is obtained by taking a given wave function in the meson rest-frame. This is then boosted into a light-cone wave function using the Brodsky–Huang–Lepage prescription, in which the expressions for the off-shellness in the centre-of-mass and light-cone frames are equated [27] or, equivalently, the expressions for the invariant mass of the $q\bar{q}$ pair in the centre-of-mass and light-cone frames are equated [28].

The simplest version of this approach assumes a simple Gaussian wave function in the meson rest-frame. Alternatively, Nemchik *et al* [23] have supplemented this by adding a hard 'Coulomb' contribution to give an improved description of the rest-frame wave function at small r. However, as discussed in [24], this results in an unphysical singularity in the light-cone wave functions at $r = 0$, $z = \frac{1}{2}$. In what follows, therefore, we confine ourselves to the simple case of a boosted Gaussian wave function. We refer to [23,24] for details of this procedure. Here we simply state the result, which is that the meson light-cone wave functions are given by (8.51), (8.50) with $\delta = 1$ and the scalar wave functions $\phi_\lambda(r, z)$:

$$\phi_\lambda(r, z) = \mathcal{N}_\lambda \, 4z(1 - z)\sqrt{2\pi R^2} \exp\left(-\frac{m_f^2 R^2}{8z(1 - z)}\right) \exp\left(-\frac{2z(1 - z)r^2}{R^2}\right).$$

$$(8.58)$$

Like the DGKP wave function discussed above, for a given quark mass the boosted Gaussian wave function contains two free parameters which can be determined from the constraints (8.52)–(8.54).

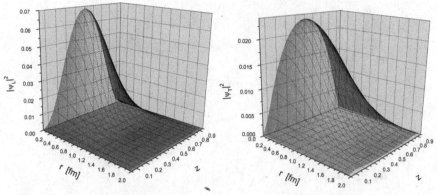

Figure 8.5. The ρ wave functions $|\Psi^L|^2$ (left) and $|\Psi^T|^2$ (right) in the boosted Gaussian model with the light quark mass $m_{u,d}^2 = 0.08$ GeV^2.

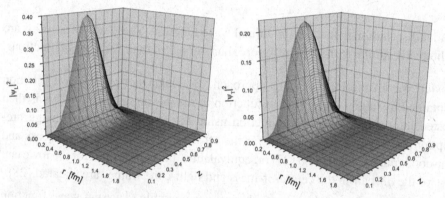

Figure 8.6. The J/ψ wave functions $|\Psi^L|^2$ (left) and $|\Psi^T|^2$ (right) in the boosted Gaussian model with the charm quark mass $m_c = 1.4$ GeV.

The behaviour of the resulting ρ wave functions is shown in figure 8.5. On comparing this with figure 8.4 one sees that they are qualitatively similar to those obtained in the DGKP case but that the peaks are a little less pronounced.

In figure 8.6, the corresponding wave functions for the J/ψ meson are shown. As expected, the larger charm mass ensures that these wave functions are more tightly peaked around $z = \frac{1}{2}$ and $r = 0$.

8.4 The dipole cross section

Different versions of the colour-dipole model are characterized by different forms for the dipole cross section (8.21). In choosing them, authors are guided by perturbative QCD for small dipoles, and hadron dominance for large dipoles. The free parameters, which are invariably needed, are usually determined by fitting to the

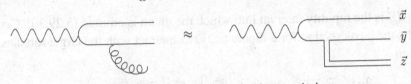

Figure 8.7. Dipole evolution via gluon emission,

copious and high-precision DIS data, and the resulting dipole cross section is then used to make predictions for other processes. The most stringent test comes from comparison to the DDIS data, for reasons mentioned earlier. In what follows, we will focus mainly on models which have provided successful predictions for DDIS. First, however, we will discuss the theoretical expectations for small dipoles, where perturbative QCD is expected to be appropriate.

8.4.1 Small dipoles and QCD

Physically, a small dipole created far upstream of the target proton may evolve to a more complicated state through partonic radiation. To leading order in the number of colours, the original dipole may branch into two dipoles as a consequence of gluon emission, as illustrated in figure 8.7. These two dipoles may then also radiate additional gluons and by the time the fluctuation reaches the target proton the original dipole may have evolved into a system of many dipoles. The amount of evolution clearly depends upon the evolution time of the dipole system and this in turn is controlled by the overall centre-of-mass energy or, more conveniently, the rapidity $Y = \frac{1}{2} \log\{(E + p_z)/(E - p_z)\} \approx \frac{1}{2} \log (s/s_0)$ in the diffractive limit.

That we can describe the dipole evolution in terms of a classical branching process needs some justification. It follows from the assumption that the radiation of a gluon always takes place off a long-lived state, so that the emission is independent of the prior history of the state; that is the timescale of emission is much less than the lifetime of the state which is doing the emitting. The lifetime of the relevant emitter (a dipole) is determined by the rapidity interval into which the emission can take place and provided this is large enough we are justified in our use of this dipole branching picture. We therefore can view the evolution of the original dipole as an evolution in rapidity, and given the probability that a particular dipole branches into a pair of dipoles we can immediately write down the appropriate evolution equation. The probability that a dipole whose ends are located at transverse positions $(\vec{x}, \vec{y})$ will emit a gluon at transverse position $\vec{z}$, thereby creating two dipoles at $(\vec{x}, \vec{z})$ and $(\vec{y}, \vec{z})$, is

$$dP = \frac{N_c \alpha_s}{\pi} dY \frac{d^2\vec{z}}{2\pi} \frac{(\vec{x} - \vec{y})^2}{(\vec{x} - \vec{z})^2 (\vec{y} - \vec{z})^2}, \tag{8.59}$$

where dY is the rapidity interval into which the gluon is emitted [5,29,30]. Consequently, the cross section for a dipole $(\vec{x}, \vec{y})$ to interact with the target satisfies the equation

$$\frac{\partial\sigma(\vec{x}, \vec{y}, Y)}{\partial Y} = \frac{N_c \alpha_s}{\pi} \int \frac{d^2\vec{z}}{2\pi} \frac{(\vec{x} - \vec{y})^2}{(\vec{x} - \vec{z})^2(\vec{y} - \vec{z})^2}$$

$$\times (\sigma(\vec{x}, \vec{z}, Y) + \sigma(\vec{y}, \vec{z}, Y) - \sigma(\vec{x}, \vec{y}, Y)). \qquad (8.60)$$

The first two terms express the fact that the original dipole may branch into two, one or other of which may then scatter off the proton, whilst the last term reflects the fact that after emission the original dipole is destroyed. Note that the possibility that both newly-created dipoles might scatter off the proton is not included in this equation. We might expect this approximation to break down if the dipole cross section is too large. This is indeed the case and we shall return to this issue in the following subsection.

Given the dipole cross section at some $Y = Y_0$, (8.60) allows us to compute it at all other values of Y under the assumption that $Y, Y_0 \gg 1$. In fact, (8.60) is none other than the dipole form of the Balitsky, Fadin, Kuraev, Lipatov (BFKL) equation [31,32] which correctly sums all of the leading logarithms in energy to all orders, that is all terms $\sim (\alpha_s Y)^n$. (Note that (8.60) does not actually depend upon us making the leading N_c approximation.) In summing multi-gluon emissions into the dipole cross section we are effectively including some of the higher-Fock components in the photon wave function, in particular those components which contain leading logarithms in energy.

Making contact with the discussion in section 8.2.2, the dipole cross section of (8.60) is simply

$$\sigma(\vec{x}, \vec{y}, Y) = \frac{\text{Im}\tau(\vec{b}, s; z, \vec{r})}{s} \qquad (8.61)$$

with $\vec{x} - \vec{y} = \vec{r}$ and $\frac{1}{2}(\vec{x} + \vec{y}) = \vec{b}$.

It is worth remarking here that, for non-diagonal processes such as the electroproduction of vector mesons, there is an asymmetry in the dipole evolution. This asymmetry is beyond the leading logarithmic approximation and as such it is only appropriate to speak of a truly universal dipole cross section within this approximation. Beyond leading logarithms, it is more appropriate to use a formalism which builds in the off-diagonal nature of scattering amplitudes from the start, such as is provided by the so-called skewed parton distribution functions: see chapter 9.

Let us now turn to another way to view the dipole cross section in perturbative QCD. Equations (8.27) and (8.28) determine the structure function $F_2(x, Q^2)$ in terms of the photon wave function and the dipole cross section. Now the Q^2 evolution equations of Dokshitzer, Gribov, Lipatov, Altarelli and Parisi (DGLAP) tell us that [33]

$$\frac{\partial F_2(x, Q^2)}{\partial \log Q^2} \approx \sum_q e_q^2 \frac{\alpha_s}{2\pi} \int_x^1 dz\, G(x/z, Q^2) P_{qg}(z), \qquad (8.62)$$

where $G(x, Q^2)$ is the gluon momentum density, $P_{qg}(z) = \frac{1}{2}(z^2 + (1-z)^2)$ and e_q is the electric charge of a quark, q, in units of the electron charge. This approximation is reasonable in the small-x region where the gluon density is dominant. Formally, it is correct in the double leading logarithmic approximation (DLLA) where all terms proportional to $(\alpha_s \log Q^2 \log(1/x))^n$ are summed. Equating the dipole prediction for $\partial F_2/\partial \log Q^2$ (using (8.27), (8.28), (8.45)) with the DGLAP prediction of (8.62), implies that

$$\sigma(s, r) \approx \frac{\pi^2 \alpha_s}{3} r^2 G(x, A^2/r^2). \qquad (8.63)$$

The constant A is not determined in DLLA, although numerical studies indicate that $A \approx 3$.

The equivalence of the BFKL and DGLAP approaches in the double logarithmic region is evident in that (8.60), with the aid of (8.63), reduces to

$$\frac{\partial G(x, 1/r^2)}{\partial \log(1/x)} \approx \frac{N_c \alpha_s}{\pi} \int_{b>r} \frac{db^2}{b^2} G(x, 1/b^2) \qquad (8.64)$$

in the double logarithmic approximation, which corresponds to the region where the emitted dipole is much larger than the parent dipole. This is just the DGLAP evolution of the gluon in the double logarithmic approximation.

Equation (8.63) informs us that as $r \to 0$ so $\sigma(s, r) \to r^2$ modulo logarithmic corrections. This property of the dipole cross section means that small dipoles tend to cut through the target without hindrance and goes by the name of 'colour transparency': see chapter 11. It is a key property that is included in all models of the dipole cross section.

8.4.2 Non-linear dynamics and saturation

For small enough dipoles, perturbative QCD anticipates that the dipole cross section should rise with increasing s. Equation (8.60) is a linear evolution equation: dipoles branch to produce more dipoles and the result is a dipole cross section which grows exponentially with rapidity, that is $\sigma \sim \exp(\lambda Y)$ where $\lambda = (N_c \alpha_s/\pi) 4 \log 2$. Eventually, as the dipole cross section rises, non-linear effects will necessarily become relevant. In the last section we noticed this effect when, in writing down (8.60), we acknowledged that it ignored the possibility that *both* newly-created dipoles could scatter off the proton.

If we assume that the two dipoles scatter independently off the proton, then we can immediately generalize (8.60). In terms of the S-matrix $S(\vec{x}, \vec{y}, Y)$ for scattering a dipole $(\vec{x}, \vec{y})$ off a proton at rapidity Y, independent scattering implies that

$$\frac{\partial S(\vec{x}, \vec{y}, Y)}{\partial Y} = \frac{N_c \alpha_s}{\pi} \int \frac{d^2 \vec{z}}{2\pi} \frac{(\vec{x} - \vec{y})^2}{(\vec{x} - \vec{z})^2 (\vec{y} - \vec{z})^2}$$
$$\times (S(\vec{x}, \vec{z}, Y) S(\vec{y}, \vec{z}, Y) - S(\vec{x}, \vec{y}, Y)). \qquad (8.65)$$

The first term now expresses the possibility that the newly-created dipoles can each either scatter off the proton or pass through it without any scattering. Substituting for $1 - S(\vec{x}, \vec{y}, Y) = \frac{1}{2}\sigma(\vec{x}, \vec{y}, Y)$, which follows from (8.21), (8.61) and the fact that

$$\sigma(s, r) = 2 \int d^2 \vec{b} \, (1 - S(\vec{x}, \vec{y}, Y)), \qquad (8.66)$$

we can write (8.65) as

$$\frac{\partial \sigma(\vec{x}, \vec{y}, Y)}{\partial Y} = \frac{N_c \alpha_s}{\pi} \int \frac{d^2 \vec{z}}{2\pi} \frac{(\vec{x} - \vec{y})^2}{(\vec{x} - \vec{z})^2 (\vec{y} - \vec{z})^2}$$
$$\times \left(\sigma(\vec{x}, \vec{z}, Y) + \sigma(\vec{y}, \vec{z}, Y) - \sigma(\vec{x}, \vec{y}, Y)\right.$$
$$\left. - \frac{1}{2}\sigma(\vec{x}, \vec{z}, Y)\sigma(\vec{y}, \vec{z}, Y)\right). \qquad (8.67)$$

This is the Balitsky–Kovchegov (BK) equation [34,35]. The assumption of independent scattering is a strong one but can be expected to capture at least some of the essential physics. To go beyond this approximation takes us into the domain of the 'colour glass condensate' (CGC), which is a subject whose details lie beyond the scope of this chapter. We refer to the review articles [36–38] for the details and here restrict ourselves to providing a very brief overview that we present at the end of this subsection.

The non-linear term in the BK equation has the effect of reducing (or 'saturating') the dipole cross section at large Y. Although we focus on the BK equation here, we ought to stress that the idea of saturation through non-linear gluon dynamics is not new [39]. For sufficiently small dipoles, the non-linear effects are unimportant. Conversely, dipoles larger in size than some 'saturation radius', $r_s(x)$, feel the non-linear dynamics. We expect that $r(x)$ ought to push to lower values as x decreases, reflecting the rapid rise of the dipole cross section with decreasing x. Detailed analysis using the next-to-leading order (NLO) BFKL equation to determine where non-linear effects should become important indicates that $r(x)^2 \sim x^\lambda$, where $\lambda \approx 0.3$ [40]. However, the question of precisely when saturation dynamics becomes relevant is one that has eluded theoretical efforts to date. One of the predictions of the BK equation is that, for $r > r_s(x)$, the dipole cross section should only depend upon the geometrical scaling variable $r/r_s(x)$.

It is interesting to note that although the BK equation guarantees that the scattering probability at fixed impact parameter does not violate unitarity, it does not necessarily follow that unitarity is preserved in the sense of the amplitude as a whole (that is, in the sense of the Froissart bound). There remains a perturbative tail out to large impact parameters which is power-like and which generates a growth of the cross section stronger than that implied by unitarity. For further discussion we refer to [41,42].

Many authors [16,18,43–45] have formulated dipole models which incorporate gluon saturation effects in an approximate manner. In this chapter, we will concentrate on two saturation models whose parameters have been determined by fitting to the precise DIS data.

The GW saturation model The model of Golec-Biernat and Wüsthoff (GW) [16,17] combines the approximate behaviour $\sigma \propto r^2$ at small r together with a purely phenomenological implementation of the anticipated saturation dynamics which also satisfies geometric scaling. The dipole cross section takes the attractively simple form:

$$\sigma = \sigma_0 \left[1 - \exp\left(\frac{-r^2 Q_0^2}{4(x_{mod}/x_0)^\lambda} \right) \right]. \tag{8.68}$$

Here $Q_0 = 1$ GeV and x_{mod} is a modified Bjorken variable,

$$x_{mod} = x \left(1 + \frac{4m_f^2}{Q^2} \right), \tag{8.69}$$

where m_f is the quark mass. The three free parameters x_0, σ_0 and λ were determined using a purely perturbative photon wave function, which is somewhat enhanced at large r-values by the use of a light quark mass somewhat smaller than the constituent mass. The fit to data on the F_2 structure function yielded $\sigma_0 = 29.12$ mb, $\lambda = 0.277$ and $x_0 = 0.41 \times 10^{-4}$, with quark masses 0.14 GeV for the light quarks and 1.5 GeV for the charm quark.

The resulting behaviour of the dipole cross section is illustrated in the left-hand panel of figure 8.8. (The curves marked 'FKS' will be discussed in the next subsection.) As can be seen, the dipole cross section is characterized by a rapid increase with decreasing x for small r, changing to a softer x dependence as r increases beyond the 'saturation radius'

$$r_s(x) = \frac{2}{Q_0} \left(\frac{x_{mod}}{x_0} \right)^{\lambda/2}. \tag{8.70}$$

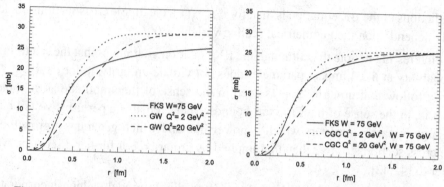

Figure 8.8. The GW dipole cross section (left) and CGC dipole cross section (right) at $W = 75$ GeV for $Q^2 = 2$ GeV^2 (dotted line) and $Q^2 = 20$ GeV^2 (dashed line). The Q^2-independent FKS dipole cross section (solid line) at the same energy is shown for comparison.

Saturation arises because of the decrease of the saturation radius with decreasing x. Let us consider a dipole of fixed transverse size r. If $r < r_s(x)$, the dipole cross section increases rapidly as x decreases. However, this rapid rise switches to a softer x dependence when x becomes so small that $r_s(x)$ itself decreases below the fixed dipole size r. In the figure, this manifests itself in the fact that the softer x dependence sets in at a lower value of r at the lower value of Q^2. In the language of the gluon density, $G(x, 1/r^2)$, saturation means that the steep rise as x decreases is tamed as Q^2 falls, and that the taming is deferred to lower x as Q^2 increases.

We note that a subsequent refinement of the GW model takes into account corrections due to DGLAP evolution at large Q^2 [46].

The CGC model   The dipole model of Iancu *et al* [43] can be thought of as a development of the GW saturation model. Though still largely a phenomenological parametrization, Iancu *et al* do argue that it contains the main features of the CGC regime, where the gluon densities are high and non-linear effects become important. In particular, they take

$$\sigma = 2\pi R^2 \mathcal{N}_0 \left(\frac{rQ_s}{2}\right)^{2[\gamma_s + \log(2/rQ_s)/(\kappa\lambda\log(1/x))]} \qquad \text{for } rQ_s \le 2$$

$$= 2\pi R^2 \{1 - \exp[-a\log^2(brQ_s)]\} \qquad \text{for } rQ_s > 2, \qquad (8.71)$$

where the saturation scale $Q_s(x) \equiv (x_0/x)^{\lambda/2}$ GeV. The coefficients a and b are uniquely determined by ensuring continuity of the cross section and its first derivative at $rQ_s = 2$. For $rQ_s < 2$ the solution matches that of the leading order BFKL equation and fixes $\gamma_s = 0.63$ and $\kappa = 9.9$. The coefficient $\mathcal{N}_0$ is strongly correlated to the definition of the saturation scale and Iancu *et al* found that the quality of fit

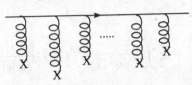

Figure 8.9. The Wilson line contributing to a high-energy quark passing through th colour field of the proton.

to F_2 data is only weakly dependent upon its value. For a fixed value of $\mathcal{N}_0$, there are therefore three parameters which need to be fixed by a fit to the data, that is x_0, λ and R. In this chapter, we take $\mathcal{N}_0 = 0.7$ and a light quark mass of $m_q = 140$ MeV, for which the fit values are $x_0 = 2.67 \times 10^{-5}$, $\lambda = 0.253$ and $R = 0.641$ fm. The CGC dipole cross section is shown in the right-hand panel of figure 8.8. It is very similar to the GW dipole cross section and we show only the predictions of the CGC model when comparing to data.

The colour glass condensate   The dipole cross section can be obtained by averaging the product of the two Wilson lines, which correspond to the quark–antiquark dipole, over all colour configurations of the proton, that is,

$$S(\vec{x}, \vec{y}, Y) = \frac{1}{N_c} \langle V^\dagger(\vec{x})_{ij} V(\vec{y})_{ji} \rangle, \tag{8.72}$$

where

$$V_{ij}^\dagger(\vec{x}) = P \, \exp\left(ig \int_{-\infty}^{\infty} dx^+ \, A_a^-(x^+, \vec{x}) t^a \right)_{ij} \tag{8.73}$$

is the Wilson line for a right-moving ($x^- \approx 0$) quark at fixed $\vec{x}$, as represented in figure 8.9. P denotes that the fields $A_a t^a$ are ordered from left to right in order of descending x^+ and the angled brackets in (8.72) indicate an average over gluon fields in the proton. In this way the essential gluon dynamics is associated with the proton, which is natural in a frame where the proton is left-moving and at much larger rapidity than the right-moving dipole. In (8.73), the interaction with the gluon field of the proton is of eikonal type and as such implicitly assumes that the gluon field is sufficiently soft, that is it is dominated by Fourier components that are small compared to the energy of the quark – we expect this to hold in the diffractive limit [47].

The field averaging at rapidity Y can be formulated as a functional integral with a weight functional $W_Y[A^-]$. Generally this functional cannot be evaluated. However, further progress can be made since the evolution of $W_Y[A^-]$ in rapidity can be computed in perturbative QCD. This is the Jalilian-Marian, Iancu, McLerran, Weigert, Leonidov, Kouner (JIMWLK) evolution equation which lies at the heart of the CGC approach [48]:

$$\frac{\partial W_Y[\alpha]}{\partial Y} = \frac{1}{2} \int_{x,y} \frac{\delta}{\delta \alpha^a(\vec{x})} \eta^{ab}(\vec{x}, \vec{y}) \frac{\delta W_Y[\alpha]}{\delta \alpha^b(\vec{y})}, \tag{8.74}$$

where

$$\eta^{ab}(\vec{x}, \vec{y}) = \frac{1}{\pi} \int \frac{d^2\vec{z}}{(2\pi)^2} \frac{(\vec{x} - \vec{z}) \cdot (\vec{y} - \vec{z})}{(\vec{x} - \vec{z})^2 (\vec{y} - \vec{z})^2}$$

$$\times \left(1 + \tilde{V}^\dagger(\vec{x})\tilde{V}(\vec{y}) - \tilde{V}^\dagger(\vec{x})\tilde{V}(\vec{z}) - \tilde{V}^\dagger(\vec{z})\tilde{V}(\vec{y})\right)^{ab} \tag{8.75}$$

and $\tilde{V}^\dagger(\vec{x})$ is defined analogously to (8.73) but is in the adjoint representation. As rapidity increases, JIMWLK accounts for the fact that more and more gluon emissions can take place and that eventually, as a result of the non-linear gluon dynamics, the proton saturates with gluons.

The dipole DIS S-matrix is thus given by

$$S(\vec{x}, \vec{y}, Y) = \frac{1}{N_c} \int [D\alpha] \, V^\dagger(\vec{x})_{ij}[\alpha] V(\vec{y})_{ji}[\alpha] \, W_Y[\alpha] \tag{8.76}$$

and hence its evolution satisfies the equation

$$\frac{\partial S(\vec{x}, \vec{y}, Y)}{\partial Y} = \frac{1}{N_c} \int [D\alpha] \, V^\dagger(\vec{x})_{ij}[\alpha] V(\vec{y})_{ji}[\alpha] \, \frac{\partial W_Y[\alpha]}{\partial Y}. \tag{8.77}$$

Substituting for the Wilson lines and performing the functional derivatives in (8.74) yields the Balitsky equation [35]:

$$\frac{\partial S(\vec{x}, \vec{y}, Y)}{\partial Y} = \frac{N_c \alpha_s}{\pi} \int \frac{d^2\vec{z}}{2\pi} \frac{(\vec{x} - \vec{y})^2}{(\vec{x} - \vec{z})^2(\vec{y} - \vec{z})^2}$$

$$\times \left(\frac{1}{N_c^2} \langle V^\dagger(\vec{x})_{ij} V(\vec{z})_{ji} \, V^\dagger(\vec{z})_{kl} V(\vec{y})_{lk} \rangle - S(\vec{x}, \vec{y}, Y)\right). \tag{8.78}$$

This equation reduces to the BK equation (8.67) if one assumes that the average over the product of Wilson loops is the product of the averages. Otherwise, the evolution equation is not closed but rather the first equation in an infinite hierarchy. The BFKL equation (8.60) emerges on expanding in the Wilson lines to lowest order in the strong coupling.

8.4.3 Regge dipole models

With the discovery of 'hard diffraction' at HERA, it became clear that diffractive photoprocesses could no longer be described by the exchange of a single Regge pole, the 'soft pomeron', characterized by the trajectory (8.1). One possible interpretation is that there is a new phenomenon – 'hard diffraction' – which becomes dominant for hard enough scales and large enough energies. If one assumes that this can also be approximated by a Regge pole, then one is led to the hypothesis of two pomerons: the soft pomeron (8.1) which dominates in hadronic diffraction and some 'soft' photoprocesses; and a second pomeron, the 'hard pomeron', which dominates for hard enough scales and large enough energies. This hypothesis has been explored by Donnachie and Landshoff [49], who obtained an excellent fit to data on the proton structure function, the charmed structure function and J/ψ production using a hard pomeron trajectory

$$\alpha_P(t) \approx 1.42 + 0.10\, t.$$

The varying energy dependence (8.9) observed in diffractive photoprocesses then arises from the varying relative importance of these two contributions.

The same hypothesis has also been used in the context of GVD models by Kerley and Shaw [50], who argued that the hard-pomeron term must be associated with long-lived fluctuations of the photon with very large invariant masses. In the context of dipole models, such heavy long-lived states are naturally associated with small dipoles, with correspondingly large components of the quark's transverse momentum leading to large expectation values for the dipole invariant mass. These ideas led FKS [15] to propose a two-component Regge model for the dipole cross section of the form

$$\sigma(s, r) = \sigma_{soft}(s, r) + \sigma_{hard}(s, r). \tag{8.79}$$

The soft term has a weaker energy dependence, $\sigma_{soft} \sim s^{0.06}$, than the hard term, $\sigma_{hard} \sim s^{0.44}$, with the hard term most important at low r ($\sigma_{hard} \sim r^2$) and the soft term dominant at large r.

The free parameters in both the dipole cross section and the photon wave function were determined by fitting to DIS and real photoabsorption data, with the FKS model using a photon wave function incorporating the enhancement factor (8.48). The resulting dipole cross section is shown in figure 8.10. As can be seen, as s increases the dipole cross section grows most rapidly for small r, where the hard term dominates, eventually exceeding the typically hadronic cross section found for dipoles of large $r \approx 1$ fm. This rise should be tamed, sooner or later, by saturation effects. However, such saturation effects are not included in the FKS model, or in any two-component Regge model. For comparison, we show the FKS dipole cross section in each of the panels in figure 8.8. The non-monotonicity at $W = 300$ GeV is a direct consequence of not including any saturation.

Other parametrizations of similar form have been presented by Donnachie and Dosch [51], who obtained successful predictions for a variety of reactions, but did not consider DDIS, and by Forshaw and Shaw [18].

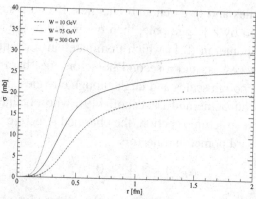

Figure 8.10. The FKS Regge dipole cross section: dashed line $W = 10$ GeV, solid line $W = 75$ GeV and dotted line $W = 300$ GeV.

The Forshaw–Shaw (FS2004) approach is identical in philosophy to the original model of FKS in that it is a sum of hard and soft components. Since we shall use the FS2004 form of the dipole cross section to make comparisons to data in the next section we shall be a little more explicit. Most importantly, FS2004 can serve either as an unsaturated Regge model or as a saturation model, depending on how one of the parameters is determined.

The dipole cross section is assumed to satisfy the simple form

$$\sigma = A_H r^2 x_m^{-\lambda_H} \quad \text{for } r < r_0 \text{ and}$$

$$= A_S x_m^{-\lambda_S} \quad \text{for } r > r_1, \tag{8.80}$$

where

$$x_m = \frac{Q^2}{Q^2 + W^2}\left(1 + \frac{4m^2}{Q^2}\right). \tag{8.81}$$

For light-quark dipoles, the quark mass m is a parameter in the fit, whilst for charm quark dipoles the mass is fixed at 1.4 GeV.

In the intermediate region $r_0 \leq r \leq r_1$ one has to interpolate linearly between the two forms of (8.80). Whether this is a Regge-inspired model or a saturation model depends entirely upon the way in which the boundary parameter r_0 is determined.

If the boundary parameter r_0 is kept constant, then the parametrization reduces to a sum of two powers, as might be predicted in a two-pomeron approach. It is plainly unsaturated, with the dipole cross section obtained at small r-values growing rapidly with increasing s at fixed Q^2 (or equivalently with decreasing x) without damping of any kind. We shall henceforth call this the *FS2004 Regge model*.

Alternatively, saturation can be introduced by adopting a device previously utilized in [47]. Instead of taking r_0 to be fixed one determines it to be the value at which the hard component is some fixed fraction of the soft component, that is

$$\sigma(s, r_0)/\sigma(s, r_1) = f, \tag{8.82}$$

and treat f instead of r_0 as a fitting parameter. This introduces no new parameters. However, the scale r_0 now moves to lower values as x decreases, and the rapid growth of the dipole cross section at a fixed value of r begins to be damped as soon as r_0 becomes smaller than r. In this sense saturation effects are modelled, albeit crudely, with r_0 the saturation radius. We shall henceforth call this the *FS2004 saturation model*.

8.5 Dipole phenomenology

It is now time to confront the dipole model predictions to the data. First, the precise data on DIS will be used to pin down the parameters of a particular dipole model. Afterwards, one is able to make predictions for other observables involving few, if any, additional parameters. Apart from wanting to examine how the dipole formalism fares against the data, we would like to exploit the universality of the formalism to test explicit theoretical ideas against the widest-possible variety of data. In particular, it is most interesting to inquire whether the data provide evidence for saturation dynamics or whether a two-component Regge model is successful.

It may well be that the Regge approach with two pomeron poles is the most appropriate description of the data with non-linear saturation effects entering at energies beyond those accessed in experiment so far. On the other hand, one might take the view that there is only a single pomeron pole with an intercept substantially above unity (or an even more complicated j-plane analytic structure) and that in order to explain the data saturation dynamics enters, at low scales there being more saturation than at higher scales. Both of these rather general approaches can at least qualitatively explain why the energy dependence of the data is as it is. The two-pomeron approach attributes the more steeply rising cross section at higher Q^2 to the increasing role of the harder pomeron whilst the saturation model approach attributes this steepening to the diminishing role of saturation effects.

To this end we shall now compare the CGC and FS2004 models with the available data.

8.5.1 Analysis of the structure function data

We shall first discuss the fits to the DIS data [52] in the kinematic range

$$0.045 \, \text{GeV}^2 < Q^2 < 45 \, \text{GeV}^2 \quad x \le 0.01. \tag{8.83}$$

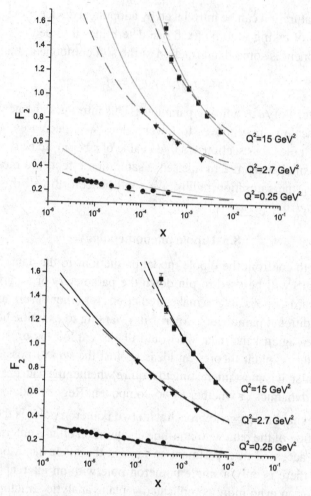

Figure 8.11. Comparison of the FS2004 dipole model to a subset of DIS data [52]. Upper: No saturation fits. FS2004 Regge dipole fit (dashed line) and (solid line) a fit of the same model to data in the restricted range $5 \times 10^{-4} < x < 10^{-2}$, extrapolated over the whole x-range $x < 0.01$. Lower: Saturation fits. FS2004 saturation fit (solid line) and the CGC dipole model (dot-dashed line).

As the parameters of the CGC fit to data have already been presented, let us turn to the FS2004 fits.

Regge fits to DIS data   The best fit obtained using the FS2004 Regge model (that is without saturation) is shown as the dashed line in figure 8.11 (upper) and the parameter values are listed in table 8.1. While the values of the Regge exponents λ_S and λ_H and of the boundary parameters r_0 and r_1 are eminently sensible, the quality

A_H	0.650	λ_H	0.338
A_S	58.42	λ_S	0.0664
r_1	4.844	r_0	0.872
m	0.223		

Table 8.1. Parameters for the FS2004 Regge model in the appropriate GeV-based units.

of the fit is not good, corresponding to a χ^2/data point of 428/156. One might worry that this merely demonstrates a failure of this admittedly simple parametrization. However, we have attempted to fit the data with other Regge inspired models, including the original FKS parametrization, but without success.

A possible reason for this failure is suggested by figure 8.11 (upper). At fixed-Q^2, the poor χ^2 arises because the fit has much too flat an energy dependence at the larger x values for all except the lowest Q^2 value. This could be corrected by increasing the proportion of the hard term, but this necessarily would lead to a steeper dependence at the lower x values at all Q^2. This interpretation is confirmed by the solid curve in figure 8.11 (upper), which shows the result of fitting only to data in the range $5 \times 10^{-4} < x < 10^{-2}$, and then extrapolating the fit to lower x values, corresponding to higher energies at fixed Q^2. As can be seen, this leads to a much steeper dependence at these lower x values than is allowed by the data at all Q^2. An obvious way to solve this problem is by introducing saturation at high energies, to dampen this rise. We shall discuss this in the next subsection.

Another way to alleviate the problem would be to multiply the dipole cross section by a factor $(1 - x)^\delta$. Such a correction is clearly beyond the control of the Regge approach and is anticipated in a partonic approach. However, even with δ as large as 20, the fit is still not satisfactory. The problem could also be eliminated by restricting the fitted region to $x \leq 10^{-3}$. In our view, there is no justification for this as the non-diffractive contributions are already small at $x \approx 0.01$, the $(1 - x)^\delta$ correction providing an example of this.

Saturation fits to DIS   The best-fit parameter values for what we refer to as our FS2004 saturation model are listed in table 8.2 corresponding to a χ^2/data point of 155/156. The corresponding fits to the data are shown in figure 8.11 (lower). Also shown are the very similar results obtained using the CGC model [45].

It is clear from these results that the introduction of saturation into the FS2004 model immediately removes the tension between the soft and hard components which is so disfavoured by the data. However, it is important to note that this conclusion relies on the inclusion of the data in the low-Q^2 region: both the Regge and saturation

A_H	0.836	λ_H	0.324
A_S	46.24	λ_S	0.0572
r_1	4.48	f	0.129
m	0.140		

Table 8.2. Parameters for the FS2004 saturation model in the appropriate GeV-based units.

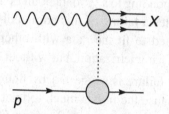

Figure 8.12. The DDIS process

models yield satisfactory fits if we restrict the fit to $Q^2 \geq 2\,\text{GeV}^2$, with χ^2/data point values of 78/86 and 68/86 respectively.

8.5.2 *Diffractive deep inelastic scattering*

We are now ready to compare the predictions of the FS2004 and CGC models with data on DDIS.

Kinematics of diffraction   The generic diagram of the DDIS process (8.6) is shown in figure 8.12. In addition to the standard DIS variables x and Q^2, there are two other kinematic variables: t, the squared four-momentum transfer to the proton, and m_X^2, the invariant mass of the diffractively produced state. In addition, it is useful to introduce two further variables:

$$x_{I\!P} = \frac{q.(p - p')}{q.p} \simeq \frac{m_X^2 + Q^2}{W^2 + Q^2},$$

$$\beta = \frac{Q^2}{2q.(p - p')} = \frac{x}{x_{I\!P}} \simeq \frac{Q^2}{m_X^2 + Q^2}. \tag{8.84}$$

In the infinite momentum frame, $x_{I\!P}$ gives the fraction of the longitudinal proton momentum taken from the proton vertex by the diffractive exchange, while β gives the fraction of $x_{I\!P}$ transferred to the struck quark. Hence the product $x_{I\!P}\beta$ gives the fraction of the (longitudinal) momentum of the proton possessed by the struck quark, that is $x = x_{I\!P}\beta$. The approximate relations included in (8.84) hold good in the diffractive regime of high s.

As in DIS, the differential cross sections observed in DDIS are expressed in terms of structure functions which characterize the interaction of the exchanged photon with the hadronic system. Specifically, the transverse and longitudinal diffractive structure functions $F_T^{D(4)}$ and $F_L^{D(4)}$ are defined by

$$\frac{d^4\sigma(ep \to eXp)}{dx\,dQ^2\,dx_{I\!P}\,dt} = \frac{4\pi\alpha^2}{xQ^4}\left[\left(1 - y + \frac{y^2}{2}\right)F_T^{D(4)} - (1-y)F_L^{D(4)}\right], \quad (8.85)$$

where y is the usual DIS variable $y = q \cdot p/k \cdot p$. The sum of the transverse and longitudinal structure functions defines

$$F_2^{D(4)} = F_T^{D(4)} + F_L^{D(4)}. \quad (8.86)$$

All three structure functions $F_i^{D(4)}$ are expressed in terms of the associated virtual photon cross sections by a relation analogous to (8.27) in inclusive DIS:

$$F_i^{D(4)}(t, x_{I\!P}, \beta, Q^2) = \frac{Q^2}{4\pi^2\alpha}\frac{d\sigma_i(\gamma^*p \to Xp)}{dx_P\,dt}, \quad (8.87)$$

where $i = 2, T, L$. We shall be interested in the structure functions

$$F_2^{D(3)}, F_T^{D(3)}, F_L^{D(3)}, \quad (8.88)$$

which are obtained by integrating the corresponding $F_i^{D(4)}$ over the momentum transfer $|t|$. As for exclusive processes, this is often done using the empirical parametrization

$$\frac{d\sigma}{dt} = \frac{d\sigma}{dt}\bigg|_{t=0}\exp(-B|t|). \quad (8.89)$$

Dipole model predictions   The contribution due to quark–antiquark dipoles to $F_{L,T}^{D(3)}$ can be obtained from a momentum space treatment, as described in [17,53]. Here we shall simply quote the result.

First we introduce the definitions

$$\Phi_{0,1} \equiv \left(\int_0^\infty r\,dr\,K_{0,1}(\epsilon r)\sigma(x_p, r)J_{0,1}(kr)\right)^2, \quad (8.90)$$

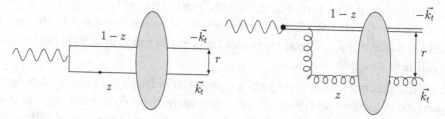

Figure 8.13. The $q\bar{q}$ and $q\bar{q}g$ contributions to $F_2^{D(3)}$.

where $k^2 = m_X^2 z(1-z) + m_f^2$. Note the shift in notation for the dipole cross section to emphasize that the relevant energy is that determined by $x_{I\!P}$. The diffractive structure functions are then given by

$$x_{I\!P} F_{q\bar{q},L}^{D(3)}(Q^2, \beta, x_{I\!P}) = \frac{3Q^6}{32\pi^4 \beta B} \cdot \sum_f e_f^2 \cdot 2 \int_{z_0}^{1/2} dz z^3 (1-z)^3 \Phi_0, \qquad (8.91)$$

and

$$x_{I\!P} F_{q\bar{q},T}^{D(3)}(Q^2, \beta, x_{I\!P}) = \frac{3Q^4}{128\pi^4 \beta B} \cdot \sum_f e_f^2 \cdot 2 \int_{z_0}^{1/2} dz z(1-z)$$
$$\times \left\{ \epsilon^2 [z^2 + (1-z)^2] \Phi_1 + m_f^2 \Phi_0 \right\} \qquad (8.92)$$

for the longitudinal and transverse components respectively. The lower limit of the integral over z is given by $z_0 = (1/2)(1 - \sqrt{1 - 4m_f^2/m_X^2})$ (which ensures that $k^2 > 0$) and B is the slope parameter.

If we are to confront the data at low values of β, corresponding to large invariant masses m_X, it is necessary to also include a contribution from the higher Fock state $q\bar{q}g$. We can estimate this contribution using an effective 'two-gluon dipole' approximation due to Wüsthoff [17,53], as illustrated in figure 8.13. This gives

$$x_{I\!P} F_{q\bar{q}g,T}^{D(3)}(Q^2, \beta, x_{I\!P}) = \frac{81\beta\alpha_s}{512\pi^5 B} \sum_f e_f^2 \int_\beta^1 \frac{d\theta}{(1-\theta)^3} \left[\left(1 - \frac{\beta}{\theta}\right)^2 + \left(\frac{\beta}{\theta}\right)^2 \right]$$
$$\times \int_0^{(1-\theta)Q^2} dk_t^2 \log\left(\frac{(1-\theta)Q^2}{k_t^2}\right)$$
$$\times \left[\int_0^\infty u\, du\, \sigma(u/k_t, x_{I\!P}) K_2\left(\sqrt{\frac{\theta}{1-\theta}} u^2\right) J_2(u) \right]^2,$$
$$(8.93)$$

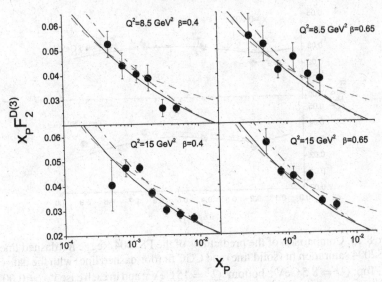

Figure 8.14. Comparison of the predictions of the FS2004 Regge fit (dashed line), the FS2004 saturation fit (solid line) and CGC fit (dot-dashed line) with the data on $F_2^{D(3)}$. Preliminary data from [55].

where, following [54], we have inserted a missing factor of $\frac{1}{2}$ compared with the expression in [17]. The light-cone variable

$$z = \frac{\theta}{1-\theta}\frac{k_t^2}{Q^2}$$

and $u = k_t r$. Note in particular that the normalization of this component is rather uncertain.

Comparison with experiment The predictions obtained from the above formulae for DDIS involve no adjustment of the dipole cross sections used to describe the F_2 data. However, we are free to adjust the forward slope for inclusive diffraction, B, within the range acceptable to experiment, which means that the overall normalization of $F_2^{D(3)}$ is free to vary slightly. We choose the rather low value of $B = 4.5 \text{ GeV}^{-2}$ in what follows. We are also somewhat free to vary the value of α_s used to define the normalization of the $q\bar{q}g$ component, which is important at low values of β. Rather arbitrarily we take $\alpha_s = 0.1$.

In principle, if evidence for saturation is seen in the DIS data, then it should also be detectable at some level in DDIS data.

As we have seen, there is a characteristic difference in the predictions of the Regge and saturation models for the energy dependence of the DIS structure function F_2 at

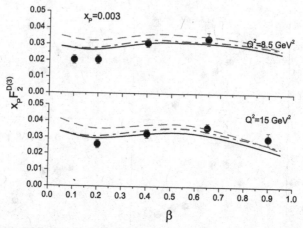

Figure 8.15. Comparison of the predictions of the FS2004 Regge fit (dashed line), the FS2004 saturation fit (solid line) and CGC fit (dot-dashed line) with the data on $F_2^{D(3)}$. Top: $Q^2 = 8.5\,\text{GeV}^2$; bottom: $Q^2 = 15\,\text{GeV}^2$; and in each case $x_{I\!P} = 0.003$. Preliminary data from [55].

fixed Q^2. If DIS and DDIS are described by the same dipole cross section, then there should be corresponding differences in the predictions for the energy dependence of the diffractive structure function $F_2^{D(3)}$ at fixed Q^2 and fixed diffractively produced mass m_X. In other words, it should be seen in the $x_{I\!P}$ dependence at fixed Q^2 and fixed β.

In figure 8.14 we show the predictions of our new FS2004 Regge and saturation models for the $x_{I\!P}$ dependence of the structure function $F_2^{D(3)}$ at fixed Q^2 and β, together with the corresponding predictions of the CGC model. In doing so, we have chosen to focus on β values in the intermediate range where the predictions are relatively insensitive to the $q\bar{q}g$ term and to the large-r behaviour of the photon wave function, which are both rather uncertain. There is, however, still some freedom in the choice of slope parameter, and hence in the absolute normalization. As can be seen there is, as expected, a characteristically different energy dependence of the Regge model and the two saturation models. There is a hint that the saturation models are preferred, but more accuracy is required in order to make a more positive statement.

In figure 8.15 we show the β dependence at fixed $x_{I\!P}$ and Q^2. Although this is unlikely to exhibit saturation effects in a transparent way, it is nonetheless a significant test of the dipole models discussed, because different size dipoles enter in markedly different relative weightings to the DIS case. For $\beta \geq 0.4$, where the $q\bar{q}$ dominates, it is clear that the Regge dipole and both saturation models are

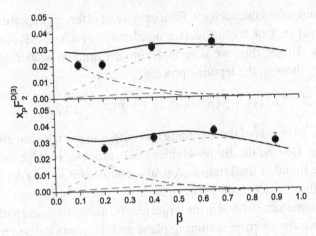

Figure 8.16. Breakdown of the various contributions to $F_2^{D(3)}$ (in the CGC model). Solid line is the total contribution, short dashed line is the $q\bar{q}$ contribution (light quarks only), dash-dotted line is the $q\bar{q}g$ contribution and the long dashed line is the $c\bar{c}$ contribution. Top: $Q^2 = 8.5$ GeV^2; bottom: $Q^2 = 15$ GeV^2; and in each case $x_{I\!P} = 0.003$. Preliminary data are from [55].

compatible with the data, given the uncertainty in normalization associated with the value of the slope parameter mentioned above. At the lowest β values, where the model-dependent $q\bar{q}g$ component is dominant, there is a discrepancy between the data and the predictions of all three models, which can only be removed by reducing the value of α_s well below the value assumed.

Finally, figure 8.16 explicitly shows the contribution from the light quark–antiquark dipoles. The majority of the remaining contribution comes from the $q\bar{q}g$ term and there is a small contribution from charm quark–antiquark dipoles. As can be seen, in the intermediate- to high-β region, where the $q\bar{q}g$ contribution is negligible, the predictions agree well with the data. However, the $q\bar{q}g$ contribution can be seen to dominate, as expected, for small β, corresponding to diffractively produced states with invariant mass $m_X^2 > Q^2$. The breakdown is qualitatively similar in other dipole models.

8.5.3 Deeply virtual Compton scattering

DVCS on protons, $\gamma^* p \to \gamma p$, is seen as an important reaction for the study of diffraction in QCD, despite the obvious difficulty in measuring it. In the standard QCD approach the amplitude is described at large Q^2 by skewed parton distributions, corresponding to operator products evaluated between protons of unequal momenta, as discussed in chapter 9. Furthermore, it offers an opportunity to access directly the real part of the diffractive amplitude, which is difficult to measure in other ways. To see this, we note that virtual Compton scattering is accessed experimentally through the leptonic process:

$$e^{\pm}(k) + p(p) \rightarrow e^{\pm}(k') + p(p') + \gamma(q'), \tag{8.94}$$

where the four-momenta of the incoming and outgoing particles are given in brackets. As well as DVCS, the Bethe–Heitler (BH) process, in which the photon is radiated by the initial or final state lepton, also contributes to the DVCS amplitude. In particular, interference terms between these two contributions can be detected by measuring asymmetries defined in the target rest-frame with respect to the azimuthal angle ϕ between the electron scattering plane and the plane defined by the virtual photon and the outgoing proton momenta. Because the BH amplitude is itself real, for the case of unpolarized electron scattering from an unpolarized target, these interference terms are sensitive to the real part of the diffractive amplitude.

We will not discuss the asymmetries further, because they have not yet been measured, but refer to Balitsky and Kuchina [56] for their detailed definitions and to Freund and McDermott [57] for predictions of their magnitudes in the dipole model. However, data integrated over ϕ are available for unpolarized particles. In this case the interference term between the two processes vanishes in the limit of large Q^2, so that the differential cross section can be written as:

$$\frac{d^2\sigma}{dy \, dQ^2} = \frac{d^2\sigma_{DVCS}}{dy \, dQ^2} + \frac{d^2\sigma_{BH}}{dy \, dQ^2}.$$

Here $y \approx (Q^2 + W^2)/S$ is the fraction of the incoming electron energy carried by the virtual photon, where $S = (k + p)^2$ is the square of the lepton–proton centre-of-mass energy and we have neglected the lepton and proton masses. The BH contribution is essentially known in terms of the Dirac and Pauli form factors (for example see (18) and (27) of [58]) and can be easily calculated and subtracted from the total to leave the DVCS cross section

$$\frac{d^2\sigma_{DVCS}}{dW \, dQ^2} = \frac{\alpha_{em}}{\pi Q^2 W} [1 + (1 - y)^2] \, \sigma(\gamma^* p \rightarrow \gamma p) \tag{8.95}$$

in a form convenient for comparing to dipole predictions for $\sigma(\gamma^* p \rightarrow \gamma p)$.

Dipole predictions   The imaginary part of the DVCS amplitude at zero momentum transfer is given by

$$\text{Im} \, A_{DVCS}(W^2, Q^2, t = 0) = \int dz \, d^2r \, \Psi_\gamma^{T*}(r, z, Q^2) \sigma(s, r) \Psi_\gamma^T(r, z, 0). \tag{8.96}$$

We have here made explicit the dependence of the photon wave function upon the photon virtuality. Thus dipole models provide predictions for this process with no adjustable parameters beyond those used to describe DIS and DDIS. However, before comparing these predictions with experiment, some general comments are worth making.

Firstly, only transverse virtual photons contribute to this amplitude, because the final photon is necessarily transverse, and the dipoles scatter without changing helicity. In addition, (8.96) indicates that DVCS is a good probe of the transition between soft and hard regimes in the dipole model. This is because DVCS is more sensitive to large dipoles than DIS at the same Q^2, because of the presence of the real-photon wave function in the amplitude. Finally we note that although (8.96) contains no adjustable parameters, a value of the badly-determined slope parameter B must be assumed in order to turn the predictions for the forward amplitude into predictions for the cross section. For this reason, the normalization is uncertain.

In figure 8.17, we compare the predictions of the FS2004 Regge and saturation models and the CGC model with H1 data [59], using a value $B = 7\,\mathrm{GeV}^{-2}$ for the slope parameter. Bearing in mind the normalization uncertainty, the agreement is excellent for all three models, although significant differences between the saturated and unsaturated models appear when the predictions are extrapolated to higher energies, as one would expect.

8.5.4 Vector-meson production

We next consider the diffractive vector-meson production reactions, that is

$$\gamma^*(q) + p(p) \to V(q') + p(p') \qquad V = \rho, \phi \text{ or } J/\psi. \qquad (8.97)$$

In these processes the choice of vector meson, as well as of different photon virtualities, allows one to explore contributions from dipoles of different transverse sizes [60]. The process also has the advantage that there is a wide range of available data, including data on the ratio of the contributions from longitudinal and transverse photons, obtained from the angular distribution of the vector-meson decay products. It ought also to provide important information on the poorly-known light-cone wave functions of the vector mesons.

Here we will again confine ourselves to the CGC model and the FS2004 Regge and saturation models. We will show predictions corresponding to the boosted Gaussian wave functions discussed in section 8.3, without adjustment of parameters to fit the data. In comparing with data, one must always bear in mind the uncertainties in the wave functions, and also in the empirical slope parameter B. A more detailed study of these issues can be found in [24].

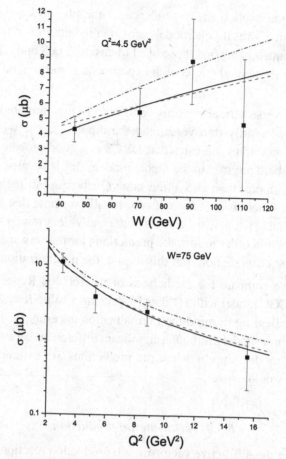

Figure 8.17. Comparison of the predictions of the FS2004 Regge fit (dashed line), the FS2004 saturation fit (solid line) and CGC fit (dot-dashed line) with the DVCS data [59]. The upper plot shows the W dependence at $Q^2 = 4.5\,\text{GeV}^2$. The lower plot shows the Q^2 dependence at $W = 75\,\text{GeV}$.

Production of ρ and ϕ   In figure 8.18 we show the predictions of the three models for ρ production. A correction for the real parts is included, as discussed in [24], and it is typically less than 20%. The parametrization of [61] is used for the B parameter, that is (in appropriate GeV-based units)

$$B = 0.60 \left(\frac{14}{(Q^2 + M_V^2)^{0.26}} + 1 \right). \tag{8.98}$$

Note that the flatness of the CGC prediction at $Q^2 = 0$ is a consequence of using the modified x variable of (8.69). The original CGC fit was performed without this

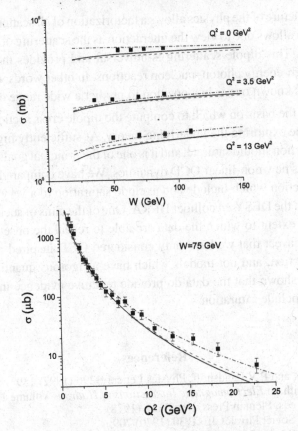

Figure 8.18. Comparison of the predictions of the FS2004 Regge fit (dashed line), the FS2004 saturation fit (solid line) and CGC fit (dot-dashed line) to the ρ meson data. The upper plot shows the W dependence at $Q^2 = 0\,\text{GeV}^2$, $Q^2 = 3.5\,\text{GeV}^2$ and $Q^2 = 13\,\text{GeV}^2$ [62]. The lower plot shows the Q^2 dependence at $W = 75\,\text{GeV}$ [63].

modification but to higher-Q^2 DIS data. Some modification is clearly needed if we are to try to extrapolate to lower Q^2 otherwise the cross section diverges. The quark-mass correction is sensible and does not spoil the agreement with the DIS data.

The data on ϕ meson production are qualitatively very similar to those for the ρ.

8.6 Conclusions

In this chapter we have shown how the high-energy interactions of a photon with a hadron can be described within the framework of the colour-dipole model. The space-time structure of the physics allows a factorization of all scattering amplitudes in a way which allows one to view the interaction as the scattering of a colour dipole off the proton. This dipole scattering is universal and provides the unifying link between all high-energy photon–nucleon reactions. In other words, once the dipole cross section is known one is in a position to predict a wide range of observables.

QCD provides the basis on which to compute the dipole cross section and we have spent some time exploring the underlying theory. At sufficiently high energies the dipole cross section should saturate, and it is one of the principal goals of experiment to probe for this new non-linear QCD dynamics. We have compared models of the dipole cross section which include and exclude saturation to a very wide range of data collected at the DESY *ep* collider HERA. One of the aims of such a comparison is to assess the extent to which the data are able to reveal the onset of saturation. Subject to the caveat that we have only considered QCD inspired models for the dipole cross section, and not models which have a rigorous quantitative basis in QCD, we have shown that the data do provide tentative evidence in favour of the models which include saturation.

References

[1] A Donnachie and P V Landshoff, Physics Letters B296 (1992) 139
[2] D W G S Leith, in *Electromagnetic Interactions of Hadrons* Volume 1, A Donnachie and G Shaw eds, Plenum Press, New York (1978)
[3] V N Gribov, Soviet Physics JETP 30 (1970) 709;
    K Gottfried, *Proceedings of the 1971 International Symposium on Electron and Photon Interactions*, Cornell University, Ithaca, New York (1972)
[4] N N Nikolaev and B G Zakharov, Zeitschrift für Physik C49 (1991) 607; *ibid* C53 (1992) 331
[5] A H Mueller, Nuclear Physics B415 (1994) 373;
    A H Mueller and B Patel, Nuclear Physics B425 (1994) 471
[6] S J Brodsky, H C Pauli and S S Pinsky, Physics Reports 301 (1998) 299 and references therein
[7] H J Miettinen and J Pumplin, Physical Review D18 (1978) 1696
[8] A Hebecker and P V Landshoff, Physics Letters B419 (1998) 393
[9] M Abramowitz and I A Stegun, *Handbook of Mathematical Functions*, Dover, New York (1970)
[10] A Donnachie and G Shaw, in *Electromagnetic Interactions of Hadrons* Vol 2, A Donnachie and G Shaw eds, Plenum Press, New York (1978)
[11] G Shaw, Physical Review D47 (1993) R3676, Physics Letters B228 (1989) 125;
     P Ditsas and G Shaw, Nuclear Physics B113 (1976) 246
[12] H Fraas, B J Read and D Schildknecht, Nuclear Physics B86 (1975) 346
[13] L Frankfurt, V Guzey and M Strikman, Physical Review D58 (1998) 094039
[14] G Cvetic, D Schildknecht, B Surrow and M Tentyukov, European Physical Journal C20 (2001) 77
[15] J R Forshaw, G Kerley and G Shaw, Physical Review D60 (1999) 074012
[16] K Golec-Biernat and M Wüsthoff, Physical Review D59 (1999) 014017

[17] K Golec-Biernat and M Wüsthoff, Physical Review D60 (1999) 114023

[18] J R Forshaw and G Shaw, Journal of High Energy Physics 0412 (2004) 052

[19] D O Caldwell *et al*, Physical Review Letters 40 (1978) 1222

[20] S J Brodsky and G P Lepage, Physical Review D22 (1980) 2157

[21] S Munier, A H Mueller and A Stasto, Nuclear Physics B603 (2001) 427

[22] H G Dosch, T Gousset, G Kulzinger and H G Pirner, Physical Review D55 (1997) 2602

[23] J Nemchik, N N Nikolaev, E Predazzi and B G Zakharov, Zeitschrift für Physik C75 (1997) 71

[24] J R Forshaw, R Sandapen and G Shaw, Physical Review D69 (2004) 094013

[25] H Halperin and A Zhitnitsky, Physical Review D56 (1997) 184

[26] M Wirbel, B Stech and M Bauer, Zeitschrift für Physik C29 (1985) 637

[27] S J Brodsky, T Huang and G P Lepage, SLAC-PUB-2540 (1980)

[28] A Donnachie, J Gravelis and G Shaw, Physical Review D63 (2001) 114013

[29] N N Nikolaev and B D Zakharov, JETP Letters 59 (1994) 6

[30] E Avsar, G Gustafson and L Lönnblad, hep-ph/0503181

[31] E A Kuraev, L N Lipatov and V S Fadin, Soviet Physics JETP 45 (1977) 199;
I Balitsky and L N Lipatov, Soviet Journal of Nuclear Physics 28 (1978) 822

[32] J R Forshaw and D A Ross, *Quantum Chromodynamics and the Pomeron*, Cambridge University Press, Cambridge (1997)

[33] V N Gribov and L N Lipatov, Soviet Journal of Nuclear Physics 15 (1972) 438;
Y L Dokshitzer, Soviet Physics JETP 46 (1977) 641;
G Altarelli and G Parisi, Nuclear Physics B126 (1977) 298

[34] Yu V Kovchegov, Physical Review D60 (1999) 034008

[35] I Balitsky, Nuclear Physics B463 (1996) 99

[36] A H Mueller, Lectures given at Cargese Summer School on QCD Perspectives on Hot and Dense Matter, Cargese, France, 2001, hep-ph/0111244

[37] E Iancu, A Leonidov and L D McLerran, Lectures given at Cargese Summer School on QCD Perspectives on Hot and Dense Matter, Cargese, France, 2001, hep-ph/0202270

[38] H Weigert, arXiv:hep-ph/0501087

[39] L V Gribov, E M Levin and M G Ryskin, Physics Reports 100 (1983) 1

[40] D N Triantafyllopoulos, Nuclear Physics B648 (2003) 293

[41] A Kovner and U A Wiedemann, Physical Review D66 (2002) 051502; Physical Review D66 (2002) 034031; Physics Letters B551 (2003) 311

[42] E Ferreiro, E Iancu, K Itakura and L McLerran, Nuclear Physics A710 (2002) 373

[43] E Iancu, K Itakura and S Munier, Physics Letters B590 (2004) 199

[44] A I Shoshi, F D Steffen and H J Pirner, Nuclear Physics A709 (2002) 131;
A Kovner and U Wiedemann, Physical Review D66 (2002) 034031;
S Munier, Physical Review D66 (2002) 114012;
D Kharzeev, E Levin and L McLerran, Physics Letters B561 (2003) 93;
E Gotsman, E Levin, M Lublinsky and U Maor, European Physical Journal C27 (2003) 411;
L Favart and M V T Machado, European Physical Journal C29 (2003) 365;
H Kowalski and D Tearney, Physical Review D68 (2003) 114005;
J Bartels, E Gotsman, E Levin, M Lublinsky and U Maor, Physics Letters B556 (2003) 114; Physical Review D68 (2003) 054008;
E Levin, hep-ph/0408039

[45] M F McDermott, L Frankfurt, V Guzey and M Strikman, European Physical Journal C16 (2000) 641

[46] J Bartels, K Golec-Biernat and H Kowalski, Physical Review D66 (2002) 014001
[47] A Hebecker, Physics Reports 331 (2000) 1
[48] J Jalilian-Marian, A Kovner, A Leonidov and H Weigert, Nuclear Physics B504 (1997) 415; Physical Review D59 (1999) 014014;
     H Weigert, Nuclear Physics A703 (2002) 823;
     E Iancu, A Leonidov and L D McLerran, Nuclear Physics A692, 583 (2001); Physics Letters B510 (2001) 133
[49] A Donnachie and P V Landshoff, Physics Letters B437 (1998) 408
[50] G Kerley and G Shaw, Physical Review D56 (1997) 7291
[51] A Donnachie and H G Dosch, Physics Letters B502 (2001) 74
[52] S Chekanov et al, ZEUS Collaboration, European Physical Journal C21 (2001) 442;
     C Adloff et al, H1 Collaboration, European Physical Journal C21 (2001) 33
[53] M Wüsthoff, Physical Review D56 (1997) 4311
[54] J R Forshaw, G Kerley and G Shaw, Nuclear Physics A675 (2000) 80
[55] H1 Collaboration in Proceedings of the International Europhysics Conference on High Energy Physics, Aachen (2003), European Physical Journal C33, Supplement 1 (2004)
[56] I Balitsky and E Kuchina, Physical Review D62 (2000) 074004
[57] A Freund and M F McDermott, Physical Review D65 (2002) 074008; ibid 091901
[58] A Belitsky et al, Nuclear Physics B593 (2001) 289
[59] C Adloff et al, H1 Collaboration, Physics Letters B517 (2001) 47
[60] B Z Kopeliovich, J Nemchick, N N Nikolaev and B G Zakharov, Physics Letters B309 (1993) 179
[61] B Mellado, unpublished result given in [64]
[62] J Breitweg et al, ZEUS Collaboration, European Physical Journal C6 (1999) 603
[63] C Adloff et al, H1 Collaboration, European Physical Journal C13 (2000) 371; Nuclear Physics B468 (1996) 3
[64] A Caldwell and M Soares, Nuclear Physics A696 (2001) 125

9

Generalized parton distributions

V Burkert and M Diehl

Generalized parton distributions (GPDs) have been recognized as a versatile tool to investigate and describe the structure of hadrons at the quark–gluon level. They are closely related to conventional parton distributions and also to hadronic form factors, but contain information that cannot be accessed with either of these quantities. Important areas where GPDs can provide new insight are the spatial distribution of quarks and gluons within a hadron and the contribution of quark orbital angular momentum to the nucleon spin. In this chapter we present the basics of the theory of GPDs, the dynamics they encode and the efforts of phenomenology and experiment to measure them in exclusive scattering processes. We do not attempt to give a comprehensive account of the vast literature and refer to the reviews [1–5] for more detailed discussion and references.

Experimental access to GPDs is provided in suitable hard scattering processes with exclusive final states, especially in processes initiated by a highly virtual photon. Recall that the cross section for inclusive deep inelastic scattering (DIS) is related to the amplitude for forward Compton scattering, $\gamma^* p \to \gamma^* p$, via the optical theorem. In the Bjorken limit of large $Q^2 = -q^2$ at fixed $x_B = Q^2/(2pq)$, this amplitude factorizes into a short-distance process involving quarks and gluons and the usual parton distributions which encode the structure of the target hadron at quark–gluon level. At leading order in α_s one then obtains the handbag diagram in figure 9.1(a) with $p' = p$ and $q' = q$. In appropriate kinematics this type of factorization generalizes to the non-forward amplitude which can appear in exclusive processes. One such example is $ep \to ep\gamma$, called deeply virtual Compton scattering (DVCS), where the photon with momentum q' is real and one takes the limit of large Q^2 at fixed x_B and fixed squared momentum transfer $t = (p - p')^2$. The long-distance dynamics is now encoded in generalized parton distributions which

Electromagnetic Interactions and Hadronic Structure, eds Frank Close, Sandy Donnachie and Graham Shaw. Published by Cambridge University Press. © Cambridge University Press 2007

(a)

(b)

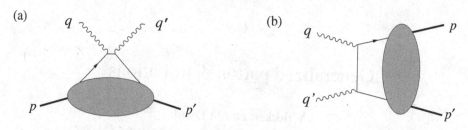

Figure 9.1. (a) Handbag diagram for the Compton amplitude in Bjorken kinematics. A second diagram is obtained by interchanging the photon vertices. (b) The crossed-channel diagram for two-photon annihilation into a hadron–antihadron pair.

involve the different hadron states before and after the scattering. Not only can the hadron momenta differ but also their polarization, and one may also have the transition from one hadron to another. Each of these degrees of freedom opens a way to study important aspects of hadron structure, as we will see in the following.

Crossing symmetry relates Compton scattering with the annihilation of two photons into a hadron–antihadron pair. An analogous pattern of factorization, shown in figure 9.1(b), is found in the limit where at least one of the photons goes far off-shell at fixed invariant mass $s = (p + p')^2$ of the produced hadrons. This leads to quantities describing the hadronization of a quark–antiquark pair into the final state. These generalized distribution amplitudes are connected to GPDs by s–t crossing. At the same time they are direct extensions of the light-cone distribution amplitudes for, say, a single-pion state which occur in two-photon annihilation $\gamma^* \gamma \to \pi^0$.

9.1 Properties and physics of GPDs

9.1.1 Definitions and basics

The usual parton distributions are expectation values of quark or gluon operators for a given hadron state. Their generalizations describe the matrix elements of the same operators, but between different hadron states. To describe partons it is useful to introduce light-cone coordinates $v^{\pm} = (v^0 \pm v^3)/\sqrt{2}$ and the transverse part $\vec{v} = (v^1, v^2)$ of a four-vector v, where we define the scalar product such that $\vec{v}^2 \geq 0$. Note that a plus-momentum p^+ becomes proportional to the longitudinal momentum of a particle in the infinite-momentum frame, where $p^3 \to \infty$ and where the parton picture is most immediate. Fractions of plus-momentum (often called 'momentum fractions' for ease of language) are invariant under boosts in the z-direction. For the generalized quark distributions in the nucleon one can

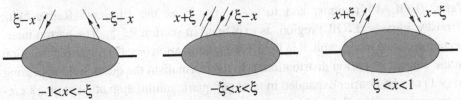

Figure 9.2. Partonic representation of a GPD in different regions of x. Momentum fractions x and ξ refer to the average hadron momentum $P = \frac{1}{2}(p + p')$.

define [1]

$$F^q = \frac{1}{2} \int \frac{dz^-}{2\pi} e^{ixP^+z^-} \langle p' | \bar{q} \left(-\tfrac{1}{2}z\right) \gamma^+ q \left(\tfrac{1}{2}z\right) | p \rangle \Big|_{z^+=0, \vec{z}=\vec{0}}$$

$$= \frac{1}{2P^+} \left[H^q(x,\xi,t) \bar{u}(p')\gamma^+ u(p) + E^q(x,\xi,t) \bar{u}(p') \frac{i\sigma^{+\alpha} \Delta_\alpha}{2m} u(p) \right],$$

$$\tilde{F}^q = \frac{1}{2} \int \frac{dz^-}{2\pi} e^{ixP^+z^-} \langle p' | \bar{q} \left(-\tfrac{1}{2}z\right) \gamma^+ \gamma_5 \, q \left(\tfrac{1}{2}z\right) | p \rangle \Big|_{z^+=0, \vec{z}=\vec{0}}$$

$$= \frac{1}{2P^+} \left[\tilde{H}^q(x,\xi,t) \bar{u}(p')\gamma^+ \gamma_5 u(p) + \tilde{E}^q(x,\xi,t) \bar{u}(p') \frac{\gamma_5 \Delta^+}{2m} u(p) \right],$$

(9.1)

where we omit the polarization dependence of the hadron states and spinors for brevity. As a shorthand for different quark flavour combinations we will use $H^{u+d} = H^u + H^d$ etc. We write

$$P = \tfrac{1}{2}(p + p'), \quad \Delta = p' - p, \quad t = (p' - p)^2, \tag{9.2}$$

and m for the nucleon mass. The plus-momentum transfer to the proton is described by the skewness parameter $\xi = (p - p')^+/(p + p')^+$. As a consequence of Lorenz invariance the GPDs $H, E, \tilde{H}, \tilde{E}$ only depend on t and on the momentum fractions x and ξ relative to the average hadron momentum P. These definitions hold in light-cone gauge $A^+ = 0$. In other gauges a Wilson line, or gauge link, has to be inserted between the antiquark and quark fields $\bar{q}(-\frac{1}{2}z)$ and $q(\frac{1}{2}z)$ as explained in chapter 7.

There are three distinct regions in x for a given value of ξ, as shown in figure 9.2. We assume $\xi > 0$ here, which is the case of relevance in the physical processes studied so far. For $\xi < x < 1$, GPDs describe the emission of a quark by the incoming hadron and its absorption by the outgoing one. For $-1 < x < -\xi$ one has emission and absorption of an antiquark. In the central region $-\xi < x < \xi$, which has no analogue in the usual parton distributions, one has the emission of a quark–antiquark pair from the initial hadron. Following the pattern of evolution one refers to $\xi < x < 1$ and $-1 < x < -\xi$ as Dokshitzer, Gribov, Lipatov, Altrarelli,

Parisi (DGLAP) regions, and to $-\xi < x < \xi$ as the Efremov, Radyushkin, Brodsky, Lepage (ERBL) region, as explained in section 9.1.5. The parton interpretation can be made explicit in light-cone quantization (see [7] for an introduction in the context of parton distributions). In this formalism the quark field appearing in (9.1) can be Fourier expanded in terms of quark annihilation and antiquark creation operators, b and $d^\dagger$, which come with factors $e^{-iz^-k^+}$ and $e^{iz^-k^+}$ respectively. Here k^+ is constrained to be non-negative, as $k^0 + k^3$ must be for a positive-energy state on mass shell. The three regions in figure 9.2 then select the terms $dd^\dagger$, db, $b^\dagger b$, respectively.

In the forward limit $p = p'$ the GPDs H^q and $\tilde{H}^q$ become the usual parton distribution functions (see chapter 7). For $x > 0$ one has

$$H^q(x,0,0) = q(x), \qquad H^q(-x,0,0) = -\bar{q}(x),$$
$$\tilde{H}^q(x,0,0) = \Delta q(x), \qquad \tilde{H}^q(-x,0,0) = \Delta\bar{q}(x) \qquad (9.3)$$

in terms of the unpolarized and polarized quark distributions $q(x)$ and $\Delta q(x)$ and their analogues $\bar{q}(x)$ and $\Delta\bar{q}(x)$ for antiquarks. E^q and $\tilde{E}^q$ are multiplied with $p' - p$ in their definitions, so that they decouple in the forward limit and are invisible in processes where the usual parton densities are accessed. The combinations $H^q(x,\xi,t) - H^q(-x,\xi,t)$ and $\tilde{H}^q(x,\xi,t) + \tilde{H}^q(-x,\xi,t)$ describe the exchange of positive C parity in the t-channel; in the forward limit they are the sum of quark and antiquark distributions. In turn, $H^q(x,\xi,t) + H^q(-x,\xi,t)$ and $\tilde{H}^q(x,\xi,t) - \tilde{H}^q(-x,\xi,t)$ correspond to negative C parity in the t channel. The relevant combinations for E are as for H, and those for $\tilde{E}$ are as for $\tilde{H}$.

For gluons we have definitions similar to those in the quark sector:

$$
\begin{aligned}
F^g &= \frac{1}{P^+} \int \frac{dz^-}{2\pi} e^{ixP^+z^-} \langle p' | G^{+\mu}\left(-\tfrac{1}{2}z\right) G_\mu{}^+\left(\tfrac{1}{2}z\right) |p\rangle \Big|_{z^+=0,\,\vec{z}=\vec{0}} \\
&= \frac{1}{2P^+}\left[H^g(x,\xi,t)\,\bar{u}(p')\gamma^+ u(p) + E^g(x,\xi,t)\,\bar{u}(p')\frac{i\sigma^{+\alpha}\Delta_\alpha}{2m}u(p) \right], \\
\tilde{F}^g &= -\frac{i}{P^+} \int \frac{dz^-}{2\pi} e^{ixP^+z^-} \langle p' | G^{+\mu}\left(-\tfrac{1}{2}z\right) \tilde{G}_\mu{}^+\left(\tfrac{1}{2}z\right) |p\rangle \Big|_{z^+=0,\,\vec{z}=\vec{0}} \\
&= \frac{1}{2P^+}\left[\tilde{H}^g(x,\xi,t)\,\bar{u}(p')\gamma^+\gamma_5 u(p) + \tilde{E}^g(x,\xi,t)\,\bar{u}(p')\frac{\gamma_5\Delta^+}{2m}u(p) \right],
\end{aligned}
$$

$$(9.4)$$

where $\tilde{G}^{\alpha\beta} = \frac{1}{2}\varepsilon^{\alpha\beta\gamma\delta}G_{\gamma\delta}$ is the dual field-strength tensor ($\varepsilon_{0123} = 1$). For symmetry reasons H^g, E^g are even and $\tilde{H}^g$, $\tilde{E}^g$ are odd functions of x. This is because the field strength operator $G^{\alpha\beta}$ both creates and annihilates gluons (in contrast to the quark field q, which annihilates quarks but creates antiquarks). In the forward limit one has

$$H^g(x, 0, 0) = xg(x), \qquad \tilde{H}^g(x, 0, 0) = x\Delta g(x) \qquad (9.5)$$

for $x > 0$, whereas E^g and $\tilde{E}^g$ decouple again. Notice the extra factors of x compared with (9.3).

Using time-reversal symmetry one can show that for both quarks and gluons $H(x, \xi, t) = H(x, -\xi, t)$, that is H is an even function of ξ. Together with the relation $[H(x, \xi, t)]^* = H(x, -\xi, t)$, which directly follows from the definitions (9.1) or (9.4), this constrains H to be real-valued. Analogous relations hold for E, $\tilde{H}$, $\tilde{E}$. One can also define GPDs for the transition from one hadron to another. Transition GPDs provide a tool to study the parton structure of hadrons not readily available as targets (and give information quite different from the fragmentation functions into these hadrons).

The $p \rightarrow \Delta(1232)$ transition GPDs have been studied in [3,8]. Note that transition GPDs are in general not even in ξ, since time-reversal symmetry relates different transitions such as $p \rightarrow \Delta$ and $\Delta \rightarrow p$. Isospin invariance can be used to express the $p \rightarrow n$ and $n \rightarrow p$ transition GPDs through the difference of u and d quark GPDs in the proton. GPDs for other transitions within the ground-state baryon octet, such as $p \rightarrow \Lambda$, are related to those in the proton by approximate flavour SU(3) symmetry, see [3,5]. Here it will depend on a particular case whether these transitions can provide additional information on the nucleon GPDs or rather on details of SU(3) symmetry breaking.

The outgoing hadron state in matrix elements defining GPDs need not be a single hadron, but can also be a multi-hadron state. The distributions then depend on its invariant mass and on further variables describing its internal structure [3,9]. A case of particular practical relevance is the transition $N \rightarrow N\pi$, where N denotes a nucleon. Time-reversal invariance transforms a state $|N\pi\rangle_{out}$ into $|N\pi\rangle_{in}$ and hence does not force such transition GPDs to be real-valued; their complex phases reflect the strong interactions in the $N\pi$ system.

9.1.2 Spin structure

As one can see from the operators in their definitions, the distributions H and E for a spin-$\frac{1}{2}$ target involve the sum of positive and negative quark or gluon helicities, whereas $\tilde{H}$ and $\tilde{E}$ involve their difference. Referring to the parton spins one often calls H, E 'unpolarized' and $\tilde{H}$, $\tilde{E}$ 'polarized' GPDs. In both cases there is no net helicity transfer on the parton side: in the DGLAP regions the emitted and the absorbed partons have the same helicity and in the ERBL region the helicities of the two emitted partons couple to zero. To discuss the spin dependence on the hadron side it is useful to choose states that have definite light-cone helicity [6], which is for instance invariant under boosts along the z-axis. For $p^+ \gg m$, light-cone helicity differs from the usual helicity only by effects of order m/p^+, which is reminiscent to the near-equivalence of plus-momentum and ordinary momentum for particles moving fast in the positive z-direction. Inserting the appropriate spinors (given in [4] for example) into the matrix elements (9.1) one obtains

$$F_{++} = \sqrt{1-\xi^2}\left[H - \frac{\xi^2}{1-\xi^2}E\right], \quad F_{-+} = e^{i\varphi}\frac{\sqrt{t_0-t}}{2m}E,$$

$$\tilde{F}_{++} = \sqrt{1-\xi^2}\left[\tilde{H} - \frac{\xi^2}{1-\xi^2}\tilde{E}\right], \quad \tilde{F}_{-+} = e^{i\varphi}\frac{\sqrt{t_0-t}}{2m}\xi\tilde{E}, \qquad (9.6)$$

where the first index in $F_{\lambda'\lambda}$ denotes the light-cone helicity of the outgoing proton and the second index that of the incoming proton (for better legibility we label fermion helicities only by their sign). The other spin combinations are given as $F_{--} = F_{++}$, $F_{+-} = -[F_{-+}]^*$, $\tilde{F}_{--} = -\tilde{F}_{++}$, $\tilde{F}_{+-} = [\tilde{F}_{-+}]^*$ by parity invariance. The smallest kinematically allowed value of $-t$ at given ξ is

$$-t_0 = 4\xi^2 m^2/(1-\xi^2), \qquad (9.7)$$

and the azimuthal angle φ is given by $e^{i\varphi}|\vec{D}| = D^1 + iD^2$ with

$$\vec{D} = \frac{\vec{p}'}{1-\xi} - \frac{\vec{p}}{1+\xi}, \qquad t = t_0 - (1-\xi^2)\vec{D}^2. \qquad (9.8)$$

The special role played by this vector will again appear in section 9.1.7. We see in (9.6) that H and $\tilde{H}$ conserve the proton helicity. In turn, E and $\tilde{E}$ are responsible for proton helicity flip. The factors $e^{i\varphi}$ indicate that in these transitions one unit of orbital angular momentum compensates the mismatch of helicity transferred on the parton and on the hadron side, thus ensuring angular-momentum conservation along z. In sections 9.1.3 and 9.1.8 we will see from other points of view that E and $\tilde{E}$ involve orbital angular momentum in an essential way.

In addition to the distributions discussed so far there are GPDs describing the helicity flip of quarks or gluons in the DGLAP regions, transferring a net helicity of 1 in the case of quarks and of 2 in the case of gluons. They become diagonal in parton polarization if one changes basis from states of definite helicity to states of definite transversity (see [7] for example). The relevant operators for quark and gluon transversity GPDs are $\bar{q}\sigma^{+i}q$ and $G^{+i}G^{j+} + G^{+j}G^{i+} - g^{ij}G^{+\alpha}G_\alpha{}^+$, where i, j are transverse indices. For spin-$\frac{1}{2}$ targets, four transversity GPDs for each parton species parametrize the dependence on the initial and final hadron polarizations [10]. For quarks one of these GPDs tends to the usual quark transversity distribution in the forward limit (see chapter 7). All others require orbital angular momentum to balance a helicity mismatch between the partons and hadrons and hence decouple for equal hadron momenta.

A general way to count the number of GPDs for given spin of the hadrons uses the interpretation of parton distributions as parton–hadron scattering amplitudes, see [7,10] for example. Rewriting

$$\int \frac{dz^-}{2\pi} \, e^{ixP^+z^-} \langle p' | \bar{q} \left(-\tfrac{1}{2}z\right) \gamma^+ q \left(\tfrac{1}{2}z\right) |p\rangle \Big|_{z^+=0,\, \vec{z}=\vec{0}}$$

$$= \int \frac{dk^- \, d^2\vec{k}}{(2\pi)^4} \left[\int d^4z \, e^{i(kz)} \langle p' | T\bar{q} \left(-\tfrac{1}{2}z\right) \gamma^+ q \left(\tfrac{1}{2}z\right) |p\rangle \right]_{k^+=xP^+}, \qquad (9.9)$$

the second line is identified as a Green's function with two external hadrons and two (off-shell) external quarks, integrated over the minus and transverse momentum of the quarks (and hence over their off-shellness). The time-ordering prescription has been inserted when going to the second line, which is possible because the fields are taken at light-like distance (see section 3.4 of [4] for references).

Using this connection one obtains the number of GPDs for a parton species as the number of independent helicity amplitudes for parton–hadron scattering when the constraints of parity invariance are included. We have seen in section 9.1.1 that time-reversal symmetry relates distributions at ξ and $-\xi$ and thus does not reduce their number, except for $\xi = 0$. With a spin-0 target one gets for each quark flavour and for gluons one unpolarized GPD (given by the same matrix element as $F^{q,g}$ in (9.1) and (9.4)) and one transversity GPD. The nine parton helicity conserving GPDs in a spin-1 target like the deuteron have been classified in [11].

9.1.3 Sum rules, form factors and the nucleon spin

Taking x moments of GPDs one obtains matrix elements of local operators with twist 2, just as for the usual parton densities. These matrix elements are readily parametrized in terms of form factors, which depend only on t. Of special importance are the contributions of a given quark flavour to the usual Dirac and Pauli nucleon form factors,

$$\langle p' | \bar{q}(0)\gamma^\mu q(0) |p\rangle = F_1^q(t)\, \bar{u}(p')\gamma^\mu u(p) + F_2^q(t)\, \bar{u}(p')\frac{i\sigma^{\mu\alpha}\Delta_\alpha}{2m} u(p), \qquad (9.10)$$

and the form factors parametrizing the quark part of the energy–momentum tensor,

$$\langle p' | T_q^{(\mu\nu)}(0) |p\rangle = A^q(t)\, \bar{u}(p')P^{(\mu}\gamma^{\nu)}u(p) + B^q(t)\, \bar{u}(p')\frac{P^{(\mu} i\sigma^{\nu)\alpha}\Delta_\alpha}{2m} u(p)$$

$$+ C^q(t)\, \frac{\Delta^{(\mu}\Delta^{\nu)}}{m} \bar{u}(p')u(p), \qquad (9.11)$$

where $t^{(\mu\nu)} = \frac{1}{2}(t^{\mu\nu} + t^{\nu\mu}) - \frac{1}{4}g^{\mu\nu} t^{\alpha}{}_{\alpha}$ denotes symmetrization of indices and subtraction of the trace for any tensor $t^{\mu\nu}$. The trace part of $T^{\mu\nu}$ has twist 4 and is not accessible from the distributions we consider here. The form factors in (9.10) and (9.11) are related to the lowest Mellin moments of H^q and E^q by

$$
\int_{-1}^{1} dx \, H^q(x, \xi, t) = F_1^q(t), \qquad \int_{-1}^{1} dx \, x H^q(x, \xi, t) = A^q(t) + 4\xi^2 C^q(t),
$$

$$
\int_{-1}^{1} dx \, E^q(x, \xi, t) = F_2^q(t), \qquad \int_{-1}^{1} dx \, x E^q(x, \xi, t) = B^q(t) - 4\xi^2 C^q(t).
$$

$$(9.12)$$

Due to Lorentz invariance the ξ dependence on the right-hand sides of (9.12) can only originate from factors of Δ^μ in the form-factor decompositions. These are readily generalized to higher moments [1], and the x^n moments of H^q and E^q turn out to be polynomials in ξ of order $n + 1$ (which in addition must be even due to time-reversal invariance). This 'polynomiality' property puts important constraints on the x and ξ dependence of the original distributions. The terms with the highest power cancel in the sum of $H + E$ as seen in (9.12). Analogous sum rules hold for the polarization-dependent distributions, where $\int_{-1}^{1} dx \, \tilde{H}^q$ and $\int_{-1}^{1} dx \, \tilde{E}^q$ give, for instance, the axial and pseudoscalar nucleon form factors. In the axial sector there is, however, no analogue of the form factors C^q [4], so that the x^n moments of $\tilde{H}^q$ and $\tilde{E}^q$ are polynomials in ξ of order n. Finally there are similar sum rules for the gluon GPDs, where in particular $\int_0^1 dx \, H^g$ and $\int_0^1 dx \, E^g$ are related to the gluon part of the energy–momentum tensor. Notice the correspondence of x^n moments for quarks with x^{n-1} moments for gluons, which reflects the different factors of x in the forward limits (9.3) and (9.5).

The Dirac and Pauli form factors are well measured, at least for small to moderate t (see chapter 2), and can be used as constraints on the hitherto unknown GPDs. In contrast, the energy–momentum form factors (9.11) cannot be directly measured, and it is hoped that information about GPDs will provide access to them. Widespread interest in GPDs has in fact been triggered by Ji's sum rule [12]

$$
2\langle J_q^3 \rangle = A^q(0) + B^q(0) = \lim_{t \to 0} \int_{-1}^{1} dx \, x [H^q(x, \xi, t) + E^q(x, \xi, t)], \qquad (9.13)
$$

where $\langle J_q^3 \rangle$ is the *total* angular momentum along z carried by quarks and antiquarks of flavour q in a proton polarized in the positive z-direction. This comprises not only the contribution from quark helicity, which can be obtained from the moments $\int_0^1 dx \, [\Delta q(x) + \Delta \bar{q}(x)]$ of the usual polarized quark densities (see chapter 7), but also the part due to the orbital angular momentum carried by the quarks. A direct evaluation of the integral in (9.13) from measurements of hard exclusive processes will be very difficult (see section 9.3.9). On the other hand, the relation (9.13) provides the only known route to $\langle J_q^3 \rangle$ and thus how the total spin of the nucleon is made up at the quark–gluon level. It can also be used at the theory level to calculate $\langle J_q^3 \rangle$, for instance using lattice QCD (see section 9.2.6).

In analogy to (9.13) the total angular momentum carried by gluons is given as $2\langle J_g^3 \rangle = \lim_{t \to 0} \int_0^1 dx \, [H^g(x, \xi, t) + E^g(x, \xi, t)]$. Note that for vanishing t and ξ the corresponding integrals over xH^q or H^g alone give the momentum fraction along z for each parton species. Combining the sum rules for total momentum and for total angular momentum, one thus finds that $\int_0^1 dx \, E^g + \sum_q \int_{-1}^1 dx \, x E^q$ must vanish when both t and ξ are zero.

9.1.4 Generalized distribution amplitudes

Generalized distribution amplitudes [13,14] parametrize matrix elements of the same operators as GPDs. The most detailed investigations have been made for the two-pion system, where one has

$$\Phi^q(z, \zeta, s) = \int \frac{dx^-}{2\pi} \, e^{i(z - \frac{1}{2})(p+p')^+ x^-} \left\langle \pi^+ \pi^- \left| \bar{q} \left(-\tfrac{1}{2}x\right) \gamma^+ q \left(\tfrac{1}{2}x\right) \right| 0 \right\rangle \Big|_{x^+=0, \, \vec{x}=\vec{0}}$$

(9.14)

and an analogous definition for gluons. Here p and p' are the respective momenta of π^+ and π^-, and $s = (p + p')^2$ is their invariant mass. The momentum fraction of the quark is denoted by z and $\zeta = p^+/(p + p')^+$ is the momentum fraction of the π^+ relative to the total hadron momentum. Moments in $(2z - 1)$ of the generalized distribution amplitudes are polynomials in $(2\zeta - 1)$, whose coefficients are form factors of twist-2 operators in the time-like domain, for example the time-like pion form factor, given by $\int_0^1 dz \, \Phi^u(z, \zeta, s) = (2\zeta - 1) F_\pi(s)$. By analytic continuation these form factors are connected with the form factors parametrizing the x moments of the GPDs in the pion; this is the crossing relation between GPDs and generalized distribution amplitudes. For the same reason we discussed in connection with $N \to N\pi$ GPDs at the end of section 9.1.1, generalized distribution amplitudes are not real-valued but have dynamical phases reflecting the strong interactions within the multi-hadron system. The phases of two-pion distribution amplitudes are simply related with the phases of elastic $\pi\pi$ scattering by Watson's theorem, provided that s is below the inelastic threshold [15]. Two-pion distribution amplitudes reflect in various ways the dynamics of the two-pion system and its interplay with quarks and gluons, see [4] for an overview. Examples are their connection with chiral perturbation theory and their connection with the distribution amplitudes of resonances such as the ρ or f_2 mesons, which are defined just as in (9.14) but with $|\pi\pi\rangle$ replaced by the single-meson state.

9.1.5 Evolution

Like the usual parton distributions, GPDs depend on a factorization scale μ^2, which reflects the separation of parton scattering and hadronic matrix element in process amplitudes and physically describes the resolution scale at which partons are probed. For a non-singlet combination of GPDs like H^{u-d} the evolution equations have the form

$$\mu^2 \frac{d}{d\mu^2} H^{u-d}(x, \xi, t) = \int_{-1}^{1} \frac{dx'}{|\xi|} V\left(\frac{x}{\xi}, \frac{x'}{\xi}\right) H^{u-d}(x', \xi, t), \qquad (9.15)$$

whereas in the singlet sector the mixing of quark and gluon distributions is described by a matrix equation, just as for the usual parton densities. Note that due to their spin structure, transversity GPDs do *not* mix quarks and gluons under evolution and in this sense allow one to probe 'primordial' quarks or gluons in a hadron, which have not been 'generated' by the usual perturbative parton-splitting processes.

The evolution kernels $V(x, x')$ have been studied in detail to leading [13,16] and next-to-leading [17] logarithmic accuracy. Two different patterns of evolution can be distinguished. In the DGLAP regions the change in μ^2 of a GPD at x only depends on GPDs at momentum fractions $|x'| \geq |x|$. Here, evolution to higher scales μ^2 shifts parton momenta to smaller values and corresponds to momentum loss of partons by radiation. For $\xi \to 0$ one recovers the well-known DGLAP evolution equations and kernels for the usual parton densities. In the ERBL region, all values of x' contribute on the right-hand sides of the evolution equations. This region thus acts as a sink of partons evolving 'out' of the DGLAP regions. Inside the ERBL region there is a redistribution of parton momenta, described by the kernels $V(x, x')$ in the domain $|x|, |x'| \leq 1$, where they coincide with the usual ERBL kernels governing the scale evolution of meson distribution amplitudes [18,19]. This is just one example of the close connection between distribution amplitudes and GPDs in the region $-\xi < x < \xi$. Another one will be discussed in section 9.2.1. The evolution of GPDs makes a deep connection manifest between the evolution of parton densities and of meson distribution amplitudes [13,16]. The link providing this connection is that both types of quantities are matrix elements of the same quark–gluon operators, and one can indeed formulate their evolution equations as the renormalization group equations of these operators.

9.1.6 Double distributions and the D-term

The matrix elements in the definitions of GPDs can be parametrized in a different way which involves double distributions [13,20,2]. For the quark vector current one

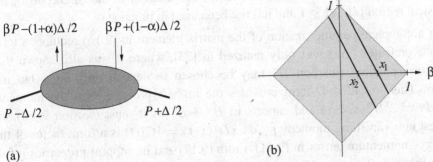

Figure 9.3. (a) Momenta associated with the partons and hadrons in a double distribution. (b) Support region of double distributions, and lines of integration to obtain GPDs for $\xi < x_1 < 1$ and $-\xi < x_2 < \xi$ according to (9.17).

has

$$
\langle p' | \bar{q} \left(-\tfrac{1}{2}z\right) (z\gamma) q \left(\tfrac{1}{2}z\right) | p \rangle \Big|_{z^2=0}
$$

$$
= \bar{u}(p')(z\gamma) u(p) \int d\beta \, d\alpha \, e^{-i\beta(Pz)+i\alpha(\Delta z)/2} \, f^q(\beta, \alpha, t)
$$

$$
+ \bar{u}(p') \frac{i\sigma^{\mu\alpha} z_\mu \Delta_\alpha}{2m} u(p) \int d\beta \, d\alpha \, e^{-i\beta(Pz)+i\alpha(\Delta z)/2} \, k^q(\beta, \alpha, t)
$$

$$
- \bar{u}(p') \frac{\Delta z}{2m} u(p) \int d\alpha \, e^{i\alpha(\Delta z)/2} \, D^q(\alpha, t). \tag{9.16}
$$

The exponents in the Fourier transform can be associated with the momentum flow shown in figure 9.3(a), where β appears similar to the momentum fraction x in a forward parton density and $\frac{1}{2}(1 + \alpha)$ similar to the momentum fraction z in a meson distribution amplitude. The support region of the double distributions f^q and k^q is the rhombus $|\alpha| + |\beta| \leq 1$ [13,20], and the support region of D^q is $|\alpha| \leq 1$. Time-reversal invariance forces f^q and k^q to be even and D^q to be odd in α. Comparing with (9.1) one finds the reduction formula

$$
H^q(x, \xi, t) = \int d\beta \, d\alpha \, \delta(x - \beta - \xi\alpha) \, f^q(\beta, \alpha, t) + \text{sign}(\xi) D^q\left(\frac{x}{\xi}, t\right),
$$

$$
E^q(x, \xi, t) = \int d\beta \, d\alpha \, \delta(x - \beta - \xi\alpha) \, k^q(\beta, \alpha, t) - \text{sign}(\xi) D^q\left(\frac{x}{\xi}, t\right),
$$

$$
\tag{9.17}
$$

where the lines of integration in the β–α plane are shown in figure 9.3(b). One readily checks that the representation (9.17) automatically satisfies the polyno-miality constraints (section 9.1.3), which explains the prominent role played by double distributions in constructing ansätze for GPDs. The inversion of (9.17) has been given in [21,22], but it requires analytic continuation of the GPDs outside the physical region $|x|, |\xi| \leq 1$ and has not been used in practice.

That a complete parametrization of the matrix element in (9.16) requires a term with a prefactor (Δz) was only realized in [23], where it was also shown that the corresponding function D^q may be chosen to depend only on α but not on β. This so-called D-term provides the highest power ξ^{2n} in the moments $\int_{-1}^{1} dx \, x^{2n-1} H^q(x, \xi, t)$ and cancels in $H^q + E^q$ as it must (section 9.1.3). Its lowest non-vanishing moment $\int_{-1}^{1} dx \, x D^q(x, t) = 4C^q(t)$ is a form factor of the energy–momentum tensor in (9.11). From (9.17) and its support properties it follows that the D-term only contributes to the ERBL region in GPDs. We finally remark that there is an analogous D-term for H^g and E^g in the gluon sector, but not for $\tilde{H}^{q,g}$ or $\tilde{E}^{q,g}$. For further discussion we refer to [21,24].

9.1.7 Impact parameter and the spatial structure of hadrons

A physically appealing property of GPDs is that they contain information about the spatial distributions of partons in a hadron, as first realized in [25,26] and reviewed in [4,27]. To obtain this formulation one changes basis from the usual momentum eigenstates of hadrons to states

$$|p^+, \vec{b}\rangle = \int \frac{d^2p}{16\pi^3} e^{-i\vec{b}\cdot\vec{p}} |p^+, \vec{p}\rangle, \qquad (9.18)$$

which are localized in the transverse plane while still having definite plus-momentum. Remarkably, the localization of a particle in two dimensions *is* possible to arbitrary accuracy, whereas localization in all three dimension is limited within the Compton wavelength. A spatial interpretation of GPDs in all three dimensions has been proposed in [28,29]. Some insight into why two-dimensional localization is possible can be gained by forming wave packets [25]; on the more formal level the states (9.18) are eigenstates of a two-dimensional position operator [6]. The mixed representation in 'impact parameter' $\vec{b}$ and plus-momentum p^+ is well suited for a parton interpretation, since one can ensure that all partons move fast by boosting to a frame where p^+ is large.

To understand what a definite position $\vec{b}$ means for a hadron with internal structure, we consider an important symmetry on the light-cone. A subgroup of the Lorentz group called 'transverse boosts' transforms any four-vector k according to

$$k^+ \to k^+, \qquad \vec{k} \to \vec{k} - k^+\vec{v}, \qquad (9.19)$$

with a given transverse vector $\vec{v}$. (The transformation law for k^- is easily obtained from the invariance of k^2.) These transformations are analogous to the Galilean

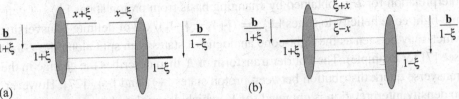

Figure 9.4. Representation of a GPD in impact parameter space. Plus-momentum fractions refer to the average proton momentum P and are indicated above or below lines. The region $\xi < x < 1$ is shown in (a), and the region $-\xi < x < \xi$ in (b).

transformations in non-relativistic mechanics, where k^+ is replaced by the mass of the particle. By Noether's theorem the light-cone analogue of the conserved centre of mass $\vec{r} = \sum m_i \vec{r}_i / \sum m_i$ for a multi-particle system thus is the 'centre of momentum' $\vec{b} = \sum k_i^+ \vec{b}_i / \sum k_i^+$. One can explicitly show that $\vec{b}$ for a proton state $|p^+, \vec{b}\rangle$ is the centre of momentum of the partons in the proton. Inserting (9.18) into the definition (9.1) one obtains

$$\int \frac{d^2 D}{(2\pi)^2} e^{i\vec{b}\cdot\vec{D}} \left[H - \frac{\xi^2}{1-\xi^2} E \right] = \mathcal{N}^{-1} \frac{1+\xi^2}{(1-\xi^2)^{5/2}}$$

$$\times \left\langle p'^+, \frac{\vec{b}}{1-\xi} \middle| \int \frac{dz^-}{4\pi} e^{ixP^+z^-} \bar{q}\left(-\tfrac{1}{2}z\right) \gamma^+ q\left(\tfrac{1}{2}z\right) \middle| p^+, \frac{\vec{b}}{1+\xi} \right\rangle_{z^+=0,\,\vec{z}=\vec{0}}$$

$$(9.20)$$

if both proton states have positive light-cone helicity, where $\mathcal{N}$ is a normalization factor. The GPDs depend on $\vec{D}$ via t as given in (9.8). The corresponding matrix element with proton helicity flip is obtained from the Fourier transform of $(D^1 + iD^2) E$ according to (9.6). The physical content of (9.20) is illustrated in figure 9.4. The transverse distance between the extracted parton or parton pair from the centre of momentum of the hadron, up to factors of $(1 \pm \xi)^{-1}$, is specified by $\vec{b}$. The shift in the centre of momentum between the initial and the final hadron reflects the change in plus-momentum [30].

At $\xi = 0$ the interpretation becomes yet simpler, and the Fourier transformed GPDs give the *density* of partons with definite plus-momentum x and transverse distance $\vec{b}$ from the proton centre. Integrating over $\vec{b}$ one sets $t = 0$ in momentum space and thus recovers the usual parton densities, where all transverse information is lost. The average $\vec{b}^2$ of partons at given x is obtained as

$$\langle \vec{b}^2 \rangle = 4 \frac{\partial}{\partial t} \log H(x, 0, t) \Big|_{t=0}. \qquad (9.21)$$

The Fourier transforms of H and $\tilde{H}$ readily give the sum and difference of partons with positive or negative helicity in a proton with helicity along z. A density interpretation for E is obtained by changing basis from proton states $|\pm\rangle_z$ of definite light-cone helicity to states $|\pm\rangle_x = (|+\rangle_z \pm |-\rangle_z)/\sqrt{2}$ of definite transversity, which may be seen as the light-cone analogue of states with spin along the x-axis (see [7] for example). The Fourier transform of E then describes the change in the transverse quark distribution between proton states $|+\rangle_x$ and $|-\rangle_x$ [27]. However, no density interpretation is obtained for $\tilde{E}$, which decouples at $\xi = 0$.

The interpretation as a *two-dimensional* density carries over to the Fourier transforms of form factors like F_1, F_2 or A^q, B^q, which respectively describe the transverse distribution of quark charge or of longitudinal quark momentum in the nucleon. Note that F_1, F_2 belong to a conserved current and hence have no dependence on a factorization scale μ^2. They may be interpreted at large μ^2, where quarks can be resolved, but their information is not specific to short-distance degrees of freedom. In contrast, A^q and B^q do refer to a specific resolution scale; evolution redistributes longitudinal momentum between quarks and gluons.

We finally remark that the information about transverse hadron structure in GPDs is different from that in k_T-dependent parton distributions (see chapter 7), which are more easily interpreted in transverse momentum and not in transverse position space [30]. Connections between the two types of distributions have been proposed in the context of spin structure [31].

9.1.8 The wave function representation

A detailed interpretation of the information encoded in GPDs is obtained by writing them in terms of light-cone wave functions. In the formalism of light-cone quantization and light-cone gauge (see section 9.1.1) one obtains parton states by acting on the vacuum with the creation operators for quarks, antiquarks and gluons appearing in the Fourier decomposition of quark and gluon fields at $z^+ = 0$. One proceeds by expanding a physical hadron state on to these parton states. The coefficients of this Fock-state expansion are the light-cone wave functions, which thus contain the most detailed description of the hadron structure at quark–gluon level. This formalism makes the ideas of the parton model explicit and provides valuable insights, although there are subtle unresolved issues, for instance concerning zero-modes, renormalization and the choice of gauge (see [4] for references).

Inserting the Fock-state expansion into the definitions (9.1) or (9.4) and expressing the fields in terms of parton creation and annihilation operators, one obtains a representation of GPDs as a product of wave functions $\Psi_{in} \Psi_{out}^*$ for the incoming and outgoing hadron [32]. This is to be summed over all possible configurations of spectator partons, including their number, keeping fixed the momentum fractions and quantum numbers of the parton emitted and reabsorbed in the DGLAP regions, or those of the parton pair emitted in the ERBL region. For equal hadron momenta and spins, one obtains squared wave functions $|\Psi|^2$ and thus the interpretation of usual parton distributions as classical probabilities. In non-forward kinematics one has instead the interference of different wave functions and thus correlates different parton configurations in the hadron. One can also Fourier transform the wave function representation from transverse momentum to transverse position space. This leads to the picture shown in figure 9.4, where the blobs represent wave functions $\tilde{\Psi}_{in}$ and $\tilde{\Psi}^*_{out}$ describing partons of definite impact parameter and plus-momentum. In the case $\xi = 0$ one then has squared wave functions $|\tilde{\Psi}|^2$ and again an interpretation as classical probabilities, as discussed just above.

According to (9.6) the GPDs E and $\tilde{E}$ describe matrix elements with opposite helicities of the initial and the final hadron. On the other hand, the helicities of all partons in their respective wave functions are the same. This implies that at least in one of the wave functions Ψ_{in} and Ψ^*_{out} the parton helicities do not add up to the helicity of the proton, which requires orbital angular momentum among the partons. We thus see that the integral $\int_{-1}^1 dx\, x E^q(x, 0, 0)$ appearing in Ji's sum rule (9.13) is just one particular piece of information on orbital angular momentum in the nucleon.

The polynomiality conditions on GPDs are non-trivial in the wave function representation. This reflects that Lorentz invariance, which underlies these conditions, is not manifest in the light-cone formalism, where certain coordinate directions are singled out. The independence of $\int_{-1}^1 dx\, H^q(x, \xi, t)$ on ξ, for instance, requires cancellation of the ξ dependence from the individual contributions of the regions $|x| > \xi$ and $|x| < \xi$, where wave functions with different parton numbers are involved. Relations among the light-cone wave functions for states with different particle number are indeed produced by the equations of motion of the theory. Examples of such relations can be studied in field-theoretical models which permit explicit calculations in perturbation theory (see [4] for references), but little is known about these relations in general.

Non-trivial relations between different light-cone wave functions are also required for the behaviour of GPDs at the transition point $x = \xi$ between the DGLAP and the ERBL regions, where they must be continuous although they may be non-analytic in x [33]. Physically the point $x = \xi$ is rather intriguing since it involves one parton with finite and another with vanishing plus-momentum, as can be seen in figure 9.2. In several field-theory models one indeed finds quark GPDs that are continuous at $x = \xi$ whereas the first derivative in x has a jump (see references in [4]). For gluon GPDs there are indications that also the first derivative in x is continuous [34], but no conclusive analysis has yet been given.

The wave function representation has been used in [35] to construct models of GPDs at large x, say above 0.5, where one can expect that Fock states with only a limited number of partons dominate. One then needs to model a limited number of wave functions, which can be tested against further observables.

9.1.9 Positivity constraints

In the DGLAP regions the wave function representation has the structure of a scalar product in the functional space of wave functions. The corresponding Schwartz inequalities provide constraints on GPDs which can be written in a variety of forms. One type of constraint has a combination of GPDs bounded from above by the conventional parton distributions, for instance [36]

$$(1 - \xi^2)\left(H^q - \frac{\xi^2}{1 - \xi^2} E^q\right)^2 + \frac{t_0 - t}{4m^2} (E^q)^2 \leq q\left(\frac{x + \xi}{1 + \xi}\right) q\left(\frac{x - \xi}{1 - \xi}\right),$$

(9.22)

where we have omitted the arguments x, ξ, t of H^q and E^q for brevity and where $\xi < x < 1$ is required. Powerful positivity constraints can also be derived using the impact-parameter space representation. The bound [37]

$$(E^q(x, 0, 0))^2 \leq m^2 (q(x) + \Delta q(x)) (q(x) - \Delta q(x))$$

$$\times 4 \frac{\partial}{\partial t} \log(H^q(x, 0, t) \pm \tilde{H}^q(x, 0, t))|_{t=0}$$

(9.23)

for $x > 0$ (valid with both signs on the right-hand side) involves the average squared impact parameter of partons in the second line and is especially restrictive when x becomes large (see section 9.2.5). The validity of all these constraints is subject to the same caveats as the positivity of forward parton densities. They can be derived in a region where the non-forward Compton amplitude is well approximated by the expression (9.34) of leading twist and leading order in α_s. Positivity constraints hence require the factorization scale μ^2 of the distributions to be sufficiently large.

9.2 Dynamics and models

9.2.1 Chiral dynamics

Dynamics associated with chiral symmetry and its breaking plays a role for different aspects of GPDs. The most prominent example is the contribution of pion t-channel exchange to the isovector combination $\tilde{E}^{u-d}$. Extrapolating to $t = m_\pi^2$ one obtains [38,39]

$$\tilde{E}^{u-d}(x, \xi, t) \xrightarrow{t \to m_\pi^2} \theta(|x| < |\xi|) \frac{1}{2|\xi|} \phi_\pi\left(\frac{x + \xi}{2\xi}\right) \frac{4m^2 g_A}{m_\pi^2 - t},$$

(9.24)

where $g_A \approx 1.26$ is the nucleon axial charge and ϕ_π the twist-2 distribution amplitude of the pion, normalized to $\int_0^1 dz \, \phi_\pi(z) = 1$ with z being the momentum fraction of the quark in the pion. Since m_π^2 is so small one expects this pion-pole contribution to remain prominent for small negative values of t; in this sense pion exchange is

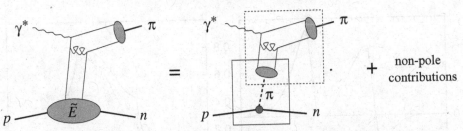

Figure 9.5. Pion exchange in $\gamma^* p \to \pi^+ n$. The full box is the pion-pole contribution to $\tilde{E}$, and the dashed box gives the pion form factor at leading order in $1/Q^2$ when t is analytically continued to m_π^2.

special compared with any other exchange in the t-channel. The physical picture of (9.24) is illustrated in figure 9.5, where isospin invariance has been used to connect $\tilde{E}^{u-d}$ with the transition GPD from proton to neutron. An off-shell pion is emitted by the nucleon and annihilates into a $q\bar{q}$ pair which is probed in a hard scattering process.

Heavy-baryon chiral perturbation theory has been applied to the x moments of the u and d quark GPDs of the nucleon in [40] to evaluate non-analytic terms in the small-t behaviour, which originate in the chiral logarithms generated at one-loop level. In the same calculation one also obtains the non-analytic part of the dependence on m_π^2, which is important for the extrapolation of calculations in lattice QCD (see section 9.2.6).

Predictions of chiral symmetry and its breaking can be obtained in the limit where the momentum of a pion becomes soft. Such soft-pion theorems have been used in [8] to express the $N \to N\pi$ transition GPD in soft-pion kinematics through the nucleon GPDs. The results were confirmed in [41,42] but have been challenged in [43,44] and this important issue remains to be clarified. Soft-pion theorems as well as non-analytic behaviour in t and m_π^2 have also been considered for the GPDs in a pion and for the two-pion distribution amplitude [15,45,46].

In [47] the pion-cloud contribution to the gluon distribution in the nucleon has been obtained as the convolution of the distribution of pions in the nucleon with the gluon distribution in a pion. For this to be meaningful the pion virtuality should only be of order m_π^2, which requires $-t$ of order m_π^2 and gluon momentum fractions below m_π/m. In terms of impact parameters one then probes gluons at large distances of order $1/m_\pi$.

9.2.2 The large-N_c limit

The limit of SU(N_c) gauge theory for a large number N_c of colours provides important insights into QCD with $N_c = 3$. Of particular practical relevance for GPDs is

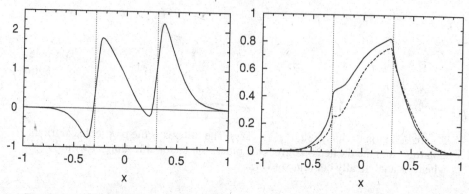

Figure 9.6. The GPD flavour combinations H^{u+d} (left), E^{u-d} (right, solid) and $\tilde{H}^{u-d}$ (right, dashed) at $\xi = 0.3$ and $t = -0.35$ GeV^2, as obtained in the chiral quark–soliton model [3].

that a hierarchy of sizes is obtained for different spin–flavour combinations in the nucleon. In kinematics where $t/m^2 \sim 1/N_c^2$ and $x, \xi \sim 1/N_c$ one obtains [3]

$$H^{u+d} \sim N_c^2, \qquad H^{u-d} \sim N_c, \qquad \tilde{H}^{u+d} \sim N_c, \qquad \tilde{H}^{u-d} \sim N_c^2,$$
$$E^{u+d} \sim N_c^2, \qquad E^{u-d} \sim N_c^3, \qquad \tilde{E}^{u+d} \sim N_c^3, \qquad \tilde{E}^{u-d} \sim N_c^4,$$

$$(9.25)$$

as well as $D^{u+d} \sim N_c^2$ and $D^{u-d} \sim N_c$ for the quark D-term. Some consequences of the relations (9.25) are known to work rather well, such as the relative size and sign of the forward parton densities u, d and Δu, Δd at $x \sim 1/3$, as well as the small ratio F_2^{u+d}/F_2^{u-d} of isoscalar and isovector Pauli form factors at small t. The large-N_c limit further relates the GPDs for the $N \to \Delta(1232)$ transition with those of the nucleon [3].

9.2.3 The chiral quark–soliton model

The chiral quark–soliton model is an effective theory of QCD at low scales, implementing the chiral dynamics and the $1/N_c$ expansion of QCD. For a detailed account in the context of GPDs and further references see [3]. The model gives a fair description of nucleon form factors at small t and of the forward quark densities taken at the scale $\mu \sim 600$ MeV, which plays the role of an ultraviolet cutoff in the effective theory. The nucleon in this model can be represented as consisting of N_c 'valence' quarks in a bound-state level together with 'sea' quarks in the negative-energy Dirac continuum, thus including both quark and antiquark degrees of freedom. Gluon distributions are found to be parametrically suppressed and have not been studied so far.

The leading flavour combinations of GPDs in (9.25) have been calculated within this model in [39,48]. The x dependence of H^{u+d}, E^{u-d} and $\tilde{H}^{u-d}$ has considerable

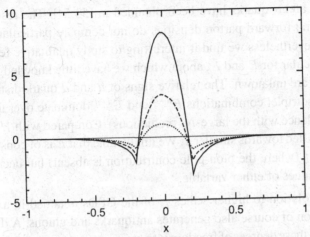

Figure 9.7. Predictions of the chiral quark–soliton model [39] for $\tilde{E}^{u-d}$ at $\xi = 0.2$ and $t = -0.5\ \text{GeV}^2$ (solid), $t = -0.7\ \text{GeV}^2$ (dashed) and $t = -1\ \text{GeV}^2$ (dotted). The curves result from the positive pion-pole term (9.24) and a negative non-pole contribution.

structure due to the interplay of the 'valence' and 'sea' contributions, as seen in figure 9.6. Neither H^{u+d} nor $\tilde{H}^{u-d}$ is found to be compatible with a factorized form $F(t)f(x, \xi)$ for the t dependence, which is often used as a simplification in phenomenology [3]. The pion-pole contribution (9.24) is obtained analytically in the model and generally dominates E^{u-d}, but other contributions are not negligible already at moderate t, both in the ERBL and in the DGLAP regions, as is seen in figure 9.7. Dedicated studies of the nucleon helicity-flip distribution E have been presented for E^{u-d} in [49] and for the non-leading flavour combination E^{u+d} in [50].

An isosinglet D-term of considerable size has been extracted from the model in [51] and is often used in phenomenological studies. Extrapolated to $t = 0$ it can be written as

$$D^{u+d}(x) \approx (1 - x^2)\big[-4.0\,C_1^{3/2}(x) - 1.2\,C_3^{3/2}(x) - 0.4\,C_5^{3/2}(x)\big], \qquad (9.26)$$

in terms of Gegenbauer polynomials. Note that the coefficients in (9.26) do not refer to the low scale $\mu \sim 600\ \text{MeV}$ of the model calculation but are an estimate of the values at μ^2 of a few GeV^2, as clarified in [52].

9.2.4 Quark models

A number of dynamical studies of nucleon GPDs have been performed in constituent-quark models, with or without the inclusion of relativistic effects. The description of the nucleon as a three-quark system with corresponding Schrödinger wave functions restricts such studies to the region $x > \xi$. Results for the GPDs H,

E, $\tilde{H}$, $\tilde{E}$ of u and d quarks have been obtained in [53,54]. The resulting elastic form factors and forward parton densities do not compare particularly well with experiment. Nevertheless we find it interesting to study qualitative features of the model, in particular for E and $\tilde{E}$, about which we have little knowledge since their forward limits are unknown. The relative signs of u and d quark distributions are such that the isotriplet combinations E^{u-d} and $\tilde{E}^{u-d}$ dominate over the isosinglet ones (in accordance with the large-N_c predictions). Compared with H and $\tilde{H}$ both E and $\tilde{E}$ are shifted towards smaller x. We finally note that $\tilde{E}$ is of considerable size for zero ξ and t (where the pion-pole contribution is absent) but decreases rather fast for finite values of either variable.

Perturbative evolution to higher scales μ of the GPDs obtained in a constituent-quark description of course also generates antiquarks and gluons. A different proposal to include these degrees of freedom is to assign a substructure to the constituent quarks, parametrized by GPDs for partons within a constituent quark. This has been pursued in [55,56].

The nucleon GPDs have also been investigated in the MIT bag model [57]. Finally, a variety of studies have been performed for the GPDs of the pion, using different formalisms and different types of quark models. We do not have the space to discuss them and refer to [4] for details and references.

9.2.5 *The interplay of x and t dependence*

A very important feature of GPDs is the correlation between their dependence on t and on the momentum fractions x and ξ. (Most studies so far have considered the simplified case $\xi = 0$ for this purpose.) In physical terms, this correlation reflects that the transverse distribution of partons depends on their momentum fraction. This dependence contains valuable clues to the dynamics at work. An example of such an interplay is the contribution to GPDs from the pion cloud at transverse distances of order $1/m_\pi$, which requires parton momentum fractions to be small (see section 9.2.1).

At very small x the behaviour of parton distributions is related to Regge theory, where the simplest type of x dependence is a power in x, due to the exchange of a single Regge pole. Extending this form to finite t one obtains a power behaviour like $x^{-\alpha(t)}$, where Regge phenomenology suggests a linear form $\alpha(t) = \alpha(0) + \alpha't$ for $-t$ not too large: see chapter 5. This results in a factor $\exp[\alpha't \log(1/x)]$ correlating the x and t dependence of GPDs. According to (9.21) the average squared impact parameter $\langle \vec{b}^2 \rangle$ of partons then grows like $\log(1/x)$ for decreasing x. The physical picture behind this is Gribov diffusion, which views repeated parton radiation at small x as a diffusion process in impact-parameter space.

As x becomes large, the spectator partons in figure 9.4 have less and less weight in the average $\sum x_i \vec{b}_i$, so that the position of the struck parton will tend to be close to the centre-of-momentum of the hadron [27]. For growing x, the $\vec{b}$ distribution should thus become more narrow and hence the t distribution more flat. If one requires that the distance $\vec{b}/(1-x)$ between the struck parton and the centre of momentum of the *spectators* stays finite on average, one needs $\langle \vec{b}^2 \rangle$ to vanish at least like $(1-x)^2$ [58]. A simple form of t dependence that satisfies this requirement and also exhibits Regge behaviour at small x is $\exp[tf(x)]$ with $f(x) \to A(1-x)^2$ for $x \to 1$ and $f(x) \to \alpha' \log(1/x)$ for $x \to 0$. Such parametrizations have been explored in more detail in [59,60].

9.2.6 Lattice QCD

We have seen in section 9.1.3 that x moments of GPDs are matrix elements of local operators and are parametrized in terms of form factors. Such quantities can be evaluated in lattice QCD, and detailed studies [61,62] have investigated the form factors of the quark energy–momentum tensor (9.11). Further progress will be needed, particularly in the extrapolation to physical values of the quark masses and in the calculation of 'disconnected graphs', where the external current attaches to a quark line which is not connected with the initial and final hadron state. Such contributions cancel in the difference of u and d quark distributions but may well be important in their sum. Nevertheless, it is interesting to note characteristic features of present results. The relative size of $u + d$ and $u - d$ combinations for the form factors A, B, C is for instance found to follow the predictions of the large-N_c limit given in section 9.2.2, as seen in figure 9.8. Evaluating $A^q + B^q$ at $t = 0$ one gets in particular the total quark angular momentum, and on subtracting the moments $\int_0^1 dx \, [\Delta q(x) + \Delta \bar{q}(x)]$ of the polarized quark densities calculated on the lattice, results for the orbital angular momentum $\langle L_q^3 \rangle$ are obtained. Both [61] and [62] found a very small $\langle L_{u+d}^3 \rangle$, but given the uncertainties just discussed no definite conclusions can be drawn. Closer investigation suggests that a small $\langle L_{u+d}^3 \rangle$ may result from cancellations between $\langle L_u^3 \rangle$ and $\langle L_d^3 \rangle$ [63].

The t dependence of the moments with different powers x^n has been compared for $n = 0, 1, 2$ in [64,65] and a clear flattening of the t slope with increasing n was observed. Such a trend is in line with our discussion at the end of the previous subsection, and it is interesting that it sets in already for moments in which the average values of x are below 0.5.

A different line of development is 'transverse lattice gauge theory', where lattice methods for the transverse directions are combined with the formalism of discrete light-cone quantization. In [66] this was used to evaluate the quark GPD of the pion

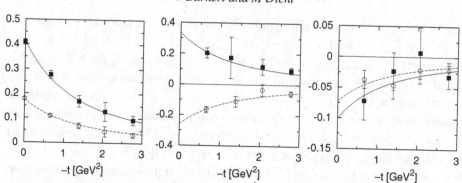

Figure 9.8. The form factors A^q (left), B^q (middle) and C^q (right) of the energy–momentum tensor for u quarks (full squares) and d quarks (open circles), obtained in a quenched lattice calculation [61] with quark masses corresponding to $m_\pi = 870$ MeV. The renormalization scale is $\mu = 2$ GeV.

directly in the mixed representation of momentum fraction and impact parameter. For increasing $\vec{b}$ the x dependence was found to be strongly suppressed at large x and strongly enhanced at small x.

9.2.7 Ansätze for phenomenology

The methods we have discussed so far address particular aspects and features of GPDs, but none of them provides sufficient input for evaluating scattering amplitudes of hard processes, where distributions in the full range of x are needed for several quark flavours and possibly gluons. Phenomenological studies so far have used ansätze for GPDs, whose typical form we now describe. Detailed accounts are given in [3,67]. In the forward limit the distributions H and $\tilde{H}$ reduce to the usual parton densities, which are rather well measured and can be used as an input. For definiteness let us concentrate on H^u and H^d. The t dependence has so far mostly been simplified as a factorizing form $H^q(x, \xi, t) = H^q(x, \xi) F_1^q(t)/F_1^q(0)$ with the u and d quark contributions to the Dirac form factor taken from experiment. This ansatz satisfies the sum rules (9.12), but according to our discussion in sections 9.2.3 and 9.2.5 can only be a crude approximation.

The simplest model one can make for the ξ dependence is to neglect it, setting $H^q(x, \xi) = q(x)$ for $x > 0$ and $H^q(x, \xi) = -\bar{q}(-x)$ for $x < 0$. This so-called 'forward model' is consistent with general symmetries, but it is physically not plausible near $x = 0$ for finite ξ, where it gives a singular behaviour in x. In the forward limit this singularity corresponds to partons with vanishing plus-momentum. For finite ξ the point $x = 0$ describes the emission of a quark–antiquark pair with equal momentum fractions ξ, for which there is no reason to expect a singular behaviour.

To ensure polynomiality in ξ of all x moments, a more sophisticated treatment of the ξ dependence is conveniently achieved by constructing a model at the level of double distributions (section 9.1.6). A widely used ansatz for the double distribution is to set $f^q(\beta, \alpha, t) = f^q(\beta, \alpha) F_1^q(t)/F_1^q(0)$ with

$$f^q(\beta, \alpha) = \frac{q(\beta) - \bar{q}(\beta)}{1 - \beta} \rho_{val}\left(\frac{\alpha}{1 - \beta}\right) + \frac{\bar{q}(\beta)}{1 - \beta} \rho_{sea}\left(\frac{\alpha}{1 - \beta}\right),$$

$$f^q(-\beta, \alpha) = -\frac{\bar{q}(\beta)}{1 - \beta} \rho_{sea}\left(\frac{\alpha}{1 - \beta}\right) \tag{9.27}$$

for $\beta > 0$, where the 'profile functions' ρ are normalized to $\int_{-1}^{1} d\alpha \, \rho(\alpha) = 1$ [68,3]. Their form is taken as $\rho(\alpha) \propto (1 - \alpha^2)^b$ with a parameter b that may be chosen differently in ρ_{val} and ρ_{sea}. For $b = \infty$ one gets $\rho(\alpha) = \delta(\alpha)$ and thus recovers the forward model discussed above. Analogous ansätze can be made for $\tilde{H}^q$, H^g and $\tilde{H}^g$. Common choices in the literature are $b = 1$ for quarks and $b = 2$ for gluons.

The separation into 'valence' and 'sea' parts follows the convention in the forward limit, but note that the 'valence' part in (9.27) generates a non-zero $H^q(x, \xi, t)$ in the ERBL region. Since the 'sea' part in (9.27) does not contribute to $\int_{-1}^{1} dx \, H^q = F_1^q(t)$ it may be taken with a different t dependence in the full double distribution $f^q(\beta, \alpha, t)$ [67]. A first attempt at a more realistic model for the interplay between t and the longitudinal momentum fractions has been made in [3], with a form $f^q(\beta, \alpha, t) = f(\alpha, \beta) |\beta|^{-\alpha' t}$ motivated by Regge theory (see section 9.2.5). The sum rule $\int_{-1}^{1} dx \, (e_u H^u + e_d H^d) = F_1^p(t)$ for the Dirac form factor of the proton was found to be rather well satisfied at small t for $\alpha' \approx 0.8$ GeV^2, which is quite close to the value for meson trajectories in Regge phenomenology. Such an ansatz has been developed further in [59,60].

In figure 9.9 we show GPDs obtained in the forward model and from the model (9.27) with $b = 1$. As seen in (9.17), a double-distribution ansatz misses the possible contribution from D-terms. To the ansatz just described, many phenomenological studies have added the isosinglet D-term contribution (9.26) estimated in the chiral quark–soliton model, or its extension to the SU(3) flavour-singlet combination D^{u+d+s} proposed in [3]. Figure 9.9 shows that such a term leads to considerable structure in the ERBL region.

At small x and ξ the values of β relevant in the reduction formula (9.17) are small as well, and one may replace the factors $1 - \beta$ with 1 in (9.27). For an approximate small-x behaviour like $x^{-\lambda}$ of $xq(x)$ one can perform the relevant integrations and obtain an analogous behaviour like $\xi^{-\lambda} \hat{H}(x/\xi)$ of $\xi H^q(x, \xi)$. For gluons, an approximate behaviour like $x^{-\lambda}$ of $xg(x)$ translates into a behaviour like $\xi^{-\lambda} \hat{H}(x/\xi)$ of $H^g(x, \xi)$. One finds that $H^{q,g}(\xi, \xi)$ decreases with the profile parameter b, a trend already seen in our above example at $\xi = 0.3$. The dependence of the GPDs on b

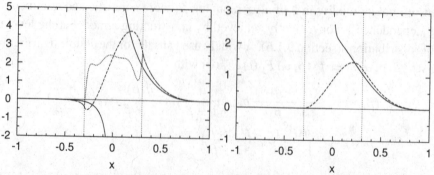

Figure 9.9. H^{u+d} (left) and $\tilde{H}^{u-d}$ (right) at $\xi = 0.3$ and $t = -0.35\,\text{GeV}^2$, obtained from the ansatz (9.27) and its polarized analogue, using $\rho_{val} = \rho_{sea}$ and the quark distributions from [69,70] at $\mu = 1$ GeV. The solid and dashed curves are for $b = \infty$ and $b = 1$, respectively, and the dotted curve shows the result for $b = 1$ plus the D-term given in (9.26). The t dependence is modelled by multiplying H^q (both the double-distribution part and the D-term) with the Dirac form factor $F_1^q(t)$ and $\tilde{H}^q$ with the nucleon axial form factor, both properly normalized.

is significantly larger for quarks than for gluons, see [4] for example. An ansatz proposed by Shuvaev *et al* [71] corresponds to $b = 1 + \lambda$ and is often used in the literature for processes at small x.

In [72] it was claimed that the ansatz we have described gives amplitudes for DVCS that are too large to be compatible with cross section data from H1 [101] and the beam spin asymmetries from HERMES [102] and CLAS [103], unless one takes $b = \infty$. The authors propose instead an ansatz consisting of the forward model in the DGLAP regions, whereas in the ERBL region they take a simple polynomial form with coefficients chosen such that the polynomiality conditions are met for the lowest two Mellin moments. Such an ansatz has the inherent problem that polynomiality of all higher moments (and hence Lorentz invariance) is violated. How important this violation is at the level of the GPDs or of observables for given kinematics has not been investigated.

The GPDs E and $\tilde{E}$ cannot be modelled using the strategies we described, since their forward limits are unknown. For $\tilde{E}$ most studies use only the pion-pole contribution (9.24). First attempts at a detailed model of E, which is crucial in the context of orbital angular momentum, have been made in [3,59,60].

9.3 Processes

9.3.1 Factorization

The measurement of GPDs in hard exclusive processes rests on factorization theorems very similar to those that connect the usual parton densities with inclusive processes like DIS or Drell–Yan lepton pair production. Detailed investigations for Compton scattering and for exclusive meson production have been made in [20,33,73]; a brief account can be found in [74].

Factorization of the Compton amplitude holds in a generalization of the Bjorken limit of DIS, namely for $|q^2| + |q'^2| \to \infty$ at fixed t and fixed q^2/W^2 and q'^2/W^2. Here four-momenta are as in figure 9.1(a) and $W^2 = (p+q)^2$ is the squared centre-of-mass energy of the proton–photon system. In practical terms, at least one of the photon virtualities as well as W^2 should be much larger than $-t$ and than the scale of soft hadronic interactions. Special cases are: (i) DVCS, which has $q'^2 = 0$ and is accessible in electroproduction $ep \to ep\gamma$ [75], and (ii) time-like Compton scattering, which has $q^2 = 0$ and can be seen in heavy lepton pair production $\gamma p \to p\,\ell^+\ell^-$ [76]. 'Double DVCS' can be studied in $ep \to ep\,\ell^+\ell^-$ and has both photons off-shell [77,78]. In general one has two scaling variables,

$$\xi = -\frac{\Delta\bar{q}}{2P\bar{q}} \approx \frac{Q^2 + Q'^2}{2W^2 + Q^2 - Q'^2}, \qquad \rho = -\frac{\bar{q}^2}{2P\bar{q}} \approx \frac{Q^2 - Q'^2}{2W^2 + Q^2 - Q'^2}, \quad (9.28)$$

with $\bar{q} = \frac{1}{2}(q+q')$ and $Q'^2 = q'^2$. The approximations are valid in the generalized Bjorken limit. In a frame where the incoming and outgoing protons have large momenta in the positive z direction (with transverse momenta of order $\sqrt{-t}$) one finds $\xi \approx (p-p')^+/(p+p')^+$, as we have used in GPDs so far. The factorization theorem states that the amplitude can be written as

$$\mathcal{A}(\gamma^* p \to \gamma^* p) = \sum_i \int_{-1}^{1} dx\, T^i(x, \rho, \xi, Q^2 - Q'^2) F^i(x, \xi, t) \quad (9.29)$$

with corrections suppressed by inverse powers of the hard scale Q or Q' (for simplicity we henceforth refer to Q as the large scale, keeping in mind that either Q or Q' may be large). Here F^i stands for any of the quark and gluon matrix elements in (9.1) and (9.4), which are parametrized by GPDs. The dependence of the hard scattering kernels T^i on $Q^2 - Q'^2$ is logarithmic and due to radiative corrections, where for legibility we have not displayed the factorization and renormalization scales. Apart from this logarithmic dependence the amplitude is independent of the photon virtualities at given ξ, ρ and t; this is the exact analogue of Bjorken scaling in DIS. We will discuss the dependence of the amplitude (9.29) on the photon helicities in the next subsection.

Analogous factorization can be shown for electroproduction $ep \to epM$ of a light meson M. The relevant limit is as in the Compton case, where q'^2 is now the squared meson mass and hence neglected along with other hadron masses. Note that one then has

$$\xi \approx x_B/(2 - x_B) \quad (9.30)$$

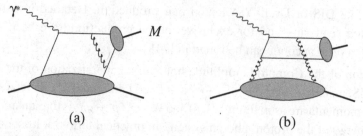

Figure 9.10. Example diagrams for hard meson production $\gamma^* p \to Mp$ with (a) quark and (b) gluon GPDs.

in terms of the usual Bjorken variable, as one has for DVCS. The factorization formula for the photon–proton amplitude is

$$\mathcal{A}(\gamma^* p \to Mp) = \frac{1}{Q} \sum_{ij} \int_{-1}^{1} dx \int_{0}^{1} dz\, T^{ij}(x, \xi, z, Q^2) F^i(x, \xi, t) \phi^j(z) \quad (9.31)$$

with corrections to it going like $1/Q^2$. Apart from a logarithmic Q^2 dependence in T^{ij}, the amplitude (9.31) falls like $1/Q$ at fixed ξ and t and requires both photon and meson to be longitudinally polarized; all other helicity transitions are suppressed by further powers of $1/Q$. Long-distance physics is parametrized by GPDs on the target-hadron side, and by light-cone distribution amplitudes ϕ^j for the $q\bar{q}$ and (if quantum numbers permit) the two-gluon content of the produced meson (see section 9.1.4). Example diagrams at leading order in α_s are shown in figure 9.10.

Let us mention some important aspects of the factorization theorems for both Compton scattering and light meson production. The outgoing hadron with momentum p', as well as the produced meson, may be replaced with multi-hadron states, described by transition GPDs or by generalized distribution amplitudes. In the generalized Bjorken limit the invariant masses of these multi-hadron states stay fixed, that is they should be small compared with the hard scale Q^2. This ensures that the systems with momenta p' and q' are well separated in rapidity.

The hard-scattering subprocesses are evaluated with external parton momenta exactly collinear and on-shell. This 'collinear approximation' corresponds to the leading term in a Taylor expansion

$$\int d^4k\, H(k) A(k) \approx \int dk^+ H(k) \Big|_{k^- = 0, \vec{k} = \vec{0}} \int dk^-\, d^2k\, A(k), \quad (9.32)$$

where H denotes the amplitude for the hard subprocess, A the amplitude represented by the lower blobs in figures 9.1(a) and 9.10, and k the loop momentum connecting them. Integrated over k^- and $\vec{k}$, the amplitude A becomes a matrix element parameterized by GPDs according to (9.9). The analogous approximation is made for the produced meson, where the plus-momenta instead of the minus-momenta of the external partons are neglected in H.

To ensure large internal virtualities in the hard scattering subprocess, the partons connecting it with the GPDs must have large plus-momenta, and those connecting it with the produced meson must have large minus-momenta. The factorization formulae (9.29) and (9.31) contain, however, momentum configurations where this is not the case, in particular the point $x - \xi = 0$ in the GPDs and the end-point regions $z = 0$ or $z = 1$ in the distribution amplitudes. An essential part of the factorization theorems is to show that such soft regions either give only power suppressed contributions to the amplitude (and hence may be included in the formulae to leading-power accuracy) or that they can be 'circumvented' by deforming the integration contours for x or z in the complex plane into a region where internal virtualities in the 'hard scattering subprocess' are truly hard. This property is not trivial: it holds, for instance, if the virtual photon in meson elec-troproduction is longitudinally polarized, but does not hold if it is transverse. It also does not hold in exclusive hadron–hadron scattering processes such as $p\bar{p} \to p\bar{p}\,\mu^+\mu^-$, which hence do not factorize (see [4] for a discussion and further references).

The power behaviour in Q^2 and the dependence on the polarization of the photon and (if applicable) the meson are generic predictions of the factorization theorems. Both can be experimentally measured, which provides model-independent tests whether the approximations leading to factorization are adequate in given kinematics.

9.3.2 Compton scattering

To leading order in $1/Q$ and in α_s the Compton amplitude in generalized Bjorken kinematics is given by the quark-handbag diagrams as shown in figure 9.1(a). In this approximation the hadronic tensor is

$$T^{\alpha\beta} = i \int d^4x\, e^{iqx} \langle p' | T J_{em}^{\alpha} \left(-\tfrac{1}{2}x\right) J_{em}^{\beta} \left(\tfrac{1}{2}x\right) | p \rangle = -g_T^{\alpha\beta} \mathcal{F} - i\varepsilon_T^{\alpha\beta} \tilde{\mathcal{F}}, \quad (9.33)$$

from which the Compton amplitude is obtained by contracting with the polarization vectors of the photons and multiplying with the squared lepton charge e^2. The only non-zero components of the transverse tensors are $g_T^{11} = g_T^{22} = -1$ and $\varepsilon_T^{12} = -\varepsilon_T^{21} = 1$, and $e J_{em}^{\alpha}(x)$ is the electromagnetic current. In (9.33) we have introduced convolutions

$$\mathcal{F}(\rho, \xi, t) = \sum_q e_q^2 \int_{-1}^1 dx \, F^q(x, \xi, t) \left(\frac{1}{\rho - x - i\varepsilon} - \frac{1}{\rho + x - i\varepsilon} \right),$$

$$\tilde{\mathcal{F}}(\rho, \xi, t) = \sum_q e_q^2 \int_{-1}^1 dx \, \tilde{F}^q(x, \xi, t) \left(\frac{1}{\rho - x - i\varepsilon} + \frac{1}{\rho + x - i\varepsilon} \right),$$

(9.34)

where the sum is over all flavours of quarks with electric charge ee_q. Following the decomposition (9.1) one obtains 'Compton form factors' $\mathcal{H}$ or $\mathcal{E}$ by replacing F^q with H^q or E^q, and $\tilde{\mathcal{H}}$ or $\tilde{\mathcal{E}}$ by replacing $\tilde{F}^q$ with $\tilde{H}^q$ or $\tilde{E}^q$ in (9.34). The relevant combinations of these form factors for definite helicities of the initial and final protons are readily obtained from (9.6). At next-to-leading order in α_s the convolutions $\mathcal{F}$ and $\tilde{\mathcal{F}}$ (and correspondingly the Compton form factors) obtain corrections which also involve the gluon GPDs H^g, E^g, $\tilde{H}^g$, $\tilde{E}^g$. Several groups have evaluated the $O(\alpha_s)$ terms; see [67,79] for detailed numerical investigations and [4] for further references.

To leading accuracy in $1/Q$ (where transverse momenta of the photons are to be neglected) the tensor (9.33) only contributes to transitions where both the photons have positive helicity or both have negative helicity. At $O(\alpha_s)$ there are in addition amplitudes where the photon helicity is flipped by two units. To ensure angular momentum conservation in the collinear hard scattering process these transitions require the gluon transversity GPDs mentioned in section 9.1.2. Unless one of the photons is on-shell there are also $O(\alpha_s)$ amplitudes where both photons have helicity 0; this is the analogue of the longitudinal structure function in DIS. Amplitudes where one photon is longitudinal and the other is transverse are $1/Q$ suppressed and will be discussed in section 9.3.4.

9.3.3 Light-meson production

Electroproduction of light mesons offers a variety of channels where different combinations of GPDs can be studied. To leading order in $1/Q$ and in α_s the amplitude for $\gamma^* p \to \rho^0 p$ with a longitudinally polarized photon is

$$\mathcal{A}_L = \frac{2e}{Q} \mathcal{F}_\rho(\xi, t),$$

$$\mathcal{F}_\rho(\xi, t) = \frac{4\pi\alpha_s}{9} \frac{f_\rho}{\sqrt{2}} \int_0^1 dz \, \frac{\phi_\rho(z)}{z(1-z)} \int_{-1}^1 dx \left(\frac{1}{\xi - x - i\varepsilon} - \frac{1}{\xi + x - i\varepsilon} \right)$$
$$\times \left[\tfrac{2}{3} F^u(x, \xi, t) + \tfrac{1}{3} F^d(x, \xi, t) + \tfrac{3}{8} x^{-1} F^g(x, \xi, t) \right],$$

(9.35)

where the meson distribution amplitude ϕ_ρ is normalized to, $\int_0^1 dz\, \phi_\rho(z) = 1$ and $f_\rho \approx 209$ MeV is the ρ decay constant. In analogy to $\mathcal{F}_\rho$ one defines convolutions $\mathcal{H}_\rho$ and $\mathcal{E}_\rho$ for the GPDs H and E. Contrary to Compton scattering, quark and gluon contributions appear here at the same order in α_s. To obtain the analogous expression for $\gamma^* p \to \pi^0 p$ one replaces $f_\rho \to f_\pi \approx 131$ MeV, $\phi_\rho \to \phi_\pi$, $F^q \to \tilde{F}^q$ and omits the term with F^g. The formulae for other cases can be found in [3,4]. By parity invariance the production of mesons with natural parity $P = (-1)^J$, such as vector mesons, selects the unpolarized GPDs. The production of mesons with unnatural parity $P = (-1)^{J+1}$, such as pseudoscalars, selects only polarized GPDs. Hence meson production allows one to study these distributions separately, unlike Compton scattering where such parity constraints do not hold because the transverse photon polarizations single out additional directions. No meson production channel is known where the polarized gluon distributions $\tilde{H}^g$ or $\tilde{E}^g$ contribute at leading order in $1/Q$ and α_s. Quark or gluon transversity GPDs do not contribute either at leading power in $1/Q$.

If the produced meson has definite C-parity, the hard scattering kernels are either even or odd in x and thus select combinations of GPDs describing definite C-parity in the t-channel (section 9.1.1). In particular the production of ρ^0, ω, ϕ selects the C-even combinations, just as Compton scattering. Assuming $\phi_\rho(z) = \phi_\omega(z) = \phi_\phi(z)$ and neglecting small SU(3) breaking effects for the meson decay constants, the appropriate combinations in $\mathcal{F}(\xi, t)$ are

$$\rho : \sqrt{\tfrac{1}{2}} \left(\tfrac{2}{3}F^u + \tfrac{1}{3}F^d + \tfrac{3}{8}x^{-1}F^g\right),$$

$$\omega : \sqrt{\tfrac{1}{2}} \left(\tfrac{2}{3}F^u - \tfrac{1}{3}F^d + \tfrac{1}{8}x^{-1}F^g\right), \quad \phi : -\left(\tfrac{1}{3}F^s + \tfrac{1}{8}x^{-1}F^g\right). \tag{9.36}$$

Combined information from these meson channels and DVCS is one of the few handles for disentangling the distributions H and E for different quark flavours and gluons. Another possibility is to use scattering on both proton and neutron targets, see section 9.4.

In contrast to neutral mesons, the production of charged mesons involves a mixture of C-even and C-odd GPD combinations. One of the few processes where the C-odd combinations of unpolarized GPDs (analogous to the valence quark distributions $q - \bar{q}$) can be accessed separately is the production of f_0 or f_2 resonances. Alternatively one can study the production of $\pi^+\pi^-$ pairs in the continuum, where in particular the interference between pairs of even and odd C-parity can be extracted from angular measurements [80,4]. First experimental results have been presented by HERMES [81]. Further issues of special interest are the physics of strangeness and flavour SU(3) violation in the production of K or K^* mesons (together with the transition from the nucleon to a Σ or Λ) and the study of SU(3) breaking and the axial anomaly in the production of η and η', see [3,4] and references therein.

An important part of charged-pion production comes from the pion-pole contribution (9.24) to $\tilde{E}^{u-d}$. Since its support is the ERBL region, it only appears in the real part of the scattering amplitude. At amplitude level one can write

$$A_L^{pole}(\gamma^* p \to \pi^+ n) = -e \frac{Q F_\pi(Q^2)}{f_\pi} \frac{2mg_A}{m_\pi^2 - t} \bar{u}(p')\gamma_5 u(p), \qquad (9.37)$$

where

$$F_\pi(Q^2) = \frac{2\pi\alpha_s}{9} \frac{f_\pi^2}{Q^2} \left[\int_0^1 dz \frac{\phi_\pi(z)}{z(1-z)} \right]^2 \qquad (9.38)$$

is the electromagnetic pion form factor to leading order in $1/Q$ and α_s (see figure 9.5). Analogous kaon-pole contributions appear in $\gamma^* p \to K^+ \Lambda$ and $\gamma^* p \to K^+ \Sigma^0$. The relation (9.37) remains valid at higher orders in α_s or in $1/Q$ and is the basis for attempts to measure F_π at moderate to large Q^2: see chapter 2. The formulation in terms of GPDs provides a tool to estimate in which kinematical region the contribution from F_π may actually dominate the longitudinal amplitude A_L, see [38] for example. Notice in particular that the pion-pole contribution (9.37) is independent of ξ at given Q^2 and t, whereas the corresponding contribution from $\tilde{H}$ should approximately follow the ξ dependence of the polarized quark densities $\Delta q(\xi)$ and hence rise when ξ becomes small. The t dependence of the contribution from $\tilde{H}$ is presently not known but should significantly differ from (9.37).

Next-to-leading order corrections in α_s for meson production were initially studied only for $\gamma^* p \to \pi^+ n$ in [82], and subsequently the calculation of the scattering kernels for other channels was completed [83]. Both studies indicate that $O(\alpha_s)$ corrections can be very large (as they can be in F_π), and further investigation will be needed to clarify for which kinematics and which observables one has sufficient theoretical control.

The $\gamma^* p$ cross section for longitudinal photons following from (9.35) is

$$\frac{d\sigma_L}{dt} = \frac{\alpha_{em}}{Q^6} \frac{4\xi^2}{1-\xi^2} (C_U + C_T \sin\beta) \qquad (9.39)$$

with

$$C_U(\xi, t) = (1 - \xi^2)|\mathcal{H}_\rho|^2 - \left(\frac{t}{4m^2} + \xi^2\right) |\mathcal{E}_\rho|^2 - 2\xi^2 \operatorname{Re}(\mathcal{E}_\rho^* \mathcal{H}_\rho),$$

$$C_T(\xi, t) = \sqrt{1 - \xi^2} \frac{\sqrt{t_0 - t}}{m} \operatorname{Im}(\mathcal{E}_\rho^* \mathcal{H}_\rho), \qquad (9.40)$$

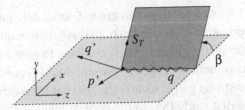

Figure 9.11. Definition of the azimuthal angle β between the hadronic plane and the transverse target spin S_T in the target rest frame. S_T is perpendicular to the z-axis, which points in the direction opposite to the virtual photon momentum.

where β is the angle between the hadron plane and the transverse target polarization S_T as defined in figure 9.11. The analogous expression for π^0 production is

$$C_U(\xi, t) = (1 - \xi^2) |\tilde{\mathcal{H}}_\pi|^2 - \frac{t}{4m^2} |\xi \tilde{\mathcal{E}}_\pi|^2 - 2\xi \, \mathrm{Re} \, (\xi \tilde{\mathcal{E}}_\pi^* \tilde{\mathcal{H}}_\pi),$$

$$C_T(\xi, t) = -\sqrt{1 - \xi^2} \, \frac{\sqrt{t_0 - t}}{m} \, \mathrm{Im} \, (\xi \tilde{\mathcal{E}}_\pi^* \tilde{\mathcal{H}}_\pi). \tag{9.41}$$

For other meson channels the same expressions hold with the appropriate decay constants and integrals over GPDs or distribution amplitudes.

We see in (9.39) that the factorization theorem predicts a $1/Q^6$ falloff for $d\sigma_L/dt$ at fixed ξ and t, up to logarithms of Q^2 due to radiative corrections. In contrast, Bjorken scaling of the Compton amplitude results in a $1/Q^4$ behaviour at fixed ξ and t of the cross section $d\sigma_T/dt$ for transverse photons in DVCS. Note that kinematical factors $\xi^2/(1 - \xi^2)$ and $1/Q^4$ in (9.39) are due to phase space and photon flux, for which we used Hand's convention [84] and neglected terms suppressed by $1/Q^2$.

The transverse target-polarization asymmetry resulting from (9.39) is sensitive to the ratio $\mathcal{E}/\mathcal{H}$ or $\tilde{\mathcal{E}}/\tilde{\mathcal{H}}$ and thus provides a way to separate GPDs with and without proton helicity flip. This asymmetry belongs to those observables where it has been argued that theoretical uncertainties should partially cancel, see [3,4] and references therein. We note that a longitudinal target polarization asymmetry does not occur with a longitudinal photon due to parity invariance.

9.3.4 Compton scattering beyond leading-power accuracy

For the Compton amplitude not only the leading term but also the first corrections in a $1/Q$ expansion have been studied in detail. At $1/Q$ accuracy there are contributions from the handbag diagrams of figure 9.1(a) where the Taylor expansion (9.32) of the hard scattering kernel in the transverse parton momentum $\vec{k}$ is carried one order further. Other contributions arise from graphs where an additional gluon (with transverse polarization) connects the blob in figure 9.1(a) with the hard scattering. The two types of contribution are not separately gauge invariant; note that the transverse derivative ∂_j, related to transverse momentum via Fourier transformation, and the transverse gluon field A_j combine to give a covariant derivative $\partial_j - ig A_j$. Careful treatment is needed to prevent double counting or the violation of electromagnetic or strong gauge invariance, and has been carried out in different formalisms, from the first study [85] to the very detailed analysis [86] (see [4] for references). Methods and results are similar to those for the spin-dependent structure function g_2, although gauge invariance is more subtle in the non-forward case.

The additional hadronic matrix elements required to calculate the Compton amplitude at the $1/Q$ level and leading order in α_s can be chosen as in (9.1), but with the non-local currents $\bar{q}\gamma^+ q$ and $\bar{q}\gamma^+\gamma_5 q$ replaced by $\bar{q}\gamma^j q$ and $\bar{q}\gamma^j\gamma_5 q$, where j is a transverse index. Using the QCD equations of motion, they can be parametrized by a part expressed in terms of the twist-2 GPDs $H, E, \tilde{H}, \tilde{E}$ (called the Wandzura–Wilczek part) and by a part involving non-local quark–antiquark–gluon operators $\bar{q}\gamma^+ G^{+j} q$ and $\bar{q}\gamma^+\gamma_5 \tilde{G}^{+j} q$ (often called the 'genuine' twist-3 part). There are indications that nucleon matrix elements of these operators might be small, and their neglect is referred to as the Wandzura–Wilczek approximation.

The $1/Q$ corrections to Compton scattering do not affect the helicity amplitudes which are already present at leading-power accuracy (see the previous subsection) but instead provide non-zero transitions between a longitudinal and a transverse photon. At leading order in α_s the new amplitudes for DVCS can be written in terms of the Compton form factors $\mathcal{H}, \mathcal{E}, \tilde{\mathcal{H}}, \tilde{\mathcal{E}}$ already appearing at leading power, of new form factors given by

$$\mathcal{H}_W = \sum_q e_q^2 \int_{-1}^{1} dx \, \frac{1}{\xi + x} \log \frac{2\xi}{\xi - x - i\varepsilon} \left(H^q(x, \xi, t) - H^q(-x, \xi, t) \right),$$

(9.42)

and its counterparts for $E, \tilde{H}, \tilde{E}$, and of convolution integrals involving four independent quark-antiquark-gluon matrix elements. If the latter are small, the amplitudes appearing at order $1/Q$ can be used to obtain additional information on the twist-2 GPDs.

The known factorization theorems do not extend to $1/Q$ accuracy, and it is not guaranteed that one can still write the Compton amplitudes in terms of perturbative hard scattering kernels and process-independent nucleon matrix elements. The results of explicit calculation at Born level, as well as at $O(\alpha_s)$ in the Wandzura–Wilczek approximation [87], are however consistent with factorization.

Not much is known about power corrections of order $1/Q^2$ in non-forward Compton scattering. No complete evaluation of target-mass corrections has been possible yet, but partial results have been obtained and suggest that such corrections are typically of order $\rho^2 m^2/Q^2$ rather than m^2/Q^2, where ρ is defined in (9.28). This would render them moderate in much of experimentally relevant kinematics. The $1/Q^2$ corrections to the Compton amplitude with photon helicity flip by two units have been evaluated to leading order in α_s in the Wandzura–Wilczek approximation [88] and may constitute an important correction to the contribution from gluon transversity, which appears at leading power in $1/Q$ but only starts at $O(\alpha_s)$.

9.3.5 *Meson production from transverse photons*

Meson production from transverse virtual photons is power suppressed by $1/Q$ compared with the leading amplitude, where the photon is longitudinal, and it is tempting to see whether collinear factorization can consistently describe the amplitude in this case. Calculation in [89,90] of the graphs of figure 9.10 with a transverse γ^* and a transverse ρ gave a result containing integrals over the meson distribution amplitude that diverge logarithmically:

$$\int_0^1 \frac{dz}{z} \int_z^1 \frac{du}{u} \, \phi_\rho(u). \tag{9.43}$$

Further one finds integrals such as

$$\int_{-1}^1 dx \, \frac{1}{(\xi - x - i\varepsilon)^2} H^{q,g}(x, \xi, t) = -\int_{-1}^1 dx \, \frac{1}{\xi - x - i\varepsilon} \frac{\partial}{\partial x} H^{q,g}(x, \xi, t), \tag{9.44}$$

which are expected to diverge at least for quark distributions (see section 9.1.8). In the regions $z \approx 0$ and $x \approx \xi$ where the divergences occur, partons become soft and the collinear approximation breaks down. The divergences disappear when the transverse momentum of the partons is not neglected in the partonic scattering amplitude; one then has for instance an inverse propagator $zQ^2 + \vec{k}^2$ instead of zQ^2. The integral over z then behaves like $\log[Q^2/\langle \vec{k}^2 \rangle]$, where $\langle \vec{k}^2 \rangle$ is a typical transverse momentum of the quark in the meson. Such a logarithmic enhancement of the amplitude for transverse photons may explain why the measured ratio of $\gamma^* p \to \rho p$ cross sections with longitudinal and transverse photons is not very large at moderate Q^2, see section 9.6.1. From the derivation of the factorization theorem [73] one expects that for a transverse photon there are contributions to the amplitude with soft partons connecting the meson and the proton side (see our discussion in section 9.3.1). Such contributions can no longer be expressed in terms of hard scattering kernels and soft matrix elements involving *either* the proton *or* the meson and entail a breakdown of factorization.

The sensitivity of amplitudes with transverse photons or mesons to soft parts of phase space has also been seen in calculations for vector meson production at small x, where the transverse parton momenta entering the partonic amplitude were kept explicitly. With suitable model assumptions on the meson wave function a rather successful phenomenology for the γ^*-ρ helicity transitions has been achieved [91]. Given our remark at the end of the previous paragraph, it remains unclear to what extent such an approach can give an adequate description of the dynamics.

9.3.6 Power corrections

The systematic evaluation of power corrections to those helicity amplitudes which receive a leading-twist contribution remains an unsolved problem. There are, however, a number of approaches to estimate specific sources of power corrections which have been tested in processes similar to the ones we are dealing with. Here and in the following we use 'leading twist' to designate the leading term in the $1/Q$ expansion of Compton scattering or meson production, given by the factorization formulae (9.29) or (9.31).

One generally expects that power corrections are more important in meson production than in Compton scattering at equal values of Q^2. This is because already at the Born level the hard scattering subgraphs in meson production involve more internal lines than those in Compton scattering; compare figures 9.1 and 9.10. At a given hard external momentum the internal virtualities in meson production are hence smaller and corrections to the approximations underlying the leading-twist factorization formulae can be more important. This trend is indeed seen when comparing power correction estimates for the transition $\gamma^*\gamma \to \pi$ and for the elastic pion form factor, which respectively have the same hard scattering graphs as DVCS and exclusive π^+ production off the proton. The form factor $F_{\gamma\pi}$ describing $\gamma^*\gamma \to \pi$ can indeed be reproduced fairly well in the leading-twist approximation down to quite low Q^2, see figure 9.12. This does not imply that power corrections are entirely negligible in the Q^2 range shown, but the data do not require them to be large. Notice that the contribution to DVCS from the pion-pole part (9.24) of $\tilde{E}$ can be expressed through the form factor $F_{\gamma\pi}(Q^2)$, just as the elastic pion form factor $F_\pi(Q^2)$ appears in the pion-pole contribution (9.37) to $\gamma^*p \to \pi^+n$.

The inclusion of the transverse parton momentum in the hard scattering kernels has been studied for the form factors $F_{\gamma\pi}(Q^2)$ and $F_\pi(Q^2)$ in [94,95], building on the modified hard scattering formalism of Sterman and collaborators. The meson structure is parametrized here by its $q\bar{q}$ light-cone wave function. Given the close connection of the relevant hard scattering kernels (see figure 9.5) this approach is readily applied to meson electroproduction as far as the meson is concerned. Taking

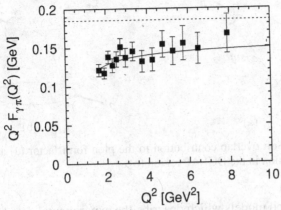

Figure 9.12. CLEO data for the scaled $\gamma-\pi$ transition form factor [92] compared with a calculation at leading twist and next-to-leading order in α_s, using a pion distribution amplitude with a fitted Gegenbauer coefficient $a_2 = -0.06$ at scale $\mu = 1$ GeV [93]. The dotted line gives the value $\sqrt{2}f_\pi$ attained for $Q^2 \to \infty$, when the running coupling goes to zero and ϕ_π takes its asymptotic form under evolution.

account of the transverse parton momentum on the proton side is more involved as it requires a model for k_T-dependent generalized parton distributions, for which there is little phenomenological guidance. Using such a model, both meson production and DVCS have been considered in [96]. A substantial suppression of the meson production cross section was found, even for Q^2 of several GeV^2, whereas for DVCS the corrections were moderate. Large suppression factors due to transverse quark momentum were also obtained for vector meson production at small x in [97,98].

A different kind of power correction in the pion form factor is shown in figure 9.13(a), where the quark struck by the photon carries most of the pion momentum whereas the spectator momentum is soft – a mechanism discussed long ago by Feynman. This gives a positive contribution to the form factor. Having one power of α_s less than the leading-twist contribution, it can be quite large at moderate Q^2, see [95]. It is rather straightforward to extend this soft overlap mechanism to electroproduction in the ERBL region, where the kinematics of the partonic subprocess are the same as in the pion form factor. One then obtains a purely-real contribution to the scattering amplitude, which was found to be quite large in $\gamma^* p \to \pi^+ n$ [96]. It is, however, not understood how to evaluate the analogous contribution in the DGLAP region, where it cannot be expressed in terms of the pion light-cone wave function (see figure 9.13(b)).

In the leading-twist approximation of DVCS, perturbation theory can be used because of the point-like coupling of the real photon to quarks. To estimate the contribution from the hadronic component of the photon to DVCS, the studies

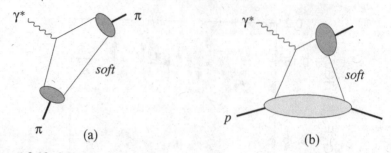

Figure 9.13. Soft overlap contribution to the pion form factor (a) and to meson electroproduction in the DGLAP region (b).

[99,100] have used models which describe the experimental data for ρ electropro-duction in certain kinematics and converted the amplitude for $\gamma_T^* p \to \rho_T p$ into a contribution to the amplitude for $\gamma_T^* p \to \gamma p$ using simple vector-meson domi-nance. Although it cannot be interpreted as a power correction in a straightforward way, the contribution in question may be seen as an indicator for effects beyond the leading-twist description of Compton scattering. In [99] this contribution was compared with DVCS data by H1 [101] and found to be at the 10–20% level in the Compton amplitude, whereas the ρ contribution estimated in [100] accounts for up to 50% of the DVCS beam spin asymmetries measured at HERMES [102] and CLAS [103].

9.3.7 Heavy-meson production

Exclusive production of charmonium or bottomonium provides another class of processes where GPDs can be accessed. Here the heavy-quark mass provides a hard scale, even in photoproduction. Neglecting heavy-quark distributions in the nucleon, the relevant graphs involve gluon GPDs as shown in figure 9.10(b). To leading order in a non-relativistic expansion, the heavy-meson wave function becomes trivial for the ground-state mesons, J/ψ and Υ, with quark and anti-quark sharing the meson four-momentum in equal parts and their relative velocity v being approximately zero. The uncertainty in the value of the charm-quark mass gives a corresponding uncertainty in the cross section for $\gamma p \to J/\psi\, p$. There is no consensus concerning the corrections to this leading-order non-relativistic expansion. In [104] the $O(v^2)$ corrections have been evaluated in the formalism of non-relativistic QCD and were found to be quite small. The studies [97,105] took into account the relative momentum between c and $\bar{c}$ directly in the hard scatter-ing kernel. Using different model wave functions for the J/ψ, moderate to large correction factors were obtained compared with the leading-order non-relativistic formula.

For photoproduction of heavy vector mesons there is no factorization theorem to all orders in perturbation theory, but the explicit calculation of the $O(\alpha_s)$ corrections to the hard scattering kernel has given a structure compatible with factorization [106]. Numerically the corrections were found to be very large, and further study will be required to understand in which kinematics this process can reliably be described in the approximation discussed here.

The description we have outlined readily generalizes to electroproduction as long as the photon virtuality is not much greater than the quark mass, and this case has indeed been included in [97,105]. In the limit where Q^2 is much bigger than the squared quark mass, it sets the hard scale and the collinear factorization scheme for light mesons becomes applicable. The structure of the meson is then encoded in its light-cone distribution amplitude.

9.3.8 A remark on small x

The connection between the production of light or heavy vector mesons at small x and the gluon distribution in the proton was realized early on [107,108]. These studies worked at leading-logarithmic accuracy in $\log(1/x)$, where the generalized gluon distribution at $t = 0$ can be replaced in the relevant loop integrals by the usual gluon distribution at a typical longitudinal momentum fraction, which is of order $(m_V^2 + Q^2)/W^2 \approx 2\xi$. We note that, taken literally, leading $\log(1/x)$ accuracy is a serious limitation for phenomenology, since even the ambiguity whether to take ξ or 2ξ as the argument of the gluon distribution has a large impact on the resulting cross section at small x. The t dependence of the generalized gluon distribution is, however, a legitimate part of the leading $\log(1/x)$ approximation, and typically parametrized via an exponential form $d\sigma/dt \propto \exp(Bt)$ for the cross section at small t.

A theoretical approach to describe processes in the leading $\log(1/x)$ approximation is based on colour dipoles, discussed in chapter 8. Within colour-dipole factorization, the usual gluon density appears as a building block when making the additional approximation of leading twist and double leading logarithm $\log Q^2 \log(1/x)$. It is important to realize that the relation between this scheme and the collinear factorization discussed in the present chapter is not always trivial (see [4] for a brief account). This is highlighted by DVCS, which in the dipole formalism is described in terms of the gluon distribution in the nucleon, whereas in collinear factorization both quark and gluon distributions are of comparable importance even at x_B of order 10^{-3}, if one chooses a factorization scale $\mu = Q$ [67]. This reflects the ambiguity in separating the process amplitude into parton distributions and other building blocks, illustrated in figure 9.14. We note also that the space-time picture of the process is different in the two schemes. The collinear factorization approach

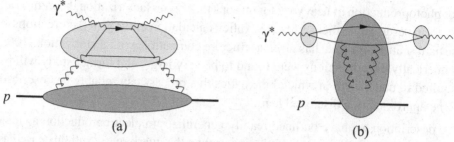

Figure 9.14. (a) Collinear factorization of a diagram for high-energy $\gamma^* p$ scattering into $\gamma^* g$ scattering and the gluon distribution. Depending on the virtuality of the internal quark lines, the same diagram can be factorized into $\gamma^* q$ scattering and the quark distribution. (b) Factorization of the same diagram into photon wave functions and the scattering of a $q\bar{q}$ dipole on the proton, as appropriate in the colour-dipole approach.

is naturally interpreted in the infinite-momentum frame, where a gluon emitted by the target scatters on the photon via a quark loop. The dipole picture may be visualized in the target rest frame, where the photon splits into a $q\bar{q}$ pair that subsequently scatters on the proton.

9.3.9 Revealing GPDs

We see in (9.34) and (9.35) that to leading order in α_s the leading-twist amplitudes for both DVCS and electroproduction of light mesons involve GPDs in the form

$$\int_{-1}^{1} dx\, H^q(x, \xi, t) \left(i\pi \delta(x - \xi) + \frac{1}{\xi - x} \right) \pm \{x \to -x\} \qquad (9.45)$$

and its analogues for E^q, $\tilde{H}^q$, $\tilde{E}^q$ and the gluon distributions, where the pole of $(\xi - x)^{-1}$ is to be regularized by Cauchy's principal-value prescription. In this approximation the imaginary part of the scattering amplitude is given directly by GPDs at the special points $x = \pm\xi$ (see section 9.1.8). Including α_s corrections one obtains integrals over x in the DGLAP regions. In contrast, the real part of the amplitude is sensitive to all x already at Born level. The most relevant region of x in the integral is of course determined by ξ, so that the dependence of scattering amplitudes on t and ξ yields rather direct information on the spatial distribution of partons with typical momentum fractions of order ξ (see section 9.1.7).

To reconstruct the dependence of GPDs on the independent momentum fractions x and ξ is a much more difficult problem. To leading order in α_s the imaginary part of the amplitude for double DVCS involves GPDs at $x = \pm\rho$, where $|\rho| < \xi$ according to (9.28). Double DVCS hence offers the unique possibility to scan GPDs in the ERBL region. Indirect information on the x dependence of GPDs resides in

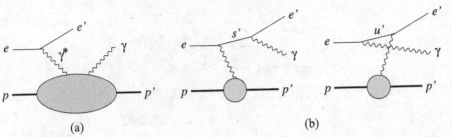

Figure 9.15. Contribution to the electroproduction of a photon from Compton scattering (a) and from the Bethe–Heitler process (b).

the logarithmic Q^2 dependence of scattering amplitudes, which is encoded in the perturbative evolution equations and hard scattering kernels. This information can of course only be used in kinematics where power corrections in $1/Q$ can be neglected, and according to our discussion in section 9.3.6 Compton scattering looks more promising in this context than meson production.

To constrain the x dependence of GPDs in practice, one needs physically well-motivated functional ansätze for these functions, as one does for the extraction of ordinary parton distributions from inclusive processes. The constraints from the forward limit, the elastic form factors, polynomiality and positivity are not trivial to fulfil, and so far only double-distribution based ansätze as discussed in section 9.2.7 have been used, where the forward limit and polynomiality are satisfied by construction.

As we have seen in sections 9.3.2 and 9.3.3, means of disentangling the spin and flavour structure of GPDs are offered by a combination of Compton scattering and suitable meson production channels, by measurements with polarized targets, and by combined data on proton and neutron targets. In the following section we will investigate the potential of Compton scattering, which in particular provides separate access to real and imaginary parts of the scattering amplitude, where, as we have just discussed, GPDs intervene in very different fashions.

9.4 Phenomenology of Compton scattering

DVCS has a rich phenomenology because it is observed in electroproduction $ep \rightarrow ep\gamma$, where it competes with the Bethe–Heitler process (see figure 9.15). The latter is calculable since the electromagnetic proton form factors are experimentally well known at the small values of momentum transfer where they are required: see chapter 2. Using the interference between the two processes, one can study the *amplitude* of virtual Compton scattering, $\gamma^* p \rightarrow \gamma p$, including its phase. As we

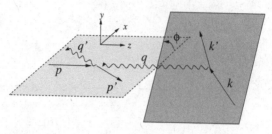

Figure 9.16. Kinematics of $ep \rightarrow ep\gamma$ in the centre of mass of the final-state photon and proton.

have seen in the previous subsection, this is most valuable for constraining the dependence of GPDs on x and ξ.

The kinematics of electroproduction are described in terms of the invariants Q^2, W^2, t entering the Compton subprocess, in addition to the azimuthal angle ϕ between the lepton and hadron plane defined in figure 9.16 and the usual inelasticity parameter $y = (qp)/(kp)$. We neglect the lepton mass throughout, and our discussion readily carries over to $\mu p \rightarrow \mu p\gamma$.

The Bethe–Heitler graphs in figure 9.15 are readily evaluated without further approximation. Expressed through the variables we have chosen, the result is quite complicated but simplifies considerably in the kinematical limit of large Q^2 at fixed t, x_B, y and ϕ, which is relevant for DVCS. For detailed discussions and formulae see [109,67,4]. Decomposing the cross section for $ep \rightarrow ep\gamma$ on an unpolarized target into contributions $d\sigma_{BH} + d\sigma_{VCS} + d\sigma_{INT}$ from the Bethe–Heitler process, Compton scattering and their interference, one finds

$$\frac{d\sigma_{BH}}{d\phi\, dt\, dQ^2\, d\log x_B} = \frac{\alpha_{em}^3}{8\pi s^2} \frac{1}{|t|} \frac{4}{\epsilon} \frac{1}{P}$$
$$\times \left[\frac{1-\xi^2}{\xi^2} \frac{t-t_0}{t} \left(F_1^2 - \frac{t}{4m^2} F_2^2 \right) + 2(F_1 + F_2)^2 + O\left(\frac{1}{Q}\right) \right], \quad (9.46)$$

with the Dirac and Pauli form factors F_1 and F_2 evaluated at t. Here $s = (k + p)^2$ is the squared ep centre-of-mass energy and $\epsilon \approx (1 - y)/(1 - y + \frac{1}{2}y^2)$ is the ratio of longitudinal to transverse photon flux in the Compton process. The factor

$$P = 1 - 2\cos\phi \sqrt{\frac{2(1+\epsilon)}{\epsilon} \frac{1-\xi}{1+\xi} \frac{t_0 - t}{Q^2}} + O\left(\frac{1}{Q^2}\right) \quad (9.47)$$

is due to the lepton propagators in the graphs of figure 9.15(b) and is proportional to the product $s'u'$ of their virtualities. Although formally $1/Q$ suppressed, the $\cos\phi$ dependent factor in (9.47) can be numerically quite important, especially in the kinematics of the HERMES and JLab experiments (see section 9.5).

For $d\sigma_{VCS}$ and $d\sigma_{INT}$ it is useful to parametrize the dynamics of $\gamma^* p \to \gamma p$ by helicity amplitudes $e^2 M_{\lambda' \mu', \lambda \mu}$. Here λ (λ') is the helicity of the incoming (outgoing) proton and μ (μ') the helicity of the incoming (outgoing) photon. With the constraints of parity invariance there are twelve independent amplitudes for DVCS, which can be chosen to correspond to the combinations $(+, +), (+, 0), (+, -)$ of the photon helicities (μ', μ). As discussed in sections 9.3.2 and 9.3.4, the amplitudes $M_{\lambda'+, \lambda+}$ appear at leading order in $1/Q$ and α_s, the twist-3 amplitudes $M_{\lambda'+, \lambda 0}$ are $1/Q$ suppressed, and $M_{\lambda'+, \lambda-}$ comes with a suppression by either α_s or $1/Q^2$. One can readily express the four amplitudes $M_{\lambda'+, \lambda+}$ in terms of the Compton form factors $\mathcal{H}, \mathcal{E}, \tilde{\mathcal{H}}, \tilde{\mathcal{E}}$ and hence of integrals over $H, E, \tilde{H}, \tilde{E}$. The four amplitudes $M_{\lambda'+, \lambda 0}$ depend on integrals over the same GPDs, and in addition on four independent quark–antiquark–gluon matrix elements. For Compton scattering on an unpolarized target one has

$$\frac{d\sigma_{VCS}}{d\phi \, dt \, dQ^2 \, d\log x_B} = \frac{\alpha_{em}^3}{8\pi s^2} \frac{1}{Q^2} \frac{1}{1 - \epsilon} \sum_{\lambda' \lambda} |M_{\lambda'+, \lambda+}|^2 + \cdots . \qquad (9.48)$$

For simplicity we have neglected the target mass in the phase space and flux factor. In order to include the target mass one should replace s^2 with $(s - m^2)^2 (1 + 4x_B^2 m^2/Q^2)^{1/2}$, both here and in (9.46) and (9.50). We have here only given the contribution from the leading amplitudes $M_{\lambda'+, \lambda+}$. The complete expression contains also terms depending on ϕ and on the lepton beam polarization P_ℓ, where $\cos\phi$ and $P_\ell \sin\phi$ accompany the interference of $M_{\lambda'+, \lambda 0}$ with $M_{\lambda'+, \lambda+}$ or $M_{\lambda'+, \lambda-}$, and $\cos 2\phi$ accompanies the interference between $M_{\lambda'+, \lambda-}$ and $M_{\lambda'+, \lambda+}$ (there is no term with $P_\ell \sin 2\phi$). The ϕ dependence of the ep cross section thus gives access to the photon helicity dependence of the Compton amplitudes and makes it possible to test their different size and Q^2 behaviour at fixed x_B and t expected from the factorization theorem. The same is possible with the interference term between the Compton and Bethe–Heitler processes, as we shall see shortly.

From (9.46) and (9.48) we see that the relative size of Bethe–Heitler and Compton cross sections is

$$d\sigma_{BH} : d\sigma_{VCS} \sim Q^2 (1 - \epsilon) : |t| \epsilon, \qquad (9.49)$$

so that Bethe–Heitler typically dominates over Compton scattering since we require $Q^2 \gg |t|$. An exception is the case $\epsilon \approx 1$, which requires small y and hence large s at given x_B and Q^2 according to the relation $Q^2 = yx_B(s - m^2)$. This is illustrated by the estimates in figure 9.17. Note that the factor P from the lepton propagators can enhance the Bethe–Heitler contribution at ϕ around $0°$, especially if ϵ is small.

Whether the Bethe-Heitler process dominates or not, information about Compton scattering can be obtained from $d\sigma_{INT}$. Especially clean access to the

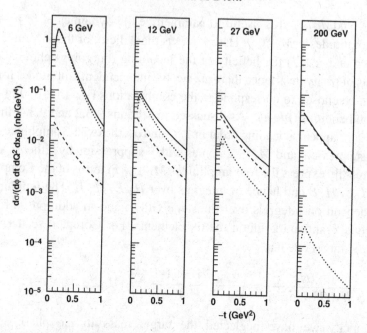

Figure 9.17. The cross section for $ep \to ep\gamma$ and its individual contributions from DVCS (dashed) and the Bethe–Heitler process (dotted) for different electron energies in the proton rest frame, given in the panels. The remaining kinematical quantities are $Q^2 = 2.5$ GeV^2, $x_B = 0.25$ and $\phi = 45°$. The curves are based on the same model as the dashed curves in figure 9.19 and will be described in section 9.5.

interference term is provided in the difference of cross sections for $e^- p \to e^- p\gamma$ and $e^+ p \to e^+ p\gamma$, where $d\sigma_{BH} + d\sigma_{VCS}$ drops out because of the lepton charge factors, whereas $d\sigma_{INT}$ remains. Furthermore, the Bethe–Heitler contribution (but not $d\sigma_{VCS}$) exactly drops out in the single-spin asymmetries for a longitudinally-polarized lepton beam and for a longitudinally- or transversely-polarized proton target. This is a consequence of parity and time-reversal invariance, see [4] for a discussion of its limits of validity. In double-spin asymmetries there is, however, a Bethe–Heitler contribution.

For an unpolarized target the interference term has the structure

$$\frac{d\sigma_{INT}}{d\phi \, dt \, dQ^2 \, d\log x_B} = -e_\ell \frac{\alpha_{em}^3}{8\pi s^2} \frac{1}{|t|} \frac{m}{Q} \frac{8\sqrt{2}}{\xi} \frac{1}{P} \left[\cos\phi \, \frac{1}{\sqrt{\epsilon(1-\epsilon)}} \operatorname{Re} \widehat{M}_{++} \right.$$

$$- \cos 2\phi \sqrt{\frac{1+\epsilon}{1-\epsilon}} \operatorname{Re} \widehat{M}_{+0} - \cos 3\phi \sqrt{\frac{\epsilon}{1-\epsilon}} \operatorname{Re} \widehat{M}_{+-}$$

$$\left. + P_\ell \sin\phi \sqrt{\frac{1+\epsilon}{\epsilon}} \operatorname{Im} \widehat{M}_{++} - P_\ell \sin 2\phi \operatorname{Im} \widehat{M}_{+0} + O\left(\frac{1}{Q}\right) \right], \qquad (9.50)$$

where $\widehat{M}_{\mu'\mu}$ is a linear combination of the Compton amplitudes $M_{\lambda'\mu',\lambda\mu}$ with known coefficients depending on ξ and t, and P is given in (9.47). The charge of the lepton beam is $e_\ell = \pm 1$, and its polarization is normalized as $-1 \leq P_\ell \leq 1$. The dependence on ϕ and ϵ is hence completely explicit in (9.50), and angular analysis can be used to separate the different terms in the interference, which involve Compton amplitudes playing very different roles in the large-Q^2 limit. In particular, the leading amplitudes $M_{\lambda'+,\lambda+}$ are those going with $\cos\phi$ and $P_\ell \sin\phi$. In this context it is important to take into account that the factor P from the Bethe–Heitler propagators provides an additional ϕ dependence of the interference term. Especially in kinematics where this is not negligible, the cleanest access to the Compton amplitudes is obtained by reweighting the cross section with P.

Upon closer inspection [67,4] and taking into account the predictions of the factorization theorem for the different Compton amplitudes, one finds that the terms denoted as $O(1/Q)$ in (9.50) are actually suppressed by $1/Q^2$ or by α_s, except for a contribution which is independent of ϕ and P_ℓ and involves the leading amplitudes $M_{\lambda'+,\lambda+}$ with kinematic coefficients of order $1/Q$. This suggests that the results extracted in an angular analysis based on (9.50) can be interpreted as definite combinations of Compton amplitudes with only moderate corrections.

With an unpolarized nucleon target one has access to a combination of Compton amplitudes and hence of GPDs. To disentangle the spin dependence parametrized by H, E, $\tilde{H}$, $\tilde{E}$, target polarization is required. The unpolarized Compton cross section is sensitive to the quadratic combination

$$\frac{1}{2} \sum_{\lambda'\lambda} |M_{\lambda'+,\lambda+}|^2 = (1-\xi^2)(|\mathcal{H}|^2 + |\tilde{\mathcal{H}}|^2) - \left(\xi^2 + \frac{t}{4m^2}\right)|\mathcal{E}|^2$$

$$- \xi^2 \frac{t}{4m^2} |\tilde{\mathcal{E}}|^2 - 2\xi^2 \, \mathrm{Re}\,(\mathcal{H}^*\mathcal{E} + \tilde{\mathcal{H}}^*\tilde{\mathcal{E}}) \tag{9.51}$$

of the integrals over GPDs we defined in section 9.3.2. Other combinations appear in the Compton contribution to the single spin asymmetry for transverse target polarization and in the double asymmetry for longitudinal lepton beam and longitudinal or transverse target polarization. A more direct separation is offered by the interference term, where GPDs appear linearly. For a polarized target the leading-twist contribution from the amplitudes $M_{\lambda'+,\lambda+}$ is

$$d\sigma_{\mathrm{INT}} \propto \cos\phi \, \mathrm{Re}\,\widehat{M}_{++} + P_\ell \sqrt{1-\epsilon^2}\,\sin\phi \, \mathrm{Im}\,\widehat{M}_{++}$$

$$+ S_L \big[\sin\phi \, \mathrm{Im}\,\widehat{M}^L_{++} + P_\ell \sqrt{1-\epsilon^2}\,\cos\phi \, \mathrm{Re}\,\widehat{M}^L_{++} \big]$$

$$+ S_T \cos\beta \big[\sin\phi \, \mathrm{Im}\,\widehat{M}^S_{++} + P_\ell \sqrt{1-\epsilon^2}\,\cos\phi \, \mathrm{Re}\,\widehat{M}^S_{++} \big]$$

$$+ S_T \sin\beta \big[\cos\phi \, \mathrm{Im}\,\widehat{M}^N_{++} - P_\ell \sqrt{1-\epsilon^2}\,\sin\phi \, \mathrm{Re}\,\widehat{M}^N_{++} \big], \tag{9.52}$$

where the target polarization vector is $(S_T \cos\beta, S_T \sin\beta, S_L)$ in the coordinate system of figure 9.11, normalized as $0 \le S_T \le 1$ and $-1 \le S_L \le 1$. The combinations of Compton amplitudes are given by

$$\widehat{M}_{++} = \sqrt{1-\xi^2}\,\frac{\sqrt{t_0-t}}{2m}\left[F_1\mathcal{H} + \xi(F_1+F_2)\tilde{\mathcal{H}} - \frac{t}{4m^2}F_2\mathcal{E}\right],$$

$$\widehat{M}^L_{++} = \sqrt{1-\xi^2}\,\frac{\sqrt{t_0-t}}{2m}\left[F_1\tilde{\mathcal{H}} + \xi(F_1+F_2)\left(\mathcal{H} + \frac{\xi}{1+\xi}\mathcal{E}\right)\right.$$
$$\left. - \left(\frac{\xi}{1+\xi}F_1 + \frac{t}{4m^2}F_2\right)\xi\tilde{\mathcal{E}}\right],$$

$$\widehat{M}^S_{++} = \left[\xi^2\left(F_1 + \frac{t}{4m^2}F_2\right) - \frac{t}{4m^2}F_2\right]\tilde{\mathcal{H}} - \left(\frac{t}{4m^2} + \frac{\xi^2}{1+\xi}\right)\xi(F_1+F_2)\mathcal{E}$$
$$+ \left[\left(\frac{t}{4m^2} + \frac{\xi^2}{1+\xi}\right)F_1 + \frac{t}{4m^2}\xi F_2\right]\xi\tilde{\mathcal{E}} - \xi^2(F_1+F_2)\mathcal{H},$$

$$\widehat{M}^N_{++} = -\frac{t}{4m^2}(F_2\mathcal{H} - F_1\mathcal{E}) + \xi^2\left(F_1 + \frac{t}{4m^2}F_2\right)(\mathcal{H}+\mathcal{E})$$
$$- \xi^2(F_1+F_2)\left(\tilde{\mathcal{H}} + \frac{t}{4m^2}\tilde{\mathcal{E}}\right) \tag{9.53}$$

in terms of integrals over GPDs. Hence, with both longitudinal and transverse target polarization, one has four independent observables to separate the four twist-2 Compton form factors. We see that the only combination where E is not kinematically suppressed compared with other GPDs is $\widehat{M}^N_{++}$. To be sensitive to this distribution, which plays a special role in the context of angular momentum, requires transverse target polarization.

So far we have focused on a proton target. Data for DVCS on both the proton and the neutron would allow separation of the GPDs for u and d quarks. At leading order in α_s the Compton amplitude involves quark combinations

$$p : \tfrac{4}{9}F^u + \tfrac{1}{9}F^d + \tfrac{1}{9}F^s, \qquad n : \tfrac{1}{9}F^u + \tfrac{4}{9}F^d + \tfrac{1}{9}F^s, \tag{9.54}$$

and their analogues for $\tilde{F}^q$, with the quark flavour referring to a proton target in both cases. The detailed phenomenology of DVCS on the proton and the neutron is different: in the t range of interest $F_1(t)$ is small but $F_2(t)$ is large for the neutron, whereas both are comparable for the proton: see chapter 2. In particular, we see in (9.53) that the unpolarized combination $\widehat{M}_{++}$ is sensitive to E in the neutron case. Measurements of DVCS on the neutron using a deuteron target have been approved at JLab [110].

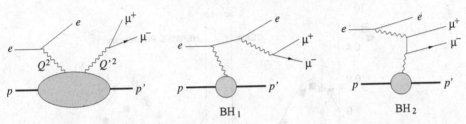

Figure 9.18. Contributions from Compton scattering and the Bethe–Heitler processes to $ep \to ep\,\mu^+\mu^-$.

9.4.1 Double DVCS

We emphasized in section 9.3.9 that double DVCS contains unique information on GPDs in the region where they describe $q\bar{q}$ emission from the target. Extraction of the Compton signal from exclusive lepton pair production $ep \to ep\,\ell^+\ell^-$ is more complex than in the case of DVCS, since there are now two different types of Bethe–Heitler processes, shown in figure 9.18. The ratio of cross sections $d\sigma_{BH_1} : d\sigma_{VCS} \sim Q^2(1-\epsilon) : |t|\epsilon$ is as in the case of DVCS, whereas for the new type of Bethe–Heitler process one has $d\sigma_{BH_2} : d\sigma_{VCS} \sim Q'^2 : |t|$ without the possibility to enhance the Compton process by going to $\epsilon \approx 1$. Using the angular distribution of the produced lepton ℓ^+ one can, however, construct observables where $d\sigma_{BH_2}$ does not contribute because of the lepton charge factors in the different processes [76]. We note that channels where the produced lepton pair has different flavour than the beam, that is $ep \to ep\,\mu^+\mu^-$ and $\mu p \to \mu p\,e^+e^-$, admit a simpler theoretical analysis than their counterparts with two identical leptons in the final state. First theoretical studies of lepton pair production have been carried out in [77,78,111], and it remains to be seen to which extent the physics potential of this process can be harvested in experiments.

9.5 Experimental results on DVCS

DVCS is the theoretically cleanest process sensitive to GPDs and offers most detailed observables. The first experimental results in fixed-target experiments [102,103] were for the lepton-beam spin asymmetry

$$A_{LU}(\phi) = \frac{1}{|P_\ell|} \frac{d\sigma^{e\uparrow} - d\sigma^{e\downarrow}}{d\sigma^{e\uparrow} + d\sigma^{e\downarrow}}. \tag{9.55}$$

Following our discussion in section 9.4 its denominator is ϕ-independent in the limit of large Q^2 and dominated by the Bethe–Heitler cross section (9.46) unless ϵ is very close to 1. In the numerator the Bethe–Heitler contribution is absent for

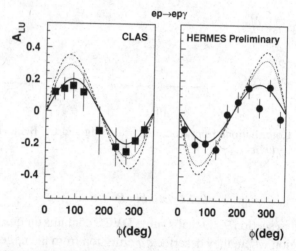

Figure 9.19. Data on the beam spin asymmetry in DVCS from CLAS [103] (left) and preliminary data from HERMES [112] (right). The solid curves show fits to the data of the form $A_{LU}(\phi) = s_1 \sin \phi$, and the remaining curves are model calculations discussed in the text. The different signs of the asymmetry reflect that CLAS used an e^- and HERMES an e^+ beam.

symmetry reasons and the Compton contribution involves twist-3 amplitudes, so that the asymmetry is approximately proportional to the combination $\operatorname{Im} \widehat{M}^{++}$ of twist-2 Compton form factors given in (9.53). At small to modest values of ξ and t this combination is dominated by $\operatorname{Im} \mathcal{H}$ due to kinematical prefactors. Hence to leading order in α_s one has direct access to $H^q(\xi, \xi, t) - H^q(-\xi, \xi, t)$ with the above approximations. For a precise interpretation of data the cross section difference $d\sigma^{e\uparrow}/d\phi - d\sigma^{e\downarrow}/d\phi$ is more suitable than the normalized asymmetry (9.55), whose denominator does receive contributions from $d\sigma_{VCS}$ and $d\sigma_{INT}$ at some level of accuracy [67]. Measurement of cross section differences is, however, experimentally more demanding.

The results on A_{LU} shown in figure 9.19 support the relevance of the large-Q^2 limit for DVCS in current experiments. The CLAS collaboration compared their data [103] with theory predictions for $Q^2 = 1.25$ GeV^2, $x_B = 0.19$, $-t = 0.19$ GeV^2 and $\epsilon = 0.3$, whereas the average kinematical values for HERMES [112] are $Q^2 = 2.5$ GeV^2, $x_B = 0.12$, $-t = 0.18$ GeV^2 and $\epsilon = 0.88$. In both measurements the asymmetry approximately follows a $\sin \phi$ behaviour, a fit to $A_{LU}(\phi) = s_1 \sin \phi + s_2 \sin 2\phi$ giving $s_2 = -0.024 \pm 0.021$ for CLAS. Preliminary CLAS data with much higher statistics find a ratio s_2/s_1 smaller than 15% [113]. This trend is in line with the large-Q^2 picture, where a $\sin 2\phi$ modulation of A_{LU} should be $1/Q$ suppressed compared with the $\sin \phi$ term.

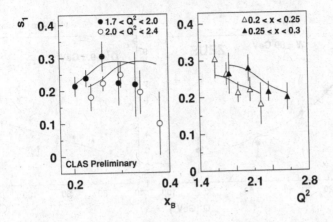

Figure 9.20. Preliminary CLAS data [113] for the coefficient s_1 of the beam spin asymmetry in a fit to the form $A_{LU}(\phi) = s_1 \sin\phi + s_2 \sin 2\phi$. The curves correspond to the same model as the dotted curves in figure 9.19.

The size of the asymmetries is described reasonably well by calculations that approximate Compton scattering to leading order in $1/Q$ and α_s (see (9.33) and (9.34)), with GPDs modelled along the lines described in section 9.2.7. The model of H^q taken for the curves in figure 9.19 assumes a factorized t dependence, adding the D-term (9.26) from the chiral quark–soliton model to the double-distribution ansatz (9.27) with $\rho_{val} = \rho_{sea} \propto (1 - \alpha^2)^b$ and the next-to-next-to-leading order (NNLO) quark densities from [114] at $\mu = 1$ GeV. The profile parameter was set to $b = 1$ for the dashed and to $b = 4$ for the dotted curves.

The measurements presented in figure 9.19 are integrated over rather large kinematical regions in x_B, Q^2 and t. In figure 9.20 we show preliminary data from CLAS for a finer binning in x_B or in Q^2. The calculation from the model just described gives a fair description of the data, except for the points at large x_B.

As we saw in section 9.4, a polarized target is necessary to disentangle GPDs with different spin structure in DVCS. For longitudinal target polarization one can define

$$A_{UL}(\phi) = \frac{1}{|S_L|} \frac{d\sigma^{p\uparrow} - d\sigma^{p\downarrow}}{d\sigma^{p\uparrow} + d\sigma^{p\downarrow}}. \tag{9.56}$$

As in the case of A_{LU} this asymmetry receives its dominant contribution from the interference term (9.52) in the large-Q^2 limit. Given the kinematical prefactors in (9.53), and taking into account that the unpolarized GPDs H are likely larger than their polarized counterparts $\tilde{H}$, one expects $\mathrm{Im}\,\widehat{M}^L_{++}$ to be dominated by $\mathrm{Im}\,\mathcal{H}$ and $\mathrm{Im}\,\tilde{\mathcal{H}}$. Together with the information from the beam spin asymmetry one can thus access $\tilde{H}^q(\xi, \xi, t) + \tilde{H}^q(-\xi, \xi, t)$ to leading accuracy in α_s.

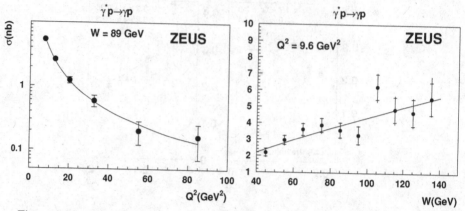

Figure 9.21. Cross section for $\gamma^* p \to \gamma p$ measured by ZEUS [117]. The curves are parametrizations $\sigma \propto Q^{-3.08}$ and $\sigma \propto W^{0.75}$, respectively.

Single-spin asymmetries provide access to the imaginary parts of Compton amplitudes. The lepton beam charge asymmetry is sensitive to their real parts and hence to GPDs in both the DGLAP and ERBL regions. In the large-Q^2 limit the term Re $\widehat{M}_{++}$ from the interference can be filtered out by the $\cos\phi$ modulation of the asymmetry

$$A_C(\phi) = \frac{d\sigma^{e^+} - d\sigma^{e^-}}{d\sigma^{e^+} + d\sigma^{e^-}}. \tag{9.57}$$

Preliminary data from HERMES [115,116] give a coefficient $c_1 = 0.11 \pm 0.04$ for a fit to $A_C = c_0 + c_1 \cos\phi$ with average kinematical values $Q^2 = 2.8$ GeV^2, $x_B = 0.12$, $-t = 0.27$ GeV^2 and $\epsilon = 0.84$. This is well in the range of present model calculations, which give a wide range of values with either sign, reflecting in particular our ignorance of GPDs in the ERBL region. In fixed-target kinematics the D-term in (9.26) gives a sizeable positive contribution to the coefficient of $\cos\phi$ in A_C. However, comparison of the model results in [51] and [67] shows that the uncertainty on the double-distribution part of GPDs is significant as well, so that one cannot constrain the D-term directly. Note, however, that by its form (9.17) the D-term gives a ξ-independent contribution to $\mathcal{H}$ at fixed Q^2 and t, and the opposite contribution to $\mathcal{E}$, and has an experimentally-accessible signature.

At high ep centre-of-mass energies $\sqrt{s}$ there is a wide kinematical region where y is small, so that the Compton and Bethe–Heitler processes are of comparable strength. Subtracting the Bethe–Heitler contribution one can then measure the Compton cross section. Such measurements have been performed by the H1 and ZEUS collaborations [101,117] at $\sqrt{s} = 300$ GeV to 318 GeV. Figure 9.21 shows the ZEUS results for the $\gamma^* p \to \gamma p$ cross section, which one can expect to be

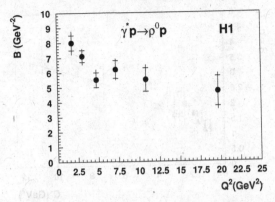

Figure 9.22. H1 data [119] on the *t*-slope in ρ production.

dominated by $|\mathcal{H}|^2$ in (9.51). A fair description of the data can be achieved with several models for GPDs [67,79,72]. We note that colour-dipole models (see section 9.3.8 and chapter 8) fare equally well. Whether the leading-log($1/x$) approximation underlying the dipole formalism or the leading-twist description discussed here is more appropriate in given small-x kinematics remains ambiguous, as it does for the inclusive structure functions of DIS. A major uncertainty for models in both approaches is the fact that the t dependence of DVCS is known only approximately [118].

9.6 Experimental results on meson production

As discussed in section 9.3, meson electroproduction probably requires rather large values of Q^2 for a leading-twist interpretation to be adequate. On the other hand, it provides valuable handles to separate different flavour and spin combinations of GPDs. Compared with DVCS it offers fewer observables that can be evaluated at leading twist, whose extraction from data is, however, more straightforward.

9.6.1 Vector mesons

Data in a wide range of kinematics are available for electroproduction of the neutral vector mesons ρ^0, ω, ϕ, both at intermediate and large W, as well as for J/ψ photo- and electroproduction. From (9.36) one expects that ρ and ω production is dominated by the gluon distribution at very small x_B and by the quark distributions at large x_B. Where the transition between these regimes takes place remains to be clarified.

Figure 9.22 shows the t-slope parameter for ρ production, obtained from an exponential fit $d\sigma/dt \propto e^{Bt}$ at small t. Similar results have been obtained in ϕ production

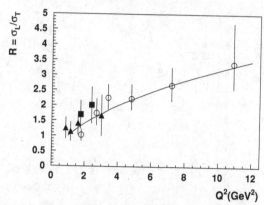

Figure 9.23. $R = \sigma_L/\sigma_T$ for exclusive ρ^0 production versus Q^2. Data are from CLAS (squares), Cornell (triangles) and H1 (circles) [121,122,119]. Average values of W are 2.1 GeV for the CLAS and Cornell data and 75 GeV in the measurement of H1. The curve is given as a guide for the eye.

[127]. The corresponding measurement for J/ψ production gives values of B between 4 and 5 GeV^{-2}, with little dependence on Q^2 between the photoproduction limit and $Q^2 = 100$ GeV^2 [128]. The steep decrease of $B(Q^2)$ for ρ production can be interpreted as an effect of the transverse quark momentum in the meson on the hard-scattering subprocess [97]. Comparison with the value for J/ψ production then suggests that this power suppressed effect becomes negligible only when $Q^2 \geq 10$ GeV^2. In the leading-twist regime $2B$ is the average impact parameter $\langle \vec{b}^2 \rangle$ of gluons according to (9.21).

The factorization theorem for light-meson production predicts that at large Q^2 both photon and meson should have longitudinal polarization. The photon helicities in the $\gamma^* p$ cross section $\sigma = \sigma_T + \epsilon \sigma_L$ can be disentangled by a Rosenbluth separation (see chapter 2), where ϵ is varied at given x_B and Q^2. However, this requires measurement at different ep centre-of-mass energies. The polarization of the produced vector meson can be inferred from its decay angular distribution, for example in $\rho \to \pi^+ \pi^-$. Helicity-changing transitions from the photon to the meson are known to be small empirically, and neglecting them one can determine the ratio $R = \sigma_L/\sigma_T$ from the measured polarization of the vector meson. This method has been used in the measurements shown in figure 9.23. Preliminary data for Q^2 up to 30 GeV^2 [120] suggest that R continues to rise with Q^2. Power counting at large Q^2 predicts R to rise like Q^2 up to logarithms (see also section 9.3.5). We see that σ_L is comparable to σ_T for Q^2 of a few GeV^2, indicating that formally $1/Q^2$ suppressed effects are not numerically small in that range.

The longitudinal cross section σ_L for ρ production has been measured by CLAS [121] for Q^2 between 1.5 and 3 GeV^2 and several bins in x_B. The data for $x_B = 0.31$

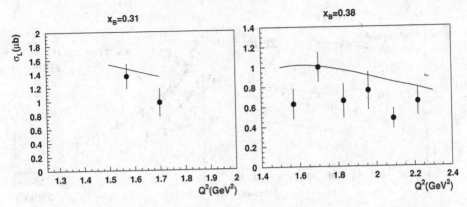

Figure 9.24. Data from CLAS [121] on σ_L for ρ production. The curves are explained in the text.

and 0.38 are shown in Figure 9.24 and compared with the calculation from [96], which is restricted to quark GPDs and takes into account power corrections from the transverse momentum of quarks in the hard scattering kernel (see section 9.3.6). The model taken for H^q uses a factorizing t dependence and the double-distribution ansatz (9.27) with $\rho_{val} = \rho_{sea}$ and $B = 5$. The bins with $x_B = 0.48$ and 0.52 in the same measurement correspond to kinematics very close to threshold, where a GPD based approach to the dynamics cannot be expected to work. Corresponding measurements at x_B around 0.1 have been made by HERMES [126]. The observed ratio of cross sections for ϕ and ρ^0 production implies that in this region of x_B both quark and gluon exchange must be relevant [129]. A consistent estimate of power corrections for this case has not yet been given. In figure 9.25 we show data from H1 and ZEUS for ρ and ϕ production. The model calculations from [98] shown in the same figure are obtained with gluon GPDs only and take into account power corrections from the transverse momentum of quarks in the meson. They use a factorized t dependence for H^g and the analogue of (9.27) for gluons with a profile parameter $b = 1$. We see that a description of the data by calculations using the leading-twist formalism and an estimate of power corrections is fair, although not perfect.

The cross sections for ρ and ϕ production at large Q^2 show a steep rise with W, in contrast to the photoproduction cross sections in these channels. A similarly steep rise is found for the cross section in DVCS (see figure 9.21) and also for J/ψ photo- and electroproduction, shown in figure 9.26. For J/ψ production the W dependence does not change significantly between $Q^2 = 0$ and 100 GeV^2 within present errors [128]. Such findings corroborate the interpretation of these processes in terms of gluon distributions, which at sufficiently large scales rise strongly with decreasing momentum fraction.

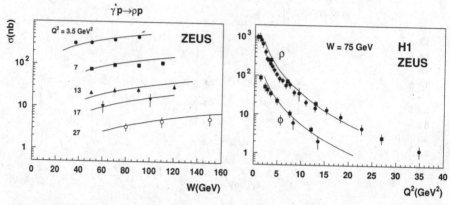

Figure 9.25. Left: cross section for $\gamma^* p \to \rho p$ measured by ZEUS [123,124]. Right: σ_L for ρ and ϕ production, extracted in [125] from data of H1 (circles) and ZEUS (squares). The curves are discussed in the text.

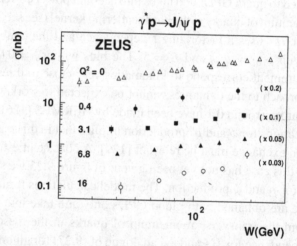

Figure 9.26. ZEUS data for J/ψ production [128] versus W for various values of Q^2 given in the plot. The electroproduction cross sections have been scaled for clarity.

9.6.2 Prospects for pseudoscalar-meson production

Separation of the cross sections for transverse and longitudinal photons can only be done by the Rosenbluth method for pseudoscalar-meson photoproduction at high Q^2. Note however that, in contrast to vector-meson production, σ_T now involves helicity flip from the photon to the meson, so that one may expect $R = \sigma_L/\sigma_T$ to be larger than for ρ production at the same Q^2. Information on the polarization of the virtual photon can be obtained in the same manner as for DVCS, namely from the distribution in the angle ϕ between the lepton and hadron planes. The unpolarized ep cross section can be written as

$$\frac{1}{\Gamma_T} \frac{d\sigma}{d\phi \, dt \, dQ^2 \, dx_B} = \frac{d\sigma_T}{dt} + \epsilon \frac{d\sigma_L}{dt} + \sqrt{2\epsilon(1+\epsilon)} \cos\phi \frac{d\sigma_{LT}}{dt} + \epsilon \cos 2\phi \frac{d\sigma_{TT}}{dt}, \tag{9.58}$$

where Γ_T is a kinematical factor. The ϕ-dependent terms go with the interference between longitudinal and transverse or between the two transverse photon polarizations, so that at large Q^2 they should be suppressed by $1/Q$ relative to σ_L. Corresponding measurements are underway at CLAS. Similar interference terms occur in the single-spin asymmetries for longitudinal-beam or longitudinal-target polarization. A non-zero value for the latter has been measured in $ep \rightarrow e\pi^+ n$ by HERMES [130] with average kinematics $Q^2 = 2.2$ GeV^2, $x_B = 0.15$, $-t = 0.46$ GeV^2 and $\epsilon = 0.95$. Preliminary HERMES results for $\sigma_T + \epsilon\sigma_L$ have been presented in [131].

Production of charged pseudoscalars is special in that it receives contributions from π^+ or K^+ exchange in the t-channel, which can be large in a wide kinematical region. As we discussed in section 9.3.3, the relative size of the contributions from GPDs $\tilde{E}$ and $\tilde{H}$ is measured in the transverse target-spin asymmetry.

9.7 GPDs and large-angle scattering processes

The concept of GPDs has also found application in processes at large invariant momentum transfer t. Although their theory is much less advanced than the one presented so far in this chapter, the developments we now briefly present have spawned a rather successful phenomenology and may be seen as a promising ansatz to understand large-t scattering in a description based on quark and gluon degrees of freedom. A more detailed account and references are given in [4,5].

To see the main physics issues in this context (which are quite different from those in processes at small t but large Q^2) consider the Dirac form factor of the proton at large t. In the asymptotic limit $-t \rightarrow \infty$ one expects that the hard scattering mechanism will dominate, where each parton in the three-quark Fock state of the proton undergoes a hard scattering (see for example [132] for a review). This can be calculated using the same collinear factorization approach we discussed in section 9.3.1, with long-distance dynamics now encoded in the leading-twist distribution amplitudes of the nucleon. Compared with analogous observables involving mesons, such as the pion form factor, the larger number of internal propagators results in a stronger 'dilution' of the external hard momentum in the hard scattering subprocess, so that much larger t is required in order for internal propagators to be predominantly far off-shell. The endpoint regions, where one or more partons in the nucleon wave function have soft momenta, play a more prominent role for baryon form factors.

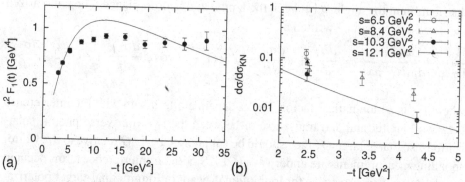

Figure 9.27. (a) Data for the proton Dirac form factor [134] and the result of modelling the soft overlap of the lowest three Fock states in the proton [35]. (b) Cross section data for wide-angle Compton scattering [135] compared to the result obtained with (9.59) and Compton form factors from the same model. The curve is for $s = 10.3$ GeV^2 and would differ by less than 20% for the other values of s, provided that $-t$ and $-u$ are above 2.3 GeV^2.

Calculations within the hard scattering approach that come close to describing the experimental data (which extends up to $-t$ of about 30 GeV^2 for the proton) typically receive their main contributions from such soft regions of phase space, where internal virtualities in the graphs are soft [133]. The approximations of the formalism then become invalid and the calculation inconsistent. A different approach is to consider precisely the contribution from endpoint configurations, where all partons in the nucleon except one are soft, as originally suggested by Feynman. This contribution may be approximated by the overlap of light-cone wave functions for the lowest few Fock states, and a rather successful phenomenology has been obtained with suitable model wave functions. In figure 9.27(a) we show results obtained with simple model wave functions for the Fock state with three quarks and the Fock states with three quarks and an additional gluon or $q\bar{q}$ pair. The same wave functions give a rather good description of polarized and unpolarized parton densities at $x \gtrsim 0.5$ (see section 9.1.8).

The preceding discussion carries over to other processes, in particular to Compton scattering at large Mandelstam invariants s, t, u, often referred to as 'wide-angle Compton scattering'. An evaluation in the hard scattering approach [136] confirms that this mechanism seriously undershoots the cross section data for s, $-t$, $-u$ of several GeV^2. For this process there is again a soft overlap contribution, where the elementary quark–photon vertex appearing in the Dirac form factor is replaced by Compton scattering on a quark, see figure 9.28. The soft overlap of the wave functions can be described in terms of generalized parton distributions: then one obtains handbag diagrams as in figure 9.1(a) with their factorized form of a hard scattering subprocess on a single quark and a soft matrix element of the nucleon [35,137].

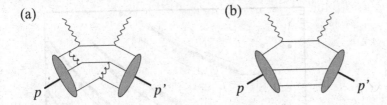

(a)   (b)

Figure 9.28. Contributions to wide-angle Compton scattering from the hard scattering (a) and the soft overlap (b) mechanism.

For the unpolarized cross section of $\gamma p \to \gamma p$ the soft handbag mechanism gives

$$\frac{d\sigma}{dt} = \frac{\pi \alpha_{em}^2}{s^2} \left[\frac{(s-u)^2}{|su|} \left(R_V^2(t) - \frac{t}{4m^2} R_T^2(t) \right) + \frac{(s+u)^2}{|su|} R_A^2(t) \right] \quad (9.59)$$

with Compton form factors

$$R_V(t) = \sum_q e_q^2 \int_{-1}^1 \frac{dx}{x} H^q(x,0,t), \qquad R_T(t) = \sum_q e_q^2 \int_{-1}^1 \frac{dx}{x} E^q(x,0,t),$$

$$R_A(t) = \sum_q e_q^2 \int_{-1}^1 \frac{dx}{x} \, \text{sign}(x) \, \tilde{H}^q(x,0,t). \quad (9.60)$$

The calculation leading to (9.59) requires the Mandelstam invariants to be large on a hadronic scale. In particular the target mass has been neglected, and the Compton subprocess was evaluated setting the momentum fraction x of the incoming quark to 1, which corresponds to this quark taking the entire plus-momentum of its parent proton. Model estimates suggest that this is not a very good approximation for moderate values of s, t, u, and to improve on it remains a major task for theory. To do this in a consistent way is not trivial since gauge invariance requires one to treat the external quarks in the hard scattering as on-shell, or to take into account additional gluons exchanged between the hard and soft subprocesses.

An important feature of the soft overlap mechanism is that in a limited but rather wide region of t it can produce an approximate power behaviour of the form factors $F_1(t)$, $R_V(t)$ and $R_A(t)$ of the form t^{-2}, which coincides with the power behaviour obtained in the leading-twist hard scattering approximation. Such a behaviour of the Compton form factors translates into an approximate power behaviour of $d\sigma/dt$ like s^{-6} at fixed t/s and thus again mimics leading-twist power behaviour, which is approximately seen in the data. In figure 9.27(b) we compare the result from (9.59) and the Compton form factors modelled in [35] with data from [135], where we required $-t$ and $-u$ to be above 2.3 GeV^2. Shown is the ratio between $d\sigma/dt$ and the Klein–Nishina cross section $d\sigma_{KN}/dt$, which up to proton mass terms is obtained from (9.59) by setting $R_V = R_A = 1$ and $R_T = 0$ (see [139]). The quality of data obtained in present experiments is illustrated in figure 9.29, where the curves

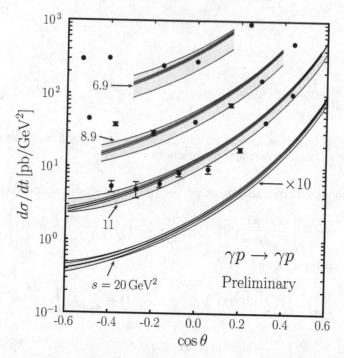

Figure 9.29. Preliminary data from Hall A at JLab [138] for wide-angle Compton scattering at different values of s given in the plot. θ is the scattering angle in the centre-of-mass frame. The theoretical curves (see text) are restricted to the region where $-t$ and $-u$ are above $2.5\,\mathrm{GeV}^2$, and the error bands reflect theoretical uncertainties as explained for figure 36 in [59]. The prediction for $s = 20\,\mathrm{GeV}^2$ has been scaled by a factor 10 and could be tested with the 12 GeV upgrade at JLab (see section 9.8).

show a calculation in the soft handbag mechanism including $O(\alpha_s)$ corrections (see below). The Compton form factors have been obtained from a parametrization of GPDs fitted to the electromagnetic form factors and the valence quark densities of the nucleon [59].

Because of the kinematical prefactors, the cross section (9.59) is mainly sensitive to the vector form factor R_V. Access to R_A and R_T is provided by polarization observables like the helicity correlation K_{LL} between the incoming photon and the outgoing proton and the correlation K_{LS} between the helicity of the incoming photon and the transverse polarization of the recoil proton in the scattering plane. With the definitions given in [140] one finds in the soft handbag approximation

$$K_{LL}\,\frac{d\sigma}{dt} = \frac{2\pi\alpha_{\mathrm{em}}^2}{s^2}\frac{s^2-u^2}{|su|}\left(R_V(t) - \frac{t}{s+\sqrt{-su}}R_T(t)\right)R_A(t),$$

$$K_{LS}\,\frac{d\sigma}{dt} = \frac{2\pi\alpha_{\mathrm{em}}^2}{s^2}\frac{s^2-u^2}{|su|}\frac{\sqrt{-t}}{2m}\left(\frac{4m^2}{s+\sqrt{-su}}R_V(t) - R_T(t)\right)R_A(t).$$

$$(9.61)$$

While K_{LL} is mainly sensitive to R_A/R_V, the ratio K_{LS}/K_{LL} depends only on R_T/R_V and thus is sensitive to the tensor form factor R_T, which describes proton helicity flip in analogy to the Pauli form factor. A variety of other polarization observables have been studied, and the soft handbag mechanism makes a number of predictions that can be experimentally tested [140]. Data from Hall A at JLab give $K_{LL} = 0.678 \pm 0.083(\text{stat}) \pm 0.040(\text{syst})$ at $s = 7$ GeV^2 and $-t = 4$ GeV^2 [141], which is well in the range of the model predictions in [140].

The $O(\alpha_s)$ corrections to the hard scattering process have also been evaluated [140]. The structure of the singularities appearing in this calculation is such that they can be absorbed by the renormalization of the Compton form factors, suggesting that a more rigorous treatment of the mechanism in the framework of factorization should be possible. After renormalization the $O(\alpha_s)$ corrections are found to be rather moderate, at the level of 10% for experimentally relevant kinematics.

The soft handbag mechanism can equally be applied to wide-angle Compton scattering with an off-shell initial photon, provided its virtuality Q^2 is not much bigger than the hard scale set by s. The relevant calculations are readily generalized from the case of real photons and the greater number of observables allows for further tests of the reaction mechanism [142]. Application of the same mechanism to the production of a meson at large s, t, u is also possible, but comes with the additional complexity of the transition from the initial photon to the meson [143]. The soft overlap mechanism for wide-angle Compton scattering has an analogue for processes in the crossed channel, meson or baryon production from two photons and photon pair production in $p\bar{p}$ collisions at large s, t, u. Its calculation leads to generalized distribution amplitudes at large s, and admits a rich phenomenology which so far has been quite successfully applied to exclusive channels in $\gamma\gamma$ annihilation (see [144] and references therein).

9.8 Future facilities for GPD physics

The knowledge of GPDs obtained by 2007 is still very limited and does not result from a dedicated experimental program. A large effort is, however, underway to improve upon existing data and to extend the kinematical reach of hard exclusive processes. Its focus is on DVCS, from which one can expect a major impact on GPD physics already at modest energies and photon virtualities. New equipment is being installed or is under construction to improve the experimental conditions for isolating the exclusive production of high-energy photons. The HERMES collaboration is building a detector to measure the recoil proton [145], the CLAS experiment is constructing a new forward-angle photon detector [146], and in Hall A at JLab a dedicated DVCS experiment is being carried out with a large photon calorimeter and a proton recoil detector [147]. At higher energies, there is an effort underway to measure DVCS and exclusive meson production in the COMPASS experiment

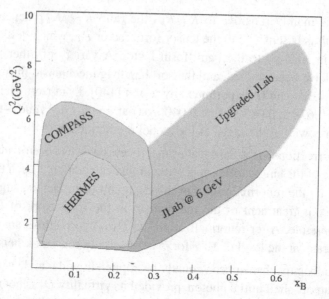

Figure 9.30. Kinematical coverage [150] for DVCS at fixed-target facilities, taking into account restrictions from both phase space and event statistics. The collider experiments at HERA cover a narrow slice at very small x_B on this linear scale and reach very high Q^2.

at CERN, which also has plans for a new recoil detection system [148]. The HERA collider experiments H1 and ZEUS will continue to take data on DVCS and vector-meson production at high Q^2 and small x_B. In the present run H1 is operating two spectrometers for high-acceptance detection of forward and very forward protons, with ξ respectively around 10^{-1} and 10^{-2} [149]. By 2007 it is expected that several of these experiments will have collected high-statistics data for beam-charge, beam-spin or target-spin asymmetries and possibly for the corresponding cross section differences in DVCS. These data will cover a significantly increased kinematical domain. On the same time scale, results on deeply virtual production of ρ^0, ρ^+, ω, ϕ, π^0, π^+ and η mesons can be expected. Figure 9.30 gives an overview of the kinematics covered in fixed-target experiments.

At the more extended time scale of 2010 and beyond, the energy upgrade of the JLab electron accelerator to 12 GeV, combined with luminosities from $\mathcal{L} = 10^{35}\,\mathrm{cm}^{-2}\,\mathrm{s}^{-1}$ for the planned CLAS upgrade to $\mathcal{L} = 10^{38}\,\mathrm{cm}^{-2}\,\mathrm{s}^{-1}$ for smaller-acceptance magnetic spectrometer setups, will increase the kinematical reach in Q^2, x_B and t, and significantly improve the statistical accuracy for all exclusive processes. As an example of the expected data quality we show in figure 9.31 projections for measuring the kinematical dependence of the beam-spin asymmetry in DVCS. Such measurements will constrain GPDs as functions of several variables to a much greater extent than currently available data, such as those shown in figure 9.20.

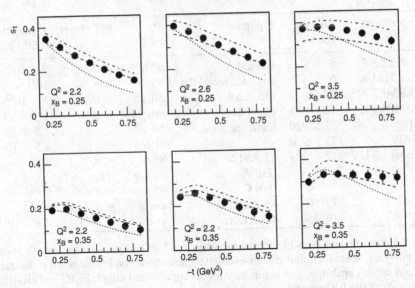

Figure 9.31. Projections for the beam-spin asymmetry in DVCS at the planned energy upgrade of the electron accelerator at JLab. The data are simulated for a beam energy of 11 GeV and the curves represent predictions with different GPD models. s_1 is obtained from a fit to $A_{LU}(\phi) = (s_1 \sin\phi + s_2 \sin 2\phi)/(1 + c_1 \cos\phi)$, where the $\cos\phi$ term has been introduced to account for the dominant ϕ dependence from the Bethe–Heitler cross section in the denominator of the asymmetry.

To exploit fully the physics potential of studying GPDs will require a dedicated machine with characteristics to cover a large range in x_B and Q^2, that is with sufficiently high $\sqrt{s}$ and luminosities that can compete with or surpass the capabilities available in present fixed-target experiments with large-acceptance detectors. Detectors should be able to measure exclusive processes in a wide range of t. These aims could be achieved with a new electron–ion collider (EIC) currently in discussion for a location in the USA [151]. The ELIC option (electron light-ion collider) would be an ideal instrument for GPD physics with exclusive processes at $x_B > 10^{-3}$ since it is designed for highest luminosities of up to 10^{35} cm^{-2} s^{-1} [152]. The option of electron–nucleus interactions (eRHIC at BNL) has lower luminosity but can reach smaller x_B values [153]. A summary of existing and possible future facilities for GPD physics is given in table 9.1.

9.9 Outlook

The theory of GPDs and exclusive processes where they occur has reached a level of sophistication similar to the description of inclusive DIS in terms of the usual parton densities. The leading-twist amplitudes for both Compton scattering and

Experiment	Energy (GeV)	Equipment	P_ℓ (%)	Pol. target	$\mathcal{L}$ (unpol.) (cm^{-2} s^{-1})
JLab/CLAS	6	em. cal.	$85/e^-$	p, d	2×10^{34}
JLab/Hall A	6	em. cal., recoil det.	$85/e^-$	^{3}He	10^{37}
HERMES	27.5	recoil det.	$50/e^\pm$	p, d	3×10^{32}
COMPASS	100–190	recoil det.	$80/\mu^\pm$	p, d	1.3×10^{32}
H1	27.5/920	forw. spectrom.	$50/e^\pm$		4.5×10^{31}
ZEUS	27.5/920		$50/e^\pm$		4.5×10^{31}
JLab@12GeV	12	CLAS12	$85/e^-$	p, d	10^{35}
		Hall A	$85/e^-$	^{3}He	10^{37}
		Hall C	$85/e^-$	p, d	10^{38}
ELIC	7/150	4π detector	yes/e^-	$p, d, ^3$He	10^{35}
eRHIC	10/250	4π detector	yes/$e^\pm$	$p, d, ^3$He	4.4×10^{32}

Table 9.1. Present and possible future facilities for GPD physics, with specification of beam energies, special equipment for exclusive measurements, lepton beam polarization, polarized target capability, and luminosity for an unpolarized target. ELIC and eRHIC are two options of the EIC project.

meson production are calculated to next-to-leading order in α_s, and the structure of $1/Q$ power corrections in Compton scattering is well understood. An outstanding task for theory remains to understand the pattern and size of next-to-leading order and power corrections in meson production sufficiently well for a quantitative interpretation of data with Q^2 of a few GeV^2. To determine the dependence of GPDs on the three variables x, ξ, t is a formidable challenge, and to date no model-independent procedure is known to achieve this in practice, with the exception of access to the ERBL region in double DVCS: see section 9.3.9. A realistic prospect is to investigate GPDs and the physics they encode with a combination of theoretical studies, non-perturbative calculations such as in lattice QCD and experimental data.

Measurements of Compton scattering have the potential to provide the most stringent and detailed constraints on GPDs. As we explained, data on meson production are more difficult to interpret quantitatively. Given their possibility to separate distributions for different quark flavours and gluons, they should, however, not be dismissed, and suitable cross section ratios may offer a practicable way to achieve this. To exploit fully the physics potential of GPDs, detailed measurements are required. Multi-dimensional spectra in x_B and t will help to disentangle the interplay of longitudinal and transverse degrees of freedom in the distribution of partons, and to lift the degeneracy among models that are able to describe overall rates. In addition, the combined dependence on Q^2 and x_B provides a crucial handle to test the power behaviour on Q^2 and its modification by logarithms predicted by the theory. To use GPDs as a tool for studying the spin structure of the nucleon, polarization measurements are mandatory. Key observables are the asymmetries for longitudinal and transverse target polarization in Compton scattering, and for transverse target polarization in meson production.

References

[1] X-D Ji, Journal of Physics G24 (1998) 1181
[2] A V Radyushkin, hep-ph/0101225
[3] K Goeke, M V Polyakov and M Vanderhaeghen, Progress in Particle and Nuclear Physics 47 (2001) 401
[4] M Diehl, Physics Reports 388 (2003) 41
[5] A V Belitsky and A V Radyushkin, hep-ph/0504030
[6] D E Soper, Physical Review D5 (1972) 1956
[7] R L Jaffe, hep-ph/9602236
[8] P A M Guichon, L Mossé and M Vanderhaeghen, Physical Review D68 (2003) 034018
[9] J Blümlein, J Eilers, B Geyer and D Robaschik, Physical Review D65 (2002) 054029
[10] M Diehl, European Physical Journal C19 (2001) 485
[11] E R Berger, F Cano, M Diehl and B Pire, Physical Review Letters 87 (2001) 142302
[12] X-D Ji, Physical Review Letters 78 (1997) 610
[13] D Müller, D Robaschik, B Geyer, F M Dittes and J Hořejši, Fortschritte der Physik 42 (1994) 101
[14] M Diehl, T Gousset, B Pire and O Teryaev, Physical Review Letters 81 (1998) 1782
[15] M V Polyakov, Nuclear Physics B555 (1999) 231
[16] J Blümlein, B Geyer and D Robaschik, Physics Letters B406 (1997) 161
[17] A V Belitsky, A Freund and D Müller, Nuclear Physics B574 (2000) 347
[18] A V Efremov and A V Radyushkin, Physics Letters B94 (1980) 245
[19] G P Lepage and S J Brodsky, Physics Letters 87B (1979) 359
[20] A V Radyushkin, Physical Review D56 (1997) 5524
[21] O V Teryaev, Physics Letters B510 (2001) 125
[22] M V Polyakov and C Weiss, Physical Review D60 (1999) 114017
[23] A V Belitsky, D Müller, A Kirchner and A Schäfer, Physical Review D64 (2001) 116002
[24] B C Tiburzi, Physical Review D70 (2004) 057504
[25] M Burkardt, Physical Review D62 (2000) 071503; Erratum, *ibid* D66 (2002) 119903
[26] J P Ralston and B Pire, Physical Review D66 (2002) 111501
[27] M Burkardt, International Journal of Modern Physics A18 (2003) 173
[28] X-D Ji, Physical Review Letters 91 (2003) 062001
[29] A V Belitsky, X-D Ji and F Yuan, Physical Review D69 (2004) 074014
[30] M Diehl, European Physical Journal C25 (2002) 223; Erratum, *ibid* C31 (2003) 277
[31] M Burkardt and D S Hwang, Physical Review D69 (2004) 074032
[32] M Diehl, T Feldmann, R Jakob and P Kroll, Nuclear Physics B596 (2001) 33; erratum, *ibid* B605 (2001) 647

[33] J C Collins and A Freund, Physical Review D59 (1999) 074009

[34] V M Braun, D Y Ivanov, A Schäfer and L Szymanowski, Nuclear Physics B638 (2002) 111

[35] M Diehl, T Feldmann, R Jakob and P Kroll, European Journal of Physics C8 (1999) 409

[36] P V Pobylitsa, Physical Review D65 (2002) 077504

[37] M Burkardt, Physics Letters B582 (2004) 151

[38] L Mankiewicz, G Piller and A Radyushkin, European Physical Journal C10 (1999) 307

[39] M Penttinen, M V Polyakov and K Goeke, Physical Review D62 (2000) 014024

[40] A V Belitsky and X-D Ji, Physics Letters B538 (2002) 289

[41] N Kivel, M V Polyakov and S Stratmann, nucl-th/0407052

[42] M C Birse, Journal of Physics G31 (2005) B7

[43] J W Chen and M J Savage, Nuclear Physics A735 (2004) 441

[44] S R Beane and M J Savage, Nuclear Physics A761 (2005) 259

[45] N Kivel and M V Polyakov, hep-ph/0203264

[46] M Diehl, A Manashov and A Schäfer, Physics Letters B622 (2005) 69

[47] M Strikman and C Weiss, Physical Review D69 (2004) 054012

[48] V Y Petrov *et al*, Physical Review D57 (1998) 4325

[49] M Wakamatsu and H Tsujimoto, Physical Review D71 (2005) 074001

[50] J Ossmann *et al*, Physical Review D71 (2005) 034011

[51] N Kivel, M V Polyakov and M Vanderhaeghen, Physical Review D63 (2001) 114014

[52] P Schweitzer, S Boffi and M Radici, Physical Review D66 (2002) 114004

[53] S Boffi, B Pasquini and M Traini, Nuclear Physics B649 (2003) 243

[54] S Boffi, B Pasquini and M Traini, Nuclear Physics B680 (2004) 147

[55] S Scopetta and V Vento, Physical Review D69 (2004) 094004

[56] S Scopetta and V Vento, Physical Review D71 (2005) 014014

[57] X-D Ji, W Melnitchouk and X Song, Physical Review D56 (1997) 5511

[58] M Burkardt, Physics Letters B595 (2004) 245

[59] M Diehl, T Feldmann, R Jakob and P Kroll, European Physical Journal C39 (2005) 1

[60] M Guidal, M V Polyakov, A V Radyushkin and M Vanderhaeghen, Physical Review D72 (2005) 054013

[61] M Göckeler *et al*, QCDSF Collaboration, Physical Review Letters 92 (2004) 042002

[62] P Hägler *et al*, LHPC and SESAM Collaborations, Physical Review D68 (2003) 034505

[63] P Hägler *et al*, LHPC and SESAM Collaborations, European Journal of Physics A24S1 (2005) 29

[64] P Hägler *et al*, LHPC and SESAM Collaborations, Physical Review Letters 93 (2004) 112001

[65] M Göckeler *et al*, QCDSF Collaboration, Nuclear Physics A755 (2005) 537

[66] S Dalley, Physics Letters B570 (2003) 191

[67] A V Belitsky, D Müller and A Kirchner, Nuclear Physics B629 (2002) 323

[68] I V Musatov and A V Radyushkin, Physical Review D61 (2000) 074027

[69] A D Martin, R G Roberts, W J Stirling and R S Thorne, European Physical Journal C4 (1998) 463

[70] E Leader, A V Sidorov and D B Stamenov, Physical Review D58 (1998) 114028

[71] A G Shuvaev, K J Golec-Biernat, A D Martin and M G Ryskin, Physical Review D60 (1999) 014015

[72] A Freund, M McDermott and M Strikman, Physical Review D67 (2003) 036001

[73] J C Collins, L Frankfurt and M Strikman, Physical Review D56 (1997) 2982

[74] J C Collins, hep-ph/9907513

[75] X-D Ji, Physical Review D55 (1997) 7114

[76] E R Berger, M Diehl and B Pire, European Physical Journal C23 (2002) 675

[77] M Guidal and M Vanderhaeghen, Physical Review Letters 90 (2003) 012001

[78] A V Belitsky and D Müller, Physical Review Letters 90 (2003) 022001

[79] A Freund and M McDermott, European Physical Journal C23 (2002) 651

[80] B Lehmann-Dronke, A Schäfer, M V Polyakov and K Goeke, Physical Review D63 (2001) 114001

[81] P di Nezza and R Fabbri, HERMES Collaboration, AIP Conference Proceedings 675 (2003) 313

[82] A V Belitsky and D Müller, Physics Letters B513 (2001) 349

[83] D Y Ivanov, L Szymanowski and G Krasnikov, JETP Letters 80 (2004) 226

[84] L N Hand, Physical Review 129 (1963) 1834

[85] I V Anikin, B Pire and O V Teryaev, Physical Review D62 (2000) 071501

[86] A V Belitsky and D Müller, Nuclear Physics B589 (2000) 611

[87] N Kivel and L Mankiewicz, Nuclear Physics B672 (2003) 357

[88] N Kivel and L Mankiewicz, European Physical Journal C21 (2001) 621

[89] L Mankiewicz and G Piller, Physical Review D61 (2000) 074013

[90] I V Anikin and O V Teryaev, Physics Letters B554 (2003) 51

[91] D Y Ivanov and R Kirschner, Physical Review D58 (1998) 114026

[92] J Gronberg et al, CLEO Collaboration, Physical Review D57 (1998) 33

[93] M Diehl, P Kroll and C Vogt, European Physical Journal C22 (2001) 439

[94] R Jakob and P Kroll, Physics Letters B315 (1993) 463; Erratum, *ibid* B319 (1993) 545

[95] R Jakob, P Kroll and M Raulfs, Journal of Physics G22 (1996) 45

[96] M Vanderhaeghen, P A M Guichon and M Guidal, Physical Review D60 (1999) 094017

[97] L Frankfurt, W Koepf and M Strikman, Physical Review D57 (1998) 512

[98] S V Goloskokov, P Kroll and B Postler, hep-ph/0308140

[99] A Donnachie, J Gravelis and G Shaw, European Physical Journal C18 (2001) 539

[100] F Cano and J-M Laget, Physics Letters B551 (2003) 317; Erratum, *ibid* B571 (2003) 250

[101] C Adloff et al, H1 Collaboration, Physics Letters B517 (2001) 47

[102] A Airapetian et al, HERMES Collaboration, Physical Review Letters 87 (2001) 182001

[103] S Stepanyan et al, CLAS Collaboration, Physical Review Letters 87 (2001) 182002

[104] M Vänttinen and L Mankiewicz, Physics Letters B440 (1998) 157

[105] M G Ryskin, R G Roberts, A D Martin and E M Levin, Zeitschrift für Physik C76 (1997) 231

[106] D Y Ivanov, A Schäfer, L Szymanowski and G Krasnikov, European Physical Journal C34 (2004) 297

[107] M G Ryskin, Zeitschrift für Physik C57 (1993) 89

[108] S J Brodsky et al, Physical Review D50 (1994) 3134

[109] M Diehl, T Gousset, B Pire and J P Ralston, Physics Letters B411 (1997) 193

[110] P Bertin et al, Jefferson Lab Experiment E-03-106

[111] A V Belitsky and D Müller, Physical Review D68 (2003) 116005
[112] F Ellinghaus, R Shanidze and J Volmer, HERMES Collaboration, AIP Conference Proceedings 675 (2003) 303
[113] H Avakian and L Elouadrhiri, CLAS collaboration, private communication
[114] A D Martin, R G Roberts, W J Stirling and R S Thorne, Physics Letters B531 (2002) 216
[115] F Ellinghaus, HERMES Collaboration, Nuclear Physics A711 (2002) 171
[116] F Ellinghaus, hep-ex/0410094
[117] S Chekanov et al, ZEUS Collaboration, Physics Letters B573 (2003) 46
[118] S Glazov, ZEUS Collaboration, *Proceedings of DIS'05*, W H Smith and S R Dean eds, AIP Conference Proceedings 792 (2005)
[119] C Adloff et al, H1 Collaboration, European Physical Journal C13 (2000) 371
[120] A Kreisel, hep-ex/0208013
[121] C Hadjidakis et al, CLAS collaboration, Physics Letters B605 (2005) 256
[122] D G Cassel et al, Physical Review D24 (1981) 2787
[123] M Derrick et al, ZEUS Collaboration, Physics Letters B356 (1995) 601
[124] J Breitweg et al, ZEUS Collaboration, European Physical Journal C6 (1999) 603
[125] B Clerbaux, hep-ph/9908519
[126] A Airapetian et al, HERMES Collaboration, European Physical Journal C17 (2000) 389
[127] S Chekanov et al, ZEUS Collaboration, Nuclear Physics B718 (2005) 3
[128] S Chekanov et al, ZEUS Collaboration, Nuclear Physics B695 (2004) 3
[129] M Diehl and A V Vinnikov, Physics Letters B609 (2005) 286
[130] A Airapetian et al, HERMES Collaboration, Physics Letters B535 (2002) 85
[131] C Hadjidakis, D Hasch and E Thomas, HERMES Collaboration, International Journal of Modern Physics A20 (2005) 593
[132] S J Brodsky and G P Lepage, in *Perturbative Quantum Chromodynamics*, A H Mueller ed, World Scientific, Singapore, (1989) p 93
[133] J Bolz et al, Zeitschrift für Physik C66 (1995) 267
[134] A F Sill et al, Physical Review D48 (1993) 29
[135] M A Shupe et al, Physical Review D19 (1979) 1921
[136] T C Brooks and L J Dixon, Physical Review D62 (2000) 114021
[137] A V Radyushkin, Physical Review D58 (1998) 114008
[138] C Hyde-Wright, A Nathan and B Wojtsekhowski, Jefferson Lab Experiment E-99-114, private communication
[139] M Diehl, T Feldmann, H W Huang and P Kroll, Physical Review D67 (2003) 037502
[140] H W Huang, P Kroll and T Morii, European Physical Journal C23 (2002) 301; Erratum, *ibid* C31 (2003) 279
[141] D J Hamilton et al, Jefferson Lab Hall A Collaboration, Physical Review Letters 94 (2005) 242001
[142] M Diehl, T Feldmann, R Jakob and P Kroll, Physics Letters B460 (1999) 204
[143] H W Huang, R Jakob, P Kroll and K Passek-Kumerički, European Physical Journal C33 (2004) 91
[144] M Diehl, Nuclear Physics B (Proceedings Supplement) 126 (2004) 271
[145] HERMES Collaboration, *The HERMES Recoil Detector*, Technical Design Report, DESY PRC 02-01, HERMES 02-003
[146] V Burkert et al, Jefferson Lab Experiment E-01-113
[147] P Bertin et al, Jefferson Lab Experiment E-00-100
[148] N d'Hose et al, hep-ex/0212047

[149] P Van Mechelen, hep-ex/0203029
[150] L Cardman *et al*, *The Conceptual Design Report for the CEBAF 12 GeV Upgrade*, Jefferson Lab (2004)
[151] R Holt *et al*, *The Electron Ion Collider: A high luminosity probe of the partonic substructure of nucleons and nuclei*, Brookhaven Report BNL-68933 (2002)
[152] L Merminga and Y Derbenev, AIP Conference Proceedings 698 (2003) 811
[153] R Milner, Proceedings of DIS 04, Štrbské Pleso, Slovakia, 14–18 April 2004, p 173

10

Quark–hadron duality

R Ent

At asymptotically high energies, the property of QCD known as asymptotic free-dom, in which quarks interact feebly at short distances, allows one to calculate hadronic observables in terms of expansions in the strong coupling constant g_s, or more commonly $\alpha_s = g_s^2/4\pi$. The small value of α_s at large momentum scales (or short distances) makes possible an efficient description of phenomena in terms of quarks and gluons or, more generally, partons.

At low-momentum scales, on the other hand, where α_s is large (see [1] for example), the effects of confinement make strongly-coupled QCD highly non-perturbative. Here, it is more efficient to work with collective degrees of freedom, the physical hadrons, that is mesons and baryons.

Despite the apparent dichotomy between the parton and hadron regimes, in nature there exist instances where the behaviour of low-energy cross sections, averaged over appropriate energy intervals, closely resembles that at asymptotically high energies, calculated in terms of quark–gluon degrees of freedom. This phenomenon is referred to as *quark–hadron duality*, and reflects the relationship between con-finement and asymptotic freedom, and the transition from perturbative to non-perturbative regimes in QCD.

The observation of such a non-trivial relationship between inclusive electron–nucleon scattering cross sections at low energy, in the region dominated by the nucleon resonances, and that in the deep-inelastic scattering (DIS) regime at high energy predates QCD. While analysing the data from the early DIS experiments at SLAC, Bloom and Gilman observed [2,3] that the inclusive structure function at low hadronic final-state mass, W, generally follows a scaling curve that describes high-W data, which the resonance structure function averages. Initial interpretations of this duality used the theoretical tools available at the time, namely finite-energy sum

Electromagnetic Interactions and Hadronic Structure, eds Frank Close, Sandy Donnachie and Graham Shaw. Published by Cambridge University Press. © Cambridge University Press 2007

rules (FESRs) (see chapter 5), or consistency relations between hadronic amplitudes inspired by the developments in Regge theory that occurred in the 1960s [4,5].

Following the advent of QCD in the early 1970s, Bloom–Gilman duality was reformulated [6,7] in terms of an operator product (or 'twist') expansion of moments of structure functions. This allowed a systematic classification of terms responsible for duality and its violation in terms of so-called 'higher-twist' operators, which describe long-range interactions between quarks and gluons. Ultimately, however, this description fell short of adequately explaining *why* particular multi-parton correlations were suppressed, and *how* the physics of resonances gave way to scaling. From the mid-1970s the subject was largely forgotten for almost two decades, as attention turned from the complicated resonance region to the more tractable problem of calculating higher-order perturbative corrections to parton distributions, and accurately describing their Q^2 evolution.

The availability of high-luminosity (polarized) beams, together with polarized targets, has allowed one to revisit Bloom–Gilman duality at a much more quantitative level than previously possible, and impressive amounts of data of unprecedented quantity and quality have now been compiled in the resonance region and beyond. One of the striking findings [8] is that Bloom–Gilman duality appears to work exceedingly well, down to Q^2 values of 1 GeV^2 or less, which is considerably lower than previously believed. In parallel, there has been a growing realization that understanding the resonance region in inelastic scattering, and the interplay between resonances and scaling in particular, represents a critical gap which must be filled if one is to fathom fully the nature of the quark–hadron transition in QCD. This has led to a resurgence of interest in questions about the origin of quark–hadron duality.

10.1 Duality in hadronic reactions

The decade or so preceding the development of QCD saw tremendous effort devoted to describing hadron interactions in terms of S-matrix theory and self-consistency relations. One of the profound discoveries of that era was the remarkable relationship between low-energy hadronic cross sections and their high-energy behaviour, in which the former on average appears to mimic certain features of the latter.

Historically, duality in strong interaction physics represented the relationship between the description of hadronic scattering amplitudes in terms of s-channel resonances at low energies, and t-channel Regge poles at high energies, as illustrated in figure 10.1. (More comprehensive discussions of Regge phenomenology can be found, for example, in the book of Collins [4], or in the more recent account by Donnachie *et al* [5]. A summary of the relevant aspects is given in chapter 5.)

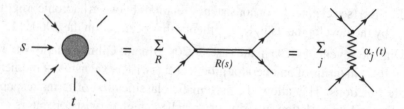

Figure 10.1. Dual descriptions of the scattering process, in terms of a sum over
s-channel resonances $R(s)$, and in terms of t-channel Reggeon exchanges $\alpha_j(t)$

At low energies, one expects the scattering amplitude to be dominated by just a
few resonance poles, R. As s increases, however, the density of resonances in each
partial wave, as well as the number of partial waves itself, increases, making it
harder to identify contributions from individual resonances. Therefore, at high s it
becomes more useful to describe the scattering amplitude in terms of a t-channel
partial-wave series, allowing the amplitude to be written as a sum of t-channel Regge
poles and cuts. At small $|t|$ the Regge trajectories are consistent with linearity:

$$\alpha(t) = \alpha(0) + \alpha' t \tag{10.1}$$

and amplitudes behave as

$$\mathcal{A}(s, t) \sim s^{\alpha(t)}, \qquad s \to \infty. \tag{10.2}$$

This implies that at large s the total cross section behaves as $\sigma \sim s^{\alpha(0)-1}$.

While the s- and t-channel partial-wave sums describe the low- and high-energy
behaviour of scattering amplitudes, respectively, an important question confronting
hadron physicists of the 1960s was how to merge these descriptions. This was especially challenging at intermediate s, where the amplitudes approach their smooth
Regge asymptotic behaviour, but some resonance structures still remain.

Progress towards synthesizing the two descriptions came with the development of
FESRs, which are generalizations of superconvergence relations in Regge theory
[9] relating dispersion integrals over the amplitudes at low energies to high-energy
parameters. The formulation of FESRs stemmed from the sum rule of Igi [10],
which used dispersion relations to express the crossing symmetric πN forward
scattering amplitude in terms of its high-energy behaviour. An implicit assumption
here is that beyond a sufficiently large energy $\nu > \bar{\nu}$, the scattering amplitude can
be represented by its asymptotic form, $\mathcal{A}_{\mathbb{R}}$, calculated within Regge theory [11].
The resulting sum rules relate functions of the high-energy parameters to dispersion
integrals which depend on the amplitude over a finite range of energies.

An important early application of FESRs was made for the case of πN scattering
amplitudes. In their seminal analysis, Dolen et al [12] observed that summing

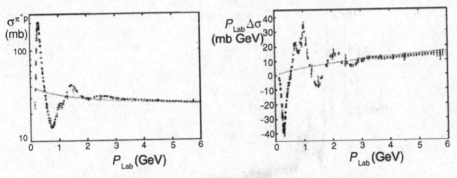

Figure 10.2. Total $\pi^+ p$ (left) and isovector-exchange πp cross section $p_{Lab}\Delta\sigma = p_{Lab}(\sigma^{\pi^+ p} - \sigma^{\pi^- p})$ (right), as a function of laboratory momentum, p_{Lab}, compared with Regge fits to high-energy data. (Adapted from [5].)

over contributions of s-channel resonances yields a result which is approximately equal to the leading (ρ) pole contribution obtained from fits to high-energy data, extrapolated down to low energies.

The original duality hypothesis embodied in the FESRs was not complete. It did not include pomeron ($\mathbb{P}$) exchange that was introduced in Regge theory to describe the behaviour of total and elastic cross sections at large s [4,5]. Since the known mesons lie on Regge trajectories with intercepts $\alpha_{\mathbb{R}}(0) < 1$, from (10.2) the resulting cross sections will obviously decrease with s. To obtain approximately constant cross sections at large s requires an intercept $\alpha_{\mathbb{P}}(0) \approx 1$. While there are no known mesons on such a trajectory, the exchange of a pomeron (which can be modelled in QCD through the exchange of two or more gluons) is introduced as an effective description of the high-energy behaviour of cross sections.

A generalization of the s- and t-channel duality to include contributions from both resonances and the non-resonant background upon which the resonances are superimposed was suggested by Harari [13] and Freund [14].

In this 'two-component duality', resonances are dual to the non-diffractive Regge pole exchanges, while the non-resonant background is dual to pomeron exchange [15],

$$A(s,t) = \sum_{res} A_{res}(s,t) + A_{bkgd}(s,t) \qquad (10.3)$$

$$= \sum_{\mathbb{R}} A_{\mathbb{R}}(s,t) + A_{\mathbb{P}}(s,t). \qquad (10.4)$$

The data on $\pi^\pm p$ scattering in figure 10.2 show pronounced resonance structures at small laboratory momenta, $p_{Lab} < 2$–3 GeV, which oscillate around the Regge fit to high-energy data, with the amplitude of the oscillations decreasing with increasing

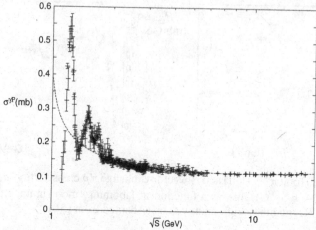

Figure 10.3. Total inclusive photoproduction cross section data for the proton as a function of the centre of mass energy, $\sqrt{s}$, compared with a parametrization (dashed curve) of the high-energy data. From [18].

momenta. Averaging the resonance data over small energy ranges thus exposes a semi-local duality between the s-channel resonances and the Regge fit. Since both the non-diffractive (isovector exchange, right) and total (left) cross sections satisfy duality, then so must the diffractive, $\mathbb{P}$-exchange component.

For the case of electroproduction, the two-component duality model has immediate application in DIS. In inclusive electroproduction from the nucleon, the behaviour of the cross sections at large $s \equiv W^2 = m_p^2 + Q^2(\omega - 1)$, where $\omega = 2m_p\nu/Q^2$, corresponds to the $\omega \to \infty$ behaviour of structure functions. Two-component duality therefore suggests a correspondence between resonances and valence quarks. At fixed Q^2, large $\omega \sim$ large s and the behaviour of the structure function is given by poles on the f_2-meson and a_2-meson Regge trajectories,

$$F_2^{val}(\omega) \sim \omega^{\alpha_{\mathbb{R}}(0)-1},\tag{10.5}$$

with the background dual to sea quarks or gluons, for which the large-ω behaviour is determined by pomeron exchange,

$$F_2^{sea}(\omega) \sim \omega^{\alpha_{\mathbb{P}}(0)-1}.\tag{10.6}$$

A dual model of DIS based on Regge calculus was developed by Landshoff and Polkinghorne [16] to describe the early DIS data. The natural extension of using Regge language to describe electroproduction can be dramatically shown in the total photon–proton cross section, $\sigma^{\gamma p}$, at low ν, as illustrated in figure 10.3. Oscillations around the high-energy behaviour can be readily seen, where the high-energy 'scaling' curve here is a fit to the large-s data by Donnachie and Landshoff [17].

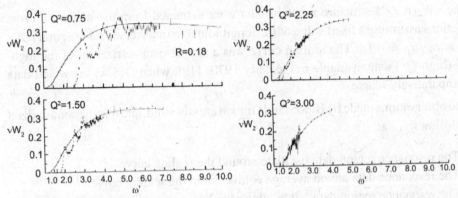

Figure 10.4. Early proton νW_2 structure function data in the resonance region, as a function of ω', compared to a smooth fit to the data in the scaling region at larger Q^2. The resonance data were obtained at the indicated kinematics, with Q^2 in GeV^2, with the longitudinal to transverse ratio $R = 0.18$. (Adapted from [3].)

They use a Regge-inspired model in which the total γp cross section is parametrized by the sum of diffractive and non-diffractive components,

$$\sigma^{\gamma p} = X \, (2m_p\nu)^{\alpha_{\mathbb{P}}-1} + Y \, (2m_p\nu)^{\alpha_{\mathbb{R}}-1}, \tag{10.7}$$

where for real photons one has $2m_p\nu = s - m_p^2$, with s the total γp centre-of-mass energy squared. The exponents $\alpha_{\mathbb{P}} = 1.0808$ and $\alpha_{\mathbb{R}} = 0.5475$ are determined by fitting the pp and $p\bar{p}$ total cross section data and the coefficients X and Y are given by $X = 0.0677$ mb and $Y = 0.129$ mb. Although the parameters were fitted to the data for $\sqrt{s} > 6$ GeV, the fit appears on average to go through the resonance data at low $\sqrt{s}$ (even at $Q^2 = 0$).

10.2 Bloom–Gilman duality

By examining the early inclusive electron–proton scattering data from SLAC, Bloom and Gilman observed [2,3] a remarkable connection between the structure function $\nu W_2(\nu, Q^2)$ in the nucleon-resonance region and that in the DIS continuum. The resonance structure function was found to be equivalent on average to the DIS one, with the averages obtained over the same range in the scaling variable

$$\omega' = \frac{2M\nu + M^2}{Q^2} = 1 + \frac{W^2}{Q^2} = \omega + \frac{M^2}{Q^2}. \tag{10.8}$$

While the physical interpretation of this modified scaling variable was not clear at the time, it did naturally allow for the direct comparison of data at higher W^2 to data at lower W^2, over a range of Q^2. The original data on the proton structure function $\nu W_2(\nu, Q^2)$ in the resonance region are illustrated in figure 10.4 for several values of Q^2 from 0.75 GeV^2 to 3 GeV^2.

The $\nu W_2(\nu, Q^2)$ structure function data were extracted from the measured cross sections assuming a fixed value of the longitudinal to transverse cross section ratio, $R = \sigma_L/\sigma_T = 0.18$. The scaling curve was a simple parametrization of the high-W (high-Q^2) data available in the early 1970s [19], when DIS was new and data comparatively scarce.

The observations made by Bloom and Gilman are still valid, and may be summarized as follows:

- The resonance region data oscillate around the scaling curve.
- The resonance data are on average equivalent to the scaling curve.
- The resonance region data 'slide' along the DIS curve with increasing Q^2.

These observations led Bloom and Gilman to make the far-reaching conclusion that *'the resonances are not a separate entity but are an intrinsic part of the scaling behaviour of νW_2'* [2].

In order to quantify their observations, Bloom and Gilman drew on the work on duality in hadronic reactions to determine a FESR equating the integral over ν of νW_2 in the resonance region, to the integral over ω' of the scaling function [2],

$$\frac{2M}{Q^2} \int_0^{\nu_m} d\nu \, \nu W_2(\nu, Q^2) = \int_1^{1+W_m^2/Q^2} d\omega' \, \nu W_2(\omega'). \qquad (10.9)$$

Here the upper limit on the ν integration, $\nu_m = (W_m^2 - m_p^2 + Q^2)/2m_p$, corresponds to the maximum value of $\omega' = 1 + W_m^2/Q^2$, where $W_m \approx 2\,\text{GeV}$, so that the integral of the scaling function covers the same range in ω' as the resonance region data. The FESR (10.9) allows the area under the resonances in figure 10.4 to be compared to the area under the smooth curve in the same ω' region to determine the degree to which the resonance and scaling data are equivalent.

A reanalysis of the resonance region and quark–hadron duality within QCD was performed by De Rújula *et al* [6,7,20], who reinterpreted Bloom–Gilman duality in terms of moments $M_2^{(n)}(Q^2)$ of the F_2 structure function. For $n = 2$ one recovers the analogue of (10.9) by replacing the νW_2 structure function on the right-hand side by the asymptotic structure function, $F_2^{asy}(x)$, so that the FESR can be written in terms of moments as

$$M_2^{(2)}(Q^2) = \int_0^1 dx \, F_2^{asy}(x). \qquad (10.10)$$

Since these moments are integrals over all x, at fixed Q^2, they contain contributions from both the DIS continuum and resonance regions. At large Q^2 the moments are saturated by the former; at low Q^2, however, they are dominated by the resonance contributions. One may expect therefore a strong Q^2 dependence in the low-Q^2

moments arising from the $1/Q^2$ power behaviour associated with the exclusive resonance channels.

Empirically, one observed only a slight difference, consistent with logarithmic violations of scaling behaviour in Q^2, between moments obtained at $Q^2 = 10$ GeV^2, and those at $Q^2 \approx 2$ GeV^2 that have a contribution of about 30% from resonances. The equivalence of the moments of structure functions at high Q^2 with those in the resonance-dominated region at low Q^2 is usually referred to as 'global duality'. If the equivalence of the averaged resonance and scaling structure functions holds over restricted regions in W, or even for individual resonances, a 'local duality' is said to exist.

Bloom and Gilman's observation that, with changing Q^2, the νW_2 structure function in the resonance region tracks a curve whose shape is the same as the scaling limit curve is expressly a manifestation of local duality, in that it occurs resonance by resonance. The scaling F_2 function becomes smaller at the larger values of the scaling variable, associated with higher values of Q^2. Therefore, the resonance transition form factors must decrease correspondingly with Q^2.

10.3 Duality in inclusive electron scattering

Since the original observations of Bloom–Gilman duality in inclusive structure functions, $F_2(x, Q^2)$ has become one of the best-measured quantities in lepton scattering, with measurements from laboratories around the world contributing to a global data set spanning over five orders of magnitude in x and Q^2. In parallel, with the advent of JLab, precise structure function data are now also available in the resonance region.

A sample of proton F_2 structure function data from JLab [8] in the resonance region is depicted in figure 10.5, where it is compared with fits to a large data set of higher-W and Q^2 data from the New Muon Collaboration (NMC) [21]. Figure 10.5 is in direct analogy to figure 10.4 above, where the Nachtmann variable $\xi = 2x/(1 + \sqrt{(1 + 4m_p^2 x^2/Q^2)})$ has replaced the more ad hoc variable ω' as a means to relate deep-inelastic data at high values of W^2 and Q^2 to data at the lower values of the resonance region, as well as to include proton target-mass corrections. It has been shown that ξ is the correct variable that systematically absorbs all target-mass corrections. These corrections are large when $Q^2 \approx m_p^2$ and need to be taken into account in studying QCD scaling violations in the nucleon [22,20].

The extraction of the F_2 structure function from cross section data can only proceed with some input for the ratio R of the longitudinal to transverse cross sections, since $F_2 \sim (d^2\sigma/d\Omega dE') \times (1 + R)/(1 + \epsilon R)$. At high Q^2 the scattering of longitudinal photons from spin-$\frac{1}{2}$ quarks is suppressed, and one expects $R \to 0$ as $Q^2 \to \infty$.

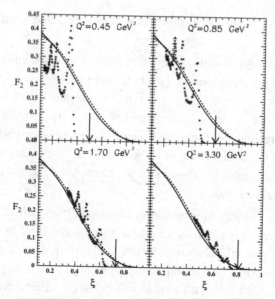

Figure 10.5. Proton $\nu W_2^p = F_2^p$ structure function data in the resonance region as a function of ξ, at $Q^2 = 0.45$, 0.85, 1.70, and 3.30 GeV^2 from JLab [8]. The resonance data were obtained with a longitudinal to transverse ratio $R = 0.20$. The arrows indicate the elastic point in ξ, corresponding to $x = 1$. The curves represent fits to DIS structure function data at the same ξ but higher (W^2, Q^2) from NMC [21] at $Q^2 = 5$ GeV^2 (dashed) and $Q^2 = 10$ GeV^2 (solid).

At low Q^2, however, R is no longer suppressed, and could be sizeable, especially in the resonance region and at large x.

The comparison shown in figure 10.5 has two shortcomings. A value of R is assumed and the presentation of the resonance data and DIS scaling curves has an inherent ambiguity. We will first address these before showing a quantification of the original Bloom–Gilman observation.

Until 2004 very little data on R existed in the region of the resonances, rendering reliable longitudinal/transverse (LT) separations impossible. The few measurements that existed below $Q^2 = 8$ GeV^2 in this region yielded $R < 0.4$, and had typical errors of 100% or more. This lack of knowledge of R was reflected in the choices of $R = 0.18$ and $R = 0.20$ for figures 10.4 and 10.5, respectively. These values simply reflected the average values of R known at the time. Precision LT-separated measurements of proton cross sections at JLab [23] have enabled detailed duality studies to be made in both of the unpolarized structure functions and their moments.

Beyond the lack of precise knowledge of R, the classic presentation of duality in electron–proton scattering, as depicted in figures 10.4 and 10.5, is also somewhat ambiguous in that resonance data at low Q^2 values are being compared to scaling curves at higher Q^2 values. It is difficult to evaluate precisely the equivalence of

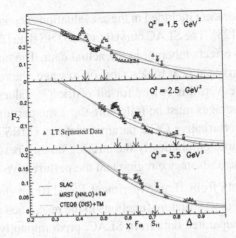

Figure 10.6. Proton F_2 structure function in the resonance region for several values of Q^2, as indicated. Data from JLab [23] compared with some parametrizations of the DIS data at the same Q^2 values (see text).

the two if Q^2 evolution is not taken into account. Furthermore, the resonance data and scaling curves, although at the same ξ or ω', are at different x and sensitive therefore to different parton distribution functions (PDFs). A more stringent test of the scaling behaviour of the resonances would compare the resonance data with fundamental scaling predictions for the *same* low-Q^2, high-x values as the data.

Such predictions are now commonly available from several groups around the world, for instance, the Coordinated Theoretical-Experimental Project on QCD (CTEQ) [24]; Martin, Roberts, Stirling, and Thorne (MRST) [25]; Glück, Reya, and Vogt (GRV) [26], and Alekhin [27]. These groups provide results from global QCD fits to a full range of hard scattering processes, such as lepton–nucleon DIS, prompt photon production, Drell–Yan measurements and jet production, to extract quark and gluon distribution functions (PDFs) for the proton. The idea of such global fitting efforts is to adjust the fundamental PDFs to bring theory and experiment into agreement for a wide range of processes. These PDF-based analyses include perturbative QCD radiative corrections which give rise to logarithmic Q^2 dependence of the structure function.

Comparison of resonance region data with PDF-based global fits allows the resonance-scaling comparison to be made at the same values of (x, Q^2), making the experimental signature of duality less ambiguous. Such a comparison is presented in figure 10.6 for LT-separated F_2 data from JLab experiment E94-110 [23], with the data bin-centred to the values $Q^2 = 1.5, 2.5$ and $3.5\,\text{GeV}^2$ indicated. The smooth curves in figure 10.6 are the perturbative QCD fits from the MRST [25] and CTEQ [24] collaborations, evaluated at the same Q^2 values as the data.

Target-mass corrections are included in these evaluations according to the prescription of Barbieri *et al* [28]. The SLAC curve is a fit to DIS data [29], which implicitly includes target-mass effects inherent in the actual data. The target-mass-corrected perturbative QCD curves appear to describe, on average, the resonance strength at each Q^2 value. Moreover, this is true for all of the Q^2 values shown, indicating that the resonance averages must be following the same perturbative Q^2 evolution which governs the perturbative QCD parametrizations (MRST and CTEQ). This demonstrates even more emphatically the striking duality between the nominally highly-non-perturbative resonance region and the perturbative scaling behaviour.

New LT-separated data from JLab experiment E94-110 for the proton transverse (F_1) and longitudinal (F_L) structure functions in the resonance region are shown in figure 10.7 [23]. LT-separated data from SLAC, predominantly in the DIS region, are also shown for comparison [30]. Where coincident, the JLab and SLAC data are in excellent agreement, providing confidence in the achievement of the demanding precision required of this type of experiment. In all cases, it is also interesting to note that the resonance and DIS data smoothly move toward one another in both x and Q^2.

The curves in figure 10.7 are from Alekhin's next-to-next-to-leading order (NNLO) analysis [27], including target-mass effects as in [20], and from the MRST NNLO analysis [25], with and without target-mass effects according to [28] included. It is clear that target-mass effects are required to describe the data. However, other than the target-mass corrections, no additional non-perturbative physics seems necessary to describe the *average* behaviour of the resonance region for $Q^2 > 1$ GeV^2. Furthermore, this is true for a range of different Q^2 values, indicating that the scaling curve describes as well the average Q^2 dependence of the resonance region. These results are analogous to those in figure 10.6 for the F_2 structure function, and are a manifestation of quark–hadron duality in the separated transverse and longitudinal channels.

Quark–hadron duality can be quantified by computing integrals of the structure function over x in the resonance region at fixed Q^2 values,

$$\int_{x_{th}}^{x_{res}} dx \, F_2(x, Q^2), \tag{10.11}$$

where x_{th} corresponds to the pion production threshold at the fixed Q^2, and $x_{res} = Q^2/(W_{res}^2 - M^2 + Q^2)$ indicates the x value at that same Q^2 where the traditional delineation between the resonance and DIS regions falls, namely $W \cong 2$ GeV. These integrals may then be compared to the analogous integrals of the 'scaling' structure function at the same Q^2 and over the same range of x. Alternatively, one can quantify local duality by computing such integrals for regions in x limited to the prominent resonance regions.

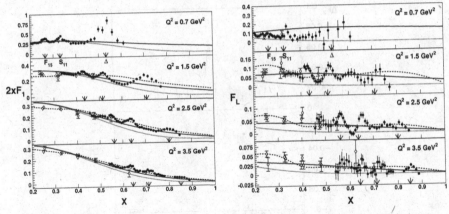

Figure 10.7. The purely transverse (left) and longitudinal (right) proton structure functions $2xF_1$ and F_L, measured in the resonance region (triangles) as a function of x, compared with existing high-precision DIS measurements from SLAC (squares). The curves are from Alekhin (dashed) [27], and from MRST [25], both at NNLO, with (dotted) and without (solid) target-mass effects included, as described in the text. The prominent resonance regions ($\Delta(1232)$, $S_{11}(1535)$, $F_{15}(1680)$) are indicated by the arrows.

The integrated perturbative strength appears equivalent to the resonance region strength to better than 5% above $Q^2 = 1 \,\mathrm{GeV}^2$. This is similarly true for all prominent resonance regions [8]. This shows unambiguously that duality is holding quite well on average in all of the unpolarized structure functions; the total resonance strength over a range in x is equivalent to the perturbative, PDF-based prediction.

If one assumes duality, it is also possible to obtain a scaling curve by averaging the resonance region data themselves. In this case one finds that the resonances oscillate around the fit to within 10%, even down to Q^2 values as low as $0.5 \,\mathrm{GeV}^2$ [8]. At low x and Q^2, such a duality-averaged curve yields a clear valence-like shape, which is in qualitative agreement with the neutrino/antineutrino data on the valence xF_3 structure function [31]. This suggests a unique sensitivity of the duality-averaged F_2 data at low Q^2 to valence quarks.

An alternative approach to quantifying the observation that the resonances average to the scaling curve has been used by Alekhin [27]. Here ΔF_2, the difference between the resonance structure function values and those of the scaling curve, is used to quantify duality, as shown in figure 10.8, where these differences are seen to oscillate around zero. Integrating ΔF_2 over the resonance region, the resonance and scaling regimes are found to be within 3% in all cases above $Q^2 = 1 \,\mathrm{GeV}^2$.

As already mentioned in section 10.2, the quantities most directly amenable to a QCD analysis are the moments of structure functions by means of the operator product expansion (OPE) [32]. According to the OPE, at large $Q^2 \gg \Lambda^2_{QCD}$

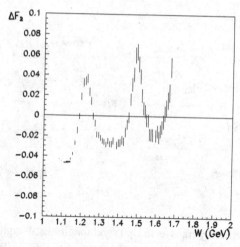

Figure 10.8. The difference ΔF_2 between proton F_2 structure function data (at $Q^2 \sim 1.5\,\mathrm{GeV}^2$) from JLab, and the scaling curve of [27] as a function of missing mass W. The integrated difference yields a value of -0.0012 ± 0.0066 for this particular W-spectrum.

the moments of the structure functions can be expanded in powers of $1/Q^2$. The coefficients in this expansion are matrix elements of quark and gluon operators corresponding to a certain *twist*, τ, defined as the mass dimension minus the spin, n, of the operator. For the second ($n = 2$) moment of the F_2 structure function, $M_2^{(2)}$ (see (10.10)) one has the expansion

$$M_2^{(2)}(Q^2) = \sum_{\tau=2}^{\infty} \frac{A_\tau^{(2)}(\alpha_s(Q^2))}{Q^{\tau-2}}, \tag{10.12}$$

where $A_\tau^{(2)}$ are the matrix elements with twist $\leq \tau$.

Asymptotically, as $Q^2 \to \infty$ the leading-twist ($\tau = 2$) terms dominate these moments. In the absence of perturbatively-generated corrections, these give rise to the Q^2 independence of the structure function moments, and hence are responsible for scaling.

The leading-twist terms correspond to diagrams such as that in figure 10.9(a), in which the virtual photon scatters incoherently from a single parton. The higher-twist terms in (10.12) are proportional to higher powers of $1/Q^2$ with coefficients which are matrix elements of local operators involving multi-quark or quark–gluon fields. Diagrammatically, these correspond to processes such as those depicted in figure 10.9(b) and (c).

The relation between the higher-twist matrix elements and duality in electron scattering was elucidated in the classic work of De Rújula, Georgi and Politzer [7,20]. The connection follows almost immediately from the definition of the moment

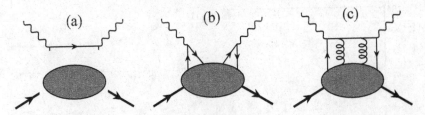

Figure 10.9. (a) Leading-twist ('handbag diagram') contribution to the structure function. (b) Higher-twist ('cat's ears') four-quark contributions. (c) Higher-twist quark–gluon interactions.

expansions in (10.12). For the F_2 structure function, the lowest moment, $M_2^{(2)}$, corresponds precisely to the Bloom–Gilman integral in (10.9), as already stated in section 10.2. At low Q^2 the moments display strong Q^2 dependence, violating both scaling and duality. In the OPE language this violation is associated with large corrections from the subleading, $1/Q^{\tau-2}$ higher-twist terms in (10.12).

At larger Q^2 the moments become independent of Q^2, as they would if they were given entirely by the scaling contribution. According to (10.12), this duality can only occur if the higher-twist contributions either are small or cancel. Duality is synonymous, therefore, with the suppression of higher twists, which in partonic language corresponds to the suppression of interactions between the scattered quark and the spectator system, as illustrated in figure 10.9(b) and (c). In other words, suppression of final-state interactions is a prerequisite for the existence of duality.

Note that since the resonances are bound states of quarks and gluons, they necessarily involve (an infinite number of) higher twists. At $Q^2 = 1$ GeV^2 approximately 70% of the cross section integral (or the $n = 2$ moment) comes from the resonance region, $W < W_{res}$. Despite this large resonant contribution, the resonances and the DIS continuum conspire to produce only about a 15% higher-twist correction at the same Q^2, as figure 10.10 (left) demonstrates. Here the total $M_2^{(2)}$ moment from proton measurements at JLab [33] is plotted versus Q^2, together with the leading-twist contribution calculated from the PDF parametrization of [25]. Remarkably, even though each bound-state resonance must be built up from a multitude of twists, when combined the resonances interfere in such a way that they closely resemble the leading-twist component.

To conclude this section, we present in figure 10.10(right) moments of new, *LT-separated*, spin-averaged, structure function data. Previous F_2 moments were constructed using assumed values for R [34,35]. One of the most remarkable features of the results in figures 10.10(right) is that the elastic-subtracted moments exhibit little or no Q^2 dependence even for $Q^2 < 1$ GeV^2. In the region where the moments

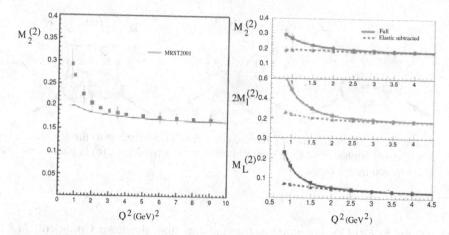

Figure 10.10. Left: total $n = 2$ moment of the proton F_2 structure function (squares) and the leading-twist contribution (solid line) [33]. Right: second ($n = 2$) moments of the F_2 (top), $2xF_1$ (centre) and F_L (bottom) structure functions, evaluated from the JLab E94-110 data [23]. The total moments are connected by solid lines, and elastic-subtracted moments by dashed lines.

are completely dominated by the nucleon resonances, the $n = 2$ moments of both of the unpolarized structure functions appear to behave just as in the DIS region at high Q^2. This phenomenological observation is even more striking when constructing similar Nachtmann moments [22]. In such moments the kinematical target-mass corrections (non-negligible at such low Q^2) are taken into account. This reduces even further the remaining Q^2 dependence of the structure function moments at low Q^2.

10.4 Scaling and duality in dynamical models

Although Bloom–Gilman duality for structure function moments at intermediate and high Q^2 can be analysed systematically within the OPE, an elementary understanding of the origins of duality for structure functions as a function of x or Q^2 is more elusive. This problem is closely related to the question of how to build up a scaling (Q^2-independent) structure function entirely out of resonances [36], each of which is described by a form factor that falls rapidly with increasing Q^2. The description of Bjorken scaling in DIS structure functions is most elegantly formulated within the QCD quark–parton model, which is justified on the basis of asymptotic freedom. On the other hand, the physical final state comprises entirely hadrons, so it must also be possible, in the general sense of quark–hadron duality, to describe the process in terms of hadronic degrees of freedom (resonances and their decays) alone.

One of the central mysteries in strong-interaction physics, and a key to the question of the origin of duality, is how scattering from bound (confined) states of quarks and gluons in QCD can be consistent with scaling, a property synonymous with scattering from free quarks.

10.4.1 Large-N_c limit

Perhaps the simplest, and most graphic, demonstration of the interplay between resonances and scaling is in QCD in the large-N_c limit. In the case of $q\bar{q}$ bound states, in this limit the hadronic spectrum consists entirely of infinitely narrow, non-interacting resonances of increasing mass. On the other hand, since no element of the perturbative QCD results for DIS depends on N_c, at the quark level one still obtains a smooth scaling structure function. Therefore in the large-N_c world duality must be invoked even in the scaling limit.

The derivation of a scaling function from large N_c resonances was demonstrated explicitly for the case of one space and one time dimension [37]. QCD in $1 + 1$ dimensions in the $N_c \to \infty$ limit, known as the 't Hooft model [38], is an exactly soluble field theory, in which all hadronic Green's functions are calculable in terms of quark degrees of freedom. (The essential simplification which allows one to solve the $1 + 1$ dimensional theory non-perturbatively is the freedom to choose gauges in which the gluon self-coupling vanishes.) In the large-N_c limit, even in lowest order, the exchange of a massless gluon between quarks corresponds to an attractive $q\bar{q}$ potential which rises linearly with r (compared with the $3 + 1$ dimensional case which gives rise to a Coulombic $1/r$ potential). Therefore confinement is an almost trivial consequence in the 't Hooft model. Furthermore, simply by power counting one can show that the theory is asymptotically free, which automatically leads to Bjorken scaling in structure functions.

In the more realistic case of QCD in $3 + 1$ dimensions one expects that $q\bar{q}$ bound states will remain narrow in the large-N_c limit, so that local duality must still be invoked. However, at finite N_c, resonances will acquire finite widths, and one can expect complications with mixing of resonant and non-resonant background contributions. Of course, confinement has not been proved in $3 + 1$ dimensions, rendering the above discussion suggestive but not rigorous. Hence, one usually resorts to quark models to learn how duality may arise in Nature.

10.4.2 Resonance parametrizations

In a more phenomenological approach, Domokos *et al* [39] showed that one could accommodate structure function scaling by summing over resonances parametrized by Q^2-dependent form factors.

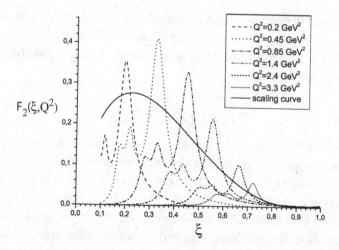

Figure 10.11. Resonance contributions to the proton F_2^p structure function versus the Nachtmann scaling variable ξ in the model of [42]. The solid line is a parametrization of DIS data [8].

Starting from a harmonic oscillator-like spectrum of nucleon excitations, the structure function F_2 was constructed by a sum of transition form factors weighted by kinematical factors. It was shown that the resonance summation, in this model, indeed yields a scaling function in the Bjorken limit. In addition, the correct $\omega' \to 1$ behaviour according to the Drell–Yan–West relation [40,41] was exhibited, with ω' the scaling variable introduced by Bloom and Gilman. Similar arguments were later used to derive scaling in spin-dependent and neutrino structure functions from sums over resonant excitations [39].

While this illustrates the compatibility of confinement and asymptotic scaling in DIS, it does not address the question of the origin of Bloom–Gilman duality at *finite* Q^2. Davidovsky and Struminsky [42] have constructed a phenomenological model of the structure functions in the resonance region in the spirit of the earlier work of [39], but with additional physical constraints for the threshold behaviour as $\mathbf{q} \to 0$, and the asymptotic behaviour as $Q^2 \to \infty$.

Summing over a total of 21 resonance states (smeared by a Breit–Wigner shape) in the isospin-$\frac{1}{2}$ and isospin-$\frac{3}{2}$ channels with masses $M_R > 2$ GeV, the total F_2 structure function is shown in figure 10.11 as a function of the Nachtmann scaling variable ξ. The $\Delta(1232)$ resonance clearly provides the largest contribution. The non-resonant background contribution here is relatively small, so that as Q^2 increases the Δ peak moves to larger ξ, following the general trend of the scaling curve. On the other hand, the higher-mass resonances lie about a factor 2 below the scaling curve at the Q^2 values shown, which reflects the absence of the non-resonant backgrounds which are relatively more important for the higher-mass resonances.

At lower ξ (higher W), a quantitative description of the data would require the inclusion of additional resonances beyond $M_R \sim 2$ GeV. This quickly becomes intractable as little phenomenological information exists on $N \to R$ transitions at high W, and indicates that a quark-level description may be more appropriate at these kinematics.

10.4.3 Harmonic oscillator model

Despite the challenges in describing the transition to scaling in terms of phenomenological form factors, it is nevertheless vital to understand how the dynamics of resonances gives way to scaling in the *preasymptotic*, finite-Q^2, region. This has been examined by several authors [36,43–46] in the context of particular dynamic models.

Isgur *et al* [36] simplified the problem by considering a spinless, charged 'quark' of mass m bound to an infinitely massive core via a harmonic oscillator potential. In this model they studied both the appearance of duality at low Q^2 and the onset of scaling at high Q^2. For the case of scalar photons, the inclusive structure function, $\mathcal{W}$, is given by a sum of squares of transition form factors (as in the models discussed above) weighted by appropriate kinematic factors [36],

$$\mathcal{W}(\nu, \mathbf{q}) = \sum_{N=0}^{N_{max}} \frac{1}{4E_0 E_N} |F_{0,N}(\mathbf{q})|^2 \delta(E_N - E_0 - \nu). \qquad (10.13)$$

The form factors $F_{0,N}$ represent transitions from the ground state to states characterized by the principal quantum number $N \equiv l + 2k$, where k is the radial and l the orbital quantum number, and the sum over states N goes up to the maximum, N_{max}, allowed at a given energy transfer ν. A related discussion which focuses on the response in the time-like region was given by Paris and Pandharipande [45].

The scaling function corresponding to the structure function in (10.13) is given by

$$S(u, Q^2) \equiv |\mathbf{q}| \, \mathcal{W} = \sqrt{\nu^2 + Q^2} \, \mathcal{W}, \qquad (10.14)$$

with dimensions [mass]$^{-2}$. The scaling variable u is defined as

$$u = \frac{1}{2m_q}(\sqrt{\nu^2 + Q^2} - \nu)\left(1 + \sqrt{1 + 4m_q^2/Q^2}\right), \qquad (10.15)$$

and takes into account both target-mass and quark-mass effects [28]. Note that the variable u in (10.15) is scaled by the quark mass, m_q, rather than the bound state mass, so that the range of u is between zero and infinity.

The structure function $S(u, Q^2)$ is shown in figure 10.12 for several finite values of Q^2, where for illustration the δ-functions have been smoothed by a Breit–Wigner

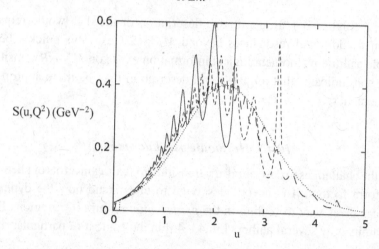

Figure 10.12. Onset of scaling for the structure function $\mathcal{S}(u, Q^2)$ as a function of u for $Q^2 = 0.5$ (solid), $Q^2 = 1$ (short-dashed), 2 (long-dashed) and 5 (dotted) GeV^2. The width for $N \geq 1$ has been arbitrarily set at $\Gamma_N = 100$ MeV with the elastic width set to $\Gamma_{N=0} = 30$ MeV. (From [36].)

shape with an arbitrary but small width, Γ_N. The resonance structure is quite evident in each of the low-Q^2 curves, with the amplitude of oscillation decreasing with increasing Q^2. As Q^2 increases, each of the resonances moves out towards higher u, as dictated by kinematics. The right-most peak in each of the curves corresponds to the elastic contribution. At $Q^2 = 0$, this is the only allowed state, and is equal to almost half of the asymptotic value of the integral over u. It remains rather prominent for $Q^2 = 0.5$ GeV^2, though most of the function is by this point built up of excited states, and it becomes negligible for $Q^2 \geq 2$ GeV^2.

For local duality to hold, the resonance 'spikes' would be expected to oscillate around the scaling curve and to average to it, once Q^2 is large enough. Remarkably, even the curves at lower Q^2 tend to oscillate around the scaling curve. The difference between the scaling function and the curve in figure 10.12 at 5 GeV^2 (dotted) is almost negligible. The asymptotic scaling function therefore straddles the oscillating resonance structure function in an apparently systematic manner. This is quite extraordinary given the very simple nature of the model, and points to the rather general nature of the phenomenon of duality.

10.4.4 Sum of squares vs square of sums

Simple models such as the one discussed above are valuable in providing physical insight into the dynamical origins of duality. However, one may wonder whether some of the qualitative features of duality and the onset of scaling could be a consequence of the restriction to scattering from a single quark charge. In general, if one neglects differences between the quark flavours, the magnitude of the structure function F_2 is proportional to the incoherent sum of the squares of the (quark and antiquark) constituent charges, $\sum_q e_q^2$. On the other hand, the summation over resonance transition form factors is implicitly driven by the coherently-summed square of constituent charges, $(\sum_q e_q)^2$, for each resonance. The basic question then arises: Why do the interference terms $\sum_{q \neq q'} e_q e_{q'}$ cancel or *how does the square of the sum become the sum of the squares?*

Close and Isgur [43] elucidated this problem by drawing attention to the necessary conditions for duality to occur for the general case of more than one quark charge. They considered a composite state made of two equal-mass scalar quarks with charges e_1 and e_2, at positions $\mathbf{r}_1$ and $\mathbf{r}_2$, respectively, interacting via a harmonic oscillator potential. This exposes the critical point that *at least one complete set of even- and odd-parity resonances must be summed over* for duality to hold [43] (see also [47]). An explicit demonstration of how this cancellation takes place in the SU(6) quark model is given below.

Similar results have also been obtained by Harrington [46], who performed a detailed study of the relationship between coherent and incoherent descriptions of the structure function within this model and the cancellation of the higher-twist interference terms. Summing over the orbital angular momentum for each N, the contributions to the structure function from a transition to the state N were shown to be proportional to $e_1^2 + e_2^2 + 2e_1 e_2 (-1)^N$, which illustrates how the contributions from alternate energy levels tend to cancel for the $e_1 e_2$ interference term.

10.4.5 SU(6) symmetry

The SU(6) spin-flavour symmetric-quark model serves as a useful basis for both visualizing the principles underpinning the phenomenon of duality and at the same time providing a reasonably close contact with phenomenology. Quark models based on SU(6) spin–flavour symmetry provide benchmark descriptions of baryon spectra, as well as transitions to excited N^* states.

In a series of classic early papers, Close, Gilman and collaborators [48–51] showed how the ratios of various DIS structure functions could be dual to sums over N^* resonances in the **56**-dimensional and **70**-dimensional representations of SU(6).

Since the nucleon ground-state wave function is totally symmetric, the only final-state resonances that can be excited have wave functions which are either totally symmetric or of mixed symmetry, corresponding to the positive parity $\mathbf{56^+}$ and negative parity $\mathbf{70^-}$ representations respectively [52]. The relative weightings of the $\mathbf{56^+}$ and $\mathbf{70^-}$ contributions are determined by assuming that the electromagnetic current is in a **35**-plet. Allowing only the non-exotic singlet **1** and **35**-plet representations in the t-channel, which corresponds to $q\bar{q}$ exchange, the reduced matrix elements for the **56**$^+$ and **70**$^-$ are constrained to be equal. In the t-channel these appear as $\gamma\gamma \to q\bar{q}$, while in the s-channel this effectively maps onto the leading-twist handbag diagram in figure 10.9(a), describing incoherent coupling to the same quark. Exotic exchanges require multi-quark exchanges, such as $qq\bar{q}\bar{q}$ in the t-channel and correspond to the 'cat's ears' diagram in figure 10.9(b). Physically, therefore, the appearance of duality in this picture is correlated with the suppression of exotics in the t-channel [48].

Although the s-channel sum was shown by Close and coworkers [48,50,51] to be dual for ratios of structure functions, this alone did not explain the underlying reason why any individual sum over states scaled. The microscopic origin of duality in the SU(6) quark model was more recently elaborated by Close and Isgur [43], who showed that the cancellations between the even- and odd-parity states found to be necessary for duality to appear, are realized through the destructive interference in the s-channel resonance sum between the **56**$^+$ and **70**$^-$ multiplets. Provided the contributions from the **56**$^+$ and **70**$^-$ representations have equal strength, this leads exactly to the scaling function proportional to $\sum_q e_q^2$.

Close and Isgur showed that in the SU(6) limit duality is also realized for F_L [43]. In general the interplay of magnetic and electric interactions will make the workings of duality non-trivial. In the $Q^2 \to 0$ limit both electric and magnetic multipoles will contribute and the interference effects can cause strong Q^2 dependence [48,49], such as that responsible for the dramatic change in sign of the lowest moment of g_1^p in the transition towards the Gerasimov–Drell–Hearn sum rule at $Q^2 = 0$. Close and Isgur suggest [43] that Bloom–Gilman duality will fail when the electric and magnetic multipoles have comparable strengths, although the precise Q^2 at which this will occur is unknown.

10.5 Duality in spin-dependent electron scattering

We have explored the transition between the parton and hadron regimes in unpolarized electron scattering and established the degree to which quark–hadron duality holds in the F_1 and F_2 structure functions. In principle, there should also exist kinematic regions in spin-dependent electron–nucleon scattering, where descriptions in terms of both hadron and parton degrees of freedom coexist. Indeed, duality in spin-dependent structure functions has been predicted from both perturbative [53] and non-perturbative QCD arguments [50,43].

The feature which most distinguishes the study of duality in spin-dependent scattering from that in spin-averaged scattering is that since spin structure functions are given by differences of cross sections, they no longer need be positive. A dramatic example of this is provided by the $\Delta(1232)$ resonance, whose contribution to the g_1 structure function of the proton is negative at low Q^2. In spin-dependent scattering several new questions for the investigation of quark–hadron duality therefore arise:

1. Does quark–hadron duality work better (or only) for positive definite quantities such as cross sections, in contrast to polarization asymmetries?
2. Is there a quantitative difference between the onset of quark–hadron duality for spin-averaged and spin-dependent scattering, and if so, to what can this be attributed?
3. Does quark–hadron duality also hold for local regions in W for spin-dependent structure functions, and if so, how do these regions differ from those in unpolarized scattering?

Expanding on the last question, the above example of the $\Delta(1232)$-resonance contribution to the polarization asymmetry is sometimes used as evidence against quark–hadron duality in spin-dependent scattering [54]. However, this argument is not complete: the $\Delta(1232)$-resonance *region* consists of both a resonant *and* a non-resonant contribution, and it is the interplay between these that is crucial for the appearance of duality [53,55]. The more relevant question is at which value of Q^2 does the $\Delta(1232)$-resonance region turn positive (in the case of the proton g_1), and whether quark–hadron duality holds at lower Q^2 if one averages over the elastic peak and other nearby resonances in addition to the $\Delta(1232)$. Clearly duality cannot be too local at low Q^2.

A large quantity of precision spin structure function data has been collected since the 1980s [56] in the DIS region ($W > 2$ GeV) over a large range of Q^2. This has allowed initial studies of the logarithmic scaling violations in the g_1 structure function, and in turn has enabled one to embark upon dedicated investigations of quark–hadron duality in spin-dependent scattering.

The first modern experiment accessing the proton g_1 spin structure function in the resonance region was SLAC experiment E143 [57]. A negative contribution in the region of the N–$\Delta(1232)$ transition was observed, and a large positive contribution for $W^2 > 2$ GeV^2. The resonance region data at $Q^2 \approx 1.2$ GeV^2 seemed to approach the DIS results, with the exception of the negative N–$\Delta(1232)$ transition region. When integrating over the region of ξ corresponding to the nucleon resonances at $Q^2 \approx 1.2$ GeV^2, one finds about 60% of the corresponding DIS strength at $Q^2 = 3.0$ GeV^2. Obviously, a large source of this missing strength lies in the $\Delta(1232)$ region, which is still negative. Indeed the integrated strength in the region $2 < W^2 < 4$ GeV^2 amounts to about 80% of the DIS strength.

More recently, the HERMES collaboration at DESY reported data on the A_1 spin asymmetry (see (3.93)) in the nucleon resonance region for $Q^2 > 1.6$ GeV^2 [58]. The resonance region data were reported to be in agreement with those measured

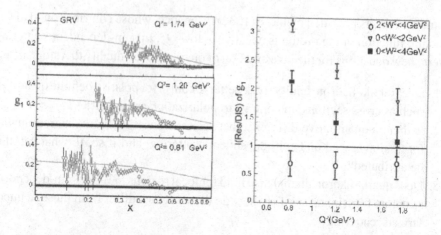

Figure 10.13. Left: Proton spin structure function g_1 from CLAS [59] in the resonance region, at three values of Q^2 indicated. The curves are the global parametrizations of the spin structure functions from [60]. Right: Ratio of the integrated strength of the g_1 data to that of the global parametrization. Both the data and the QCD parametrization are integrated for each Q^2 over the x regions corresponding to the indicated W^2 regions (with the elastic contribution included).

in the DIS region [57,56]. The average ratio of the measured A_1 asymmetry in the resonance region to the DIS power law fit is 1.11 ± 0.16 (stat.) ± 0.18 (syst.) [58]. This suggests that for $Q^2 > 1.6$ GeV^2, the description of the spin asymmetry in terms of quark degrees of freedom is, on average, also valid in the nucleon resonance region. The implication of this result is the tantalizing possibility of measuring the partonic content of A_1 at large values of x, almost up to $x = 1$, by extending such measurements into the nucleon resonance region.

The CLAS collaboration at JLab carried out inclusive polarized scattering experiments at energies of 2.6 and 4.3 GeV, using a polarized NH$_3$ target [59]. Some of the results, for $Q^2 > 0.7$ GeV^2, are shown in figure 10.13(left). In the lowest-Q^2 bins, the contribution of the $\Delta(1232)$ resonance region to g_1 is negative, whereas the contributions of the higher-mass states are positive. The negative $\Delta(1232)$ contribution obviously prevents a naive local duality interpretation at low Q^2. However, in some models [43,61] local duality is only expected to arise after averaging over the $\Delta(1232)$ and the (positive) elastic contribution. Indeed, addition of the nucleon elastic and $N-\Delta(1232)$ transition contributions would render a positive definite value for the averaged g_1.

This is further illustrated in figure 10.13(right), where we show the integrated strength of the nucleon resonance region data as compared to the integrated strength from the global structure function parametrizations [60]. Here the data have been split into two regions – the region $W^2 < 2$ GeV^2 (with the elastic contribution included), and $2 < W^2 < 4$ GeV^2 – and then integrated for each Q^2 over the x

regions corresponding to the chosen W^2. Clearly the elastic region *overcompensates* for the negative $\Delta(1232)$ region contribution, and the ratio for the region $W^2 < 2\,\text{GeV}^2$ falls as a function of Q^2. The region $2 < W^2 < 4\,\text{GeV}^2$ has $\sim75\%$ of the strength of the global QCD parametrization [60], close to the 80% found in the SLAC-E143 data [57] at $Q^2 = 1.2\,\text{GeV}^2$. The complete nucleon resonance region, with the elastic contribution included, closely resembles what one expects from the QCD parametrization at $Q^2 \approx 1.7\,\text{GeV}^2$. However, an even earlier onset is observed when both the elastic and $\Delta(1232)$ regions are left out.

The first experiment measuring the deuteron spin structure function g_1^d in the nucleon resonance region was again the SLAC experiment E143 [57], utilizing a polarized ND_3 target. The measured g_1^d structure function amounts to about half of the g_1^p structure function, leading to an almost null, but slightly negative, contribution of g_1^n. This is essentially the same behaviour as that found in the DIS data at higher W and Q^2, hinting that duality also exists for non-positive observables.

The CLAS collaboration at JLab collected g_1^d data with significantly smaller statistical uncertainties than the SLAC-E143 experiment, and better resolution in W [62]. The CLAS data do show an unambiguously positive g_1^d for $W^2 > 2\,\text{GeV}^2$, indicating that the helicity-$\frac{1}{2}$ transition amplitudes dominate even at rather low values of Q^2 ($Q^2 \approx 0.5\,\text{GeV}^2$). They conclude that the onset of local duality is slower for polarized structure functions than for unpolarized ones, as only the highest $Q^2 = 1.0\,\text{GeV}^2$ data, beyond the $\Delta(1232)$ region, show fairly good agreement with a fit to DIS data at $Q^2 = 5\,\text{GeV}^2$ [62]. For the unpolarized F_2^d structure function, local duality was observed to hold well already for $Q^2 = 0.5\,\text{GeV}^2$, from a similar comparison.

This leads us to conclude that the onset of duality in the spin-dependent structure functions occurs at larger values of Q^2 than for spin-averaged structure functions, in the region of $1 < Q^2 < 2\,\text{GeV}^2$. There are hints that duality also works for non-positive observables, and an earlier onset is definitely observed when combining the elastic and $\Delta(1232)$ regions. The evidence for quark–hadron duality in both the spin-averaged and the spin-dependent scattering process suggests that the helicity-$\frac{1}{2}$ and helicity-$\frac{3}{2}$ photoabsorption cross sections exhibit quark–hadron duality separately.

10.6 Duality in semi-inclusive reactions

In this section we generalize the duality concept to the largely unexplored domain of semi-inclusive electron scattering, $eN \rightarrow ehX$, in which a hadron h is detected in the final state in coincidence with the scattered electron. The virtue of semi-inclusive production lies in the ability to identify, in a partonic basis, individual quark species in the nucleon by tagging specific mesons in the final state, thereby enabling both the flavour and spin of quarks and antiquarks to be systematically determined.

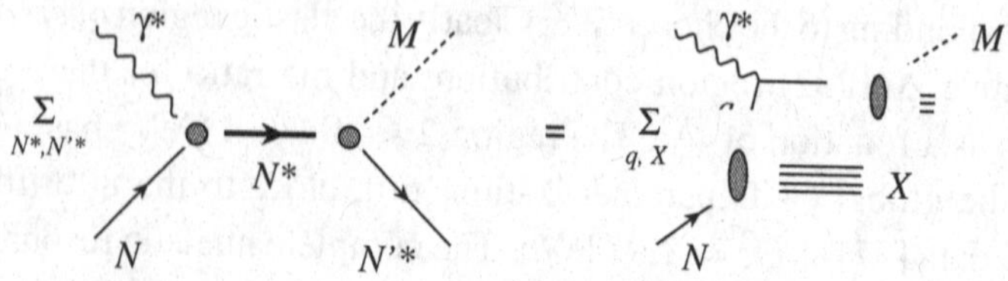

Figure 10.14. Duality between descriptions of semi-inclusive meson production
in terms of nucleon resonance (left) and quark (right) degrees of freedom [43,67].

Within a partonic description, the scattering and production mechanisms become
independent, and the cross section (at leading order in α_s) is given by a simple
product of quark distribution and quark $\rightarrow$ hadron fragmentation functions,

$$\frac{d^2\sigma}{dxdz} \sim \sum_q e_q^2 \, q(x) \, D_{q\rightarrow h}(z), \qquad (10.16)$$

where the fragmentation function $D_{q\rightarrow h}(z)$ gives the probability for a quark q to
fragment to a hadron h with a fraction z of the quark energy, $z = E_h/\nu$. Here, we
will focus on the process where a quark fragments into a pion, such that the elec-
troproduced pion carries away a large fraction, but not all, of the exchanged virtual
photon's energy. Factorization, as in (10.16), has been argued to be achievable at
lower energies, such as those available at HERMES and JLab, for pions with large
elasticity z [63,64]. Data indeed suggest that the factorization assumption may be
valid at energies accessible at HERMES and JLab with large enough z cuts [65,66].

In terms of hadronic variables the fragmentation process can be described through
the excitation of nucleon resonances, N^*, and their subsequent decays into mesons
and lower-lying resonances, which we denote by N'^*. The hadronic description
must be rather elaborate, however, as the production of a fast outgoing meson in the
current fragmentation region at high energy requires non-trivial cancellations of the
angular distributions from various decay channels [43,36]. The duality between the
quark and hadron descriptions of semi-inclusive meson production is illustrated in
figure 10.14. Heuristically, this can be expressed as [43,67]

$$\sum_{N'^*} \left| \sum_{N^*} F_{\gamma^*N\rightarrow N^*}(Q^2, W^2) \, \mathcal{D}_{N^*\rightarrow N'^*M}(W^2, W'^2) \right|^2 = \sum_q e_q^2 \, q(x) \, D_{q\rightarrow M}(z), \qquad (10.17)$$

where $D_{q\rightarrow M}$ is the quark $\rightarrow$ meson M fragmentation function, $F_{\gamma^*N\rightarrow N^*}$ is the
$\gamma^*N \rightarrow N^*$ transition form factor, which depends on the masses of the virtual
photon and excited nucleon ($W = M_{N^*}$), and $\mathcal{D}_{N^*\rightarrow N'^*M}$ is a function representing
the decay $N^* \rightarrow N'^*M$, where W' is the invariant mass of the final state N'^*.

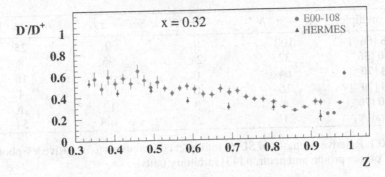

Figure 10.15. The ratio of unfavoured to favoured fragmentation function D^-/D^+ (at $z = 0.55$) as a function of x, using only deuterium data [66,68].

A dedicated experiment (E00-018) to study duality in pion electroproduction has been performed at JLab [66]. A 5.5 GeV electron beam was used to study pion electroproduction off proton and deuteron targets for Q^2 between 1.8 and 6.0 GeV^2, for $0.3 \leq x \leq 0.55$, and with z in the range 0.35–1.

Using the deuterium data only, the ratio of unfavoured to favoured fragmentation functions D^-/D^+ can be extracted. Here the favoured fragmentation function (D^+) corresponds to a pion which contains the struck quark (for example, a π^+ after a u or $\bar{d}$ quark is struck), while the unfavoured fragmentation function (D^-) describes the fragmentation of a quark not contained in the valence structure of the pion (for example, a d quark for the π^+). To a good approximation, this ratio is simply given by $D^-/D^+ = (4 - \mathcal{N}^{\pi^+}/\mathcal{N}^{\pi^-})/(4\mathcal{N}^{\pi^+}/\mathcal{N}^{\pi^-} - 1)$. Preliminary results [66] are shown in figure 10.15 in comparison with data from the HERMES experiment [68].

The first observation is that the behaviour as a function of z of D^-/D^+ measured by E00-108 closely resembles that seen in the HERMES experiment [68], albeit with slightly larger values than the HERMES ratios. However, it seems premature to draw a final conclusion from this in view of the preliminary state of the E00-108 analysis. The second observation to draw is that the D^-/D^+ ratio extracted from the JLab data shows a smooth slope as a function of z. This is quite remarkable given that the data cover the full resonance region, $0.88 < W'^2 < 4.2$ GeV^2. Apparently, there is some mechanism at work that removes the resonance excitations in the π^+/π^- ratio, and hence the D^-/D^+ ratio. This mechanism can be simply understood in the SU(6) symmetric-quark model.

Close and Isgur [43] applied the SU(6) symmetric-quark model to calculate production rates in various channels in semi-inclusive pion photoproduction, $\gamma N \rightarrow \pi X$. The pattern of constructive and destructive interference, which was a crucial feature

N'^* multiplet	$\gamma p \to \pi^+ N'^*$	$\gamma p \to \pi^- N'^*$	$\gamma n \to \pi^+ N'^*$	$\gamma n \to \pi^- N'^*$
$^2\mathbf{8}\,[\mathbf{56}^+]$	100	0	0	25
$^4\mathbf{10}\,[\mathbf{56}^+]$	32	24	96	8
$^2\mathbf{8}\,[\mathbf{70}^-]$	64	0	0	16
$^4\mathbf{8}\,[\mathbf{70}^-]$	16	0	0	4
$^4\mathbf{10}\,[\mathbf{70}^-]$	4	3	12	1
total $\mathcal{N}_N^\pi$	216	27	108	54

Table 10.1. Relative strengths of SU(6) multiplet contributions to inclusive $\pi^\pm$ photoproduction off the proton and neutron [43] (arbitrary units).

of the appearance of duality in inclusive structure functions, is also repeated in the semi-inclusive case. Defining the yields of photoproduced pions from a nucleon target as

$$\mathcal{N}_N^\pi(x,z) = \sum_{N'^*} \left| \sum_{N^*} F_{\gamma N \to N^*}(Q^2, W^2)\, \mathcal{D}_{N^* \to N'^*\pi}(W^2, W'^2) \right|^2, \quad (10.18)$$

the breakdown of $\mathcal{N}_N^\pi$ into the individual states in the SU(6) multiplets for the final W' states is shown in table 10.1 for both proton and neutron.

The results in table 10.1 suggest an explanation for the smooth behaviour of the ratio of fragmentation functions $D^-/D^+ \equiv D_d^{\pi^+}/D_u^{\pi^+}$ for a deuterium target in figure 10.15. The ratio $D^-/D^+ \approx (4 - \mathcal{N}^{\pi^+}/\mathcal{N}^{\pi^-})/(4\mathcal{N}^{\pi^+}/\mathcal{N}^{\pi^-} - 1)$, so from the relative weights of the matrix elements in table 10.1 one observes that the sum of the p and n coefficients for π^+ production is always 4 times larger than for π^- production. In the SU(6) limit, therefore, the resonance contributions to this ratio cancel exactly, leaving behind only the smooth background, as would be expected at high energies. This may account for the glaring lack of resonance structure in the resonance region fragmentation functions shown in figure 10.15.

There now exist strong hints in pion-electroproduction data that quark–hadron duality extends to semi-inclusive scattering. To convert these hints into conclusive evidence requires a new series of precision semi-inclusive experiments encompassing both the nucleon-resonance and DIS regions.

10.7 Duality in exclusive reactions

Quark–hadron duality should work better for inclusive observables than for exclusive ones, partly because perturbative behaviour appears to set in at higher Q^2 for the latter, and partly because there are more hadronic states over which to average. Although duality may be more speculative for exclusive processes, there are,

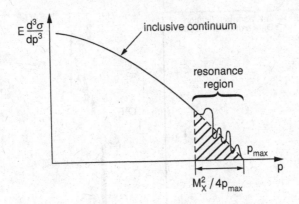

Figure 10.16. An illustration of a typical inclusive momentum spectrum of produced hadrons in the inclusive hadron production reaction $\gamma^* N \to MX$. From [69].

nevertheless, correspondence arguments that relate the exclusive cross sections at low energy to inclusive production rates at high energy.

This exclusive–inclusive connection in hadronic physics dates back to the early days of DIS and the discussion of scaling laws in high-energy processes. Bjorken and Kogut [69] articulated the correspondence relations by demanding the continuity of the dynamics as one goes from one (known) region of kinematics to another (unknown or poorly known).

For an inclusive process, such as $\gamma^* N \to MX$ where M is a meson, as the momentum (energy) of the meson increases, the invariant mass M_X of the recoiling system X enters the resonance region, as shown in figure 10.16. The correspondence argument states that the magnitude of the resonance contribution to the cross section should be comparable to the continuum contribution extrapolated from high energy into the resonance region,

$$\int_{p_{min}}^{p_{max}} dp \; E \frac{d^3\sigma}{dp^3}\bigg|_{incl} \sim \sum_{res} E \frac{d\sigma}{dp_T^2}\bigg|_{excl}, \tag{10.19}$$

where $p_{min} = p_{max} - M_X^2/4p_{max}$ and the integration region over the inclusive cross section includes contributions up to a missing mass M_X.

For inclusive electroproduction, this correspondence relation (10.19) was applied to derive the Drell–Yan–West relation [40,41] between the asymptotic behaviour of the elastic form factor and structure function in the $x \to 1$ limit. This Drell–Yan–West relation can be used to extract the proton and pion form factors from inclusive structure function data. Figure 10.17(left) shows the resulting proton magnetic form factor G_M^p extracted using the JLab scaling curve for F_2^p [8]. The extracted form

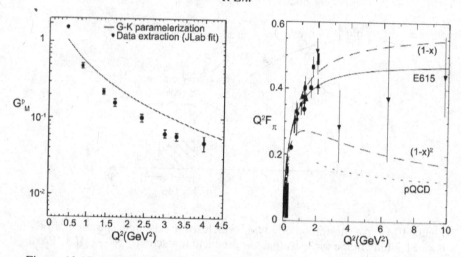

Figure 10.17. Left: proton magnetic form factor G^p_M extracted from an inelastic scaling curve using local duality, and compared with the Gari–Krumpelmann parametrization [70] of the world's G^p_M data (from [71]). Right: local duality prediction [72] for the pion form factor, using phenomenological pion structure function input from the Fermilab E615 Drell–Yan experiment [73] (solid), and the forms $F^\pi_2(x) \sim (1-x)$ and $(1-x)^2$ (dashed) [74].

factor is found to be in remarkable agreement with a parametrization of the world data on G^p_M [70]. This G^p_M parametrization is reproduced quite well, to within 30% accuracy, for Q^2 from 0.2 GeV^2 up to $\sim$4 GeV^2 [71].

Similarly, using the fit to the $F^\pi_2(x)$ data from the E615 experiment at Fermilab [73], the resulting form factor $F_\pi(Q^2)$ can be extracted, as shown in figure 10.17(right, solid curve). The agreement appears remarkably good, although there are reasons why its foundations may be questioned [72].

Soon after Bjorken and Kogut suggested the exclusive–inclusive correspondence, it was used [75] to predict the behaviour of the real Compton scattering cross section off the proton, $\gamma p \to \gamma p$, at large angles in the centre-of-mass frame. An extension of the study of duality in Compton scattering has been made to the case of one virtual and one real photon in the limit of large Q^2, known as the deeply virtual Compton scattering (DVCS) process, $ep \to e\gamma p$.

Interest in this reaction has been fostered by the realization that at high Q^2 DVCS provides access to generalized parton distributions (GPDs), with one of the important practical questions being whether the GPD formalism (see chapter 9) is applicable at intermediate energies. Here one may appeal to duality for an answer: if one can demonstrate that duality applies also to the case of DVCS, then a partonic interpretation of the scattering may be valid down to low Q^2. This problem has been investigated by Close and Zhao [44] in a generalization of the scalar constituent-quark model with a harmonic oscillator potential, a model discussed earlier for the case of the DIS structure functions [43,36]. The emergence of the scaling behaviour from duality in this model is due to the mass degeneracy between multiplets with the same N but different l, which causes a destructive interference between all but the elastic contribution.

The $\gamma p \rightarrow \pi^+ n$ reaction was considered by Afanasev *et al* [76] for $s \sim |t|$, who studied duality in the limit of fixed centre-of-mass scattering angle, θ_{cm}. Good agreement with data [77] is observed for the energy dependence at $\theta_{cm} = 90°$. However, Hoyer [78] has pointed out that at fixed angle this underestimates the measured cross section by about two orders of magnitude, due to additional diagrams involving more than a single quark in the nucleon that cannot be neglected in this limit. The appropriate limit for duality, and more generally factorization, to hold in semi-exclusive reaction is the $|t| \ll s$ limit [79].

As well as requiring an appropriate choice of kinematics, part of the apparent failure of duality in exclusive reactions also stems from the restriction to a single hadronic state. Duality arises when sufficiently many intermediate hadronic states are summed over, resulting in cancellations of non-scaling contributions. The cancellations are not exact, however, and duality violations are present at any finite kinematics.

10.8 Outlook

The historical origins of quark–hadron duality can be traced back to the 1960s, and the discovery of s- and t-channel duality in hadronic reactions. This duality reflected the remarkable relationship between low-energy hadronic cross sections and their high-energy behaviour, which, in the context of FESRs, allowed Regge parameters (describing high-energy scattering) to be inferred from the (low-energy) properties of resonances.

It was natural, therefore, that the early observations of a duality between resonance production and the high-energy continuum in inclusive electron–nucleon DIS would be interpreted within a similar framework. Bloom and Gilman found that by averaging the proton F_2 structure function data over an appropriate energy range the resulting structure function closely resembled the scaling function which described the high-energy scattering of electrons from point-like partons.

Electron scattering provides a wonderful stage for investigating the dynamical origin of quark–hadron duality. The perturbative scaling of the DIS structure functions occurs here in terms of the parton light-cone momentum fraction x, which can be accessed at different values of Q^2 and W^2, both within and outside the resonance region. Hence, both the resonance spectra and the scaling function describing the high-energy cross section can be mapped by varying Q^2 of the virtual photon.

Following the pioneering DIS experiments at SLAC in the late 1960s, the availability of (continuous wave) high-luminosity polarized beams, together with polarized targets, has allowed one to revisit Bloom–Gilman duality with unprecedented precision, and disentangle its spin, flavour, and nuclear dependence, in both local and global regions. The results have been striking: quark–hadron duality occurs at much lower Q^2 and more locally than had been expected.

Although considerable light has been shed upon the dynamical origins of quark–hadron duality, there are still important questions which need to be addressed before we come to a quantitative understanding of Bloom–Gilman duality in the structure function data. The observation of duality in spin-averaged structure functions in the region of the $\Delta(1232)$ resonance, for instance, suggests non-trivial interference effects between resonant and non-resonant (background) physics. Early descriptions of the resonance and background contributions employed the so-called two-component model of duality, in which the resonances are dual to valence quarks (associated with the exchange of Reggeons at high energy), while the background is dual to the $q\bar{q}$ sea (associated with pomeron exchange). In more modern language, this would call for a QCD-based derivation in which the properties of the non-resonant background can be calculated within the same framework as those of the resonances on top of which they sit.

It is also clear that the quark–hadron duality phenomenon is not restricted to inclusive electron–hadron scattering alone. If, as we believe, it is a general property of QCD, then it should manifest itself in other processes and in different observables. There are, in fact, predictions for quark–hadron duality in semi-inclusive and exclusive electroproduction reactions. The available evidence is scant, but it does suggest that at energy scales of a few GeV such reactions may proceed by closely mimicking a high-energy picture of free electron–quark scattering. This will be an exciting area of research for the next decade.

It is truly remarkable that in a region where we have only a few resonances, all consisting of strongly interacting quarks and gluons, the physics still ends up resembling a perturbative quark–gluon theory. Quark–hadron duality is the underlying cause of the smooth transition 'on average' from hadrons to quarks witnessed in Nature, allowing simple partonic descriptions of observables down to relatively low-energy scales.

References

[1] S Bethke, Nuclear Physics Proceedings Supplement 121 (2003) 74
[2] E D Bloom and F J Gilman, Physical Review Letters 25 (1970) 1140
[3] E D Bloom and F J Gilman, Physical Review D4 (1971) 2901
[4] P D B Collins, *An Introduction to Regge Theory and High Energy Physics*, Cambridge University Press, Cambridge (1977)

[5] A Donnachie, G Dosch, P Landshoff and O Nachtmann, *Pomeron Physics and QCD*, Cambridge University Press, Cambridge (2002)

[6] A De Rújula, H Georgi and H D Politzer, Physics Letters 64B (1976) 428

[7] A De Rújula, H Georgi and H D Politzer, Annals of Physics 103 (1975) 315

[8] I Niculescu *et al*, Physical Review Letters 85 (2000) 1186

[9] V De Alfaro, S Fubini, G Furlan and C Rossetti, Physics Letters 21 (1966) 576

[10] K Igi, Physical Review Letters 9 (1962) 76

[11] L Sertorio and M Toller, Physics Letters 18 (1965) 191; M Restignoli, L Sertorio and M Toller, Physical Review 150 (1966) 1389

[12] R Dolen, D Horn and C Schmid, Physical Review Letters 19 (1967) 402; Physical Review 166 (1968) 1768

[13] H Harari, Physical Review Letters 20 (1969) 1395; *ibid* 22 (1969) 562; *ibid* 24 (1970) 286; Annals of Physics 63 (1971) 432

[14] P G O Freund, Physical Review Letters 20 (1968) 235; P G O Freund and R J Rivers, Physics Letters 29B (1969) 510

[15] N M Queen and G Violini, *Dispersion Theory in High-Energy Physics*, Wiley, Chichester (1974)

[16] P V Landshoff and J C Polkinghorne, Nuclear Physics B28 (1971) 240

[17] A Donnachie and P V Landshoff, Physics Letters B296 (1992) 227

[18] P V Landshoff, in *Proceedings of the Workshop on the Quark-Hadron Transition in Structure and Fragmentation Functions* (Jefferson Lab, Newport News, Virginia, 2000); private communication

[19] G Miller *et al*, Physical Review D5 (1972) 528

[20] H Georgi and H D Politzer, Physical Review Letters 36 (1976) 1281; Erratum, *ibid* 37 (1976) 68; Physical Review D14 (1976) 1829

[21] M Arneodo *et al*, NMC Collaboration, Physics Letters B364 (1995) 107

[22] O Nachtmann, Nuclear Physics B63 (1973) 237

[23] Y Liang *et al*, nucl-ex/0410027 (2004)

[24] H L Lai *et al*, CTEQ Collaboration, European Physical Journal C12 (2000) 375

[25] A D Martin, R G Roberts, W J Stirling and R S Thorne, European Physical Journal C4 (1998) 463

[26] M Glück, E Reya and A Vogt, Zeitschrift für Physik C67 (1995) 433

[27] S Alekhin, Physical Review D68 (2003) 014002

[28] R Barbieri, J Ellis, M K Gaillard and G G Ross, Nuclear Physics B117 (1976) 50

[29] L W Whitlow, E M Riordan, S Dasu, S Rock and A Bodek, Physics Letters B282 (1992) 475

[30] S Dasu *et al*, Physical Review D49 (1994) 5641

[31] I Niculescu *et al*, Physical Review Letters 85 (2000) 1182

[32] K Wilson, Physical Review 179 (1969) 1499

[33] M E Christy, private communication

[34] C S Armstrong *et al*, Physical Review D63 (2001) 094008

[35] M Osipenko *et al*, Physical Review D67 (2003) 092001

[36] N Isgur, S Jeschonnek, W Melnitchouk and J W Van Orden, Physical Review D64 (2001) 054005

[37] M B Einhorn, Physical Review D14 (1976) 3451

[38] G 't Hooft, Nuclear Physics B72 (1974) 461; *ibid* B75 (1974) 461

[39] G Domokos, S Koveni-Domokos and E Schonberg, Physical Review D3 (1971) 1184; *ibid*. D3 (1971) 1191; *ibid*. D4 (1971) 2115

[40] S D Drell and T-M Yan, Physical Review Letters 24 (1970) 181

[41] G B West, Physical Review Letters 24 (1970) 1206

[42] V V Davidovsky and B V Struminsky, hep-ph/0205130

[43] F E Close and N Isgur, Physics Letters B509 (2001) 81

[44] F E Close and Q Zhao, Physical Review D66 (2002) 054001

[45] M W Paris and V R Pandharipande, Physics Letters B514 (2001) 361

[46] D R Harrington, Physical Review C66 (2002) 065205

[47] S Jeschonnek and J W Van Orden, Physical Review D65 (2002) 094038; *ibid* D69 (2004) 054006

[48] F E Close, F J Gilman and I Karliner, Physical Review D6 (1972) 2533

[49] F E Close and F J Gilman, Physics Letters 38B (1972) 541

[50] F E Close and F J Gilman, Physical Review D7 (1973) 2258

[51] F E Close, H Osborn and A M Thomson, Nuclear Physics B77 (1974) 281

[52] F E Close, *Introduction to Quarks and Partons*, Academic Press, London (1979)

[53] C E Carlson and N C Mukhopadhyay, Physical Review D58 (1998) 094029

[54] S Simula, M Osipenko, G Ricco and M Taiuti, Physical Review D65 (2002) 034017;

[55] C E Carlson and N C Mukhopadhyay, Physical Review D47 (1993) 1737

[56] K Hagiwara *et al*, Particle Data Group, Physical Review D66 (2002) 010001

[57] K Abe *et al*, E143 Collaboration, Physical Review Letters 78 (1997) 815; Physical Review D58 (1998) 112003

[58] A Airapetian *et al*, HERMES Collaboration, Physical Review Letters 90 (2003) 092002

[59] R Fatemi *et al*, CLAS Collaboration, Physical Review Letters 91 (2003) 222002

[60] M Glück, E Reya, M Stratmann and W Vogelsang, Physical Review D53 (1996) 4775; *ibid* D63 (2001) 094005

[61] F E Close and W Melnitchouk, Physical Review C68 (2003) 035210

[62] J Yun *et al*, CLAS Collaboration, Physical Review C67 (2003) 055204

[63] E L Berger, in *Proceedings of the Workshop on Electronuclear Physics with Internal Targets* (SLAC, Stanford, California, 1987).

[64] P J Mulders, in *EPIC 2000: Proceedings of the 2nd Workshop on Physics with an Electron Polarized Light Ion Collider* R G Milner ed, (MIT, Cambridge, Massachusetts, 2000) AIP Conference Proceedings 588 (2001) 75

[65] K Ackerstaff *et al*, HERMES Collaboration, Physical Review Letters 81 (1998) 5519

[66] H Mkrtchyan, private communications; JLab experiment E00-108

[67] W Melnitchouk, in *EPIC 2000: Proceedings of the 2nd Workshop on Physics with an Electron Polarized Light Ion Collider*, R G Milner ed (MIT, Cambridge, Massachusetts, 2000) AIP Conference Proceedings 588 (2001); hep-ph/0010311.

[68] P Geiger, Ph.D. Thesis, Heidelberg (1998)

[69] J D Bjorken and J B Kogut, Physical Review D8 (1973) 1341

[70] M Gari and W Krumpelmann, Physics Letters 141B (1984) 295

[71] R Ent, C E Keppel and I Niculescu, Physical Review D62 (2000) 073008; *ibid* D64 (2001) 038302

[72] W Melnitchouk, European Physics Journal A17 (2003) 223

[73] J S Conway *et al*, E615 Collaboration, Physical Review D39 (1989) 92

[74] W Melnitchouk, Physical Review D67 (2003) 077502

[75] D M Scott, Physical Review D10 (1974) 3117

[76] A Afanasev, C E Carlson and C Wahlquist, Physical Review D62 (2000) 074011

[77] R L Anderson *et al*, Physical Review Letters 30 (1973) 627; Physical Review D14 (1976) 679

[78] P Hoyer, in *Exclusive Processes at High Momentum Transfer* (Newport News, Virginia, 2002), hep-ph/0208190

[79] S J Brodsky, M Diehl, P Hoyer and S Peigne, Physics Letters B449 (1999) 306

11

Colour Transparency

G A Miller

The strong interaction between hadrons and nuclei leads to the phenomenon of shadowing. However, in the special situation of high-momentum-transfer coherent processes, these interactions can be turned off, causing the shadowing to disappear and the nucleus to become quantum-mechanically transparent. This phenomenon is known as colour transparency. In more technical language, colour transparency is the vanishing of initial- and final-state interactions, predicted by QCD to occur in high-momentum-transfer quasi-elastic nuclear reactions. These are coherent reactions in which one adds different contributions to obtain the scattering amplitude. Under such conditions the effects of gluons emitted by small colour-singlet systems vanish. Thus colour transparency is also known as colour coherence. The name 'colour transparency' is rather unusual. One might think that it concerns transparent objects that have colour, but it is really about how a medium can be transparent to objects without colour. This chapter provides a pedagogic review that defines the phenomenon and the conditions necessary for it to occur, assesses the role of colour transparency in strong interaction physics and reviews experimental and theoretical progress.

11.1 Point-like configurations

Strong interactions are strong: when hadrons hit nuclei they generally break up the nucleus or themselves. Indeed, a well-known classical formula states that the intensity of a beam of hadrons falls exponentially with the penetration distance. It is remarkable that QCD admits the possibility that, under certain conditions, the strong interactions can effectively be turned off and hadronic systems can move freely through a nuclear medium.

Electromagnetic Interactions and Hadronic Structure, eds Frank Close, Sandy Donnachie and Graham Shaw.
Published by Cambridge University Press. © Cambridge University Press 2007

Consider the elastic scattering of a small colour-singlet $q\bar{q}$ system by a nucleon. At sufficiently high energy, the separation b of the quarks (in a direction transverse to the momentum) does not change. The lowest-order perturbative contribution is given by two-gluon exchange [1–3] and the remarkable feature is that, in the limit that b approaches 0, the cross section vanishes because colour-singlet point particles do not exchange coloured gluons. This feature can be expressed in a concise form as

$$\lim_{b\to 0} \sigma(b^2) \propto b^2. \tag{11.1}$$

This reduced interaction is often termed 'colour screening' and is an unusual feature of QCD. In chiral theories of the interactions of hadronic systems, small relative distances between constituents (corresponding to small sizes) imply large relative momenta and therefore very strong interactions. Thus experimental verification of the behaviour of the cross section (11.1) and its consequences supports the belief that QCD is the correct theory of the strong interaction.

This chapter is concerned with the diverse consequences of the notion that small-sized colour-singlet objects do not interact strongly. There are interesting questions to ask and answer. In particular, how can one make a small-sized colour singlet, ensure that it propagates as a colour singlet and carefully measure the effects of vanishing interactions?

It is natural to suppose that small-sized colour-singlet objects are made in processes involving high momentum transfer. Candidate reactions include diffractive excitation of pions or photons [4] into two jets [5] moving at large relative momenta, electroproduction of vector mesons [6,7] and quasi-elastic reactions such as (p, pp), $(e, e'p)$ and $(e, e'\pi)$ [6,7]. High momentum transfer is usually associated with small wavelengths, but more is involved here. An object of small size, without soft-colour and pion fields, is supposed to be produced in the midst of a high-momentum-transfer process. This was suggested originally as a consequence of perturbative QCD applied at very high Q^2 [6,7]. The relevance of perturbative QCD for experimentally accessible values of Q^2 was questioned [8,9], but later it was shown [10,11] how including the effects of gluon radiation [12] extends the kinematic region of applicability of perturbative calculations. However, many theoretical and experimental results make it clear that perturbative treatments are usually not complete unless energies very much greater than 10 GeV and momentum transfers well in excess of 10 GeV^2 are involved, so it is relevant to ask if strong QCD can lead to small-sized objects. Studies of popular hadronic models have shown [13,14] that point-like configurations can be produced for momentum transfers as low as 1–2 GeV^2.

Producing a point-like configuration is not sufficient in itself. One must be able to identify that the important components contributing to the invariant amplitude are those of small size. This introduces the requirement of coherence. A reaction must involve a colour-singlet object before and after the interaction. Ensuring this requires a detailed experimental knowledge of the kinematics. For inclusive reactions, the influences of detailed quantum-mechanical-interference effects are lost, classical considerations dominate and the colour transparency idea becomes irrelevant. Another difficulty is that a very small-sized colour-singlet object (consisting of u, d or s quarks) is not in a physical state. It can be thought of as a wave packet consisting of a coherent superposition of physical states, undergoing time evolution that changes the relative phases of the components and consequently the object's size. Any change of the size of an object of zero spatial extent must be an increase, so the system must expand and become large eventually. The rest-frame time scale for these evolutions of sizes is of the order of hadron size divided by the speed of light, but in the laboratory this is multiplied by a time dilation factor that can be large enough to allow the influence of the vanishing interaction (11.1).

Taking these considerations into account, one may see that the magical disappearance of the strong scattering amplitude arises in three steps.

- A high-momentum-transfer reaction creates a colour-singlet point-like configuration that moves with high momentum through a nucleus.
- In coherent processes, the interaction between the point-like configuration and the nucleon will be strongly suppressed because of cancellations between gluon emission amplitudes arising from different quarks.
- The point-like configuration is not an eigenstate of the Hamiltonian. It can be regarded as a wave packet undergoing time evolution, which necessarily changes (and therefore increases) the size of the colour-singlet object, turning it into a normal-sized strongly interacting configuration. Therefore the observation of colour transparency requires that the point-like configuration escape the nucleus before it expands.

The first item addresses the interesting dynamical question of whether a point-like configuration ever exists. According to perturbative QCD, such objects are the origin of high-momentum-transfer hadronic form factors, but the validity of perturbative QCD is not a necessary condition. The second item is on firmer ground because the suppression of interactions is one of the essential ingredients needed to understand how Bjorken scaling occurs for deep-inelastic scattering at small x [6,15–17]. Furthermore, for high-energy scattering (11.1) applies and smaller objects generally do have smaller cross sections [18]. The third item leads to complications in observing the influence of colour transparency. One needs very high energies, the ability to handle the expansion in a quantitative manner, or the use of double scattering to isolate the interaction between the point-like configuration and the nucleon.

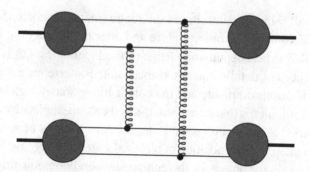

Figure 11.1. Perturbative interactions of small-sized colour-singlet objects. The heavy lines represent incident and outgoing colour-singlet objects. This shows one of the eight graphs of the same order. Both gluons can originate on either the quark or the antiquark, and graphs with crossed gluons must be included.

11.2 Small-sized objects have small cross sections

The general feature of QCD that small objects have small cross sections at high energy is supported by strong theoretical arguments and diverse experimental evidence. This is the basis for supposing that colour transparency is an observable phenomenon.

For example, consider the interaction between two colour-singlet systems. Evaluating the lowest-order perturbative two-gluon exchange contribution gives the result that if the size of either colour-singlet object vanishes, its scattering amplitude also vanishes [1–3]. This is because colour-singlet point particles do not exchange coloured gluons.

Figure 11.1 shows the high-energy, small-angle scattering of two mesons in the centre-of-mass frame, with the momentum of one of the mesons in the 3-direction. For very-high-energy scattering one may treat the transverse separations between the quark and antiquark of each meson as fixed quantities b, B. In the centre-of-mass frame, one meson has very large $P^+ = (P^0 + P^3)$ and the other a very large $P^- = (P^0 - P^3)$. In this situation, both the plus and minus components of the momenta of the exchanged gluons can be neglected and only the transverse components enter into the intermediate gluon and fermion propagators. This means that the scattering amplitude must be a function of b and B only. The quarks and antiquarks couple to the gluon with opposite signs, so that if either of b, B vanishes then the scattering amplitude vanishes. Suppose b is small, and B has a typical hadron size ~ 1 fm, then the cross section is of the form (11.1). The quadratic behaviour is interpreted as arising from two dipole factors, $\sim b$, that originate from each of the exchanged gluons.

The original work [1–3] succeeded in explaining the approximately constant nature of hadronic cross sections at high energy, provided a reasonably successful two-gluon exchange model of the pomeron and offered a natural explanation of the various sizes of hadronic cross sections. This explanation was later verified in a quantitative manner [18] for meson–nucleon scattering under the assumption that (11.1) applies over much of the meson volume. A dramatic confirmation of the idea that small objects have small cross sections is discussed in Section 11.6.

The idea that small-sized $q\bar{q}$ pairs interact with strongly-reduced cross sections is now understood to be an important part of lepton–nucleon and lepton–nucleus deep-inelastic scattering. For small values of Bjorken x the deep-inelastic scattering arises (as seen in the laboratory frame) from terms in which the virtual photon breaks up into a $q\bar{q}$ pair. Gribov [19] noticed that if the pair–target interactions are treated as those of an ordinary vector meson then the cross section for transversely-polarized virtual photons depends on $\log 1/x$. Bjorken denoted this result a disaster [20] because it badly violates scaling that requires a $1/Q^2$ behaviour. ($1/Q^2 \log Q^2$ gives the usual scaling violations caused by the influence of asymptotic freedom.) Bjorken handled the problem by reducing the value of the cross section for $q\bar{q}$ pairs moving at high relative transverse momentum. The relation between this reduced interaction and colour neutrality was noticed, using a $U(1)$ treatment [21]. The basic idea [6,17,15] is that high relative momentum implies that the transverse size of the pair is small and that the interactions of such a pair are strongly inhibited. The reduced interactions of small-sized quark–antiquark pairs play an important role in understanding the nuclear shadowing observed in deep-inelastic scattering [6,16,22].

These ideas led naturally to the dipole approach to deep-inelastic scattering and to hadronic interactions [23], discussed in chapter 8. In this approach the scattering amplitude is actually expressed as an integral over configurations of all sizes. Quantitative agreement with data for structure functions and photon–target interactions is achieved.

The original perturbative two-gluon exchange model has deficiencies when applied to elastic scattering because low-momentum gluons are important. In particular, the slope of $d\sigma/dt$ at $t = 0$ diverges for the exchange of massless gluons. The model was extended to include the effects of gluon–gluon interactions [24] and the non-perturbative role of gluon–target scattering. Essential new features were found by considering the relevant box diagrams [25,5]. The cross section depends on the gluon distribution of the target, $g_T(x, Q^2)$ [26]:

$$\sigma(b^2) = \frac{\pi^2}{3} \left(b^2 \alpha_s \left(Q^2_{\text{eff}} \right) x g_T \left(x, Q^2_{\text{eff}} \right) \right)_{x=\lambda/sb^2, Q^2_{\text{eff}}=\lambda/b^2}, \tag{11.2}$$

where λ is the proportionality factor that relates the high relative momentum to $1/b$ and depends upon the particular process and wave functions involved. For the situations of interest to this review $\lambda \approx 9$ [27].

The result (11.2) is consistent with the original idea that the point-like configuration has no interactions, but the entrance of the gluon distribution function introduces significant interesting differences from the original work [1–3]. The previously-mentioned divergence of the slope of elastic cross sections exists. Furthermore, g_T increases as x decreases to very small values and this causes interesting energy dependence. A detailed derivation and discussion of the accuracy of (11.2) is presented in [28]. Rigorous factorization theorems justifying the application of (11.2) to hard electroproduction of mesons have been derived [29].

The small-sized interactions of point-like configurations of $q\bar{q}$ pairs seems reasonably well verified. This physics is closely analogous to the reduction of the electromagnetic strength of an electric dipole of small size. That there will be similar cancellations of the field of gluons emitted by small-sized colour-neutral qqq configurations [3] is the remarkable prediction derived from QCD according to its SU(3) nature of colour.

11.3 Generic colour transparency

The issues related to colour transparency are: formation of a point-like configuration in a high-momentum-transfer reaction, suppression of the soft interactions of the point-like configuration with the target, and avoidance of the expansion of the point-like configuration into a large blob-like configuration. The experimental measurements must be done with kinematics sufficiently controlled that the separation between the hard and soft processes can be well defined. For example, in the $(e, e'p)$ reaction, the electron–proton interaction must be known to occur at a much larger momentum transfer than that of any ejectile–nucleus interactions.

Interesting special features that bring colour transparency to life for any given reaction are discussed below. Here we provide a general framework.

First consider a general high-momentum-transfer process in which an isolated nucleon in a state i is converted into a final state f. Then the invariant amplitude can be represented as

$$\mathcal{M}_{fi} = \langle f | \widehat{\mathcal{F}} | i \rangle, \tag{11.3}$$

in which the operator that induces the reaction is denoted as $\widehat{\mathcal{F}}$. For example, in electron–proton scattering at large Q^2, $\widehat{\mathcal{F}}$ is the electromagnetic current operator, i contains an incident photon and a proton at rest in the laboratory frame and f represents a high-momentum proton. In the case of pp elastic scattering, $\widehat{\mathcal{F}}$ represents the high-momentum part of the interaction and any soft interactions are subsumed within the states i, f. Other examples involve an incident pion or virtual photon. In that case the necessary high momentum originates in the wave function of the incident particle and $\widehat{\mathcal{F}}$ represents the soft-scattering interaction with the target.

An incident photon may fluctuate into a $q\bar{q}$ pair, with a time scale controlled by the uncertainty principle. For systems that propagate rapidly in the laboratory, this time scale is known as the coherence length, l_c:

$$l_c = \frac{2\nu}{Q^2 + M_{q\bar{q}}^2}, \tag{11.4}$$

because the $q\bar{q}$ fluctuation may propagate over this distance. Here ν is the photon energy and Q^2 the negative of the square of its invariant mass. If the mass of the $q\bar{q}$ system is of the order of Q, then $l_c \sim 1/(2M_N x)$. Thus for low values of x, such as those that occur in vector-meson production, l_c can be large and soft interactions of the $q\bar{q}$ fluctuation may occur. For large values of x, such as for the $(e, e'p)$ nuclear reaction, a point-like photon interacts with the target.

The intermediate state $\widehat{\mathcal{F}}|i\rangle$ may or may not be a point-like configuration. Its overlap with the physical final state will necessarily be small at large momentum transfer, but it is not possible to examine the microscopic nature of the intermediate state without using nuclear targets. Therefore consider the high-momentum-transfer process occurring on any nucleon bound in a target nucleus. The initial and final nuclear states are denoted by I, F and the invariant amplitude is given by

$$\mathcal{M}_{FI} = \langle F|(1 + T_S G_0)\widehat{\mathcal{F}}(1 + G_0 T_S)|I\rangle, \tag{11.5}$$

in which residual interactions (not contained in the states $|I, F\rangle$) involving nucleons, other than the struck one, are denoted as T_S and G_0 represents the non-interacting Green's function. The expression (11.5) is general and applies to any process. For example, in vector-meson production by virtual photons of low x, the term $G_0 T_S$ may represent the propagation of the incident photon as a $q\bar{q}$ pair. For the (p, pp) reaction, $G_0 T_S$ and $T_S G_0$ represent the soft initial- and final-state interactions.

We examine (11.5) under two separate limiting conditions. First, suppose that the energies are high enough so that any point-like configuration can propagate without changing its properties. Then G_0 can be regarded as a constant. Given the dipole nature of high-energy low-momentum transfer interactions (11.1) and (11.2), $T_S \propto b^2$. The quantities $\langle f|b^2\widehat{\mathcal{F}}|i\rangle$ and $\langle f|\widehat{\mathcal{F}}b^2|i\rangle$ control the initial- and final-state interactions. The Q^2 dependence of this matrix element is crucial. If $\langle f|b^2\widehat{\mathcal{F}}|i\rangle$ approaches 0 with increasing Q^2 much faster than $\langle f|\widehat{\mathcal{F}}|i\rangle$ does, then final-state interactions will not be important. This suggests that we define an effective size to determine whether or not a point-like configuration is formed:

$$b^2(Q^2) \equiv \frac{\langle f|b^2\widehat{\mathcal{F}}|i\rangle}{\langle f|\widehat{\mathcal{F}}|i\rangle}. \tag{11.6}$$

The vanishing of $b^2(Q^2)$ for any given reaction is a necessary condition for colour transparency to occur. Note that the effective size (11.6) is an off-diagonal matrix element. This is relevant for observing the given initial and final state, while the diagonal matrix element involving the putative point-like configuration state $\langle i|\widehat{\mathcal{F}}^{\dagger}b^2\widehat{\mathcal{F}}|i\rangle$ has no relevance.

If one treats the Green's function G_0 as a constant then $b^2(Q^2)$ is essentially the ratio of the second term to the first term of the multiple-scattering series (11.5). However, G_0 does not act as a constant, because the wave-packet state $\widehat{\mathcal{F}}|i\rangle$ is not an eigenstate of the Hamiltonian and therefore undergoes time evolution. This can only increase the wave-packet size and restore soft quark–gluon fields. The time, τ, for a quantum fluctuation from a point-like configuration of bare mass M_X to a relevant object of normal hadronic size of mass M_1 is given by

$$\tau = 2P/\left(M_X^2 - M_1^2\right), \qquad (11.7)$$

where P is the momentum of the wave packet. The expansion time is controlled by an undetermined parameter M_X. Equation (11.7) involves a simplification because any high momentum transfer would produce a set of configurations of different masses. For sufficiently large energies, τ is long enough that the object can leave the nucleus while small enough to avoid final state interactions and colour transparency occurs. But what is M_X? Purely theoretical arguments do not give a value, but we can obtain a lower limit for nucleon final states. In that case, the mass M_1 is that of the nucleon, and the lowest possible value one can think of for the value of M_X is the sum of the nucleon and pion masses. The use of (11.7) with $M_X = M_N + m_\pi$, $M_1 = M_N$ leads to $\tau \leq (E/1\ \text{GeV})$ fm, which, for the experimentally relevant value of $E \approx$ 5 GeV, is about 5 fm or smaller than the diameter of the aluminium nucleus. More realistic estimates [30,31,13,32] use larger values of $M_X \sim 1.4-1.6$ GeV and obtain smaller distances. The sadly-unavoidable conclusion is that, unless the energy is very high, the point-like configuration will expand. One must account for this effect.

Thus the conditions for colour transparency to occur are that the effective size (11.6) be vanishingly small and the fluctuation time (11.7) be huge. If both conditions are satisfied, then $T_S G_0 \widehat{\mathcal{F}} \sim T_S \widehat{\mathcal{F}} \approx 0$, and the same holds for $\widehat{\mathcal{F}} G_0 T_S$. Then one finds

$$\mathcal{M}_{FI} \approx \langle F|\widehat{\mathcal{F}}|I\rangle \equiv \mathcal{M}_{FI}(PWBA), \qquad (11.8)$$

in which PWBA refers to the plane-wave Born approximation. Two different situations are relevant. Suppose the nuclear process is coherent, then

$$\mathcal{M}_{FI} \approx A\mathcal{M}_{fi}, \qquad (11.9)$$

and the sum over all the nucleons is represented as a schematic factor A. The other relevant situation occurs when the nucleus is excited by the removal of a single nucleon (quasi-elastic reaction), and one sums over all such excited states. Then

$$\sum |\mathcal{M}_{FI}|^2 \approx A|\mathcal{M}_{fi}|^2. \tag{11.10}$$

The much sought-after limit of perfect colour transparency is expressed by either of (11.9) or (11.10). The A dependence of the cross section can be either A^2 or A depending on the particular nuclear process studied. More generally, the ratio of nuclear to nucleon cross sections is measured and defined to be the transparency $T(A, Q^2)$, with perfect transparency corresponding to a value of unity. Experimental measurements weight the numerator and denominator with the relevant acceptance, and theorists must match these in their calculations.

11.4 Simple examples

Colour transparency provides a number of simple examples that require a good understanding of quantum-mechanical-interference phenomena.

11.4.1 Quantum-mechanical invisibility

The effects of the reduced interaction (11.1) and (11.2) are interesting in the context of our understanding of visibility and invisibility. Consider a very energetic $q\bar{q}$ system made of very massive quarks (onium) incident on a slab of nuclear matter of thickness L [33]. Under the standard semi-classical theory applicable for short wavelengths, the probability $P_{sc}(L)$ for the onium passing through the slab while remaining in its ground state and without losing energy is

$$P_{sc}(L) = e^{-L/\lambda}, \tag{11.11}$$

where $\lambda = 1/(\sigma_0 \rho)$ is the path length, with σ_0 the onium–nucleon cross section and ρ the nuclear density. But the onium wave function entails a probability for the pair to have a transverse separation:

$$\Psi^2(b) = \frac{1}{\pi \langle b^2 \rangle} e^{-b^2/\langle b^2 \rangle}, \tag{11.12}$$

using a simple model. At sufficiently high energies each separation propagates without changing and each interacts with a different strength according to (11.1) and (11.2). Using this leads to a probability amplitude given by

$$\int d^2 b \, \Psi^2(b) e^{-(b^2/2\langle b^2 \rangle) L/\lambda}, \tag{11.13}$$

so that the true survival probability $P(L)$ of the ground state is

$$P(L) = \frac{1}{(1 + L/2\lambda)^2}. \tag{11.14}$$

For $\lambda \gg L$, the expressions (11.11) and (11.14) coincide. More generally, the quantum-mechanically computed probability (11.14) can be much larger than (11.11) because the small-sized components of onium have reduced interactions that allow propagation through a dense medium. The inclusive cross section computed by taking the sum of the individual cross sections corresponds to a transmission probability that varies as $1/L$ [32].

The analogy with optics is clear. Light cannot propagate through an opaque object (as in (11.11)) and the object casts a shadow. But quantum mechanically (11.14) holds and propagation through the medium is allowed. The classically opaque object loses it shadow. Only visible objects cast shadows, so (11.1) and (11.2) correspond to quantum-mechanical invisibility.

11.4.2 Existence of point-like configurations

Is a point-like configuration formed in a high-momentum-transfer reaction? This is the dynamical question of interest. The use of general principles is not sufficient to provide an answer.

It is worthwhile to examine simple examples to understand how it is that different dynamical interactions lead to different answers. Therefore the effective size $b^2(Q^2)$ of (11.6) is evaluated using simplifying assumptions following [13,14]. The first example is the non-relativistic interaction of two equal-mass quark systems. The form factor is given by

$$F(Q^2) = \int d^3r \; \psi^*(r) e^{\frac{iq\cdot r}{2}} \psi(r), \tag{11.15}$$

where $Q^2 = \mathbf{q} \cdot \mathbf{q}$ and $Q = |\mathbf{q}|$. The quantity $e^{\frac{iq\cdot r}{2}} \psi(r) \equiv \chi_Q(\mathbf{r})$ is a wave packet formed in a high-momentum-transfer reaction. The measurement of $F(Q^2)$ only determines a given integral involving $\chi_Q(\mathbf{r})$. To see whether or not this quantity represents a point-like configuration one examines the effective size (11.6) which, for this example, takes the form

$$b^2(Q^2) = \frac{1}{F(Q^2)} \int d^3r \; \psi^*(r) b^2 \chi_Q(\mathbf{r}), \tag{11.16}$$

as $\mathbf{q} \cdot \mathbf{r} = 0$. This leads to the result

$$b^2(Q^2) = -16 \frac{d \log F(Q^2)}{dQ^2}. \tag{11.17}$$

The form factor $F(Q^2)$ falls as a power of Q^2 unless the binding potential is an analytic function of r^2 [14]. An example is binding with an attractive Coulomb potential for which

$$b^2(Q^2) = \frac{32a_0^2}{16 + Q^2 a_0^2} \sim \frac{8}{(Q/2)^2}, \tag{11.18}$$

where a_0 is the Bohr radius and the second expression is the limit for very large values of Q^2. This vanishing of $b^2(Q^2)$ shows that colour transparency is possible for systems bound by Coulomb forces. Alternatively, if the binding potential is a harmonic oscillator, $F(Q^2) = e^{-Q^2 R^2/6}$ with R^2 as the mean square radius, then the effective size $b^2(Q^2) = 2R^2/3$ is a constant. If linear forces govern the dynamics there will be no colour transparency.

Relativity is very important at high momentum transfer, so we consider a relativistic system of two light quarks, spinless for simplicity, labelled as 1,2. Light-front dynamics [34–37] can be used to treat the system in any reference frame. Define the components of the total momentum vector P^μ as $P^\pm = (P^0 \pm P^3)$, and $\vec{P}$. Spatial coordinates are given by $x^+ = x^0 + x^3$, $x^- = x^0 - x^3$, $\vec{x}$. The coordinate x^- is canonically conjugate to the plus-component of the momentum that is closely related to the Bjorken scaling variable. The evolution operator for the light-front 'time' variable x^+ is P^-, so that the relativistic Schrödinger equation is

$$P^- \Psi = \frac{\vec{P}^2 + \widehat{M^2}}{P^+} \Psi, \tag{11.19}$$

where $\widehat{M^2}$ is the mass-squared operator for the system:

$$\widehat{M^2} |n\rangle = M_n^2 |n\rangle, \tag{11.20}$$

where the spectrum is discrete for a confined system and the ground state is represented by $n = 1$. The use of relative coordinates

$$\alpha \equiv p_1^+/P^+, \quad \vec{p} = (1 - \alpha)\vec{p}_1 - \alpha \vec{p}_2, \tag{11.21}$$

where p_1 and p_2 are the quark four-momenta, leads to the unique result that in *any* reference frame the Schrödinger equation (11.19) can be expressed as

$$\left(\frac{\vec{p}^2 + m^2}{\alpha(1 - \alpha)} + V \right) |n\rangle = M_n^2 |n\rangle, \tag{11.22}$$

where V is the effective quark–quark interaction operator. This use of light-cone variables allows the separation of centre-of-mass and relative variables, so that the wave function of any two-quark system of total four-momentum P^μ can be expressed as $e^{-iP \cdot x} \langle \vec{p}, \alpha | \phi \rangle$, in which momentum-space $\vec{p}, \alpha$ are used for the internal variables and coordinate-space x is used to describe the motion of the centre of mass. In deep-inelastic scattering with Bjorken scaling, the variable $\alpha = Q^2/(2M_N\nu) = x$.

Form factors can be computed easily using light-front dynamics if one employs a reference frame in which $q^+ = 0$ and $Q^2 = -q_\perp^2$. The boosts are kinematic and the form factor for the transition between states n and n' is

$$F_{n'n}(Q^2) = \int d^2b \int_0^1 d\alpha \frac{\langle n'|\vec{b}, \alpha\rangle e^{i\vec{q}_\perp \cdot \vec{b}\alpha}\langle \vec{b}, \alpha|n\rangle}{\alpha(1-\alpha)} \equiv \langle n'|e^{i\vec{q}_\perp \cdot \vec{b}\alpha}|n\rangle, \quad (11.23)$$

where $\vec{b}$ is conjugate to $\vec{p}$. The effective size analogous to (11.6),

$$b_{n'n}^2(Q^2)F_{n'n}(Q^2) = \langle n'|b^2|n\rangle, \quad (11.24)$$

can be vanishingly small for large values of Q^2 [13,14].

11.4.3 Expansion of point-like configurations

The existence of a point-like configuration is not sufficient to guarantee the observation of colour transparency, as we have seen in section 11.3. Here we provide a simple treatment of the dynamics governing expansion.

Suppose that a state $|\phi\rangle$ of total four-momentum P^μ is created in a point-like configuration, $|PLC\rangle$, at some initial time defined as $t = 0$. We may treat it as a coherent expansion of complete eigenstates of $\widehat{M^2}$:

$$\langle \vec{p}, \alpha|\phi(t)\rangle = \sum_n \langle \vec{p}, \alpha|n\rangle e^{-i\sqrt{P^2+M_n^2}\,t}\langle n|PLC\rangle, \quad (11.25)$$

for the relativistic two-quark system of the previous subsection. The condition for avoiding expansion is that the phase factors for states n that contribute importantly to the sum be approximately the same. This occurs if τ of (11.7) is larger than the radius of the nucleus.

Note that the interaction V generally causes evolution in *all* of the three variables $\vec{p}, \alpha$. Two-dimensional approaches [17,38] that keep the value of α fixed are not complete. Indeed, the failure to maintain the influence of confinement in three spatial dimensions is responsible for the difficulties in end-point behaviour encountered in [38]. The expression (11.25) uses a hadronic basis, but one may also use only quark degrees of freedom.

The point-like configuration is not the system that one encounters. Instead it is $\sum_n |n\rangle e^{-i\sqrt{P^2+M_n^2}\,t}\langle n|PLC\rangle$. Suppose this system interacts with a strength governed by b^2, and the end product is detected as the ground state. Then the experimentally-relevant overlap $b^2(t)$ is

$$b^2(t) = \langle 1|b^2 \sum_n |n\rangle e^{-i\sqrt{P^2+M_n^2}\,t}\langle n|PLC\rangle. \tag{11.26}$$

The use of completeness and the definition of a point-like configuration lead to the result that $b^2(t=0) = 0$. Jennings and Miller developed a simple harmonic oscillator model [39,31] in which the matrix element $\langle 1|b^2|n\rangle$ is non-zero when n corresponds to the ground state or states of energy M_2. Then

$$b^2(t) = e^{-iE_1t}\langle 1|b^2|1\rangle\langle 1|PLC\rangle\left(1 - e^{i\frac{(M_1^2-M_2^2)}{2P}t}\right), \tag{11.27}$$

if $P \gg M_2$. The real part of the quantity in parentheses, $2\sin^2\left(\frac{M_1^2-M_2^2}{4P}t\right)$, controls the strength of the final-state interaction.

11.4.4 Point-like configurations in the nuclear ground state

The existence of a point-like configuration as a significant part of the nucleon wave function has consequences for physics other than form factors. Quarks within a point-like configuration have high relative momentum and are therefore much more likely to carry a large plus-component of momentum (α of (11.21)). Therefore deep-inelastic scattering from a free proton at large x should be strongly influenced by the point-like configuration. The EMC effect [40] is the observation that structure function of a bound nucleon is smaller than that of a free nucleon by about 10–15% for large values of x, so it is natural to consider the effects of nuclear binding on the structure of the nucleon.

For simplicity consider the nucleon $|N\rangle$ to be a superposition of a configuration of normal size $|N_0\rangle$ and an orthogonal point-like configuration that occurs with small probability. Then

$$|N\rangle \approx |N_0\rangle + \epsilon_0|PLC\rangle, \tag{11.28}$$

where $\epsilon_0 = v/\Delta E$ with v the interaction that connects the components and $\Delta E > 0$ the energy difference between the unperturbed configurations. Based on the nucleon spectrum, we might expect ΔE to be about 500–600 MeV. Now suppose the nucleon is bound within a nucleus: its different components respond to the nuclear force in different ways. The point-like configuration would not feel the attractive interactions with the medium but the average-sized configuration would. For an attractive potential energy $-U$, the energy denominator is increased from ΔE to $\Delta E + U$, and the PLC amplitude is suppressed by the nuclear binding [41–43]:

$$\epsilon = \epsilon_0 \left(1 - \frac{U}{\Delta E} \right). \tag{11.29}$$

A typical nuclear binding potential is of the order of 60 MeV deep, so the probability amplitude is reduced by 10%.

The effect of this suppression can be large for processes in which the interference effect between the point-like configuration and the normal-sized configuration provides the dominant contribution. Deep-inelastic scattering at high x is one example, and the resulting suppression is one of the explanations of the EMC effect [41]. Another is that the nucleon form factor of a bound nucleon would be suppressed compared to the free one at high Q^2. Experiments [44] that search for nuclear modifications of the proton form factor could detect this effect of a point-like configuration at larger values of Q^2.

Another issue concerns the convergence of the series that determines the effective interaction between nucleons. The effects of colour transparency would help to suppress the short-distance interaction between bound nucleons and could ultimately help theorists to better understand convergence.

11.5 The best case: dijet production by high-energy pions

Fermilab experiment E791 [45] has obtained an unambiguous observation of colour transparency in the diffractive dissociation into dijets of pions of momentum $P = 500$ GeV scattering coherently from carbon and platinum targets: $\pi A \to q\bar{q} A$. The $q\bar{q}$ pair (or resulting dijet) state has a relative transverse momentum $\vec{\kappa}$ and the quark carries a plus component of momentum zP. Then the mass m_f of the diffractively produced state is

$$m_f^2 = \frac{m_q^2 + \vec{\kappa}^2}{z(1-z)}. \tag{11.30}$$

The process begins when, long before the incident pion hits the target, it fluctuates into a $q\bar{q}$ pair. The coherence length (11.4) for this to occur is very large because of the large value of P. A two-gluon exchange interaction with a single nucleon then brings the virtual pair into reality ultimately as two jets. Detailed arguments show [46] that the dominant term is the diagram T_1 of figure 11.2 in which the high relative momentum arises from gluon exchange between the quark and antiquark in the simplest configuration of the initial state pion, and in which the two-gluon exchange interaction is soft. The special result here is that the invariant amplitude depends only on the $q\bar{q}$ components of the initial-pion wave function.

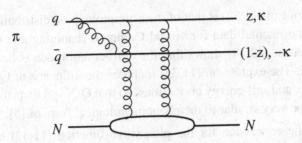

Figure 11.2. Dominant graph, T_1: the large momentum transfer arises from the initial-π wave function. The figure shows one of the eight of this class of diagrams. Both gluons can originate on either the quark or the antiquark, and graphs with crossed gluons must be included.

11.5.1 Dominant amplitude for $\pi N \to N + dijet$

Consider the forward ($t = t_{min} \approx 0$) amplitude, $\mathcal{M}$, for coherent dijet production on a nucleon $\pi N \to N + $ dijet [5]:

$$\mathcal{M}_{fi} \approx T_1 = \langle f(\vec{\kappa}, z), N' \mid \widehat{\mathcal{F}} \mid \pi, N \rangle, \tag{11.31}$$

where $\widehat{\mathcal{F}}$ represents the interaction with the target nucleon. The initial $|\pi\rangle$ and final $|f(\vec{\kappa}, z)\rangle$ states represent the physical states that generally involve all manner of multi-quark and gluon components.

The dominant Feynman diagrams in figure 11.2 can be rewritten as a product of a high-momentum component of a light-cone pion wave function with the amplitude for the scattering of a quark–antiquark dipole by a target. Its evaluation consists of three parts: (i) determining the relevant part of the pion wave function, (ii) identifying the interaction with the target (in this case with the gluon field of the target) that causes the pion to dissociate into a $q\bar{q}$ pair and (iii) computing the overlap matrix element (11.31).

The high-transverse-momentum component of the pion light-cone wave function, $\chi_\pi(\vec{p}, \alpha)$, arises from the perturbative one-gluon-exchange acting on the strong QCD light-cone pion wave function [47,46]:

$$\chi(\vec{p}, \alpha) = \frac{4\pi C_F \alpha_s(\vec{p}^{\,2})}{\vec{p}^{\,2}} \sqrt{3} f_\pi \alpha(1 - \alpha) \left(\log \frac{Q^2}{\Lambda_{QCD}^2} \right)^{C_F/\beta}, \tag{11.32}$$

where $C_F = N_c^2 - 1/2N_c = \frac{4}{3}$, $\alpha_s(\vec{p}^{\,2}) = 4\pi/(\beta \log(\vec{p}^{\,2}/\Lambda_{QCD}^2))$, $\beta = 11 - \frac{2}{3}n_f$, and $f_\pi = 93$ MeV. The argument of the strong coupling constant α_s is $\vec{p}^{\,2}$ in leading logarithm approximation and Q is the renormalization scale of the wave function [47].

The α dependence of (11.32) is that of the asymptotic pion distribution amplitude. Analysis of experimental data for virtual Compton scattering and the pion form factor performed in [48,49] shows that the correct amplitude is not far from the asymptotic one. The expression (11.32) includes the influence of QCD evolution as well as vertex and self-energy corrections, so that QCD predicts the dependence of $\chi_\pi(\vec{p}, \alpha)$ to be very similar to the phenomenological form of [5].

If the asymptotic expression for the pion wave function (11.32) is used in the scattering amplitude (11.31) and one evaluates only the diagrams of figure 11.2, one obtains the dominant term T_1. To proceed it is necessary to specify the soft scattering operator $\widehat{\mathcal{F}}$. The transverse distance operator $\vec{b} = (\vec{b}_q - \vec{b}_{\bar{q}})$ is canonically conjugate to the transverse relative-momentum operator. At sufficiently small values of $b = |\vec{b}|$, the leading-twist effect and the dominant term at large s arise from diagrams in which the pion fragments into two jets as a result of interactions with the two-gluon component of the gluon field of a target, see figure 11.2. The perturbative QCD determination of this interaction involves a diagram similar to the gluon-fusion contribution to the nucleon sea-quark content observed in deep-inelastic scattering. One calculates the box diagram for large values of the relative momentum using the wave function of the pion instead of the vertex for $\gamma^* \to q\bar{q}$. This leads [46] to a purely-imaginary scattering amplitude $\widehat{\mathcal{F}}$ that is essentially of the form (11.2), except that the function g_N becomes the skewed or generalized gluon distribution [46] (see chapter 9), because the mass difference between the pion and the final two-jet state necessitates that the reaction requires a non-zero momentum transfer to the target.

The most important effect shown in (11.2) is the b^2 dependence, which shows the diminished strength of the interaction for small values of b. To proceed we need the scattering operator $\widehat{\mathcal{F}}$ corresponding to (11.2):

$$\widehat{\mathcal{F}}(b^2) = is \frac{\sigma_0}{\langle b^2 \rangle} b^2, \tag{11.33}$$

in which σ_0 is treated as a constant. This scattering operator is independent of the longitudinal momenta and has a simple dependence on the transverse momentum operator, so that the evaluation of the matrix element (11.31) leads to an expression in which the arguments of χ_π turn out to be the final-state variables $(\vec{\kappa}, z)$. The approximate pion wave function, valid for large relative momenta, is given by (11.32). The b^2 operator appearing in (11.33) can be taken to act on the pion wave function as $-\delta(\vec{p} - \vec{\kappa})\nabla_p^2$. Then combining (11.32) and (11.33) and evaluating (11.31) leads to the QCD result [46]:

$$T_1 = -4is \frac{\sigma_0}{\langle b^2 \rangle} \frac{4\pi C_F \alpha_s(\kappa^2)}{\kappa^4} \left(\log \frac{\kappa^2}{\Lambda_{QCD}^2} \right)^{C_F/\beta} a_0 z(1-z) \tag{11.34}$$

of the same form as the 1993 result [5]. Derivatives of the logarithmic terms of (11.32) are ignored because similar terms are neglected in taking $\widehat{\mathcal{F}}$ to be proportional to b^2. Such corrections are of the order of the inverse of $\log(\kappa^2/\Lambda_{QCD}^2)$. The κ dependence of T_1 leads to $d\sigma(\kappa)/d\kappa^2 \sim \kappa^{-8}$ or $d\sigma(\kappa)/d\kappa \sim \kappa^{-7}$, which is to be compared with the experimental result [45] that the power fall-off is 6.5 ± 2 for $\kappa > 1.8$ GeV.

11.5.2 Nuclear dependence

The amplitude is dominated by a process in which the pion becomes a $q\bar{q}$ pair, of essentially zero transverse extent and having the very large longitudinal momentum of the incident pion, well before hitting the nuclear target. This point-like configuration can move through the entire nucleus without expanding. The dominant contribution occurs when the $q\bar{q}$ pair interacts with one nucleon and passes undisturbed through the remainder of the nuclear matter. The nuclear amplitude $\mathcal{M}_{FI}$ takes the form of (11.9) $\mathcal{M}(A) \approx A\,\mathcal{M}_{fi}$ if the momentum transfer t to the nucleus vanishes. The effects of nuclear modifications of the skewed gluon distribution function [46] can modify this simple dependence. There also is an unusual $q\bar{q}$-nucleus rescattering term, proportional to the nuclear radius divided by κ^2, that interferes constructively with the dominant amplitude [5] as a result of the b^2 dependence of the scattering operator (11.33). The usual effect of rescattering is destructive interference associated with shadowing.

Suppose one measures the production of a dijet state of mass m_f (11.30) and relative transverse momentum $\vec{\kappa}$. Then the differential cross section for a nuclear target, of radius R_A, obtained from the amplitude (11.9) takes the form

$$\frac{d\sigma(A)}{dt} = A^2\frac{d\sigma(N)}{dt}e^{tR_A^2/3}, \tag{11.35}$$

where $d\sigma(N)/dt$ is the cross section for a nucleon target. For very small values of t the A dependence is a spectacular A^2. In reality, one hopes to measure the integral

$$\sigma(A) = \int dt \frac{d\sigma(A)}{dt} \approx \frac{3}{R_A^2}A^2\frac{d\sigma(N)}{dt}\bigg|_{t=0}. \tag{11.36}$$

A typical procedure is to parametrize $\sigma(A)$ as

$$\sigma(A) = \sigma_1 A^\alpha \tag{11.37}$$

in which σ_1 is a constant independent of A. The nominal dependence is given by $\alpha = \frac{4}{3}$, but one finds $\alpha \approx 1.45$ for the two specific targets, platinum ($A = 195$) and carbon ($A = 12$), used in E791.

The strong A dependence is supposed to be the unique signature of a coherent nuclear diffraction process. Another mechanism causing a similar A dependence is that of electromagnetic production by the Primakoff process in which the final state is produced from a pion well outside the nucleus so that there would be no initial-state interactions. However, the Primakoff process is governed by the fine structure constant $\alpha = 1/137$ and is real in contrast with the purely-imaginary dominant term. It is therefore expected to provide a very small effect. A detailed evaluation [46] showed that the Primakoff amplitude leads to a negligible correction for E791, but would be more important for higher energies available at colliding-beam facilities.

The ideal procedure would be to measure the jet momenta precisely enough so as to be able to identify the final nucleus as the target ground state. While it is feasible to consider this for electron–nucleus colliders, it is impossible for high-energy fixed-target experiments, so another technique must be used. E791 [45] isolated the dependence of the coherent elastic peak on the momentum transfer to the target, t, as the distinctive property of the coherent processes. This is shown in figure 11.3. The coherent slope of these curves is consistent with the radii measured in electron scattering experiments. It is immediately apparent that the peak is much narrower for the larger nucleus. The main advantage of the E791 experiment that enabled the data of figure 11.3 to be obtained is the excellent resolution of the transverse momentum.

The extraction of the coherent signal involves separating the coherent contributions from the incoherent ones and the background, as in figure 11.3. The theory for this is detailed in [46]. The result is a modified version of the quantity σ_1 of the total cross section (11.37):

$$\sigma_1 \approx \frac{3(A-1)^2}{r_N^2 + R_A^2} \frac{d\sigma_N}{dt}\bigg|_{t=0}. \tag{11.38}$$

Here the nucleon radius r_N^2 takes into account the slope of the elementary cross section, assuming that it is determined solely by the nucleon vertex. The result (11.38) differs by a factor of $(A-1)^2/A^2$ from the A dependence predicted previously for coherent processes (11.35). Using $A = 12$ and 195, the nuclear radii mentioned above, $r_N = 0.83$ fm and parametrizing the ratio of cross sections obtained from (11.38) by $\sigma \propto A^\alpha$, gives

$$\alpha = 1.54, \tag{11.39}$$

instead of $\alpha = 1.45$. The experimental result [45] is

$$\alpha \approx 1.55 \pm 0.05, \tag{11.40}$$

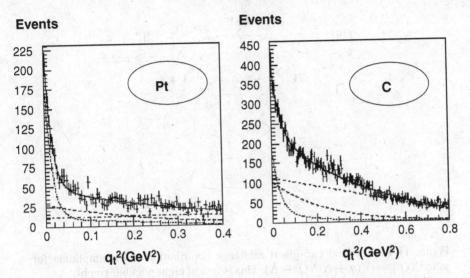

Figure 11.3. Coherent diffractive peaks observed in pion–nucleus interactions: $q_t^2 \equiv -t$. (From [45].) Here $1.5 \leq \kappa \leq 2.0$ GeV for the platinum and carbon targets. The curves are fits of Monte Carlo simulations to the data: coherent dissociation (dotted line), incoherent dissociation (dashed line), background (dashed-dotted line), and total fit (solid line).

which is remarkably close to the theoretical value shown in (11.39). Under the naive treatment of diffraction, the incident pion would be strongly absorbed by a large nucleus and α would be about $\frac{2}{3}$. Having $\alpha = 1.55$ instead of 0.67 causes the ratio of the platinum to carbon cross sections to be enhanced by about a factor of 12. If one takes into account that effects of colour fluctuations increase diffractive cross sections [50], the ratio of 12 becomes a still startling enhancement factor of 7 [46]. The effects of colour transparency have been observed.

11.6 Vector-meson production

We have seen in section 11.5 that colour transparency was discovered in the π–dijet reaction. Meson electroproduction on nuclei also involves a $q\bar{q}$ final state and so it is natural to use it to search for effects of colour transparency [7,6].

A QCD factorization theorem [29], proved for exclusive meson production by longitudinally polarized photons on nucleons, shows that contributions involving $q\bar{q}$ point-like configurations dominate the process in the Bjorken limit. The combination of this feature with the colour screening of QCD, (11.1) and (11.2), leads immediately to the prediction that colour transparency should occur for both coherent and incoherent nuclear processes [51,52].

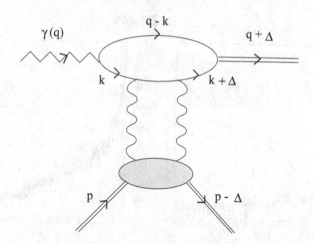

Figure 11.4. A typical two-gluon exchange contribution to the amplitude for
$\gamma^*(q)N(p) \to V(q + \Delta)N(p - \Delta)$. This is one of eight possible graphs.

We shall first discuss how point-like configurations and colour screening enter
for a nucleon target. This will be followed by a discussion of nuclear colour
transparency.

11.6.1 Production on nucleons

Consider the differential cross section for $\gamma_L^*(q) + N(p) \to V(q + \Delta) +$
$N(p - \Delta)$ in the near-forward region at kinematics for which Bjorken scaling
applies. A detailed perturbative QCD calculation of this [51] has been made, and
we follow this reference. The process, illustrated in figure 11.4, takes place sequen-
tially in time, in three steps. The process begins when the virtual photon breaks up
into a quark–antiquark pair with a lifetime τ_i given by

$$\tau_i = q^+ \tau_i^{-1} = Q^2 + \frac{\vec{k}^2 + m^2}{\alpha(1 - \alpha)} \approx Q^2, \qquad (11.41)$$

where m is the current quark mass and α and $\vec{k}$ are the quark longitudinal-momentum
fraction and transverse momentum. Then the quark–antiquark pair is scattered by
the target proton, with the cross section of (11.2). The process is completed when
the final-state vector meson is formed after a time $\tau_f \geq \tau_i$, given by

$$q^+ \tau_f^{-1} = \frac{\vec{k}^2 + m^2}{\alpha(1 - \alpha)}. \qquad (11.42)$$

The amplitude $\mathcal{M}_{fi}$ can therefore be written as a product of three factors: the wave
function, determined by QED, giving the amplitude for the virtual photon to break

into a quark–antiquark pair; the amplitude for $q\bar{q}$-target scattering given by (11.2); and the wave function giving the amplitude for the scattered quark–antiquark pair to become a vector meson, V. The only target dependence is caused by differing gluon distributions. See chapter 8 for a related discussion.

The invariant amplitude $\mathcal{M}_{fi}$ is given, using the notation of (11.23), by [52]:

$$\mathcal{M}_{fi} = \frac{\pi^2}{3}\alpha_s(Q_{\text{eff}}^2)xg_N(x, Q_{\text{eff}}^2, t)\langle V|b^2 e^{i\alpha\vec{\Delta}_\perp\cdot\vec{b}}|\gamma_L^*\rangle, \tag{11.43}$$

with $\vec{\Delta}_\perp$ as the transverse momentum transfer and $t = -\Delta_\perp^2$. Contributions to the integral are dominated by terms with $\alpha \approx \frac{1}{2}$ for the wave functions of [52].

The main features of the perturbative QCD analysis [51] are consistent with those obtained by Donnachie and Landshoff (DL) [53] and Cudell [54]. The pomeron is represented by the effective exchange of two non-perturbative gluons and the ρ wave function is expressed as a non-relativistic vertex function. In [51] the two-gluon aspect of the QCD pomeron is directly related to the proton's gluon structure function (11.2) and the relativistic structure of the vector meson is treated. Both perturbative QCD [51] and the DL model predict that the dominant amplitude involves a longitudinally-polarized photon and vector meson, and the leading cross section falls as Q^{-6}. That this is compatible with data is shown in [54] and chapter 7.

Reproduction of the measured magnitudes of cross sections for large values of Q^2 at HERA energies was achieved [52] using relativistic ρ-meson wave functions that include the effects of a high-momentum tail. This indicates that perturbative QCD is relevant for describing the interaction of small-sized $q\bar{q}$ pairs with nucleons. Furthermore, the cross section for vector-meson production by transverse photons is predicted to be suppressed by extra powers of Q^2, and this is consistent with the current data [55,56] for ρ-meson production. This suppression is a crucial feature of the perturbative QCD predictions. Indeed, experiments at HERA (see the review [57]) have convincingly confirmed the basic predictions of the perturbative QCD calculation. At high Q^2 there is a rapid increase with energy (corresponding to a decrease in x) and the cross section for longitudinally-polarized photons dominates over that for transverse ones.

Another central feature of the predictions of [51] is that, for fixed small values of x and $Q^2 \to \infty$, the slope of the cross section for small values of t is determined solely by the slope of the target gluon distribution, B_g. If the t dependence is parameterized as $d\sigma/dt = Ae^{B(Q^2)t}$, (11.43) yields

$$B(Q^2) = B_g + \langle V|b^4\alpha^2|\gamma_L^*\rangle/\langle V|b^2|\gamma_L^*\rangle, \tag{11.44}$$

in which the second term (with a fixed value of α) is of the form of the effective size (11.6). The predicted Q^2 dependence of B, depends on the Q^2-dependent

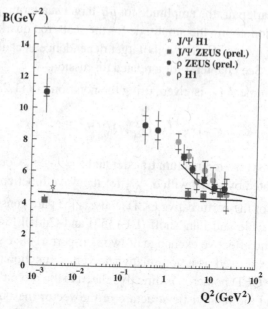

Figure 11.5. Q^2 dependence [52] of the slope parameter B of (11.44). The solid curve shows that the slope for ρ production becomes the same as the one for J/Ψ production at large values of Q^2. The data are from [58]. Figure courtesy of M Strikman [59].

effective size of the $q\bar{q}$ pair, not the size of the meson V. A comparison between the predicted value of B and data is shown in figure 11.5. The agreement between theory and experiment provides a dramatic confirmation that the wave function of a longitudinally-polarized virtual photon acts as a point-like configuration.

11.6.2 Hard coherent processes with nuclei

The dominance of configurations of small size is a key feature of perturbative QCD predictions for forward diffractive leptoproduction of longitudinally-polarized vector mesons with $1/2m_N x \gg 2R_A$ and for the coherent diffraction of pions into two jets. Thus, even for a nuclear target, colour screening inherent in the cross sections (11.1) and (11.2) implies that the coherent $q\bar{q}$ point-like system cannot interact. In leading-logarithmic approximation and in the light-cone gauge only two gluons connect the pion–dijet system and photon–vector-meson system with the nucleus, as illustrated in figures 11.2 and 11.4. Thus, in the appropriate kinematic region, the hadronic system propagating through the nucleus suffers no initial-state or final-state absorption, and the nuclear dependence of the $\pi + A \rightarrow$ dijet $+ A$ and $\gamma^* A \rightarrow V A$ forward amplitudes will be approximately additive in the nucleon number A. We can understand this remarkable feature of QCD from the space-time arguments given above. The final-state vector meson or dijet is formed over a long time given by (11.42) from a compact $q\bar{q}$ pair which does not attain its final physical size and normal strong interactions until it is well outside the target nucleus. Although the vector meson (or two jets) suffers no final-state interactions, the forward amplitude is not strictly additive in nuclear number because the gluon distribution itself can be shadowed at small values of x. Thus one predicts [51,32] an identical nuclear dependence for the forward vector-meson diffractive leptoproduction cross sections, diffractive production of dijets by pions, the longitudinal structure functions $F_A^L(x, Q^2)$, and the square of the gluon structure functions:

$$\frac{\frac{d\sigma}{dt}(\pi A \to \text{jet} - \text{jet} + A)\big|_{t=0}}{\frac{d\sigma}{dt}(\pi N \to \text{jet} - \text{jet} + N)\big|_{t=0}} =$$

$$\frac{\frac{d\sigma}{dt}(\gamma^* A \to V A)\big|_{t=0}}{\frac{d\sigma}{dt}(\gamma^* N \to V N)\big|_{t=0}} = \left[\frac{F_A^L(x, Q^2)}{F_N^L(x, Q^2)}\right]^2 = \frac{g_A^2(x, Q^2)}{g_N^2(x, Q^2)}. \quad (11.45)$$

The nuclear gluon distribution is expected to be more strongly shadowed than the nuclear quark structure functions at intermediate Q^2 because of the larger colour charge of the gluon. Numerical estimates [6,60,61] lead to $g_A(x, Q_0^2)/Ag_N(x, Q_0^2) \sim 0.85$ for $A = 12$ and 0.6 for $A = 200$ at $x = 0.01$ and $Q^2 = 4$ GeV^2. However, at fixed $x \sim 0.01$–0.03, shadowing substantially decreases with Q^2 due to scaling violation effects [60], which should lead to an effective increase of transparency for ρ leptoproduction at fixed x with increasing Q. At higher energies, when the gluon distributions are very large, one expects the onset of colour opacity. This can be checked by studying ultra-peripheral collisions at the LHC [62].

Evidence for the influence of colour transparency has been obtained in a Fermilab experiment [63] from the A dependence of the cross section for coherent J/ψ photoproduction which finds $\sigma \propto A^{1.40}$. The coefficient in the exponential arises from a calculation of the form of (11.35)–(11.37), using the nuclear radii of the particular targets, assuming the absence of initial- and final-state interactions. More recently, evidence for colour transparency effects has been reported, with moderate statistics, in incoherent ρ^0-meson production in deep-inelastic muon–nucleus scattering [56]. The data appear to indicate the onset of colour transparency predicted by perturbative QCD for incoherent $\gamma^* A \to \rho N(A - 1)'$ reactions. However, it is important to distinguish whether the increase in transparency with increasing Q^2 arises from the influence of an increase in the coherence length (11.4) or from the effects of colour transparency [64]. At small values of ν, $l_c \ll R_A$ any observed vector meson is produced inside the nucleus and typically travels through only about half of the nuclear thickness to escape. For large values of ν, $l_c \gg R_A$, the vector meson exists as a quantum fluctuation of the virtual photon before it encounters the nucleus and must propagate through the entire nuclear thickness. This variation in $q\bar{q}$-nuclear interactions with l_c has been observed by a HERMES experiment [65]. Another HERMES experiment [66] reported a rise of nuclear transparency with Q^2 (with fixed values of l_c) for both coherent and incoherent ρ^0 production. A JLab Laboratory experiment [67] plans a measurement of incoherent ρ^0 electro-production. The Q^2 dependence of the nuclear transparency ratio for two different targets with fixed coherence lengths will be measured. A correct interpretation of vector-meson experiments depends on the ability to separate the effects of changes in the coherence length from those of colour transparency [68,38]. Observing that the amplitude for coherent nuclear forward diffractive ρ leptoproduction is approximately proportional to A would provide dramatic evidence for colour transparency.

11.7 Quasi-elastic nuclear reactions

We have seen in sections 11.5 and 11.6 that the general concepts of colour transparency in the perturbative domain are now firmly established for high-energy processes: the presence of point-like configurations in vector mesons and pions and the form of the small-size $q\bar{q}$ dipole–nucleon interaction at high energies are well established experimentally. It is important to observe the influence of colour screening and colour transparency in nucleon–nucleon interactions. The unique feature of QCD is the prediction that small colour-neutral qqq configurations undergo reduced interactions due to the cancellation of amplitudes involving different quarks.

As discussed above, colour transparency is the vanishing or near-vanishing of initial-and/or final-state interactions in large Q^2 nuclear quasi-elastic reactions. This phenomenon depends on the formation of a small-sized object, a point-like configuration, in two-body projectile–nucleon collisions. High energies are needed to avoid the expansion of the point-like configuration, but one also needs the experimental resolution to be sufficiently good to ensure that no extra pions are produced in the interaction. Also it is difficult to enforce the requirement of exclusivity or semi-exclusivity that maintains the coherent nature of the reaction. At lower energies, one must take account of the expansion of the point-like configuration. The combination of uncertainties regarding the experimental kinematics and the treatment of the expansion of the point-like configuration has led to doubts about the meaning of some of the experimental results.

11.7.1 Time development

A point-like configuration produced in the interior of the nucleus expands as it moves through the nucleus, as described in section 11.3 and subsection 11.4.3.

Including this effect has been shown to be a crucial element in accurate calculations [69,39,31,70–72] for laboratory beam momenta less than 20 GeV, for which quasi-elastic measurements are possible.

We discuss some of the attempts to include this effect, using the $(e, e'p)$ reaction as the simplest example. If we denote by $\mathcal{M}_i$ the amplitude to remove a proton from the shell-model orbital i and detect an outgoing proton of momentum $\mathbf{p}$, the general expression (11.5) can be written as

$$\mathcal{M}_{fi} = \langle \mathbf{p} \, |(1 + T_S G_0) J | i \rangle, \tag{11.46}$$

where J is the electromagnetic-current operator. The soft ejectile–nucleus scattering amplitude T_S is obtained by solving the equation $T_S = U + U G_0 T_S$, where U represents the interaction between the ejected object and the nucleus.

The old-fashioned approach is to treat the ejectile as a proton. Then the operator U is the optical potential U^{opt}, with

$$U^{opt} \approx -i\sigma \rho(R), \tag{11.47}$$

where σ is the proton–nucleon total cross section and $\rho(R)$ is the nuclear density, R being the distance from the nuclear centre. For simplicity, the finite range of the nucleon–nucleon interaction and the usually small real part of U^{opt} are neglected. The term $-i\sigma$ represents the forward proton–nucleus scattering amplitude via the optical theorem. If the proton wave function is obtained from U^{opt}, the proton wave is said to be distorted (from the plane-wave approximation). The use of such a wave function in computing $\mathcal{M}_\alpha$ is the distorted-wave impulse approximation, DWIA, where the 'impulse' refers to the use of the free nucleon–nucleon cross section. The notation DWIA is used by nuclear physicists, while the high-energy limits of such calculations are termed as Glauber calculations.

The new feature that arises at large Q^2 is that the ejected object may be a point-like configuration that does not interact. If the energies are not sufficiently high, the ejectile expands as it moves through the nucleus and the soft interactions must be included. Farrar *et al* [69] argued that the square of the transverse size is roughly proportional to the distance travelled, Z, from the point of hard interaction where the point-like configuration is formed. Thus the scattering amplitude 'σ' that appears in the optical potential is replaced by one that grows as the ejectile moves in the Z-direction

$$\sigma^{PLC}(Z) = \left(\sigma_{hard} + \frac{Z}{\tau}[\sigma - \sigma_{hard}] \right) \theta(\tau - Z) + \sigma \theta(Z - \tau), \tag{11.48}$$

where the expansion time τ of (11.7) was originally obtained with $M_X^2 - M_1^2 = 1.4 \text{ fm}^{-2}$. Using $\sigma^{PLC}(Z)$ is theoretically justified for small times, when the system is small enough so that the leading-logarithmic approximation to perturbative QCD can be applied [69,7,6,73]. For very large times, $\sigma^{PLC}(Z)$ reverts to the standard hadronic cross section. In between these times, one may expect that (11.48) serves as a smooth interpolation formula.

The time development of the point-like configuration can also be obtained by modelling the ejectile–nucleus interaction as $\hat{U} = -i\sigma(b^2)\rho(R)$ (recall (11.1)) in which an interpolation formula such as $\sigma(b^2) = \sigma_0 b^2/\langle b^2 \rangle$ can be used. Then the average of $\sigma(b^2)$ is consistent with the measured projectile–nucleon cross section.

One may assume a baryonic basis and compute the relevant matrix elements of $\sigma(b^2)$ [39,31,70]. Jennings and Miller [39,31] used a harmonic-oscillator basis and suggested that the Lippmann–Schwinger equation $T_S = \hat{U}(1 + G_0 T_S)$ be solved by computing the term of first order in $\hat{U}$ and exponentiating the result. Then U^{opt} is modified by replacing σ by an effective cross section, σ_{eff}, given by,

$$\sigma_{eff}(Z) = \sigma\left(1 - e^{-iZ/\tau}\right), \qquad (11.49)$$

where τ is determined from the mass of the lowest radial excitation of the ground state. Using the frozen approximation ($\tau \to \infty$) leads to $\sigma_{eff} = 0$ and colour transparency, but at low energies the old-fashioned optical potential emerges. Exponentiation is an accurate approximation [74].

A sum-rule approach [75] used measured matrix elements for deep-inelastic scattering to represent the hard scattering operator, measured diffractive-dissociation cross sections to represent the soft scattering operator and some model assumptions. The result can be expressed as

$$\sigma_{eff}(Z) = \int_{(M+m_\pi)^2}^{\infty} dM_X^2 \rho\left(M_X^2, Q^2\right)\left(1 - e^{-i(M_X^2 - M_N^2)Z/2P}\right), \qquad (11.50)$$

where P is the magnitude of the total three-momentum of the ejectile and the function ρ represents the product of measured matrix elements. All values of M_X^2 contribute, so that some colour-transparency effects appear at low values of P, but obtaining a nearly-complete cancellation requires higher values of P than does the model (11.49). For small values of Z, σ_{eff} is approximately linear in Z as is σ^{PLC} of (11.48) [32], so numerical differences between the results of using (11.48) and (11.50) are much smaller than other uncertainties. Other approaches are discussed in the review [76]. The minimum requirements for any successful treatment of expansion are that the point-like configuration should not interact for small values of Z, but interact with full strength for large values. Any search for colour transparency at an energy for which two such treatments give very different answers should be terminated.

11.7.2 *Electromagnetic interactions with nuclei*

The use of the $(e, e'p)$ reactions has an appealing simplicity. Either the struck proton is, or is not, transformed into a point-like configuration and it should be straightforward to observe or rule out the effects of colour transparency. An intrepid group of experimentalists made pioneering measurements in the NE18 experiment at SLAC [77,78]. These experiments were later redone [79], with greatly improved statistical and systematic uncertainties.

It is necessary to discuss kinematics. For a given Q^2 and a virtual photon striking a proton at rest, the kinetic energy of the outgoing nucleon, T_p is equal to the virtual photon energy and for elastic scattering $(x = 1)$ is

$$T_p = \frac{Q^2}{2M_N}. \tag{11.51}$$

This is only 4 GeV for the largest measured value of Q^2. We may estimate an upper limit on the expansion time (11.7) by using the lowest possible value of M_X in (11.7). This is that of the Roper resonance $P_{11}(1440)$ (because interactions of the form of (11.1) and (11.2) do not flip spin), so that $\tau \leq 4$ fm. Expansion effects mask the expected small effect that colour transparency yields as it turns on with increasing Q^2.

In the JLab experiment [79], coincident detection of the momenta of the recoil electron and ejected proton enabled the energy transfer, $\nu = E_e - E_e'$, to be determined. The separation energy, E_m, needed to remove the nucleon is $E_m = \nu - T_p - T_{A-1}$, where T_p and T_{A-1} are the kinetic energies of the final-state proton and $(A - 1)$ recoil nucleus, both of which could be determined. The momentum of the detected proton is $\mathbf{p}$ and $\mathbf{q}$ is the three-momentum transfer. These quantities determine the missing momentum $\mathbf{p}_m = \mathbf{p} - \mathbf{q}$. Under the plane-wave impulse approximation (PWIA), the missing momentum $\mathbf{p}_m$ is the same as the initial momentum of the proton within the nucleus. In a non-relativistic PWIA formalism, the cross section can be written as a product

$$\frac{d^4\sigma}{dE_e'd\Omega_e'dE_pd\Omega_p} = K\sigma_{ep}S(E_m, \mathbf{p}_m), \tag{11.52}$$

where $dE_e', d\Omega_e', dE_p$ and $d\Omega_p$ are the phase-space factors of the electron and proton, $K = |\mathbf{p}|E_p$, and σ_{ep} is the off-shell electron–proton cross section. The spectral function $S(E_m, \mathbf{p}_m)$ is the joint probability to find a proton of momentum $\mathbf{p}_m$ and separation energy E_m within the nucleus.

The definition of the transparency ratio is the same as in the earlier $A(e, e'p)$ colour-transparency experiment [77,78], namely the ratio of the cross section measured on a nuclear target to the cross section for $(e, e'p)$ scattering in PWIA. This ratio

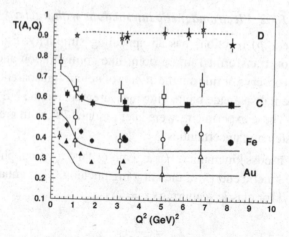

Figure 11.6. Transparency for $(e, e'p)$ quasi-elastic scattering from D (stars), C (squares), Fe (circles), and Au (triangles). Data from [79] are represented by the large solid stars, squares, and circles, respectively. Earlier JLab data (small solid squares, circles, and triangles) are from [88]. Previous SLAC data (large open symbols) are from [77] and [78]. Previous Bates data [83] (small open symbols) occur at the lowest Q^2 on C, Ni and Ta targets. The solid curves shown from $0.2 < Q^2 < 8.5\ \text{GeV}^2$ represent the results of Glauber calculations from [82] for C, Fe and Au and [84,85] for the D.

is defined as a transparency $T(A, Q)$, with

$$T(A, Q) = \frac{\int_V d^3 p_m dE_m Y_{exp}(E_m, \mathbf{p}_m)}{\int_V d^3 p_m dE_m Y_{PWIA}(E_m, \mathbf{p}_m)}, \tag{11.53}$$

where the integral is over the phase space V defined by the cuts $E_m < 80$ MeV and $|\mathbf{p}_m| < 300$ MeV. $Y_{exp}(E_m, \mathbf{p}_m)$ and $Y_{PWIA}(E_m, \mathbf{p}_m)$ are the corresponding experimental and simulation yields.

Spectral functions, generated from an independent-particle model, are used as input to the simulation. This procedure overestimates the experimental yield because the repulsive effects of nucleon–nucleon correlations are not included. To account for this, a correlation correction was applied by multiplying $T(A, Q)$ by factors of 1.11 for ^{12}C and 1.22 for ^{56}Fe, as determined from calculations in [80] and [81]. The data and related calculations are shown in figure 11.6. Experimental systematic uncertainties of $\pm 2.3\%$ are included with the statistical errors, but the effects of model-dependent systematic uncertainties on the simulations (typically about 5%) are not displayed. The solid curves shown from $0.2 < Q^2 < 8.5\ \text{GeV}^2$ represent the results of Glauber calculations from [82]. In the case of deuterium, the dashed curve is a Glauber calculation from [84,85]. The only energy dependence observed arises from the energy dependence of the elementary nucleon–nucleon cross section σ. This causes the fall in $T(A, Q)$ as Q^2 is increased from low values of Q^2 to about 2 GeV^2. No evidence for the effects of colour transparency is obtained.

A great variety of calculations were performed in response to the prospect of learning about colour transparency, including [86], [31], [75], [71], [42] and [87]. For precise comparisons between different calculations it is necessary to compute $T(A, Q^2)$ using the same kinematic cuts and the same ground-state wave functions. Typically, there is an unavoidable model dependence of about 5%, arising from using different bound state wave functions, nucleon–nucleon cross sections and nuclear densities so that detecting colour transparency requires the observation of a much larger effect.

A JLab experiment measured the nuclear transparency of the process $\gamma n \to \pi^- p$ in ^{4}He [89] for photons of energy between 1.6 and 4.5 GeV. Some barely perceptible deviations from the DWIA Glauber calculations are seen. A higher-energy experiment could provide a more fruitful search for the effects of colour transparency.

In principle, pion transparency can be determined by measuring pion electroproduction from nuclear targets [90]. Measurable effects of about 40% have been predicted [91,92] for the kinematics of this experiment.

11.7.3 Rescattering conquers time development

The practical problem in looking for colour-transparency effects in experiments at values of Q^2 from about one to a few GeV^2 is that the assumed point-like configuration expands while propagating through the nucleus. Observing colour transparency at intermediate values of Q^2 requires finding a way to suppress the expansion effects. One might use the lightest nuclear targets, where the propagation distances are small. But then the transparency is close to unity so any effects of colour transparency must be small. However, suppose a final-state interaction is specifically required to produce a given final state. Then colour-coherent effects would manifest themselves as a consequent decrease of the probability to reach the final state, and one could observe an effect decreasing from the Glauber value to zero. Thus, the measured cross section is to be compared with a vanishing quantity so that the relevant ratio of cross sections runs from unity to infinity. The first calculations [93] showed that substantial colour-transparency effects could be observable in $(e, e'pp)$ reactions on ^{4}He and ^{3}He targets.

To be specific, consider the cross section for the reaction $(ed \to e'pn)$ (in which the proton carries the large momentum of the photon) for various momenta of magnitude p_s of the recoiling neutron. For small values of p_s, easily accommodated by the deuteron wave function, the Glauber rescattering mechanism should provide

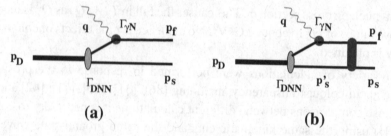

(a)   **(b)**

Figure 11.7. (a) Impulse approximation diagram. (b) Rescattering diagram.

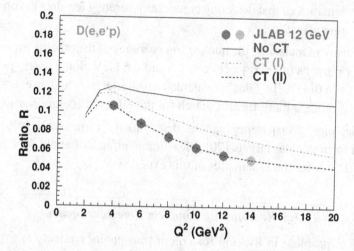

Figure 11.8. The ratio (11.54) of the cross section at 400 MeV missing momentum to that at 200 MeV. The solid line shows the eikonal approximation with no colour transparency. The dashed and solid lines represent the expansion of (11.48) obtained using $\Delta M^2 = 0.7$ or 1.1 GeV^2. The vanishing of rescattering effects produced by colour transparency causes the ratio to drop with increasing Q^2 [94]. The full circles show projected data for the 12 GeV upgrade of JLab. See [95] for details.

a small decrease in the cross section in the usual shadowing way by interfering with the impulse approximation term, figure 11.7(a). But rescattering effects, figure 11.7(b), are the dominant source of large values of p_s. Specific calculations [84,94, 95] showed that it is possible to separate the two kinematic regions by choosing two momentum intervals for the recoiling neutron: (300–500 MeV) for the rescattering region and (0–200 MeV) for the shadowing region. The rescattering mainly changes the momentum of the recoiling neutron by giving its momentum a component transverse to the direction of the virtual photon. Therefore both effects are enhanced if the momentum of the recoiling neutron is transverse to the direction (z) of the virtual photon. This corresponds to $\alpha = (E_s - p_s^z)/m \approx 1$. Figure 11.8 shows the expected Q^2 dependence of the ratio

$$R = \frac{\sigma(p_s = 400\,\mathrm{MeV})}{\sigma(p_s = 200\,\mathrm{MeV})}. \tag{11.54}$$

Another idea involves pionic degrees of freedom. The pion cloud of a nucleon is suspected to be responsible for many observed features, such as the relatively large slope of the neutron electric form factor [96]. One can consider processes in which pion exchanges cause the recoiling neutron to be detected as a $\Delta(1232)$ in the $d(e, e'p)\Delta$ reaction. An example is the quasi-elastic production of the Δ^{++} in electron scattering, the $(e, e'\Delta^{++})$ reaction. The initial singly-charged object is knocked out of the nucleus by the virtual photon and converts to a Δ^{++} by emitting or absorbing a charged pion. But pionic coupling to small-sized systems is suppressed, so this cross section for quasi-elastic production of Δ^{++}s should fall faster with increasing Q^2 than the predictions of conventional theories. Significant effects of this new 'chiral transparency' [13] have been predicted [97].

Another interesting variation of the rescattering idea is in the planned measurement of coherent vector-meson production from the deuteron [98]. Double scattering is necessary to produce a vector-deuteron final state at large scattering angles. This would be suppressed if the vector meson is produced as a point-like configuration. Relatively small propagations distances are involved, so that expansion confusion is avoided [99,100].

11.7.4 *Proton-nucleus interactions*

The first dedicated experiment seeking the effects of colour transparency was performed at Brookhaven National Laboratory [101]. The cross sections for the process $A(p, pp)(A-1)$ were measured for beam momenta p_L of 6, 10 and 12 GeV. The experimental setup used [102] was that for proton–proton elastic scattering at a centre-of-mass angle of $90°$. Very stimulating and intriguing results were obtained. In contrast with the prediction of the Glauber formalism that the ratio of the nuclear to nucleon cross sections is small and independent of energy, the ratio rises rapidly as p_L goes from 6 to 10 GeV and falls at larger values of p_L. The rise seems consistent with the onset of the effects of colour transparency, but the fall-off was a surprise.

The original experiment is subject to several questions. For a hydrogen target, the correct identification of an elastic-scattering event requires detecting only the outgoing momentum of one proton and the angle of the other. This is not sufficient for nuclear scattering because of the influence of the Fermi motion of the bound proton, so the observed reactions do not correspond exactly to pp elastic scattering at $90°$ in the centre of mass. Furthermore, the data were plotted as a function of an

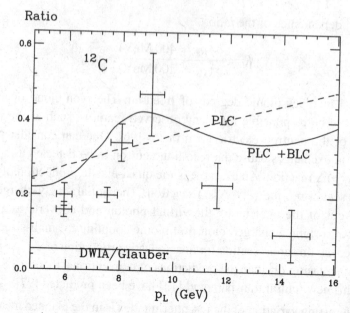

Figure 11.9. The transparency ratio of cross section to the partial-wave Born approximation (PWBA) as a function of beam momentum p_L and $k_z \approx 0$. The data are those of [103–105]. The standard DWIA, or Glauber, calculation is represented by the lower solid line. The effect of colour transparency using a point-like configuration (PLC) only is shown as the dashed line, while accounting for the interplay between a point-like configuration and a blob-like configuration (BLC), using the model of [106], and including necessary Fermi motion and time-evolution effects according to [107] is shown as the upper solid line.

effective beam momentum, P_{eff}, related to the invariant mass of the two outgoing nucleons. This is an incorrect treatment of the Fermi momentum of the initial bound proton as that of a freely moving proton. It is more correct (and substantially different) to display the data as a function of $-k_z$, the component of the momentum of the struck nucleon (in the beam direction) calculated in the PWIA.

The DWIA describes similar data at intermediate energies of $E_p = 1$ GeV with an accuracy of better than 20%; see [27] for reviews. However, the BNL data are generally considerably above the DWIA/Glauber results of [101,69,31,108,42], as shown in figure 11.9. The relatively-large value of $d\sigma/d\sigma^B$ indicates the presence of a large transparency effect, but the drop at high energy led to new ideas [109,106].

Understanding these ideas requires knowing some features of the pp elastic scattering data. The energy dependence of the 90° angular distribution is of the form of $R(s)/s^{10}$, where R(s) oscillates between 1 and 3 over the energy range of the BNL experiment. Ralston and Pire [106] suggested that the energy dependence is caused by an interference between a hard amplitude, which produces a point-like configuration, and a soft one, which involves a large blob-like configuration. This makes sense because it is difficult to imagine that the intermediate state can be treated as a point-like configuration with all six quarks in the same location. The idea of Ralston and Pire is that the blob-like configuration is due to the Landshoff process [106]. Another mechanism is that of Brodsky and de Teramond [109], in which the two-baryon system couples to charmed quarks (there is a small six-quark configuration and a blob-like configuration that is a six-quark plus $c\bar{c}$ object). This suggestion originated from the observation that the mass scale of the rapid energy variation in A_{NN} [102] and in the measured transparency matches that of the charm threshold. In both [106] and [109] the observed nuclear effect is claimed to result from the suppression of the blob-like configuration in the nucleus. The observed ratio is that of a pp cross section in the nucleus which varies smoothly with s and the free pp cross section which has a bump. However, it is necessary to include the effects of wave-packet expansion and proton Fermi motion [75,107,110,111] to make quantitative calculations.

The interplay between the contributions of the point-like configuration and the blob-like configuration leads to a drop in the transparency ratio. A description of the drop obtained from these ideas was successful in reproducing the early data [107], but many questions remained because of the uncertainty in the kinematics.

To answer the questions regarding kinematics, a new detector was developed so that the momenta of both outgoing protons could be detected [103–105]. These newer experimental results largely confirmed the original ones. No effects other than colour transparency can account for the rise in the ratio. However, it is difficult to account completely for the fall-off. The new data and the 1997 predictions of this author, applying the theory of [107], are shown in figure 11.9. Qualitative agreement is achieved, but the drop in the ratio is not as steep as that of the data. Calculations that describe the data [112,113] neglect the essential influences of expansion and Fermi motion, so we cannot accept the results. It seems clear that including the effects of colour transparency improves the agreement between theory and experiment. However, the remaining disagreement indicates a lack of knowledge in handling the details of calculating the (p, pp) cross section, such as the influence of nucleon–nucleon correlations [108]), and perhaps some experimental difficulties remain. The original incorrect use of P_{eff} was maintained in the newer work. A general remark is that the electron-scattering experiments should be easier to understand.

A final comment concerns the relation between the searches for the effects of colour transparency using (p, pp) and $(e, e'p)$ reactions. In the former, the emerging protons have half the energy of the incident proton, and the effects of expansion are very important. Thus colour-transparency effects originate in the initial state for incident momenta greater than 6 GeV. According to the kinematic relation (11.51), one should see the effects of colour transparency at $Q^2 \approx 12 \text{ GeV}^2$ for the $(e, e'p)$ reaction. The present maximum value of Q^2 for which $(e, e'p)$ data exist is 8.1 GeV^2 [79]. Thus there is no contradiction between the BNL data that show signs of colour transparency and $(e, e'p)$ data that do not. A discovery or contradiction will be revealed when the energy of JLab is increased.

11.8 Assessment and outlook

The basic origin of colour transparency is that small-sized objects have small-sized cross sections: recall (11.1) and (11.2). This occurs because of cancellations among the effects of gluons emitted from different constituents of point-like configurations. Point-like colour-singlet objects do not exchange gluons. The experimental evidence confirming this ranges from the measured sizes of hadronic cross sections to the observation of scaling in deep inelastic lepton–nucleon scattering.

The small size can be observed under suitable experimental conditions that maintain the coherence needed to produce a small-sized object or point-like configuration and operate at high-enough energy so that the object remains point-like as it moves through the entire nucleus. The effect of colour transparency has been observed in the coherent diffractive dissociation of pions into jets [45]: a ratio of nuclear cross sections is about seven times higher than the one expected from the standard DWIA/Glauber theory. Furthermore, the predictions of perturbative QCD for the process $\gamma p \rightarrow \rho^0 p$ have been verified. This makes it natural for colour transparency to be expected as soon as nuclear targets are used for vector-meson production at high energies. Indeed, there is already some evidence [63] and studies have found tantalizing hints [56], but nothing as strong as a factor of 7 has been seen.

So far, the influence of point-like configurations and colour transparency has been detected most clearly in reactions involving $q\bar{q}$ pairs. Observing colour transparency in reactions involving baryonic point-like configurations would verify a remarkable feature of the SU(3) nature of QCD that the gluon fields coherently emitted by a small-sized colour-neutral qqq system will also be cancelled. The first attempts were the pioneering BNL (p, pp) experiments. Tantalizing evidence was seen at laboratory momenta from 6 to 10 GeV [101,103–105]. However, there are difficulties: (i) in understanding the experimental kinematics, (ii) with the theory of treating a six-quark object as a point-like configuration and (iii) with treating various details of the reaction theory. The hope of using a simpler process spurred efforts to use the $(e, e'p)$ reaction along with quasi-elastic kinematics. No evidence for colour transparency was observed for values of Q^2 up to about 8 GeV^2 [79]. However, even the largest value corresponds to an outgoing proton kinetic energy of only about 4 GeV, significantly lower than used at BNL.

Establishing the existence of colour transparency for three-quark systems remains an important goal. Increasing the beam energy at JLab to enable experiments at values of Q^2 of greater than 10 GeV^2, or making careful kinematically-complete measurements of the $ed \rightarrow e'pn$ reaction, would be the best way to achieve this.

What would be the implications of a successful observation of baryonic colour transparency? The first is that one of the unusual predictions of QCD would have been verified. But there are a host of other implications related to the physics of nuclei because the point-like configurations of a bound nucleon would be suppressed compared to those of a free one. This provides one of the explanations of the EMC effect [40], predicts a nuclear modification of the nucleon form factor and could help theorists to understand better the interaction between nucleons separated by small distances.

Colour transparency is one of the novel and unusual consequences of QCD. Experimental verification that this phenomenon exists for both mesons and baryons is an important part of the general program of testing QCD and learning how the strong interaction really works. New phenomena such as the disappearance of the pion–nucleus interaction at energies of 500 GeV have been discovered already, and analogous features of baryon–nuclear interactions could be discovered using precision electron scattering experiments at much lower energies.

References

[1] F E Low, Physical Review D12 (1975) 163
[2] S Nussinov, Physical Review Letters 34 (1975) 1286
[3] J F Gunion and D E Soper, Physical Review D15 (1977) 2617
[4] G Bertsch *et al*, Physical Review Letters 47 (1981) 297
[5] L Frankfurt, G A Miller and M Strikman, Physics Letters B304 (1993) 1
[6] L Frankfurt and M Strikman, Physics Reports 160 (1988) 235
[7] S J Brodsky and A H Mueller, Physics Letters 206B (1988) 685
[8] A V Radyushkin, Acta Physica Polonica B15 (1984) 403;
    G P Korchemski and A V Radysushkin, Soviet Journal of Nuclear Physics 45 (1987) 910 and references therein
[9] N Isgur and C H Llewellyn Smith, Physical Review Letters 52, (1984) 1080; Physics Letters B217 (1989) 535
[10] J Botts and G Sterman, Nuclear Physics B325 (1989) 62; Physics Letters B224 (1989) 20; *ibid* B227 (1989) 501; J Botts, Physical Review D44 (1991) 2768
[11] H-N Li and G Sterman, Nuclear Physics B381 (1992) 129
[12] V V Sudakov, Soviet Physics JETP 3 (1956) 65
[13] L Frankfurt, G A Miller and M Strikman, Comments on Nuclear and Particle Physics 21 (1992) 1
[14] L Frankfurt, G A Miller and M Strikman, Nuclear Physics A555 (1993) 752
[15] L Frankfurt and M Strikman, Physical Review Letters 66 (1991) 2289
[16] L Frankfurt and M Strikman, Nuclear Physics B316 (1989) 340

[17] N N Nikolaev, Communications in Nuclear and Particle Science 21 (1992) 41

[18] B Povh and J Hufner, Physical Review Letters 58 (1987) 1612

[19] V N Gribov, Soviet Physics JETP 30 (1970) 709

[20] J D Bjorken, SLAC-PUB-0571 *Invited paper at the Amer. Phys. Soc. New York Meeting, Feb 3, 1969*

[21] J D Bjorken, J B Kogut and D E Soper, Physical Review D3 (1971) 1382

[22] V Barone *et al*, Zeitschrift für Physik C58 (1993) 541

[23] A Donnachie, G Dosch, P Landshoff and O Nachtmann, *Pomeron Physics and QCD*, Cambridge University Press (2002)

[24] L N Lipatov, Soviet Journal of Nuclear Physics 23 (1976) 338;
E A Kuraev, L N Lipatov and V S Fadin, Soviet Physics JETP 45 (1977) 199;
I Balitskii and L N Lipatov, Soviet Journal of Nuclear Physics 28 (1978) 822

[25] B Blattel, G Baym, L Frankfurt and M Strikman, Physical Review Letters 70 (1993) 896

[26] S Eidelman *et al* (Particle Data Group), Physics Letters 'B592 (2004) 1

[27] S L Belostotsky *et al*, in *Proceedings of the International Symposium on Modern Developments in Nuclear Physics*, O Sushkov ed, World Scientific (1987);
J Saudinos and C Wilkin, ARNPS 24 (1974) 341;
G J Igo, Reviews of Modern Physics 50 (1978) 523;
R J Glauber, Physical Review, 100 (1955) 242

[28] L Frankfurt, A Radyushkin and M Strikman, Physical Review D55 (1997) 98

[29] J C Collins, L Frankfurt and M Strikman, Physical Review D56 (1997) 2982

[30] G R Farrar, H Liu, L Frankfurt and M I Strikman, Physical Review Letters 62 (1989) 1095

[31] B K Jennings and G A Miller, Physical Review D44 (1991) 692

[32] L Frankfurt, G A Miller and M Strikman, ARNPS 44 (1994) 501

[33] B Z Kopeliovich, L I Lapidus and A B Zamolodchikov, JETP Letters 33 (1981) 595

[34] L Frankfurt and M Strikman, Physics Reports 76 (1981) 215

[35] L Frankfurt and M Strikman, Progress in Particle and Nuclear Physics 27 (1991) 135

[36] S J Brodsky, H C Pauli and S S Pinsky, Physics Reports 301 (1998) 299

[37] G A Miller, Progress in Particle and Nuclear Physics 45 (2000) 83

[38] B Z Kopeliovich, J Nemchik, A Schafer and A V Tarasov, Physical Review C65 (2002) 035201

[39] B K Jennings and G A Miller, Physics Letters B236 (1990) 209

[40] J J Aubert *et al*, EMC Collaboration, Physics Letters 123B (1983) 275

[41] L Frankfurt and M Strikman, Nuclear Physics B250 (1985) 147

[42] L Frankfurt, M Strikman and M Zhalov, Physical Review C50 (1994) 2189

[43] M R Frank, B K Jennings and G A Miller, Physical Review C54 (1996) 920

[44] S Strauch *et al*, Jefferson Laboratory E93-049 Collaboration, Physical Review Letters 91 (2003) 052301

[45] E M Aitala *et al*, E791 Collaboration, Physical Review Letters 86 (2001) 4768 and 4773

[46] L Frankfurt, G A Miller and M Strikman, Physical Review D65 (2002) 094015

[47] S J Brodsky and G P Lepage, Physical Review D22 (1980) 2157

[48] I V Musatov and A V Radyushkin, Physical Review D56 (1997) 2713

[49] P Kroll and M Raulfs, Physics Letters B387 (1996) 848

[50] L Frankfurt, G A Miller and M Strikman, Physical Review Letters 71 (1993) 2859

[51] S J Brodsky, L Frankfurt, J F Gunion, A H Mueller and M Strikman, Physical Review D50 (1994) 3134

[52] L Frankfurt, W Koepf and M Strikman, Physical Review D57 (1998) 512

[53] A Donnachie and P V Landshoff, Nuclear Physics B311 (1989) 509

[54] J R Cudell, Nuclear Physics B336 (1990) 1

[55] P Amaudruz *et al*, NMC Collaboration, Zeitschrift für Physik C54 (1992) 239

[56] M R Adams *et al*, E665 Collaboration, Physical Review Letters 74 (1995) 1525

[57] H Abramowicz and A Caldwell, Reviews of Modern Physics 71 (1999) 1275

[58] A Levy, for the H1 and Zeus Collaborations, hep-ex/0301022

[59] L Frankfurt and M Strikman, Nuclear Physics (Proc. Supp.) 79 (1909) 671; L Frankfurt, M Strikman, C Weiss and M Zhalov, hep-ph/0412260

[60] L Frankfurt, M Strikman and S Liuti, Physical Review Letters 65 (1990) 1725

[61] L Frankfurt, V Guzey, M McDermott and M Strikman, Journal of High Energy Physics 202 (2002) 27

[62] L Frankfurt, M Strikman and M Zhalov, Physics Letters B540 (2002) 220

[63] M D Sokoloff *et al*, Physical Review Letters 57 (1986) 3003

[64] J Hufner and B Kopeliovich, Physics Letters B403 (1997) 128

[65] K Ackerstaff *et al*, HERMES Collaboration, Physical Review Letters 82 (1999) 3025

[66] A Airapetian *et al*, HERMES Collaboration, Physical Review Letters 90 (2003) 052501

[67] Jefferson Laboratory experiment E-02-110, K Hadifi, M Holtrop and B Mustapha, spokespersons

[68] T Falter, K Gallmeister and U Mosel, Physical Review C67 (2003) 054606; erratum *ibid* C68 (2003) 019903

[69] G R Farrar, H Liu, L Frankfurt and M I Strikman, Physical Review Letters 61 (1988) 686

[70] B Z Kopeliovich and B G Zakharov, Physics Letters B264 (1991) 434

[71] O Benhar *et al*, Physical Review Letters 69 (1992) 881

[72] N N Nikolaev, Nuclear Physics A567 (1994) 781

[73] Yu L Dokshitzer, V A Khoze, A H Mueller and S I Troyan, in *Basics of Perturbative QCD*, J Tran Thanh Van ed, Editions Frontières, (1991)

[74] W R Greenberg and G A Miller, Physical Review D47 (1993) 1865

[75] B K Jennings and G A Miller, Physical Review Letters 70 (1992) 3619; Physics Letters B274 (1992) 442

[76] P Jain, B Pire and J P Ralston, Physics Reports 271 (1996) 67

[77] N C R Makins *et al*, Physical Review Letters 72 (1994) 1986

[78] T G O'Neill *et al*, Physics Letters B351 (1995) 87

[79] K Garrow *et al*, Physical Review C66 (2002) 044613

[80] J W Van Orden, W Truex and M K Banerjee, Physical Review C21 (1980) 2628

[81] X Ji, private communications to [79]

[82] V R Pandharipande and S C Pieper, Physical Review C45 (1992); H Gao, V R Pandharipande and S C Pieper, private communication to [79]

[83] G Garino *et al*, Physical Review C45 (1992) 780

[84] L Frankfurt *et al*, Zeitschrift für Physik A352 (1995) 97

[85] M Sargsian, private communication to [83]

[86] L Frankfurt, M Strikman and M Zhalov, Nuclear Physics A515 (1990) 599

[87] W R Greenberg and G A Miller, Physical Review C49 (1994) 2747

[88] D Abbott *et al*, Physical Review Letters 80 (1998) 5072

[89] D Dutta *et al*, Jefferson Laboratory E94-0104 Collaboration, Physical Review C68 (2003) 021001

[90] Jefferson Laboratory Experiment, E01-107, K Garrow and R Ent spokespersons

[91] B Kundu, J Samuelsson, P Jain and J P Ralston, Physical Review D62 (2000) 113009

[92] G A Miller and M Strikman, in *Proceedings of the Workshop on Options for Color Coherence/Transparency*, Jefferson laboratory (1995)

[93] K Egiian *et al*, Nuclear Physics A580 (1994) 365

[94] M M Sargsian, International Journal of Modern Physics E10 (2001) 405

[95] M M Sargsian *et al*, Journal of Physics G29 (2003) R1

[96] A W Thomas, S Théberge and G A Miller, Physical Review D22 (1980) 2838; *ibid* D23 (1981) 2106E; *ibid* D24, (1981) 216; Physics Letters 91B (1980) 192: G A Miller, Physical Review C66 (2002) 032201

[97] L Frankfurt, T S H Lee, G A Miller and M Strikman, Physical Review C55 (1997) 909

[98] Jefferson Laboratory experiment E02-012, L Kramer, F Klein and S Stepanyan, spokespersons.

[99] L Frankfurt *et al*, Nuclear Physics A622 (1997) 511

[100] L Frankfurt, G Piller, M Sargsian and M Strikman, European Physical Journal A2 (1998) 301

[101] A S Carroll *et al*, Physical Review Letters 61 (1988) 1698

[102] E A Crosbie *et al*, Physical Review D23 (1981) 600

[103] I Mardor *et al*, Physical Review Letters 81 (1998) 5085

[104] A Leksanov *et al*, Physical Review Letters 87 (2001) 212301

[105] J Aclander *et al*, Physical Review C70 (2004) 015208

[106] J P Ralston and B Pire, Physical Review Letters 61 (1988) 1823

[107] B K Jennings and G A Miller, Physics Letters B318 (1993) 7

[108] T S H Lee and G A Miller, Physical Review C45 (1992) 1863

[109] S J Brodsky and G F De Teramond, Physical Review Letters 60 (1988) 1924

[110] B K Jennings and B Kopeliovich, Physical Review Letters 70 (1993) 3384

[111] A Bianconi, S Boffi and D E Kharzeev, Physics Letters B305 (1993) 1; Nuclear Physics A565 (1993) 767 ; Physical Review C49 (1994) R1243

[112] P Jain and J P Ralston, Physical Review D48 (1993) 1104

[113] V V Anisovich *et al*, Physics Letters B292 (1992) 169

Index